{ VIRTUAL FIELD TRIPS GEOLOGY }

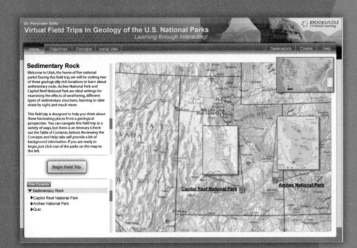

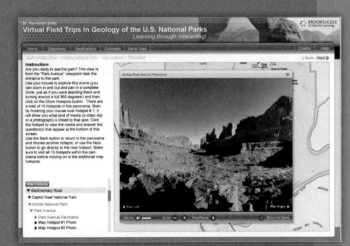

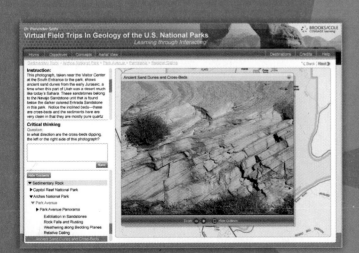

Bringing the field to you.

The Virtual Field Trips are concept-based modules that teach students geology by using famous locations throughout the United States. Sideling Hill Syncline, Grand Canyon, and Hawaii Volcanoes National Parks are included, as well as many others. Designed to be used as homework assignments or lab work, the modules use a rich array of multimedia to demonstrate concepts. High definition videos, images, animations, quizzes, and Google Earth layers work together in **Virtual Field Trips** to bring the concepts to life.

Packaged with GEOL.

- Sedimentary Rocks
- Geologic Time
- Desert Environment
- Running Water
- Hydrothermal Activity

Additional Bundle Options*

- Metamorphism and Metamorphic Rocks
- Mass Wasting Processes
- Glaciers and Glaciation
- Igneous Rock Textures
- Earthquakes & Seismicity
- Volcano Types
- Plate Tectonics
- Mineral Resources
- Groundwater
- Shorelines and Shoreline Process

*For availability and pricing information, contact your local Brooks/Cole representative.

BROOKS/COLE
CENGAGE Learning™

GEOL
Reed Wicander, James S. Monroe

Publisher: Yolanda Cossio

Earth Science Editor: Laura Pople

Developmental Editor: Anna Lustig

Assistant Editor: Samantha Arvin

Editorial Assistant: Kristina Chiapella

Media Editor: Alexandria Brady

Marketing Manager: Nicole Mollica

Marketing Coordinator: Kevin Carroll

Marketing Communications Manager:
   Belinda Krohmer

Content Project Manager: Hal Humphrey

Art Director: John Walker

Print Buyer: Karen Hunt

Rights Acquisitions Account Manager, Text:
   Bob Kauser

Rights Acquisitions Account Manager, Image:
   Deanna Ettinger

Production Service: Lachina Publishing Services

Cover Design: Red Hanger Design

Cover Image: Stephen Matera/Getty Images

Compositor: Lachina Publishing Services

For product information and technology assistance, contact us at
Cengage Learning Customer & Sales Support, 1-800-354-9706

For permission to use material from this text or product,
submit all requests online at **cengage.com/permissions**
Further permissions questions can be emailed to
permissionrequest@cengage.com

Library of Congress Control Number: 2010921361

ISBN-13: 978-0-538-49453-3

ISBN-10: 0-538-49453-0

Brooks Cole
20 Davis Drive
Belmont, CA 94002-3098
USA

Cengage Learning is a leading provider of customized learning solutions with office locations around the globe, including Singapore, the United Kingdom, Australia, Mexico, Brazil, and Japan. Locate your local office at:
international.cengage.com/region

Cengage Learning products are represented in Canada by Nelson Education, Ltd.

To learn more about Brooks/Cole visit **www.cengage.com/brookscole**.

Purchase any of our products at your local college store or at our preferred online store **www.CengageBrain.com**.

Printed in the United States of America
1 2 3 4 5 6 7 14 13 12 11 10

Brief Contents

Contents

3 MINERALS: THE BUILDING BLOCKS OF ROCKS 46

4 IGNEOUS ROCKS AND INTRUSIVE IGNEOUS ACTIVITY 64

5 VOLCANOES AND VOLCANISM 82

6 WEATHERING, SOIL, AND SEDIMENTARY ROCKS 102

7 METAMORPHISM AND METAMORPHIC ROCKS 132

8 EARTHQUAKES AND EARTH'S INTERIOR 148

9 THE SEAFLOOR 174

10 DEFORMATION, MOUNTAIN BUILDING, AND THE CONTINENTS 192

11 MASS WASTING 214

15 THE WORK OF WIND AND DESERTS 302

16 SHORELINES AND SHORELINE PROCESSES 320

17 GEOLOGIC TIME: CONCEPTS AND PRINCIPLES 342

18 EARTH HISTORY 366

19 LIFE HISTORY 400

VIRTUAL FIELD TRIPS: LET'S GO!

GOALS OF THE TRIPS

1. Engage students with panoramas, videos, photographs, and Google Earth.
2. Help teach geologic concepts in a clear way.
3. Allow students to check their learning, with critical thinking questions throughout.
4. Help students get excited about the Earth and its processes.
5. Have fun!

Integration of Field Trips with the GEOL text

Students are directed to further their knowledge of key concepts using the web-based Virtual Field Trips. The Virtual Field Trips sections in the book help students focus on what they should pay particular attention to while completing their field trip. Used as road maps to the field trips, the text sections include top objectives, integrated points from the text and field trip, photos, and critical thinking questions. Here is a list of the modules that are included in the GEOL package:

Sedimentary Rocks

Found In: Chapter 6, Weathering, Soil, and Sedimentary Rocks
Locations: Arches and Capitol Reef National Parks
Main concepts covered:
- Sandstones, shales, and volcanic ash in the field
- Lithology and weathering profiles
- Bedding planes, strata, and unconformities
- Formation of arches in sandstones
- Story of the Delicate Arch

Running Water

Found In: Chapter 12, Running Water
Location: Zion National Park
Main concepts covered:
- River erosion, cut banks, and slot canyons
- River deposition and point bars
- River abrasion and rounded pebbles
- Incised meanders and downcutting

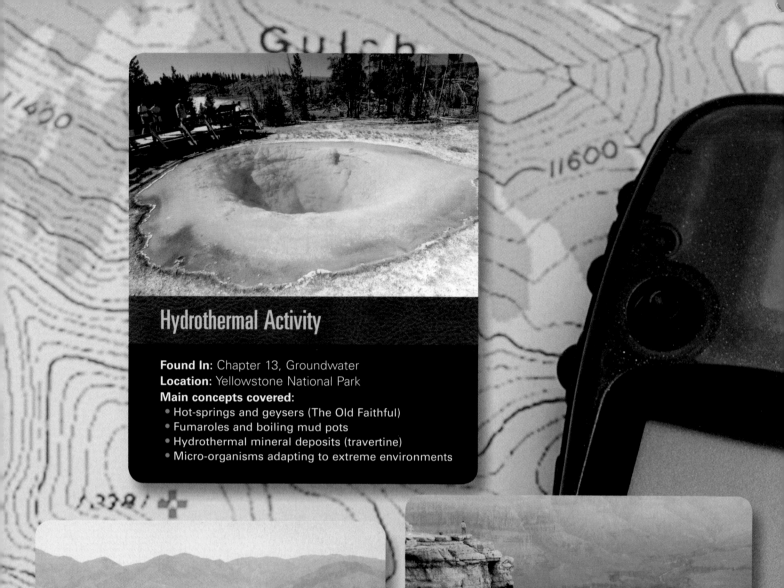

Hydrothermal Activity

Found In: Chapter 13, Groundwater
Location: Yellowstone National Park
Main concepts covered:
- Hot-springs and geysers (The Old Faithful)
- Fumaroles and boiling mud pots
- Hydrothermal mineral deposits (travertine)
- Micro-organisms adapting to extreme environments

Desert Environments

Found In: Chapter 15, The Work of Wind and Deserts
Location: Death Valley National Park
Main concepts covered:
- Geological features of an arid landscape
- Chemical sedimentary deposits & evaporites
- Formation and migration of sand dunes
- Features of an alluvial fan
- Borax–mining in Death Valley

Geologic Time

Found In: Chapter 17, Geologic Time: Concepts and Principles
Locations: Grand Canyon and Capitol Reef National Parks
Main concepts covered:
- Stratigraphic section and correlation of units
- Laws of relative dating (Superposition, Original Horizontality, Cross-cutting Relationships, & Inclusions)
- Unconformities and missing time
- Ancient sedimentary environments and primary and secondary structures preserved in rocks (cross-beds, ripple marks, trace fossils)
- Index fossils & correlation

CHAPTER 1
UNDERSTANDING EARTH: A DYNAMIC AND EVOLVING PLANET

A picturesque view of the Yosemite Valley with the Merced River, Yosemite National Park, USA. The large granite batholiths looming above the river were emplaced deep below Earth's surface millions of years ago and exposed as overlying rocks were removed through weathering and erosion.

Introduction

"Earth is a complex, dynamic planet that has changed continuously since its origin some 4.6 billion years ago."

A major benefit of the space age has been the ability to look back from space and view our planet in its entirety. Every astronaut has remarked in one way or another on how Earth stands out as an inviting oasis in the otherwise black void of space. We are able to see not only the beauty of our planet, but also its fragility. We can also decipher Earth's long and frequently turbulent history by reading the clues preserved in the geologic record.

LEARNING OUTCOMES

After reading this unit, you should be able to do the following:

LO1 Define geology

LO2 Understand the impact the formulation of theories has had on the study of geology

LO3 Explain how geology relates to the human experience

LO4 Explain how geology affects our everyday lives

LO5 Describe global geologic and environmental issues facing humankind

LO6 Describe the origin of the universe and solar system, and Earth's place in them

LO7 Explain why Earth is a dynamic and evolving planet

LO8 Describe the rock cycle

LO9 Understand organic evolution and its role in the history of life

LO10 Describe geologic time and uniformitarianism

LO11 Explain how the study of geology benefits us

A major theme of this book is that Earth is a complex, dynamic planet that has changed continuously since its origin some 4.6 billion years ago. By viewing Earth as a whole—that is, thinking of it as a system—we not only see how its various components are interconnected, but we can also better appreciate its complex and dynamic nature. The system concept makes it easier for us to study a complex subject such as Earth, because it divides the whole into smaller components that we can easily understand, without losing sight of how the components fit together as a whole.

A **system** is a combination of related parts that interact in an organized manner. An automobile is a good example of a system. Its various components or subsystems, such as the engine, transmission, steering, and brakes, are all interconnected in such a way that a change in any one of them affects the others.

We can examine Earth in the same way we view an automobile—that is, as a system of interconnected components that interact and affect each other in many ways. The principal subsystems of Earth are the *atmosphere*, *biosphere*, *hydrosphere*, **lithosphere**, **mantle**, and **core** (Figure 1.1). The complex interactions among these subsystems result in a dynamically changing planet in which matter and energy is continuously recycled into different forms.

We must also not forget that humans are part of the Earth system, and our activities can produce changes with potentially wide-ranging consequences. When people discuss and debate such environmental issues as acid rain, the greenhouse effect and global warming, and the depleted ozone layer, it is

system A combination of related parts that interact in an organized fashion; Earth systems include the atmosphere, hydrosphere, biosphere, and solid Earth.

lithosphere Earth's outer, rigid part, consisting of the upper mantle, oceanic crust, and continental crust.

mantle The thick layer between Earth's crust and core.

core The interior part of Earth beginning at a depth of 2,900 km that probably consists mostly of iron and nickel.

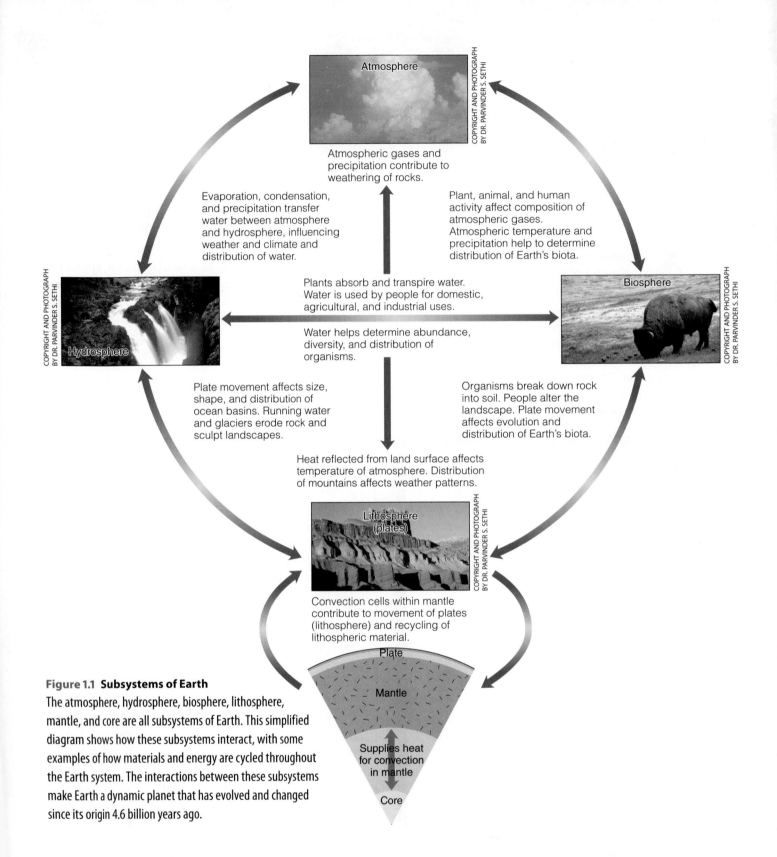

Figure 1.1 Subsystems of Earth

The atmosphere, hydrosphere, biosphere, lithosphere, mantle, and core are all subsystems of Earth. This simplified diagram shows how these subsystems interact, with some examples of how materials and energy are cycled throughout the Earth system. The interactions between these subsystems make Earth a dynamic planet that has evolved and changed since its origin 4.6 billion years ago.

The following labels appear in the figure:

Atmosphere

Atmospheric gases and precipitation contribute to weathering of rocks.

Evaporation, condensation, and precipitation transfer water between atmosphere and hydrosphere, influencing weather and climate and distribution of water.

Plant, animal, and human activity affect composition of atmospheric gases. Atmospheric temperature and precipitation help to determine distribution of Earth's biota.

Hydrosphere

Biosphere

Plants absorb and transpire water. Water is used by people for domestic, agricultural, and industrial uses.

Water helps determine abundance, diversity, and distribution of organisms.

Plate movement affects size, shape, and distribution of ocean basins. Running water and glaciers erode rock and sculpt landscapes.

Organisms break down rock into soil. People alter the landscape. Plate movement affects evolution and distribution of Earth's biota.

Heat reflected from land surface affects temperature of atmosphere. Distribution of mountains affects weather patterns.

Lithosphere (plates)

Convection cells within mantle contribute to movement of plates (lithosphere) and recycling of lithospheric material.

Plate

Mantle

Supplies heat for convection in mantle

Core

important to remember that these are not isolated issues but are part of the larger Earth system. Furthermore, remember that Earth goes through time cycles that are much longer than humans are used to. Although they may have disastrous short-term effects on the human species, global warming and cooling are also part of a longer-term cycle that has resulted in many glacial advances and retreats during the past 1.8 million years.

LO1 What Is Geology?

Geology, from the Greek *geo* and *logos*, is defined as the study of Earth, but now must also include the study of the planets and moons in our solar system. It is generally divided into two broad areas—physical geology and historical geology. *Physical geology* is the study of Earth materials, such as **minerals** and **rocks**, as well as the processes operating within Earth and on its surface. *Historical geology* examines the origin and evolution of Earth and its continents, oceans, atmosphere, and life.

Nearly every aspect of geology has some economic or environmental relevance. Many geologists are involved in exploration for mineral and energy resources, using their specialized knowledge to locate the natural resources on which our industrialized society is based.

Other geologists use their expertise to help solve environmental problems. Finding adequate sources of groundwater for the ever-burgeoning needs of communities and industries is becoming increasingly important, as is the monitoring of surface and underground water pollution and its cleanup. Geologic engineers help find safe locations for dams, waste-disposal sites, and power plants, as well as designing earthquake-resistant buildings.

Geologists are also engaged in making short- and long-range predictions about earthquakes and volcanic eruptions, and the potential destruction that may result from them.

LO2 Geology and the Formulation of Theories

The term **theory** has various meanings. In colloquial usage, it means a speculative or conjectural view of something—hence, the widespread belief that scientific theories are little more than unsubstantiated wild guesses. In scientific usage, however, a theory is a coherent explanation for one or several related natural phenomena supported by a large body of objective evidence. From a theory, scientists derive predictive statements that can be tested by observations and/or experiments so that their validity can be assessed. The law of universal gravitation is an example of a theory that describes the attraction between masses (an apple and Earth in the popularized account of Newton and his discovery).

Theories are formulated through the process known as the **scientific method**. This method is an orderly, logical approach that involves gathering and analyzing facts or data about the problem under consideration. Tentative explanations, or **hypotheses**, are then formulated to explain the observed phenomena. Next, the hypotheses are tested to see whether what was predicted actually occurs in a given situation. Finally, if one of the hypotheses is found, after repeated tests, to explain the phenomena, then the hypothesis is proposed as a theory. Remember, however, that in science, even a theory is still subject to further testing and refinement as new data become available.

The fact that a scientific theory can be tested and is subject to such testing separates it from other forms of human inquiry. Because scientific theories can be tested, they have the potential for being supported or even proven wrong. Accordingly, science must proceed without any appeal to beliefs or supernatural explanations, not because such beliefs or explanations are necessarily untrue, but because we have no way to investigate them. For this reason, science makes no claim about the existence or nonexistence of a supernatural or spiritual realm.

Each scientific discipline has certain theories that are of particular importance. In geology, the formulation of plate tectonic theory has changed the way geologists view Earth. Geologists now view Earth from a global perspective in which all

---

geology The science concerned with the study of Earth materials (minerals and rocks), surface and internal processes, and Earth history.

mineral A naturally occurring, inorganic, crystalline solid that has characteristic physical properties and a narrowly defined chemical composition.

rock A solid aggregate of one or more minerals, as in limestone and granite, or a consolidated aggregate of rock fragments, as in conglomerate, or masses of rocklike materials, such as coal and obsidian.

theory An explanation for some natural phenomenon that has a large body of supporting evidence. To be scientific, a theory must be testable (e.g., plate tectonic theory).

scientific method A logical, orderly approach that involves gathering data, formulating and testing hypotheses, and proposing theories.

hypothesis A provisional explanation for observations that is subject to continual testing. If well-supported by evidence, a hypothesis may be called a *theory*.

of its subsystems and cycles are interconnected, and Earth history is seen to be a continuum of interrelated events that are part of a global pattern of change.

LO3 How Does Geology Relate to the Human Experience?

You would probably be surprised at the extent to which geology pervades our everyday lives and the numerous references to geology in the arts, music, and literature.

In the field of music, Ferde Grofé's *Grand Canyon Suite* was no doubt inspired by the grandeur and timelessness of Arizona's Grand Canyon and its vast rock exposures (Figure 1.2). The rocks on the Island of Staffa in the Inner Hebrides provided the inspiration for Felix Mendelssohn's famous *Hebrides Overture.*

References to geology abound in *The German Legends of the Brothers Grimm.* Jules Verne's novel *Journey to the Center of the Earth* describes an expedition into Earth's interior. There is even a series of mystery books by Sarah Andrews that features the fictional geologist Em Hansen, who uses her knowledge of geology to solve crimes.

Geology has also played an important role in the history and culture of humankind. Empires throughout history have risen and fallen on the distribution and exploitation of natural resources. Wars have been fought for the control of such natural resources as oil and gas, and valuable minerals such as gold, silver, and diamonds.

LO4 How Does Geology Affect Our Everyday Lives?

The most obvious connection between geology and our everyday lives is when natural disasters strike. Less apparent, but equally significant, are the connections between geology and economic, social, and political issues.

Consider for example just how dependent we are on geology in our daily routines (Figure 1.3). Much of the electricity for our appliances comes from the

Many sketches and paintings depict rocks and landscapes realistically. Leonardo da Vinci's *Virgin of the Rocks* and *Virgin and Child with Saint Anne,* Giovanni Bellini's *Saint Francis in Ecstasy* and *Saint Jerome,* and Asher Brown Durand's *Kindred Spirits* are just a few examples by famous painters.

burning of coal, oil, natural gas, or uranium consumed in nuclear-generating plants. Geologists locate the coal, petroleum (oil and natural gas), and uranium. The copper or other metal wires through which electricity travels are manufactured from materials found as the result of mineral exploration. The concrete foundation (concrete is a mixture of clay, sand, or gravel, and limestone), drywall (made largely from the mineral gypsum), and windows (the mineral quartz is the principal ingredient in the manufacture of glass) of the buildings we live and work in owe their very existence to geologic resources.

Figure 1.2 Geology and Art
Ferde Grofé's Grand Canyon Suite was inspired by the beauty of Arizona's Grand Canyon, where sedimentary rock layers grandly document some of Earth's past history.

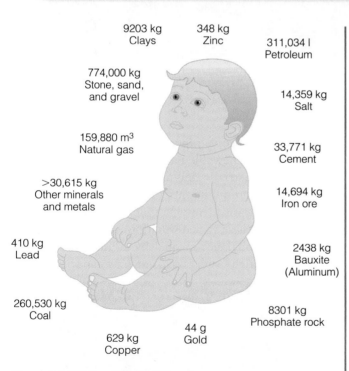

9203 kg
Clays

348 kg
Zinc

311,034 l
Petroleum

774,000 kg
Stone, sand,
and gravel

14,359 kg
Salt

159,880 m³
Natural gas

33,771 kg
Cement

>30,615 kg
Other minerals
and metals

14,694 kg
Iron ore

410 kg
Lead

2438 kg
Bauxite
(Aluminum)

260,530 kg
Coal

8301 kg
Phosphate rock

629 kg
Copper

44 g
Gold

Figure 1.3 Lifetime Mineral Usage
According to the Mineral Information Institute in Golden, Colorado, the average Amercan born in 2006 has a life expectancy of 77.8 years and will need 1,672,393 kg of minerals, metals, and fuels to sustain his or her standard of living over a lifetime. That is an average of 21,496 kg of mineral and energy resources per year for every man, woman, and child in the United States.

When we go to work, the car or public transportation we use is powered and lubricated by some type of petroleum by-product and is constructed of metal alloys and plastics. And the roads or rails we ride over come from geologic materials, such as gravel, asphalt, concrete, or steel. All of these items are the result of processing geologic resources.

As individuals and societies, we enjoy a standard of living that is obviously directly dependent on the consumption of geologic materials. We therefore need to be aware of how our use and misuse of geologic resources may affect the environment and develop policies that not only encourage management of our natural resources, but also allow for continuing economic development among all the world's nations.

LO5 Global Geologic and Environmental Issues Facing Humankind

Most scientists would argue that overpopulation is the greatest environmental problem facing the world today. The world's population reached 6.7 billion in 2007, and projections indicate that this number will grow by at least another billion people during the next

Figure 1.4 Offshore Oil Drilling
With increasing demand for energy, offshore drilling for oil and natural gas has increased in recent years.

two decades, bringing Earth's human population to more than 7.7 billion. Although this may not seem to be a geologic problem, remember that these people must be fed, housed, and clothed, and all with a minimal impact on the environment. Much of this population growth will be in areas that are already at risk from such hazards as earthquakes, tsunami, volcanic eruptions, and floods. Adequate water supplies must be found and protected from pollution. Additional energy resources will be needed to help fuel the economies of nations with ever-increasing populations (Figure 1.4). New techniques must be developed to reduce the use of our dwindling nonrenewable resource base and to increase our recycling efforts so that we can decrease our dependence on new sources of these materials.

The problems of overpopulation and how it affects the global ecosystem vary from country to country. For many poor and nonindustrialized countries, the problem is too many people and not enough food. For the more developed and industrialized countries,

it is too many people rapidly depleting both the nonrenewable and renewable natural resource base. And in the most industrially developed countries, it is people producing more pollutants than the environment can safely recycle on a human time scale. The common thread tying these varied situations together is an environmental imbalance created by a human population exceeding Earth's short-term carrying capacity.

Overpopulation is the greatest environmental problem facing the world today.

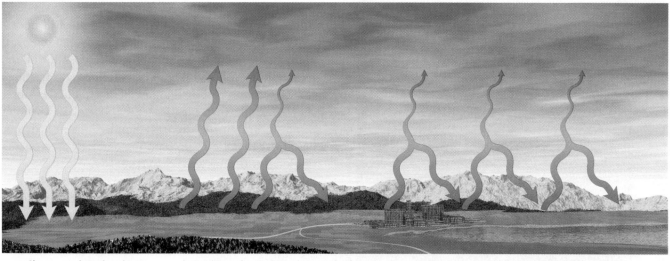

a. Short-wavelength radiation from the Sun that is not reflected back into space penetrates the atmosphere and warms Earth's surface.

b. Earth's surface radiates heat in the form of long–wavelength radiation back into the atmosphere, where some of it escapes into space. The rest is absorbed by greenhouse gases and water vapor and reradiated back toward Earth.

c. Increased concentrations of greenhouse gases trap more heat near Earth's surface, causing a general increase in surface and atmospheric temperatures, which leads to global warming.

Figure 1.5 The Greenhouse Effect and Global Warming

An excellent example of how Earth's various subsystems are interrelated is the relationship between the greenhouse effect and global warming. As a byproduct of respiration and the burning of organic material, carbon dioxide is a component of the global ecosystem and is constantly being recycled as part of the carbon cycle. The concern in recent years over the increase in atmospheric carbon dioxide levels is related to its role in the greenhouse effect.

The recycling of carbon dioxide between Earth's **crust** and atmosphere is an important climate regulator, because carbon dioxide and other gases, such as methane, nitrous oxide, chlorofluorocarbons, and water vapor, allow sunlight to pass through them but trap the heat reflected back from Earth's surface. This retention of heat is called the *greenhouse effect*. It results in an increase in the temperature of Earth's surface and, more importantly, its atmosphere, thus producing global warming (Figure 1.5).

Because of the increase in human-produced greenhouse gases during the past 200 years, many scientists are concerned that a global warming trend has already begun and will result in severe global climatic shifts. Presently, most climate researchers use a range of scenarios for greenhouse gas emissions when predicting future warming rates. State-of-the-art climate model simulations published in the *2007 Fourth Intergovernmental Panel on Climate Change* show a predicted increase in global average temperature from 2000 to 2100 of 1–3°C under the best conditions, to a 2.5–6.5° rise under business as usual conditions. These predicted increases in

crust Earth's outermost layer; the upper part of the lithosphere that is separated from the mantle by the Moho; divided into continental and oceanic crust.

Figure 1.6 Withered Corn Crop as a Result of Drought Conditions

temperatures are based on various scenarios that explore different global development pathways.

Regardless of which scenario is followed, the global temperature change will be uneven, with the greatest warming occurring in the higher latitudes of the Northern Hemisphere. As a consequence of this warming, rainfall patterns will shift dramatically. This will have a major effect on the largest grain-producing areas of the world, such as the American Midwest. Drier and hotter conditions will intensify the severity and frequency of droughts, leading to more crop failures and higher food prices (Figure 1.6). With such shifts in climate, Earth's deserts may expand, with a resulting decrease in valuable crop and grazing land.

Continued global warming will result in a rise in mean sea level, as icecaps and glaciers melt and contribute their water to the world's oceans. It is predicted that at the current rate of glacial melting, sea level will rise 21 cm by around 2050, thus increasing the number of people at risk from flooding in coastal areas by approximately 20 million!

Big Bang A model for the evolution of the universe in which a dense, hot state was followed by expansion, cooling, and a less-dense state.

We would be remiss, however, if we did not point out that some scientists are not convinced that the global warming trend is the direct result of increased human activity related to industrialization. They indicate that although the level of greenhouse gases has increased, we are still uncertain about their rate of generation and rate of removal, and whether the rise in global temperatures during the past century resulted from normal climatic variations through time or from human activity. Furthermore, they conclude that even if a general global warming trend occurs during the next hundred years, it is not certain that the dire predictions made by proponents of global warming will come true.

Earth, as we know, is a remarkably complex system, with many feedback mechanisms and interconnections throughout its various subsystems and cycles. It is very difficult to predict all of the consequences that global warming would have for atmospheric and oceanic circulation patterns and its ultimate effect on Earth's biota.

LO6 Origin of the Universe and Solar System, and Earth's Place in Them

How did the universe begin? What has been its history? What is its eventual fate, or is it infinite? These are just some of the basic questions people have asked and wondered about since they first looked into the nighttime sky and saw the vastness of the universe beyond Earth.

ORIGIN OF THE UNIVERSE: DID IT BEGIN WITH A BIG BANG?

Most scientists think that the universe originated about 14 billion years ago in what is popularly called the **Big Bang**. The Big Bang is a model for the evolution of the universe in which a dense, hot state was followed by expansion, cooling, and a less-dense state.

According to modern *cosmology* (the study of the origin, evolution, and nature of the universe), the universe has no edge and therefore no center. Thus, when the universe began, all matter and energy were compressed into an infinitely small high-temperature and high-density state in which both time and space were set at zero. Therefore, there is no "before the Big Bang," only what occurred after it. As demonstrated by Einstein's Theory of Relativity, space and time are unalterably linked to form a space–time continuum; that is, without space, there can be no time.

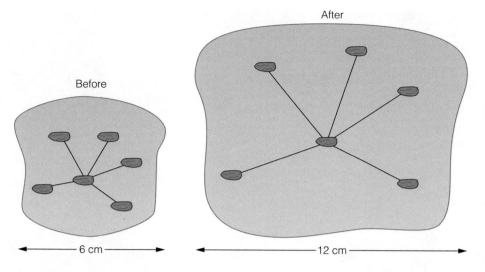

After

Before

6 cm

12 cm

Figure 1.7 The Expanding Universe
The motion of raisins in a rising loaf of raisin bread illustrates the relationship that exists between distance and speed and is analogous to an expanding universe. In this diagram, adjacent raisins are located 2 cm apart before the loaf rises. After one hour, any raisin is now 4 cm away from its nearest neighbor and 8 cm away from the next raisin over, and so on. Therefore, from the perspective of any raisin, its nearest neighbor has moved away from it at a speed of 2 cm per hour, and the next raisin over has moved away from it at a speed of 4 cm per hour. In the same way that raisins move apart in a rising loaf of bread, galaxies are receding from each other at a rate proportional to the distance between them.

One way to envision how velocity increases with increasing distance is by reference to the popular analogy of a rising loaf of raisin bread, in which the raisins are uniformly distributed throughout the loaf (Figure 1.7). As the dough rises, the raisins are uniformly pushed away from each other at velocities directly proportional to the distance between any two raisins. The farther away a given raisin is to begin with, the farther it must move to maintain the regular spacing during the expansion, and hence the greater its velocity must be. In the same way that raisins move apart in a rising loaf of bread, galaxies are receding from each other at a rate proportional to the distance between them, which is exactly what astronomers see when they observe the universe. By measuring this expansion rate, astronomers can calculate how long ago the galaxies were all together at a single point, which turns out to be about 14 billion years, the currently accepted age of the universe.

How do we know that the Big Bang took place approximately 14 billion years ago? Why couldn't the universe have always existed as we know it today? Two fundamental phenomena indicate that the Big Bang occurred: (1) the universe is expanding, and (2) it is permeated by background radiation.

When astronomers look beyond our own solar system, they observe that everywhere in the universe galaxies are moving away from each other at tremendous speeds. Edwin Hubble first recognized this phenomenon in 1929. By measuring the optical spectra of distant galaxies, Hubble noted that the velocity at which a galaxy moves away from Earth increases proportionally to its distance from Earth. He observed that the spectral lines (wavelengths of light) of the galaxies are shifted toward the red end of the spectrum; that is, the lines are shifted toward longer wavelengths. Galaxies receding from each other at tremendous speeds would produce such a redshift. This is an example of the *Doppler effect*, which is a change in the frequency of a sound, light, or other wave caused by movement of the wave's source relative to the observer.

Arno Penzias and Robert Wilson of Bell Telephone Laboratories made the second important observation that provided evidence of the Big Bang in 1965. They discovered that there is a pervasive background radiation of 2.7 Kelvin (K) above absolute zero (absolute zero equals –273°C; 2.7 K = –270.3°C) everywhere in the universe. This background radiation is thought to be the fading afterglow of the Big Bang.

Currently, cosmologists cannot say what it was like at time zero of the Big Bang, because they do not understand the physics of matter and energy under such extreme conditions. However, it is thought that during the first second following the Big Bang, the four basic forces—(1) *gravity* (the attraction of one body toward another), (2) *electromagnetic force* (combines electricity and magnetism into one force and binds atoms into molecules), (3) *strong nuclear force* (binds protons and neutrons together), and (4) *weak nuclear force* (responsible for the breakdown of an atom's nucleus, producing radioactive

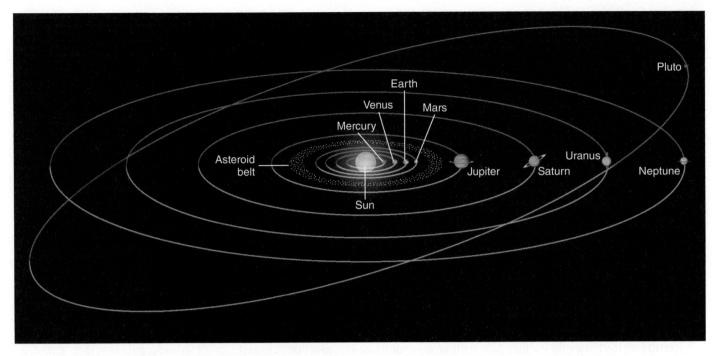

Figure 1.8 Diagrammatic Representation of the Solar System
This representation of the solar system shows the planets and their orbits around the Sun. On August 24, 2006, the International Astronomical Union downgraded Pluto from a planet to a dwarf planet. A dwarf planet has the same characteristics as a planet, except that it does not clear the neighborhood around its orbit. Pluto orbits among the icy debris of the Kuiper Belt, and therefore does not meet the criteria for a true planet.

decay)—separated and the universe experienced enormous expansion.

As the universe continued expanding and cooling, stars and galaxies began to form, and the chemical makeup of the universe changed. Initially, the universe was 100% hydrogen and helium, whereas today it is 98% hydrogen and helium and 2% all other elements by weight.

OUR SOLAR SYSTEM: ITS ORIGIN AND EVOLUTION

Our solar system, which is part of the Milky Way galaxy, consists of the Sun, eight planets, one dwarf planet (Pluto), 101 known moons or satellites (although this number keeps changing with the discovery of new moons and satellites surrounding the Jovian planets), a tremendous number of asteroids—most of which orbit the Sun in a zone between Mars and Jupiter—and millions of comets and meteorites, as well as interplanetary dust and gases (Figure 1.8). Any theory formulated to explain the origin and evolution of

solar nebula theory
A theory for the evolution of the solar system from a rotating cloud of gas.

our solar system must therefore take into account its various features and characteristics.

Many scientific theories for the origin of the solar system have been proposed, modified, and discarded since the French scientist and philosopher René Descartes first proposed, in 1644, that the solar system formed from a gigantic whirlpool within a universal fluid. Today, the **solar nebula theory** for the origin of our solar system involves the condensation and collapse of interstellar material in a spiral arm of the Milky Way galaxy (Figure 1.9).

The collapse of this cloud of gases and small grains into a counterclockwise-rotating disk concentrated about 90% of the material in the central part of the disk and formed an embryonic Sun, around which swirled a rotating cloud of material called a *solar nebula*. Within this solar nebula were localized eddies in which gases and solid particles condensed. During the condensation process, gaseous, liquid, and solid particles began to accrete into ever-larger masses called *planetesimals*, which collided and grew in size and mass until they eventually became planets.

The composition and evolutionary history of the planets are a consequence, in part, of their distance from the Sun. The **terrestrial planets**—Mercury, Venus, Earth, and Mars—so named because they are

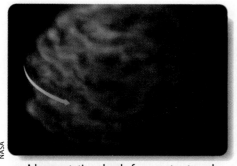

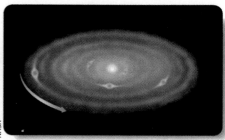

a. A huge rotating cloud of gas contracts and flattens

b. to form a disk of gas and dust with the Sun forming in the center

c. and eddies gathering up material to form planets.

Figure 1.9 Solar Nebula Theory
According to the currently accepted theory for the origin of our solar system, the planets and the Sun formed from a rotating cloud of gas.

similar to *terra*, Latin for "earth," are all small and composed of rock and metallic elements that condensed at the high temperatures of the inner nebula. The **Jovian planets**—Jupiter, Saturn, Uranus, and Neptune—so named because they resemble Jupiter (the Roman god was also called Jove), all have small rocky cores compared to their overall size, and are composed mostly of hydrogen, helium, ammonia, and methane, which condense at low temperatures.

While the planets were accreting, material that had been pulled into the center of the nebula also condensed, collapsed, and was heated to several million degrees by gravitational compression. The result was the birth of a star, our Sun.

During the early accretionary phase of the solar system's history, collisions between various bodies were common, as indicated by the craters on many planets and moons. Asteroids probably formed as planetesimals in a localized eddy between what eventually became Mars and Jupiter in much the same way that other planetesimals formed the terrestrial planets. The tremendous gravitational field of Jupiter, however, prevented this material from ever accreting into a planet. Comets, which are interplanetary bodies composed of loosely bound rocky and icy materials, are thought to have condensed near the orbits of Uranus and Neptune.

EARTH: ITS PLACE IN OUR SOLAR SYSTEM

Some 4.6 billion years ago, various planetesimals in our solar system gathered enough material together to form Earth and the other planets. Scientists think that this early Earth was probably cool, of generally uniform composition and density throughout, and composed mostly of silicates, compounds consisting of silicon and oxygen, iron and magnesium oxides, and smaller amounts

of all the other chemical elements. Subsequently, when the combination of meteorite impacts, gravitational compression, and heat from radioactive decay increased the temperature of Earth enough to melt iron and nickel, this homogeneous composition disappeared and was replaced by a series of concentric layers of differing composition and density, resulting in a differentiated planet (Figure 1.10).

This differentiation into a layered planet is probably the most significant event in Earth's history. Not only did it lead to the formation of a crust and eventually continents, but it also was probably responsible for the emission of gases from the interior that eventually led to the formation of the oceans and atmosphere.

terrestrial planets Any of the four innermost planets (Mercury, Venus, Earth, and Mars). They are all small and have high mean densities, indicating that they are composed of rock and metallic elements.

Jovian planets Any of the four planets (Jupiter, Saturn, Uranus, and Neptune) that resemble Jupiter. All are large and have low mean densities, indicating that they are composed mostly of lightweight gases, such as hydrogen and helium, and frozen compounds, such as ammonia and methane.

LO7 Why Earth Is a Dynamic and Evolving Planet

Earth is a dynamic planet that has continuously changed during its 4.6-billion-year existence. The size, shape, and geographic distribution of continents and ocean basins have changed throughout time; the composition of the atmosphere has evolved; and life-forms

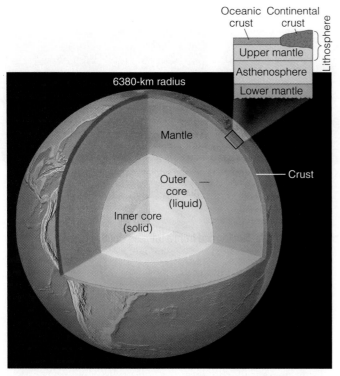

Figure 1.10 **Cross Section of Earth Illustrating the Core, Mantle, and Crust**
The enlarged portion shows the relationship between the lithosphere (composed of the continental crust, oceanic crust, and solid upper mantle) and the underlying asthenosphere and lower mantle.

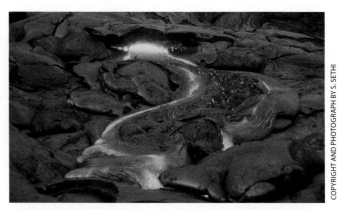

Figure 1.11 **Molten Lava Flowing Out of a Small Lava Tube in Hawaii.**

existing today differ from those that lived during the past. Mountains and hills have been worn away by erosion, and the forces of wind, water, and ice have sculpted a diversity of landscapes. Volcanic eruptions (Figure 1.11) and earthquakes reveal an active interior, and folded and fractured rocks are testimony to the tremendous power of Earth's internal forces.

Earth consists of three concentric layers: the core, the mantle, and the crust (Figure 1.10). This orderly division results from density differences between the layers as a function of variations in composition, temperature, and pressure.

The core has a calculated density of 10–13 grams per cubic centimeter (g/cm³) and occupies about 16% of Earth's total volume. Seismic (earthquake) data indicate that the core consists of a small, solid inner region and a larger, apparently liquid, outer portion. Both are thought to consist mostly of iron and a small amount of nickel.

The mantle surrounds the core and comprises about 83% of Earth's volume. It is less dense than the core (3.3–5.7 g/cm³) and is thought to be composed mostly of *peridotite*, a dark, dense igneous rock containing abundant iron and magnesium. The mantle can be divided into three distinct zones based on physical characteristics. The lower mantle is solid and forms most of the volume of Earth's interior. The **asthenosphere** surrounds the lower mantle. It has the same composition as the lower mantle, but behaves plastically and flows slowly. Partial melting within the asthenosphere generates **magma** (molten material), some of which rises to the surface, because it is less dense than the rock from which it was derived. The upper mantle surrounds the asthenosphere. The solid upper mantle and the overlying crust constitute the lithosphere, which is broken into numerous individual pieces called **plates** that move over the asthenosphere, partially as a result of underlying *convection cells* (Figure 1.12). Interactions of these plates are responsible for such phenomena as earthquakes, volcanic eruptions, and the formation of mountain ranges and ocean basins.

The crust, Earth's outermost layer, consists of two types. *Continental crust* is thick (20–90 km), has an average density of 2.7 g/cm³, and contains considerable silicon and aluminum. *Oceanic crust* is thin (5–10 km), denser than continental crust (3.0 g/cm³), and is composed of the dark igneous rocks *basalt* and *gabbro*.

PLATE TECTONIC THEORY

The recognition that the lithosphere is divided into rigid plates that move over the asthenosphere forms

asthenosphere The part of the mantle that lies below the lithosphere; it behaves plastically and flows slowly.

magma Molten rock material generated within Earth.

plate An individual segment of the lithosphere that moves over the asthenosphere.

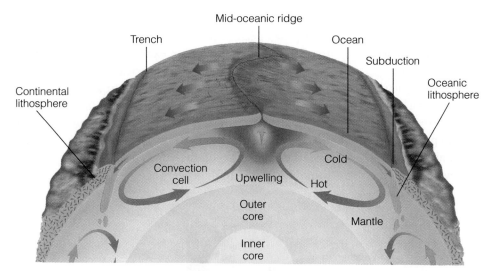

Figure 1.12 Movement of Earth's Plates
Earth's plates are thought to move partially as a result of underlying mantle convection cells, in which warm material from deep within Earth rises toward the surface, cools, and then upon losing heat, descends back into the interior as shown in this diagrammatic cross section.

the foundation of **plate tectonic theory** (Figure 1.13). Zones of volcanic activity, earthquakes, or both mark most plate boundaries. Along these boundaries, plates separate (diverge), collide (converge), or slide sideways past each other.

The acceptance of plate tectonic theory is recognized as a major milestone in the geologic sciences, comparable to the revolution that Darwin's Theory of Evolution caused in biology. Plate tectonics has provided a framework for interpreting the composition, structure, and

plate tectonic theory
The theory holding that large segments of Earth's outer part (lithospheric plates) move relative to one another.

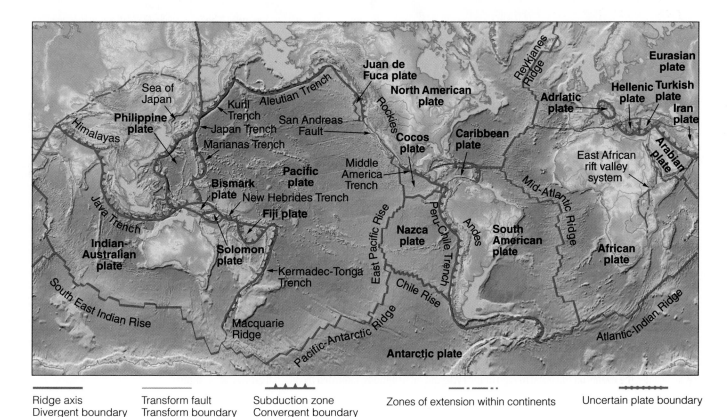

| Ridge axis
Divergent boundary | Transform fault
Transform boundary | Subduction zone
Convergent boundary | Zones of extension within continents | Uncertain plate boundary |

Figure 1.13 Earth's Plates
Earth's lithosphere is divided into rigid plates of various sizes that move over the asthenosphere.

internal processes of Earth on a global scale. It also has led to the realization that the continents and ocean basins are part of a lithosphere–atmosphere–hydrosphere system that evolved together with Earth's interior.

A revolutionary concept when it was proposed in the 1960s, plate tectonic theory has had far-reaching consequences in all fields of geology, because it provides the basis for relating many seemingly unrelated phenomena, such as the formation and occurrence of Earth's natural resources, as well as the distribution and evolution of the world's biota. Furthermore, the impact of plate tectonic theory has been particularly notable in the interpretation of Earth's history. For example, the Appalachian Mountains in eastern North America and the mountain ranges of Greenland, Scotland, Norway, and Sweden are not the result of unrelated mountain-building episodes but, rather, are part of a larger mountain-building event that involved the closing of an ancient Atlantic Ocean and the formation of the supercontinent Pangaea approximately 251 million years ago.

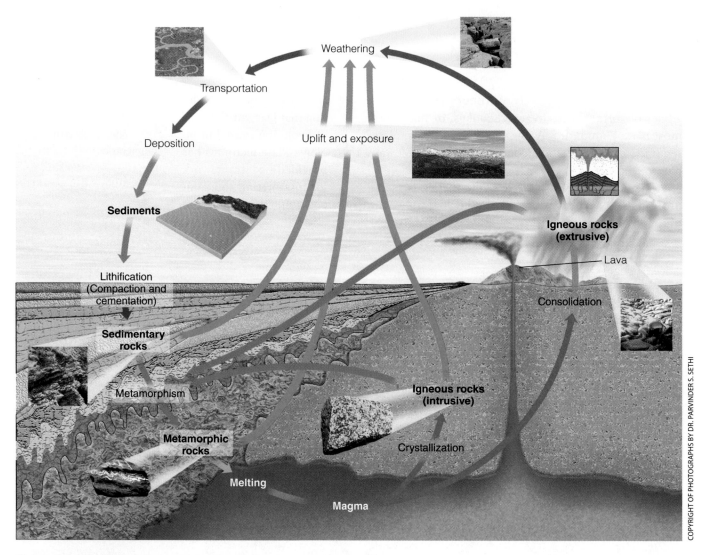

Figure 1.14 The Rock Cycle
This cycle shows the interrelationships between Earth's internal and external processes and how the three major rock groups are related. An ideal cycle includes the events on the outer margin of the cycle, but interruptions, indicated by internal arrows, are common.

LO8 The Rock Cycle

A rock is an aggregate of minerals, which are naturally occurring, inorganic, crystalline solids that have definite physical and chemical properties. Minerals are composed of elements such as oxygen, silicon, and aluminum, and elements are made up of atoms, the smallest particles of matter that retain the characteristics of an element. More than 3,500 minerals have been identified and described, but only about a dozen make up the bulk of the rocks in Earth's crust.

Geologists recognize three major groups of rocks—igneous, sedimentary, and metamorphic—each of which is characterized by its mode of formation. The **rock cycle** is a pictorial representation of events leading to the origin, destruction and/or changes, and reformation of rocks as a consequence of Earth's internal and surface processes (Figure 1.14). Furthermore, it shows that the three major rock groups are interrelated; that is, any rock type can be derived from the others.

Igneous rocks result when magma or lava crystallizes, or volcanic ejecta, such as ash, accumulate and consolidate. As magma cools, minerals crystallize, and the resulting rock is characterized by interlocking mineral grains. Magma that cools slowly beneath the surface produces *intrusive igneous rocks* (Figure 1.15a); magma that cools at the surface produces *extrusive igneous rocks* (Figure 1.15b).

Rocks exposed at Earth's surface are broken into particles and dissolved by various weathering processes. The particles and dissolved materials may be transported by wind, water, or ice and eventually deposited as *sediment*. This sediment may then be compacted or cemented (lithified) into sedimentary rock.

Sedimentary rocks form in one of three ways: (1) consolidation of mineral or rock fragments, (2) precipitation of mineral matter from solution, or (3) compaction of plant or animal remains (Figure 1.15c, d). Because sedimentary rocks form at or near Earth's surface, geologists can infer about the environment in which they were deposited, the transporting agent, and perhaps even something about the source from which the sediments were derived (see Chapter 6). Accordingly, sedimentary rocks are especially useful for interpreting Earth history.

Metamorphic rocks result from the alteration of other rocks, usually beneath the surface, by heat, pressure, and the chemical activity of fluids. For example, marble—a rock preferred by many sculptors and builders—is a metamorphic rock produced when the agents of metamorphism are applied to the sedimentary rocks limestone or dolostone. Metamorphic rocks are either *foliated* (Figure 1.15e) or *nonfoliated* (Figure 1.15f). Foliation, the parallel alignment of minerals due to pressure, gives the rock a layered or banded appearance.

How Are the Rock Cycle and Plate Tectonics Related?

Interactions between plates determine, to some extent, which of the three rock groups will form (Figure 1.16). For example, when plates converge, heat and pressure generated along the plate boundary may lead to igneous activity and metamorphism within the descending oceanic plate, thus producing various igneous and metamorphic rocks.

Some of the sediments and sedimentary rocks on the descending plate are melted, whereas other sediments and sedimentary rocks along the boundary of the nondescending plate are metamorphosed by the heat and pressure generated along the converging plate boundary. Later, the mountain range or chain of volcanic islands formed along the convergent plate boundary will be weathered and eroded, and the new sediments will be transported to the ocean, where they will be deposited.

rock cycle A group of processes through which Earth materials may pass as they are transformed from one major rock type to another.

igneous rock Any rock formed by cooling and crystallization of magma or lava or the consolidation of pyroclastic materials.

sedimentary rock Any rock composed of sediment, such as limestone and sandstone.

metamorphic rock Any rock that has been changed from its original condition by heat, pressure, and the chemical activity of fluids, as in marble and slate.

a. **Granite**, an intrusive igneous rock.

b. **Basalt**, an extrusive igneous rock.

c. **Conglomerate**, a sedimentary rock formed by the consolidation of rounded rock fragments.

d. **Limestone**, a sedimentary rock formed by the extraction of mineral matter from seawater by organisms or by the inorganic precipitation of the mineral calcite from seawater.

e. **Gneiss**, a foliated metamorphic rock.

f. **Quartzite**, a nonfoliated metamorphic rock.

Figure 1.15 Hand Specimens of Common Igneous, Sedimentary, and Metamorphic Rocks

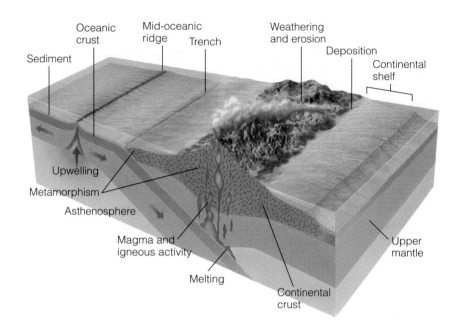

Figure 1.16 Plate Tectonics and the Rock Cycle
Plate movement provides the driving mechanism that recycles Earth materials. This block diagram shows how the three major rock groups—igneous, sedimentary, and metamorphic—are recycled through both the continental and oceanic regions. Subducting plates are partially melted to produce magma, which rises and either crystallizes beneath Earth's surface as intrusive igneous rock or spills out on the surface, solidifying as extrusive igneous rock. Rocks exposed at the surface are weathered and eroded to produce sediments that are transported and eventually lithified into sedimentary rocks. Metamorphic rocks result from pressure generated along converging plates or adjacent to rising magma.

LO9 Organic Evolution and the History of Life

Plate tectonic theory provides us with a model for understanding the internal workings of Earth and their effect on Earth's surface. The theory of **organic evolution** (whose central thesis is that all present-day organisms are related and that they have descended with modifications from organisms that lived in the past) provides the conceptual framework for understanding the history of life. Together, the theories of plate tectonics and organic evolution have changed the way we view our planet, and we should not be surprised at the intimate association between them. Although the relationship between plate tectonic processes and the evolution of life is incredibly complex, paleontological data provide indisputable evidence of the influence of plate movement on the distribution of organisms.

The publication in 1859 of Darwin's *On the Origin of Species by Means of Natural Selection* revolutionized biology and marked the beginning of modern evolutionary biology. With its publication, most naturalists recognized that evolution provided a unifying theory that explained an otherwise encyclopedic collection of biologic facts.

When Darwin proposed his theory of organic evolution, he cited a wealth of supporting evidence, including the way organisms are classified, embryology, comparative anatomy, the geographic distribution of organisms, and, to a limited extent, the fossil record.

Furthermore, Darwin proposed that *natural selection*, which results in the survival to reproductive age of those organisms best adapted to their environment, is the mechanism that accounts for evolution.

Perhaps the most compelling evidence in favor of evolution can be found in the fossil record. Just as the geologic record allows geologists to interpret physical events and conditions in the geologic past, **fossils**, which are the remains or traces of once-living organisms, not only provide evidence that evolution has occurred but also demonstrate that Earth has a history extending beyond that recorded by humans. The succession of fossils in the rock record provides geologists with a means for dating rocks and allowed for a relative geologic timescale to be constructed in the 1800s.

organic evolution
The theory holding that all living things are related and that they descended with modification from organisms that lived during the past.

fossils The remains or traces of once-living organisms.

LO10 Geologic Time and Uniformitarianism

An appreciation of the immensity of geologic time is central to understanding the evolution of Earth and its biota. Indeed, time is one of the main aspects that sets geology apart from the other sciences, except astronomy.

Most people have difficulty comprehending geologic time, because they tend to think in terms of the human perspective—seconds, hours, days, and years. Ancient history is what occurred hundreds or even thousands of years ago. When geologists talk of ancient geologic history, however, they are referring to events that happened hundreds of millions or even billions of years ago.

It is equally important to remember that Earth goes through cycles of much longer duration than the human perspective of time. Although they may have disastrous effects on the human species, global warming and cooling are part of a larger cycle that has resulted in numerous glacial advances and retreats during the past 1.8 million years.

The **geologic time scale** subdivides geologic time into a hierarchy of increasingly shorter time intervals; each time subdivision has a specific name. The geologic time scale resulted from the work of many 19th-century geologists, who pieced together information from numerous rock exposures and constructed a chronology based on changes in Earth's biota through time. Subsequently, with the discovery of radioactivity in 1895 and the development of various radiometric dating techniques, geologists have been able to assign numerical ages (also known as absolute ages) in years to the subdivisions of the geologic time scale (Figure 1.17).

One of the cornerstones of geology is the **principle of uniformitarianism**, which is based on the premise that present-day processes have operated throughout geologic time. Therefore, to understand and interpret geologic events from evidence preserved in rocks, we must first understand present-day processes and their results. In fact, uniformitarianism fits in completely with the system approach that we are following for the study of Earth.

geologic time scale A chart arranged so that the designation for the earliest part of geologic time appears at the bottom followed upward by progressively younger time designations.

principle of uniformitarianism A principle holding that we can interpret past events by understanding present-day processes, based on the idea that natural processes have always operated in the same way.

Uniformitarianism is a powerful principle that allows us to use present-day processes as the basis for interpreting the past and for predicting potential future events. We should keep in mind, however, that uniformitarianism does not exclude sudden or catastrophic events such as volcanic eruptions, earthquakes, tsunami, landslides, or floods.

What uniformitarianism means is that even though the rates and intensities of geologic processes have varied during the past, the physical and chemical laws of nature have remained the same. Although Earth is in a dynamic state of change and has been ever since it formed, the processes that shaped it during the past are the same ones operating today.

LO11 How Does the Study of Geology Benefit Us?

The most meaningful lesson to learn from the study of geology is that Earth is an extremely complex planet in which interactions are taking place between its various subsystems and have been for the past 4.6 billion years. If we want to ensure the survival of the human species, we must understand how the various subsystems work and interact with each other and, more importantly, how our actions affect the delicate balance among these systems. We can do this, in part, by studying what has happened in the past, particularly on the global scale, and by using that information to try to determine how our actions might affect the delicate balance between Earth's various subsystems in the future.

The study of geology goes beyond learning numerous facts about Earth. In fact, we don't just study geology—we *live* it. Geology is an integral part of our lives. Our standard of living depends directly on our consumption of natural resources, which formed millions and billions of years ago. However, the way we consume natural resources and interact with the environment, as individuals and as a society, also determines our ability to pass on this standard of living to the next generation.

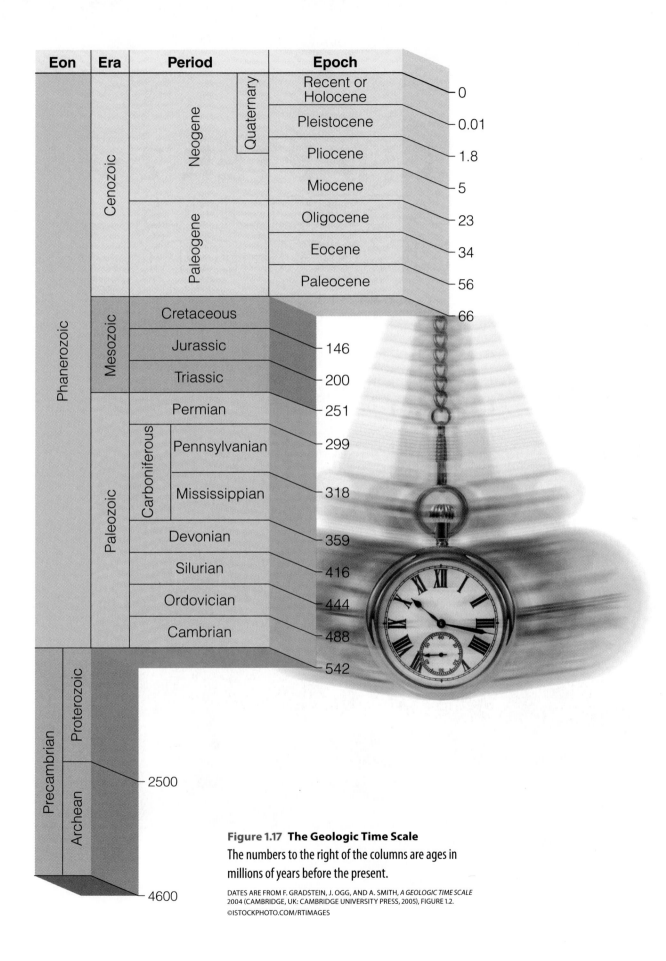

Eon	Era	Period		Epoch	
Phanerozoic	Cenozoic	Neogene	Quaternary	Recent or Holocene	0
				Pleistocene	0.01
				Pliocene	1.8
				Miocene	5
		Paleogene		Oligocene	23
				Eocene	34
				Paleocene	56
	Mesozoic	Cretaceous			66
		Jurassic			146
		Triassic			200
	Paleozoic	Permian			251
		Carboniferous	Pennsylvanian		299
			Mississippian		318
		Devonian			359
		Silurian			416
		Ordovician			444
		Cambrian			488
					542
Precambrian	Proterozoic				
	Archean				2500
					4600

Figure 1.17 The Geologic Time Scale
The numbers to the right of the columns are ages in millions of years before the present.

DATES ARE FROM F. GRADSTEIN, J. OGG, AND A. SMITH, *A GEOLOGIC TIME SCALE*
2004 (CAMBRIDGE, UK: CAMBRIDGE UNIVERSITY PRESS, 2005), FIGURE 1.2.
©ISTOCKPHOTO.COM/RTIMAGES

CHAPTER 2
PLATE TECTONICS: A UNIFYING THEORY

On May 12, 2008, a 7.9 magnitude earthquake devastated Sichuan province in China, killing approximately 69,000 people, and leaving millions homeless.

© ISTOCKPHOTO.COM/RUI PESTANA

Introduction

"The idea that Earth's past geography was different from today is not new."

Imagine it is the day after Christmas, December 26, 2004, and you are vacationing on a beautiful beach in Thailand. You look up from the book you're reading to see the sea suddenly retreat from the shoreline. Within minutes of this unusual event, a powerful tsunami will sweep over your resort and everything in its path for several kilometers inland. Within hours, the coasts of Indonesia, Sri Lanka, India, Thailand, Somalia, Myanmar, Malaysia, and the Maldives will be inundated by the deadliest tsunami in history and more than 220,000 people will die.

One year earlier, on December 26, 2003, violent shaking from an earthquake awakened hundreds of thousands of people in the Bam area of southeastern Iran. When the magnitude-6.6 earthquake was over, an estimated 43,000 people were dead, at least 30,000 were injured, and approximately 75,000 survivors were left homeless.

Now go back another 12 and a half years to June 15, 1991, when Mount Pinatubo in the Philippines erupted violently, discharging huge quantities of ash and gases into the atmosphere. Fortunately, in this case, warnings of an impending eruption were broadcast and heeded, resulting in the evacuation of 200,000 people from areas around the volcano. Unfortunately, the eruption still caused at least 364 deaths not only from the eruption, but also from the ensuing mudflows.

What do these three recent tragic events have in common? They are part of the dynamic interactions involving Earth's plates. When two plates come together, one plate is pushed or pulled under the other plate, triggering large earthquakes. If conditions are right, earthquakes can produce a tsunami.

As the descending plate moves downward and is assimilated into Earth's interior, magma is generated. Being less dense than the surrounding material, the magma rises toward the surface, where it may erupt as a volcano.

If you're like most people, you probably have only a vague notion of what plate tectonic theory is. Yet plate tectonics affects all of us. Volcanic eruptions, earthquakes, and tsunami are the result of interactions between plates. Global weather patterns and oceanic currents are caused, in part, by the configuration of the continents and ocean basins. The formation

LEARNING OUTCOMES

After reading this unit, you should be able to do the following:

LO1 Review early ideas about continental drift

LO2 Explain the evidence for continental drift

LO3 Describe Earth's magnetic field

LO4 Explain paleomagnetism and polar wandering

LO5 Explain magnetic reversals and seafloor spreading

LO6 Explain plate tectonic theory: a unifying theory

LO7 Identify the three types of plate boundaries

LO8 Describe hot spots and mantle plumes

LO9 Explain plate movement and motion

LO10 Explain the driving mechanism of plate tectonics

LO11 Recognize the role of plate tectonics in the distribution of natural resources

LO12 Recognize the role of plate tectonics in the distribution of life

© ISTOCKPHOTO.COM/GEORGE CLERK

and distribution of many natural resources are related to plate movement, and thus have an impact on the economic well-being and political decisions of nations. It is therefore important to understand this unifying theory, not only because it affects us as individuals and as citizens of nation-states, but also because it ties together many aspects of the geology you will be studying.

LO1 Early Ideas about Continental Drift

The idea that Earth's past geography was different from today is not new. The earliest maps showing the east coast of South America and the west coast of Africa probably provided people with the first evidence that continents may have once been joined together, then broken apart and moved to their present positions. As far back as 1620, Sir Francis Bacon commented on the similarity of the shorelines of western Africa and eastern South America. However, he did not make the connection that the Old and New Worlds might once have been joined together.

During the late 19th century, the Austrian geologist Edward Suess noted the similarities between the Late Paleozoic plant fossils of India, Australia, South Africa, and South America, as well as evidence of glaciation in the rock sequences of these continents. The plant fossils comprise a unique flora that occurs in the coal layers just above the glacial deposits of these southern continents. This flora is very different from the contemporaneous coal swamp flora of the northern continents, and is collectively known as the *Glossopteris* flora after its most conspicuous genus (Figure 2.1).

Suess also proposed the name Gondwanaland (or *Gondwana* as we will use here) for a supercontinent composed of the aforementioned southern continents. Abundant fossils of the *Glossopteris* flora are found in coal

Glossopteris flora A Late Paleozoic association of plants found only on the Southern Hemisphere continents and India; named for its best-known genus, *Glossopteris*.

continental drift The theory that the continents were joined into a single landmass that broke apart with the various fragments (continents) moving with respect to one another.

Pangaea The name Alfred Wegener proposed for a supercontinent consisting of all Earth's landmasses at the end of the Paleozoic Era.

Figure 2.1 Fossil *Glossopteris* Leaves
Plant fossils, such as these *Glossopteris* leaves from the Upper Permian Dunedoo Formation in Australia, are found on all five of the Gondwana continents. Their presence on continents with widely varying climates today is evidence that the continents were at one time connected. The distribution of the plants at that time was in the same climatic latitudinal belt.

beds in Gondwana, a province in India. Suess thought these southern continents were at one time connected by land bridges over which plants and animals migrated. Thus, in his view, the similarities of fossils on these continents were due to the appearance and disappearance of the connecting land bridges.

ALFRED WEGENER AND THE CONTINENTAL DRIFT HYPOTHESIS

Alfred Wegener, a German meteorologist, is generally credited with developing the hypothesis of **continental drift**. In his monumental book, *The Origin of Continents and Oceans* (first published in 1915), Wegener proposed that all landmasses were originally united in a single supercontinent that he named **Pangaea**, from the Greek meaning "all land." Wegener portrayed his grand concept of continental movement in a series of maps showing the breakup of Pangaea and the movement of the various continents to their present-day locations. Wegener amassed a tremendous amount of geologic, paleontologic, and climatologic evidence in support of continental drift; however, initial reaction of scientists to his then-heretical ideas can best be described as mixed.

Nevertheless, the South African geologist Alexander du Toit further developed Wegener's arguments and gathered more geologic and paleontologic evidence in support of continental drift. In 1937, du Toit published

Our Wandering Continents, in which he contrasted the glacial deposits of Gondwana with coal deposits of the same age found in the continents of the Northern Hemisphere. To resolve this apparent climatologic paradox, du Toit moved the Gondwana continents to the South Pole and brought the northern continents together such that the coal deposits were located at the equator. He named this northern landmass *Laurasia*. It consisted of present-day North America, Greenland, Europe, and Asia (except for India).

LO2 What Is the Evidence for Continental Drift?

What then was the evidence Wegener, du Toit, and others used to support the hypothesis of continental drift? It includes the fit of the shorelines of continents, the appearance of the same rock sequences and mountain ranges of the same age on continents now widely separated, the matching of glacial deposits and paleoclimatic zones, and the similarities of many extinct plant and animal groups whose fossil remains are found today on widely separated continents. Wegener and his supporters argued that this vast amount of evidence from a variety of sources surely indicated that the continents must have been close together in the past.

CONTINENTAL FIT

Wegener, like some before him, was impressed by the close resemblance between the coastlines of continents on opposite sides of the Atlantic Ocean, particularly South America and Africa. He cited these similarities as partial evidence that the continents were at one time joined together as a supercontinent that subsequently split apart. As his critics pointed out, though, the configuration of coastlines results from erosional and depositional processes and therefore is continuously being modified. So, even if the continents had separated during the Mesozoic Era, as Wegener proposed, it is not likely that the coastlines would fit exactly.

A more realistic approach is to fit the continents together along the continental slope where erosion would be minimal. In 1965, Sir Edward Bullard, an English geophysicist, and two associates showed that the best fit between the continents occurs at a depth of about 2,000 m (Figure 2.2). Since then, other reconstructions using the latest ocean basin data have confirmed the close fit between continents when they are reassembled to form Pangaea.

Figure 2.2 Continental Fit
When continents are placed together based on their outlines, the best fit isn't along their present-day coastlines, but rather along the continental slope at a depth of about 2000 m, where erosion would be minimal.

SIMILARITY OF ROCK SEQUENCES AND MOUNTAIN RANGES

If the continents were at one time joined, then the rocks and mountain ranges of the same age in adjoining locations on the opposite continents should closely match. Such is the case for the Gondwana continents (Figure 2.3). Marine, nonmarine, and glacial rock sequences of the Pennsylvanian to Jurassic periods are almost identical on all five Gondwana continents, strongly indicating that they were joined at one time.

Furthermore, the trends of several major mountain ranges also support the hypothesis of continental drift. These mountain ranges seemingly end at the coastline of one continent only to apparently continue on another continent across the ocean. The folded Appalachian Mountains of North America, for example, trend northeastward through the eastern United States and Canada and terminate abruptly at the Newfoundland coastline. Mountain ranges of the same age and deformational style are found in eastern Greenland, Ireland, Great Britain, and Norway. So, even though the Appalachian Mountains and their equivalent-age mountain ranges in Great Britain are currently separated by the Atlantic Ocean, they form an essentially continuous mountain

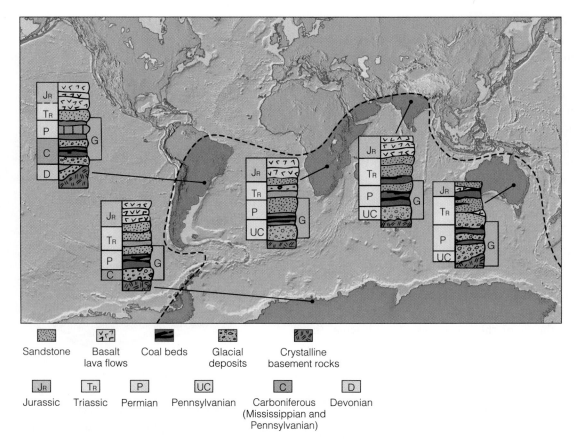

Sandstone	Basalt lava flows	Coal beds	Glacial deposits	Crystalline basement rocks

Jʀ	Tʀ	P	UC	C	D
Jurassic	Triassic	Permian	Pennsylvanian	Carboniferous (Mississippian and Pennsylvanian)	Devonian

Figure 2.3 Similarity of Rock Sequences on the Gondwana Continents

Sequences of marine, nonmarine, and glacial rocks of Pennsylvanian (UC) to Jurassic (Jʀ) age are nearly the same on all five Gondwana continents (South America, Africa, India, Australia, and Antarctica). These continents are widely separated today and have different environments and climates ranging from tropical to polar. Thus, the rocks forming on each continent are very different. When the continents were all joined together in the past, however, the environments of adjacent continents were similar and the rocks forming in those areas were similar. The range indicated by G in each column is the age range (Carboniferous–Permian) of the *Glossopteris* flora. From Robert J. Roster, *General Geology*, 5th Edition, © 1998. Reprinted by permission of Pearson Education, Inc., Upper Saddle River, NJ.

range when the continents are positioned next to each other as they were during the Paleozoic Era.

GLACIAL EVIDENCE

During the Late Paleozoic Era, massive glaciers covered large continental areas of the Southern Hemisphere. Evidence for this glaciation includes layers of till (sediments deposited by glaciers) and striations (scratch marks) in the bedrock beneath the till (Figure 2.4b). Fossils and sedimentary rocks of the same age from the Northern Hemisphere, however, give no indication of glaciation. Fossil plants found in coals indicate that the Northern Hemisphere had a tropical climate during the time that the Southern Hemisphere was glaciated.

All of the Gondwana continents except Antarctica are currently located near the equator in subtropical to tropical climates. Mapping of glacial striations

in bedrock in Australia, India, and South America indicates that the glaciers moved from the areas of the present-day oceans onto land. This would be highly unlikely because large continental glaciers (such as occurred on the Gondwana continents during the Late Paleozoic Era) flow outward from their central area of accumulation toward the sea.

If the continents did not move during the past, one would have to explain how glaciers moved from the oceans onto land and how large-scale continental glaciers formed near the equator. But if the continents are reassembled as a single landmass with South Africa located at the South Pole, the direction of movement of Late Paleozoic continental glaciers makes sense (Figure 2.4). Furthermore, this geographic arrangement places the northern continents nearer the tropics, which is consistent with the fossil and climatologic evidence from Laurasia.

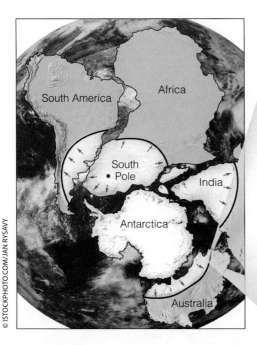

a. When the Gondwana continents are placed together so that South Africa is located at the South Pole, the glacial movements indicated by striations (red arrows) found on rock outcrops on each continent make sense. In this situation, the glacier (white area) is located in a polar climate and has moved radially outward from its thick central area toward its periphery.

b. Glacial striations (scratch marks) on an outcrop of Permian-age bedrock exposed at Hallet's Cove, Australia, indicate the general direction of glacial movement more than 200 million years ago. As a glacier moves over a continent's surface, it grinds and scratches the underlying rock. The glacial striations that are preserved on a rock's surface thus provide evidence of the direction (red arrows) the glacier moved at that time.

Figure 2.4 Glacial Evidence Indicating Continental Drift

FOSSIL EVIDENCE

Some of the most compelling evidence for continental drift comes from the fossil record (Figure 2.5). Fossils of the *Glossopteris* flora are found in equivalent Pennsylvanian and Permian-age coal deposits on all five Gondwana continents. The *Glossopteris* flora is characterized by the seed fern *Glossopteris* (Figure 2.1), as well as by many other distinctive and easily identifiable plants. Pollen and spores of plants can be dispersed over great distances by wind; however, *Glossopteris*-type plants produced seeds that are too large to have been carried by winds. Even if the seeds had floated across the ocean, they probably would not have remained viable for any length of time in saltwater.

The present-day climates of South America, Africa, India, Australia, and Antarctica range from tropical to polar and are much too diverse to support the type of plants in the *Glossopteris* flora. Wegener therefore reasoned that these continents must once have been joined so that these widely separated localities were all in the same latitudinal climatic belt (Figure 2.5).

The fossil remains of animals also provide strong evidence for continental drift. One of the best examples is *Mesosaurus*, a freshwater reptile whose fossils are found in Permian-age rocks in certain regions of Brazil and South Africa and nowhere else in the world (Figure 2.5). Because the physiologies of freshwater and marine animals are completely different, it is hard to imagine how a freshwater reptile could have swum across the Atlantic Ocean and found a freshwater environment nearly identical to its former habitat. Moreover, if *Mesosaurus* could have swum across the ocean, its fossil remains should be widely dispersed. It is more logical to assume that *Mesosaurus* lived in lakes in what are now adjacent areas of South America and Africa, but were once united into a single continent.

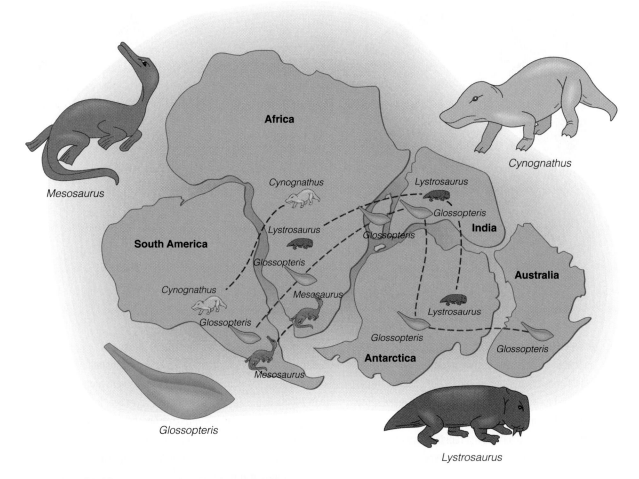

Figure 2.5 Fossil Evidence Supporting Continental Drift

Some of the plants and animals whose fossils are found today on the widely separated continents of South America, Africa, India, Australia, and Antarctica. During the Late Paleozoic Era, these continents were joined together to form Gondwana, the southern landmass of Pangaea. Plants of the *Glossopteris* flora are found on all five continents, which today have widely different climates; however, during the Pennsylvanian and Permian periods, they were all located in the same general climatic belt. *Mesosaurus* is a freshwater reptile whose fossils are found only in similar nonmarine Permian-age rocks in Brazil and South Africa. *Cynognathus* and *Lystrosaurus* are land reptiles that lived during the Early Triassic Period. Fossils of *Cynognathus* are found in South America and Africa, whereas fossils of *Lystrosaurus* have been recovered from Africa, India, and Antarctica. It is hard to imagine how a freshwater reptile and land-dwelling reptiles could have swum across the wide oceans that presently separate these continents. It is more logical to assume that the continents were once connected.

magnetism A physical phenomenon resulting from moving electricity and the spin of electrons in some solids in which magnetic substances are attracted toward one another.

magnetic field The area in which magnetic substances are affected by lines of magnetic force emanating from Earth.

Notwithstanding all of the empirical evidence presented by Wegener and later by du Toit and others, most geologists simply refused to entertain the idea that continents might have moved in the past. The geologists were not necessarily being obstinate about accepting new ideas; rather, they found the evidence for continental drift inadequate and unconvincing. In part, this was because no one could provide a suitable mechanism to explain how continents could move over Earth's surface.

LO3 Earth's Magnetic Field

Magnetism is a physical phenomenon resulting from the spin of electrons in some solids—particularly those of iron—and moving electricity. A **magnetic field** is an area in which magnetic substances such as iron are affected by lines of magnetic force emanating from Earth (Figure 2.6). The magnetic field shown in Figure 2.6 is *dipolar*, meaning that it possesses two unlike magnetic poles referred to as the north and south poles.

Earth can be thought of as a giant dipole magnet in which the magnetic poles are in close proximity to the geographic poles (Figure 2.7). This arrangement means that the strength of the magnetic field is not constant, but varies. Notice in Figure 2.7 that the lines of magnetic force around Earth parallel its surface only near the equator, just as the iron filings do around a bar magnet (Figure 2.6). As the lines of force approach the poles, they are oriented at increasingly larger angles with respect to the surface, and the strength of the magnetic field increases; it is strongest at the poles and weakest at the equator.

Another important aspect of the magnetic field is that the magnetic poles, where the lines of force leave and enter Earth, do not coincide with the geographic (rotational) poles. Currently, an 11.5° angle exists between the two (Figure 2.7). Studies of Earth's magnetic field show that the locations of the magnetic poles vary slightly over time, but that they still correspond closely, on average, with the locations of the geographic poles.

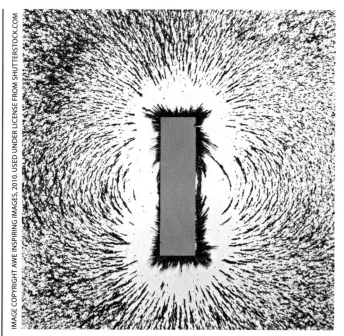

Figure 2.6 Magnetic Field
Iron filings align along the lines of magnetic force radiating from a bar magnet.

Figure 2.7 Earth's Magnetic Field

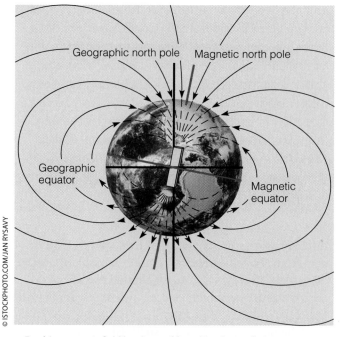

a. Earth's magnetic field has lines of force like those of a bar magnet.

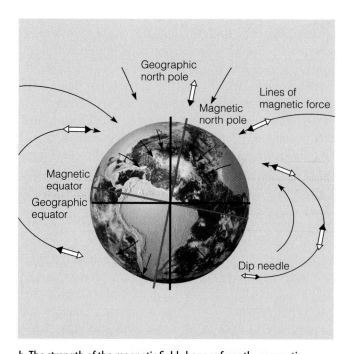

b. The strength of the magnetic field changes from the magnetic equator to the magnetic poles. This change in strength causes a dip needle (a magnetic needle that is balanced on the tip of a support so that it can freely move vertically) to be parallel to Earth's surface only at the magnetic equator, where the strength of the magnetic north and south poles are equally balanced. Its inclination or dip with respect to Earth's surface increases as it moves toward the magnetic poles, until it is at 90 degrees or perpendicular to Earth's surface at the magnetic poles.

LO4 Paleomagnetism and Polar Wandering

Interest in continental drift revived during the 1950s as a result of evidence from paleomagnetic studies, a relatively new discipline at the time. **Paleomagnetism** is the remanent magnetism in ancient rocks recording the direction and intensity of Earth's magnetic poles at the time of the rock's formation.

> *Remanent magnetism: The permanent magnetism in rocks (resulting from the orientation of the Earth's magnetic field at the time of the rocks' formation).*

When magma cools, the magnetic iron-bearing minerals align themselves with Earth's magnetic field, recording both its direction and strength. The temperature at which iron-bearing minerals gain their magnetization is called the **Curie point**. As long as the rock is not subsequently heated above the Curie point, it will preserve that remanent magnetism. Thus, an ancient lava flow provides a record of the orientation and strength of Earth's magnetic field at the time the lava flow cooled.

As paleomagnetic research progressed during the 1950s, some unexpected results emerged. When geologists measured the paleomagnetism of geologically recent rocks, they found that it was generally consistent with Earth's current magnetic field. The paleomagnetism of ancient rocks, though, showed different orientations. For example, paleomagnetic studies of Silurian lava flows in North America indicated that the north magnetic pole was located in the western Pacific Ocean at that time, whereas the paleomagnetic evidence from Permian lava flows pointed to yet another location in Asia. When plotted on a map, the paleomagnetic readings of numerous lava flows from all ages in North America trace the apparent movement of the magnetic pole (called *polar wandering*) through time (Figure 2.8).

Upon analysis, magnetic minerals from European Silurian and Permian lava flows pointed to a

paleomagnetism
Residual magnetism in rocks, studied to determine the intensity and direction of Earth's past magnetic field.

Curie point The temperature at which iron-bearing minerals in cooling magma or lava attain their magnetism.

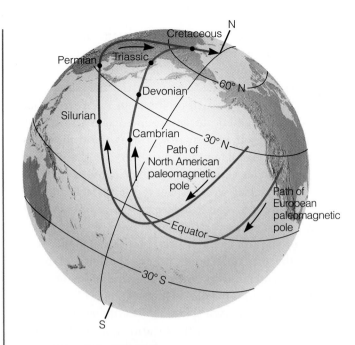

Figure 2.8 Polar Wandering
The apparent paths of polar wandering for North America and Europe. The apparent location of the north magnetic pole is shown for different periods on each continent's polar wandering path. If the continents have not moved through time, and because Earth has only one magnetic north pole, the paleomagnetic readings for the same time in the past, taken on different continents, should all point to the same location. However, the north magnetic pole has different locations for the same time in the past when measured on different continents, indicating multiple north magnetic poles. The logical explanation for this dilemma is that the magnetic north pole has remained at the same approximate geographic location during the past, and the continents have moved. From A. Cox and R. R. Doell, "Review of Paleomagnetism," *G.S.A. Bulletin*, Vol. 71, Figure 33, p. 758, with permission of the publisher, the Geological Society of America, Boulder, Colorado. USA. Copyright © 1955 Geological Society of America.

different magnetic pole location from those of the same age from North America (Figure 2.8). Furthermore, analysis of lava flows from all continents indicated that each continent seemingly had its own series of magnetic poles.

The best explanation for such data is that the magnetic poles have remained near their present locations at the geographic north and south poles and the continents have moved. When the continental margins are fit together so that the paleomagnetic data point to only one magnetic pole, we find, just as Wegener did, that the rock sequences and glacial deposits match, and that the fossil evidence is consistent with the reconstructed paleogeography.

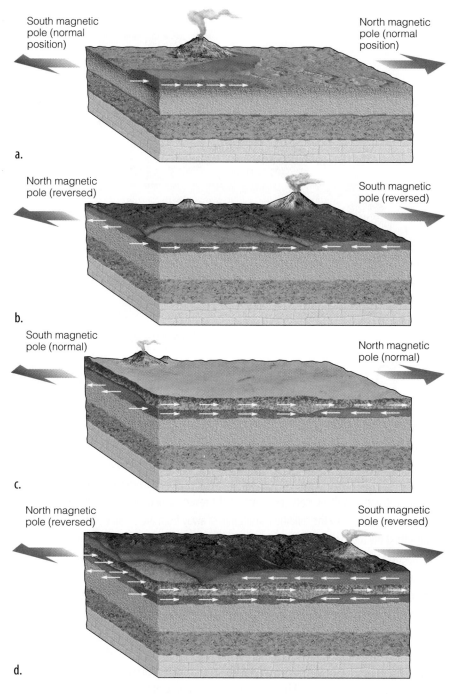

a.

b.

c.

d.

Figure 2.9 Magnetic Reversals

During the time period shown (a–d), volcanic eruptions produced a succession of overlapping lava flows. At the time of these volcanic eruptions, Earth's magnetic field completely reversed—that is, the magnetic north pole moved to the geographic south pole, and the magnetic south pole moved to the geographic north pole. Thus, the end of the needle of a magnetic compass that today would point to the North Pole would point to the South Pole if the magnetic field should again reverse. We know that Earth's magnetic field has reversed numerous times in the past because when lava flows cool below the Curie point, magnetic minerals within the flow orient themselves parallel to the magnetic field at the time. They thus record whether the magnetic field was normal or reversed at that time. The white arrows in this diagram show the direction of the north magnetic pole for each individual lava flow, thus confirming that Earth's magnetic field has reversed in the past.

LO5 Magnetic Reversals and Seafloor Spreading

Geologists refer to Earth's present magnetic field as being normal—that is, with the north and south magnetic poles located approximately at the north and south geographic poles. At various times in the geologic past, however, Earth's magnetic field has completely reversed, that is, the magnetic north and south poles reverse positions, so that the magnetic north pole becomes the magnetic south pole, and the magnetic south pole becomes the magnetic north pole. During such a reversal, the magnetic field weakens until it temporarily disappears. When the magnetic field returns, the magnetic poles have reversed their position. The existence of such **magnetic reversals** was discovered by dating and determining the orientation of the remanent magnetism in lava flows on land (Figure 2.9). Although the cause of magnetic reversals is still uncertain, their occurrence in the geologic record is well documented.

As a result of oceanographic research conducted during the 1950s, Harry Hess of Princeton University proposed, in a 1962 landmark paper, the theory of **seafloor spreading** to account for continental movement. He suggested that continents do not move through oceanic crust as do ships plowing through sea ice, but rather that the continents and oceanic crust

magnetic reversal The phenomenon involving the complete reversal of the north and south magnetic poles.

seafloor spreading The theory that the seafloor moves away from spreading ridges and is eventually consumed at subduction zones.

move together as a single unit. Thus, the theory of seafloor spreading answered a major objection of the opponents of continental drift—namely, how could continents move through oceanic crust? In fact, the continents moved with the oceanic crust as part of a litho-spheric system.

As a mechanism to drive this system, Hess revived the idea of a heat transfer system—or **thermal convection cells**—within the mantle to move the plates. According to Hess, hot magma rises from the mantle, intrudes along fractures defining oceanic ridges, and thus forms new crust. Cold crust is subducted back into the mantle at oceanic trenches, where it is heated and recycled, thus completing a thermal convection cell (see Figure 1.12).

How could Hess's hypothesis be confirmed? Magnetic surveys of the oceanic crust revealed a pattern of striped **magnetic anomalies** (deviations from the average strength of Earth's present-day magnetic field) in the rocks that are both parallel to and symmetric around the oceanic ridges (Fig-

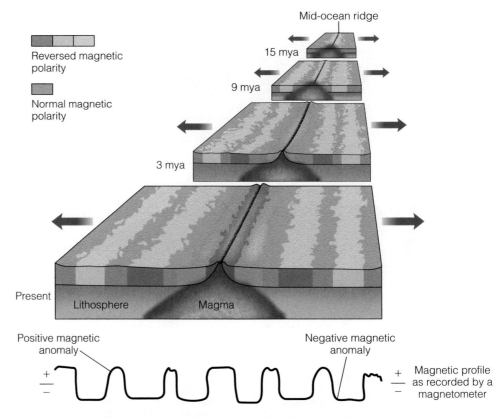

Figure 2.10 Magnetic Anomalies and Seafloor Spreading
The sequence of magnetic anomalies preserved within the oceanic crust is both parallel to and symmetric around oceanic ridges. Basaltic lava intruding into an oceanic ridge today and spreading laterally away from the ridge records Earth's current magnetic field or polarity (considered by convention to be normal). Basaltic intrusions 3, 9, and 15 million years ago record Earth's reversed magnetic field at those times. This schematic diagram shows how the solidified basalt moves away from the oceanic ridge (or spreading ridge), carrying with it the magnetic anomalies that are preserved in the oceanic crust. Magnetic anomalies are magnetic readings that are either higher (positive magnetic anomalies) or lower (negative magnetic anomalies) than Earth's current magnetic field strength. The magnetic anomalies are recorded by a magnetometer, which measures the strength of the magnetic field. Modified from Kious and Tilling, USGS and Hyndman and Hyndman, *Natural Hazards and Disasters,* Brooks/Cole, 2006, p. 15, Fig. 2.6b.

thermal convection cell A type of circulation of material involving only the asthenosphere or the entire mantle during which hot material rises, moves laterally, cools and sinks, and is reheated and continues the cycle.

magnetic anomaly Any deviation, such as a change in average strength, in Earth's magnetic field.

ure 2.10). A positive magnetic anomaly results when Earth's magnetic field at the time of oceanic crust formation along an oceanic ridge summit was the same as today, thus yielding a stronger than normal (positive) magnetic signal. A negative magnetic anomaly results when Earth's magnetic field at the time of

oceanic crust formation was reversed, therefore yielding a weaker than normal (negative) magnetic signal.

Thus, as new oceanic crust forms at oceanic ridge summits and records Earth's magnetic field at the time, the previously formed crust moves laterally away from the ridge. These magnetic stripes therefore represent times of normal and reversed polarity at oceanic ridges (where upwelling magma forms new oceanic crust), and conclusively confirm Hess's theory of seafloor spreading.

One of the consequences of the seafloor spreading theory is its confirmation that ocean basins are geologically young features whose openings and closings are partially responsible for continental movement. Radiometric dating reveals that the oldest oceanic crust is

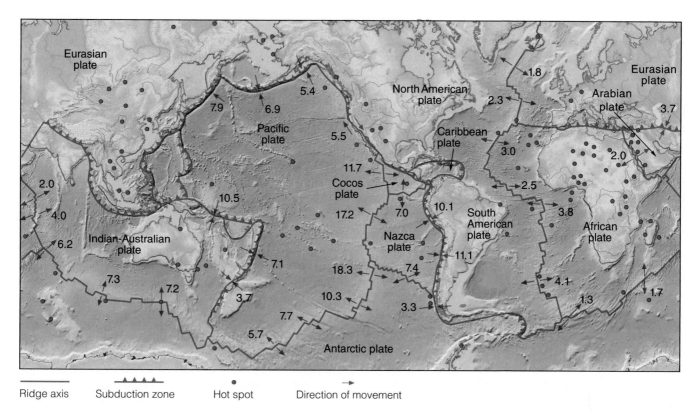

Figure 2.11 Earth's Plates
A world map showing Earth's plates, their boundaries, their relative motion and rates of movement in centimeters per year, and hot spots.

Ridge axis   Subduction zone   Hot spot   Direction of movement

somewhat younger than 180 million years old, whereas the oldest continental crust is 3.96 billion years old. Although geologists do not universally accept the idea of thermal convection cells as a driving mechanism for plate movement, most accept that plates are created at oceanic ridges and destroyed at deep-sea trenches, regardless of the driving mechanism involved.

7 PLATES

Eurasian

Indian-Australian

Antarctic

North American

South American

Pacific

African

LO6 Plate Tectonics: A Unifying Theory

Plate tectonic theory is based on a simple model of Earth. The rigid lithosphere, composed of both oceanic and continental crust, as well as the underlying upper mantle, consists of numerous variable-sized pieces called *plates* (Figure 2.11). There are seven major plates (Eurasian, Indian-Australian, Antarctic, North American, South American, Pacific, and African), and numerous smaller ones ranging from only a few tens to several hundreds of kilometers in width. Plates also vary in thickness; those composed of upper mantle and continental crust are as much as 250 km thick, whereas those of upper mantle and oceanic crust are up to 100 km thick.

The lithosphere overlies the hotter and weaker semiplastic asthenosphere. It is thought that movement resulting from some type of heat-transfer system within

plate tectonic theory
The theory holding that large segments of Earth's outer part (lithospheric plates) move relative to one another.

the asthenosphere causes the overlying plates to move. As plates move over the asthenosphere, they separate, mostly at oceanic ridges; in other areas, such as at oceanic trenches, they collide and are subducted back into the mantle.

Most geologists accept plate tectonic theory because the evidence for it is overwhelming and it ties together many seemingly unrelated geologic features and events and shows how they are interrelated. Consequently, geologists now view many geologic processes from the global perspective of plate tectonic theory in which plate interaction along plate margins is responsible for such phenomena as mountain building, earthquakes, and volcanism.

LO7 The Three Types of Plate Boundaries

Because it appears that plate tectonics has operated since at least the Proterozoic Eon, it is important that we understand how plates move and interact with each other and how ancient plate boundaries are recognized. After all, the movement of plates has profoundly affected the geologic and biologic history of this planet.

Geologists recognize three major types of plate boundaries: *divergent, convergent,* and *transform.* Along these boundaries, new plates are formed, are consumed, or slide laterally past one another. Interaction of plates at their boundaries accounts for most of Earth's volcanic eruptions and earthquakes, as well as the formation and evolution of its mountain systems.

DIVERGENT BOUNDARIES

Divergent plate boundaries or *spreading ridges* occur where plates are separating and new oceanic lithosphere is forming. Divergent boundaries are places where the crust is extended, thinned, and fractured as magma, derived from the partial melting of the mantle, rises to the surface. The magma is almost entirely basaltic and intrudes into vertical fractures to form dikes and pillow lava flows (see Figure 5.6). As successive injections of magma cool and solidify, they form new oceanic crust and record the intensity and orientation of Earth's magnetic field (Figure 2.10).

Divergent boundaries most commonly occur along the crests of oceanic ridges—for example, the Mid-Atlantic Ridge. Oceanic ridges are thus

divergent plate boundary The boundary between two plates that are moving apart.

characterized by rugged topography with high relief resulting from displacement of rocks along large fractures, shallow-depth earthquakes, high heat flow, and basaltic flows or pillow lavas.

Divergent boundaries are also present under continents during the early stages of continental breakup. When magma wells up beneath a continent, the crust is initially elevated, stretched, and thinned, producing fractures, faults, rift valleys, and volcanic activity (Figure 2.12a). As magma intrudes into faults and fractures, it solidifies or flows out onto the surface as lava flows; the latter often covering the rift valley floor (Figure 2.12b). The East African Rift Valley is an excellent example of continental breakup at this stage (Figure 2.13).

As spreading proceeds, some rift valleys continue to lengthen and deepen until the continental crust eventually breaks and a narrow linear sea is formed, separating two continental blocks (Figure 2.12c). The Red Sea separating the Arabian Peninsula from Africa (Figure 2.13) and the Gulf of California, which separates Baja California from mainland Mexico, are good examples of this more advanced stage of rifting.

As a newly created narrow sea continues to enlarge, it may eventually become an expansive ocean basin such as the Atlantic Ocean basin is today, separating North and South America from Europe and Africa by thousands of kilometers (Figure 2.12d). The Mid-Atlantic Ridge is the boundary between these diverging plates; the American plates are moving westward, and the Eurasian and African plates are moving eastward.

An Example of Ancient Rifting

What features in the geologic record can geologists use to recognize ancient rifting? Associated with regions of continental rifting are faults, dikes (vertical intrusive igneous bodies), sills (horizontal intrusive igneous bodies), lava flows, and thick sedimentary sequences within rift valleys, all features that are preserved in the geologic record. The Triassic fault basins of the eastern United States are a good example of ancient continental rifting (see Figure 18.20). These fault basins mark the zone of rifting that occurred when North America split apart from Africa. The basins contain thousands of meters of continental sediment and are riddled with dikes and sills (see Chapter 18).

Pillow lavas, in association with deep-sea sediment, are also evidence of ancient rifting. The presence of pillow lavas marks the formation of a spreading ridge in

Figure 2.12 History of a Divergent Plate Boundary

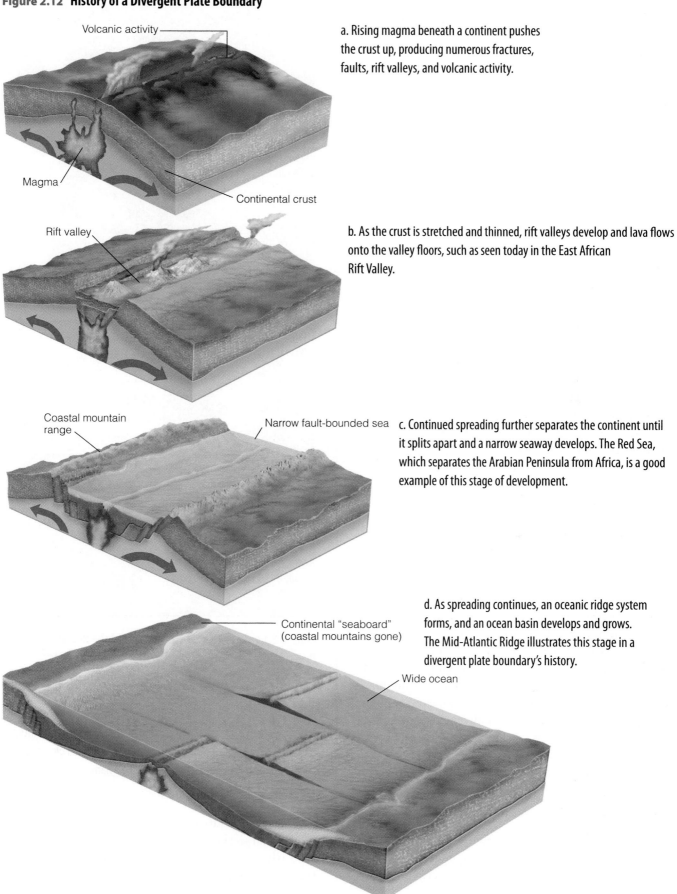

a. Rising magma beneath a continent pushes the crust up, producing numerous fractures, faults, rift valleys, and volcanic activity.

b. As the crust is stretched and thinned, rift valleys develop and lava flows onto the valley floors, such as seen today in the East African Rift Valley.

c. Continued spreading further separates the continent until it splits apart and a narrow seaway develops. The Red Sea, which separates the Arabian Peninsula from Africa, is a good example of this stage of development.

d. As spreading continues, an oceanic ridge system forms, and an ocean basin develops and grows. The Mid-Atlantic Ridge illustrates this stage in a divergent plate boundary's history.

a narrow linear sea (Figure 2.12c). Magma, intruding into the sea along this newly formed spreading ridge, solidifies as pillow lavas, which are preserved in the geologic record, along with the sediment being deposited on them.

CONVERGENT BOUNDARIES

Whereas new crust forms at divergent plate boundaries, older crust must be destroyed and recycled in order for the entire surface area of Earth to remain the same. Otherwise, we would have an expanding Earth. Such plate destruction occurs at **convergent plate boundaries** (Figure 2.14), where two plates collide and the leading edge of one plate is subducted beneath the margin of the other plate and eventually incorporated into the asthenosphere. A dipping plane of earthquake foci, called a *Benioff zone*, defines subduction zones (see Figure 8.5). Most of these planes dip from oceanic trenches beneath adjacent island arcs or continents, marking the surface of slippage between the converging plates.

Deformation, volcanism, mountain building, metamorphism, earthquake activity, and deposits of valuable minerals characterize convergent boundaries. Three types of convergent plate boundaries are recognized: *oceanic–oceanic, oceanic–continental,* and *continental–continental.*

Oceanic–Oceanic Boundaries

When two oceanic plates converge, one is subducted beneath the other along an **oceanic–oceanic plate boundary** (Figure 2.14a). The subducting plate bends downward to form the outer wall of an oceanic trench. A *subduction complex,* composed of wedge-shaped slices of highly folded and faulted marine sediments and oceanic lithosphere scraped off the descending plate, forms along the inner wall of the oceanic trench. As the subducting plate descends into the mantle, it is heated and partially melted, generating magma commonly of andesitic composition (see Chapter 4). This magma is less dense than the surrounding mantle rocks and rises to the surface of the nonsubducted plate to form a curved chain

convergent plate boundary The boundary between two plates that move toward each other.

oceanic–oceanic plate boundary A convergent plate boundary along which two oceanic plates collide and one is subducted beneath the other.

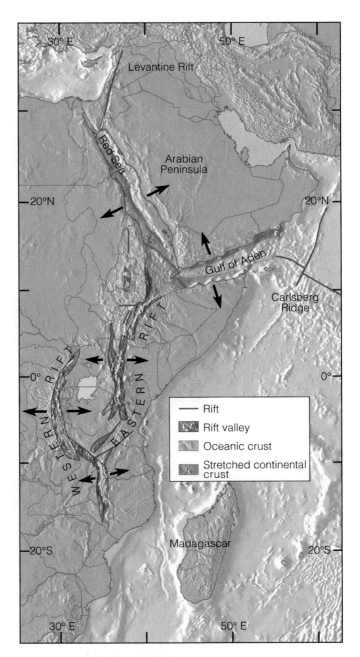

Figure 2.13 East African Rift Valley and the Red Sea—Present-Day Examples of Divergent Plate Boundaries
The East African Rift Valley and the Red Sea represent different stages in the history of a divergent plate boundary. The East African Rift Valley is being formed by the separation of eastern Africa from the rest of the continent along a divergent plate boundary. The Red Sea represents a more advanced stage of rifting, in which two continental blocks (Africa and the Arabian Peninsula) are separated by a narrow sea.

of volcanic islands called a *volcanic island arc* (any plane intersecting a sphere makes an arc). This arc is nearly parallel to the oceanic trench and is separated from it by a distance of up to several hundred

Figure 2.14 Three Types of Convergent Plate Boundaries

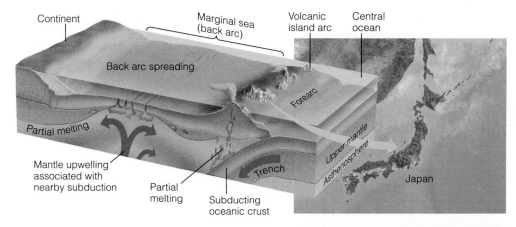

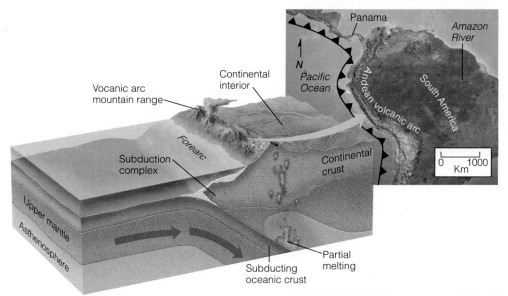

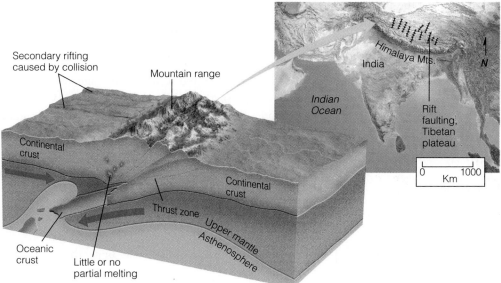

a. Oceanic–oceanic plate boundary. An oceanic trench forms where one oceanic plate is subducted beneath another. On the nonsubducted plate, a volcanic island arc forms from the rising magma generated from the subducting plate. The Japanese Islands are a volcanic island arc resulting from the subduction of one oceanic plate beneath another oceanic plate.

b. Oceanic–continental plate boundary. When an oceanic plate is subducted beneath a continental plate, an andesitic volcanic mountain range is formed on the continental plate as a result of rising magma. The Andes Mountains in Peru are one of the best examples of continuing mountain building at an oceanic–continental plate boundary.

c. Continental–continental plate boundary. When two continental plates converge, neither is subducted because of their great thickness and low and equal densities. As the two continental plates collide, a mountain range is formed in the interior of a new and larger continent. The Himalayas in central Asia resulted from the collision between India and Asia approximately 40 to 50 million years ago.

kilometers—the distance depends on the angle of dip of the subducting plate (Figure 2.14a).

In those areas where the rate of subduction is faster than the forward movement of the overriding plate, the lithosphere on the landward side of the volcanic island arc may be subjected to tensional stress and stretched and thinned, resulting in the formation of a *back-arc basin*. This back-arc basin may grow by spreading if magma breaks through the thin crust and forms new oceanic crust (Figure 2.14a). A good example of a back-arc basin associated with an oceanic–oceanic plate boundary is the Sea of Japan between the Asian continent and the islands of Japan.

Most present-day active volcanic island arcs are in the Pacific Ocean basin and include the Aleutian Islands, the Kermadec–Tonga arc, and the Japanese (Figure 2.14a) and Philippine Islands. The Scotia and Antillean (Caribbean) island arcs are in the Atlantic Ocean basin.

Oceanic–Continental Boundaries

When an oceanic and a continental plate converge, the denser oceanic plate is subducted under the continental plate along an **oceanic–continental plate boundary** (Figure 2.14b). Just as at oceanic–oceanic plate boundaries, the descending oceanic plate forms the outer wall of an oceanic trench.

The magma generated by subduction rises beneath the continent and either crystallizes as large intrusive bodies, called *plutons*, before reaching the surface or erupts at the surface to produce a chain of andesitic volcanoes, also called a *volcanic arc*. An excellent example of an oceanic–continental plate boundary is the Pacific Coast of South America where the oceanic Nazca plate is currently being subducted under South America (Figure 2.14b; see also Chapter 10). The Peru–Chile Trench marks the site of subduction, and the Andes Mountains are the resulting volcanic mountain chain on the nonsubducting plate.

oceanic–continental plate boundary A convergent plate boundary along which oceanic lithosphere is subducted beneath continental lithosphere.

continental–continental plate boundary A convergent plate boundary along which two continental lithospheric plates collide.

Continental–Continental Boundaries

Two continents approaching each other are initially separated by an ocean floor that is being subducted under one continent. The edge of that continent displays the features characteristic of oceanic–continental convergence. As the ocean floor continues to be subducted, the two continents come closer together until they eventually collide. Because continental lithosphere, which consists of continental crust and the upper mantle, is less dense than oceanic lithosphere (oceanic crust and upper mantle), it cannot sink into the asthenosphere. Although one continent may partially slide under the other, it cannot be pulled or pushed down into a subduction zone (Figure 2.14c).

When two continents collide, they are welded together along a zone marking the former site of subduction. At this **continental–continental plate boundary**, an interior mountain belt is formed consisting of deformed sediments and sedimentary rocks, igneous intrusions, metamorphic rocks, and fragments of oceanic crust. In addition, the entire region is subjected to numerous earthquakes. The Himalayas in central Asia, the world's youngest and highest mountain system, resulted from the collision between India and Asia that began 40 to 50 million years ago and is still continuing (Figure 2.14c; see Chapter 10).

Recognizing Ancient Convergent Plate Boundaries

Igneous rocks provide one clue to ancient subduction zones. The magma erupted at the surface, forming island arc volcanoes and continental volcanoes, is of andesitic composition. Another clue is the zone of intensely deformed rocks between the deep-sea trench where subduction is taking place and the area of igneous activity. Here, sediments and submarine rocks are folded, faulted, and metamorphosed into a chaotic mixture of rocks termed a *mélange*.

During subduction, pieces of oceanic lithosphere are sometimes incorporated into the mélange and accreted onto the edge of the continent. Such slices of oceanic crust and upper mantle are called *ophiolites* (Figure 2.15). They consist of a layer of deep-sea sediments that include graywackes (poorly sorted sandstones containing abundant feldspars and rock fragments, usually in a clay-rich matrix), black shales, and cherts (see Chapter 6). These deep-sea sediments are underlain by pillow lavas, a sheeted dike complex, massive gabbro (a dark intrusive igneous rock), and layered gabbro, all of which form the oceanic crust. Beneath the gabbro is peridotite (a dark intrusive igneous rock composed of the mineral olivine), which probably represents the upper mantle. The presence of ophiolites in an outcrop or drilling core is a key indicator of plate convergence along a subduction zone.

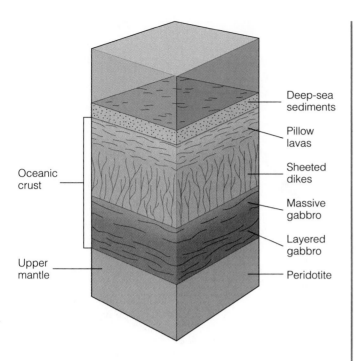

Figure 2.15 Ophiolites
Ophiolites are sequences of rock on land consisting of deep-sea sediment, oceanic crust, and upper mantle. Ophiolites are one feature used to recognize ancient convergent plate boundaries.

TRANSFORM BOUNDARIES

The third type of plate boundary is a **transform plate boundary**. These mostly occur along fractures in the seafloor, known as *transform faults,* where plates slide laterally past one another roughly parallel to the direction of plate movement. Although lithosphere is neither created nor destroyed along a transform boundary, the movement between plates results in a zone of intensely shattered rock and numerous shallow-depth earthquakes.

Transform faults "transform" or change one type of motion between plates into another type of motion. Most commonly, transform faults connect two oceanic ridge segments (Figure 2.16); however, they can also connect ridges to trenches and trenches to trenches. Although the majority of transform faults are in oceanic crust and are marked by distinct fracture zones, they may also extend into continents.

One of the best-known transform faults is the San Andreas fault in California. It separates the Pacific plate from the North American plate and connects spreading ridges in the Gulf of California with the Juan de Fuca and Pacific plates off the coast of northern California (Figure 2.17). Many of the earthquakes that affect California are the result of movement along this fault (see Chapter 8).

Unfortunately, transform faults generally do not leave any characteristic or diagnostic features except for the obvious displacement of the rocks with which they are associated.

transform plate boundary A plate boundary along which plates slide past one another and crust is neither produced nor destroyed.

transform fault A fault along which one type of motion is transformed into another; commonly displaces oceanic ridges; on land, recognized as a strike-slip fault, such as the San Andreas fault.

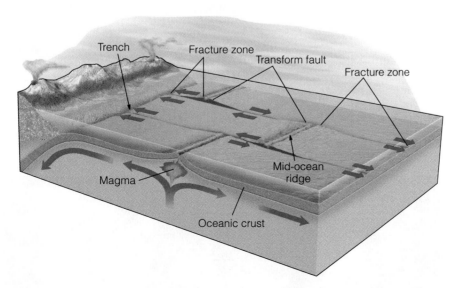

Figure 2.16 Transform Plate Boundaries
Horizontal movement between plates occurs along transform faults. Extensions of transform faults on the seafloor form fracture zones. Most transform faults connect two oceanic ridge segments.

Figure 2.17 The San Andreas Fault—A Transform Plate Boundary
The San Andreas fault is a transform fault separating the Pacific plate from the North American plate. It connects the spreading ridges in the Gulf of California with the Juan de Fuca and Pacific plates off the coast of northern California. Movement along the San Andreas fault has caused numerous earthquakes.

LO8 Hot Spots and Mantle Plumes

Before leaving the topic of plate boundaries, we should mention an intraplate feature found beneath both oceanic and continental plates. A **hot spot** (Figure 2.11) is the location on Earth's surface where a stationary column of magma, originating deep within the mantle (*mantle plume*), has slowly risen to the surface and formed a volcano. Because the mantle plumes apparently remain stationary (although some evidence suggests that they might not) within the mantle while the plates move over them, the resulting hot spots leave a trail of extinct and progressively older volcanoes called *aseismic ridges* that record the movement of the plate.

One of the best examples of aseismic ridges and hot spots is the Emperor Seamount–Hawaiian Island chain (Figure 2.18). This chain of islands and seamounts (structures of volcanic origin rising higher than 1 km above the seafloor) extends from the island of Hawaii to the Aleutian Trench off Alaska, a distance of some 6,000 km, and consists of more than 80 volcanic structures.

Mantle plumes and hot spots help geologists explain some of the geologic activity occurring within plates as opposed to activity occurring at or near plate boundaries. In addition, if mantle plumes are essentially fixed with respect to Earth's rotational axis, they can be used to determine not only the direction of plate movement, but also the rate of movement.

hot spot A localized zone of melting below the lithosphere that probably overlies a mantle plume; detected by volcanism at the surface.

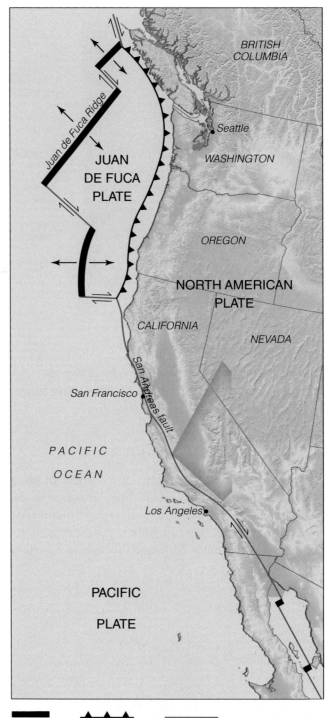

Oceanic ridge | Zone of subduction | Transform faults

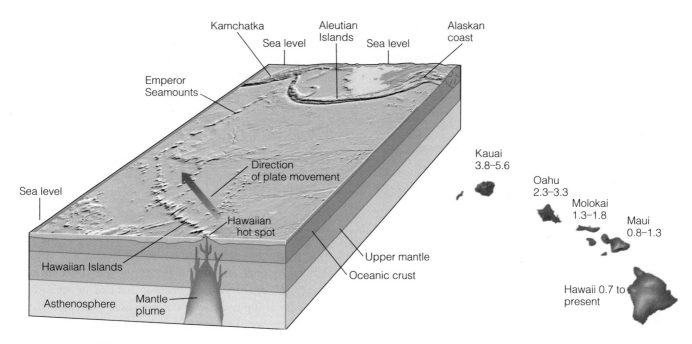

Figure 2.18 Hot Spots

A hot spot is the location where a stationary mantle plume has risen to the surface and formed a volcano. The Emperor Seamount–Hawaiian Island chain formed as a result of the Pacific plate moving over a mantle plume, and the line of volcanic islands in this chain traces the direction of plate movement. The island of Hawaii and the Loihi Seamount are the only current hot spots of this island chain. The numbers indicate the ages of the Hawaiian Islands in millions of years.

LO9 Plate Movement and Motion

Rates of plate movement can be calculated in several ways. The least accurate method is to determine the age of the sediments immediately above any portion of the oceanic crust and then divide the distance from the spreading ridge by that age. Such calculations give an average rate of movement.

A more accurate method of determining both the average rate of movement and relative motion is by dating the magnetic anomalies in the crust of the seafloor. The distance from an oceanic ridge axis to any magnetic anomaly indicates the width of new seafloor that formed during that time interval. For example, if the distance between the present-day Mid-Atlantic Ridge and anomaly 31 is 2,010 km, and anomaly 31 formed 67 million years ago (Figure 2.19), then the average rate of movement during the past 67 million years has been 3 cm per year (2,010 km, which equals 201 million cm divided by 67 million years; 201,000,000 cm/ 67,000,000 years = 3 cm/year). Thus, for a given interval of time, the wider the strip of seafloor, the faster the plate has moved. In this way, not only can the present average rate of movement and relative motion be determined (Figure 2.11), but the average rate of movement in the past can also be calculated by dividing the distance between anomalies by the amount of time elapsed between anomalies.

Geologists use magnetic anomalies not only to calculate the average rate of plate movement, but also to

Figure 2.19 Reconstructing Plate Positions Using Magnetic Anomalies

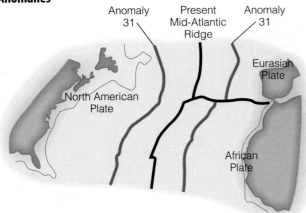

a. The present North Atlantic, showing the Mid-Atlantic Ridge and magnetic anomaly 31, which formed 67 million years ago.

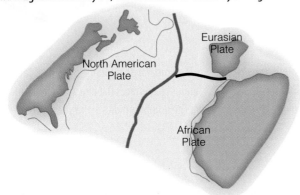

b. The Atlantic 67 million years ago. Anomaly 31 marks the plate boundary 67 million years ago. By moving the anomalies back together, along with the plates they are on, we can reconstruct the former positions of the continents.

determine plate positions at various times in the past. Because magnetic anomalies are parallel and symmetric with respect to spreading ridges, all one must do to determine the position of continents when particular anomalies formed is to move the anomalies back to the spreading ridge, which will also move the continents with them (Figure 2.19). Because subduction destroys oceanic crust and the magnetic record that it carries, we have an excellent record of plate movements since the breakup of Pangaea, but not as good an understanding of plate movement before that time.

The average rate of movement, as well as the relative motion between any two plates, can also be determined by satellite-laser ranging techniques. Laser beams from a station on one plate are bounced off a satellite (in geosynchronous orbit) and returned to a station on a different plate. As the plates move away from each other, the laser beam takes more time to go from the sending station to the stationary satellite and back to the receiving station. This difference in elapsed time is used to calculate the rate of movement and the relative motion between plates.

LO10 The Driving Mechanism of Plate Tectonics

A major obstacle to the acceptance of the continental drift hypothesis was the lack of a driving mechanism to explain continental movement. When it was shown that continents and ocean floors moved together, not separately, and that new crust is formed at spreading ridges by rising magma, most geologists accepted some type of convective heat system (convection cells) as the basic process responsible for plate motion. The question of what exactly drives the plates, however, still remains.

Most of the heat from Earth's interior results from the decay of radioactive elements, such as uranium (see Chapter 17), in the core and lower mantle. The most efficient way for this heat to escape Earth's interior is through some type of slow convection of mantle rock in which hot rock from the interior rises toward the surface, loses heat to the overlying lithosphere, becomes denser as it cools, and then sinks back into the interior where it is heated, and the process repeats itself. This type of convective heat system is analogous to a pot of stew cooking on a stove (see feature box).

Two models involving thermal convection cells have been proposed to explain plate movement (Figure 2.20). In one model, thermal convection cells are restricted to the asthenosphere; in the second model, the entire mantle is involved. In both models, spreading ridges mark the ascending limbs of adjacent convection cells, and trenches are present where convection cells descend back into Earth's interior. The convection cells therefore determine the location of spreading ridges and trenches, with the lithosphere lying above the thermal convection cells. Thus, each plate corresponds to a single convection cell and moves as a result of the convective movement of the cell itself.

Although most geologists agree that Earth's internal heat plays an important role in plate movement, there are problems with both models. The major problem associated with the first model is the difficulty of explaining the source of heat for the convection cells and why they are restricted to the asthenosphere.

Figure 2.20 **Thermal Convection Cells as the Driving Force of Plate Movement.**
Two models involving thermal convection cells have been proposed to explain plate movement.

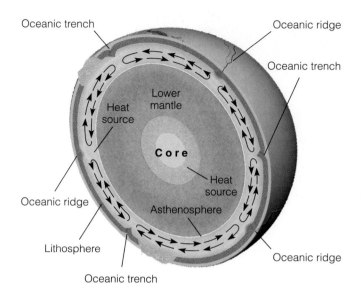

a. In the first model, thermal convection cells are restricted to the asthenosphere.

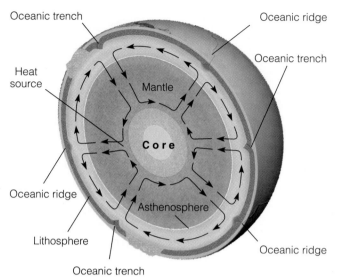

b. In the second model, thermal convection cells involve the entire mantle.

In the second model, the heat comes from the outer core, but it is still not known how heat is transferred from the outer core to the mantle. Nor is it clear how convection can involve both the lower mantle and the asthenosphere.

In addition to some type of thermal convection system driving plate movement, some geologists think that plate movement occurs partly because of a mechanism involving "slab-pull" or "ridge-push," both of which are gravity driven but still depend on thermal differences within Earth (Figure 2.21). In slab-pull, the subducting cold slab of lithosphere, being denser than the surrounding warmer asthenosphere, pulls the rest of the plate along as it descends into the asthenosphere. As the lithosphere moves downward, there is a corresponding upward flow back into the spreading ridge.

Operating in conjuction with slab-pull is the ridge-push mechanism. As a result of rising magma, the

CONVECTION IN A POT OF STEW

Heat from the stove is applied to the base of the stew pot causing the stew to heat up. As heat rises through the stew, pieces of the stew are carried to the surface, where the heat is dissipated, the pieces of stew cool, and then sink back to the bottom of the pot. The bubbling seen at the surface of the stew is the result of convection cells churning the stew. In the same manner, heat from the decay of radioactive elements produce convection cells within Earth's interior.

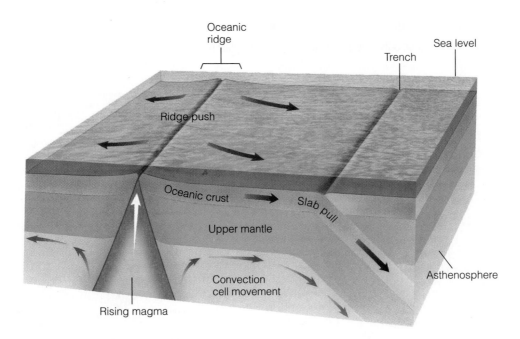

Oceanic
ridge

Sea level

Trench

Ridge push

Oceanic crust

Slab pull

Upper mantle

Asthenosphere

Convection
cell movement

Rising magma

Figure 2.21 Plate Movement Resulting from Gravity-Driven Mechanisms
Plate movement is also thought to result, at least partially, from gravity-driven "slab-pull" or "ridge-push" mechanisms. In slab-pull, the edge of the subducting plate descends into the interior, and the rest of the plate is pulled downward. In ridge-push, rising magma pushes the oceanic ridges higher than the rest of the oceanic crust. Gravity thus pushes the oceanic lithosphere away from the ridges and toward the trenches.

oceanic ridges are higher than the surrounding oceanic crust. It is thought that gravity pushes the oceanic lithosphere away from the higher spreading ridges and toward the trenches.

Currently, geologists are fairly certain that some type of convective system is involved in plate movement, but the extent to which other mechanisms, such as slab-pull and ridge-push, are involved is still unresolved. However, the fact that plates have moved in the past and are still moving today has been proven beyond a doubt. And although a comprehensive theory of plate movement has not yet been developed, more and more of the pieces are falling into place as geologists learn more about Earth's interior.

LO11 Plate Tectonics and the Distribution of Natural Resources

In addition to being responsible for the major features of Earth's crust and influencing the distribution and evolution of the world's biota, plate movement also affects the formation and distribution of some natural resources. Consequently, geologists are using plate tectonic theory in their search for petroleum and mineral deposits and in explaining the occurrence of these natural resources. It is becoming increasingly clear that if we are to keep up with the continuing demands of a global industrialized society, the application of plate tectonic theory to the origin and distribution of natural resources is essential.

Although large concentrations of petroleum occur in many areas of the world, more than 50 percent of all proven reserves are in the Persian Gulf region. The reason for this is paleogeography and plate movement. Elsewhere in the world, plate tectonics is also responsible for concentrations of petroleum.

Mineral Deposits

Many metallic mineral deposits such as copper, gold, lead, silver, tin, and zinc are related to igneous and associated hydrothermal (hot water) activity. So it is not surprising

that a close relationship exists between plate boundaries and the occurrence of these valuable deposits.

The magma generated by partial melting of a subducting plate rises toward the surface, and as it cools, it precipitates and concentrates various metallic ores. Many of the world's major metallic ore deposits are associated with convergent plate boundaries, including those in the Andes of South America, the Coast Ranges and Rockies of North America, Japan, the Philippines, Russia, and a zone extending from the eastern Mediterranean region to Pakistan. In addition, the majority of the world's gold is associated with deposits located at ancient convergent plate boundaries in such areas as Canada, Alaska, California, Venezuela, Brazil, Russia, southern India, and western Australia.

The copper deposits of western North and South America are an excellent example of the relationship between convergent plate boundaries and the distribution, concentration, and exploitation of valuable metallic ores (Figure 2.22). The world's largest copper deposits are found along this belt. The majority of the copper deposits in the Andes and the southwestern United States were formed less than 60 million years ago when oceanic plates were subducted under the North and South American plates.

Divergent plate boundaries also yield valuable ore resources. The island of Cyprus in the Mediterranean is rich in copper and has been supplying all or part of the world's needs for the past 3,000 years.

LO12 Plate Tectonics and the Distribution of Life

Plate tectonic theory is as revolutionary and far-reaching in its implications for geology as the theory of evolution was for biology when it was proposed. Interestingly, it was the fossil evidence that convinced Wegener, Suess, and du Toit, as well as many other geologists, of the correctness of continental drift. Together, the theories of plate tectonics and evolution have changed the way we view our planet, and we should not be surprised at the intimate association between them. Although the relationship between plate tectonic processes and the evolution of life is incredibly complex, paleontological data provide convincing evidence of the influence of plate movement on the distribution of organisms.

The present distribution of plants and animals is not random, but is controlled mostly by climate and geographic barriers. The world's biota occupy *biotic provinces*, which are regions characterized by a distinctive assemblage of plants and animals. Organisms within a province have similar ecological requirements, and the boundaries separating provinces are therefore natural ecological breaks. Climatic or geographic barriers are the most common province boundaries, and these are mostly controlled by plate movement.

The complex interaction between wind and ocean currents has a strong influence on the world's climates. Wind and ocean currents are thus strongly influenced by the number, distribution, topography, and orientation of continents. The distribution of continents and ocean basins not only influences wind and ocean currents, but also affects provinciality by creating physical barriers to, or pathways for, the migration of organisms. Intraplate volcanoes, island arcs, mid-oceanic ridges, mountain ranges, and subduction zones all result from the interaction of plates, and their orientation and distribution strongly influence the number of provinces and hence total global diversity. Thus, provinciality and diversity will be highest when there are numerous small continents spread across many zones of latitude.

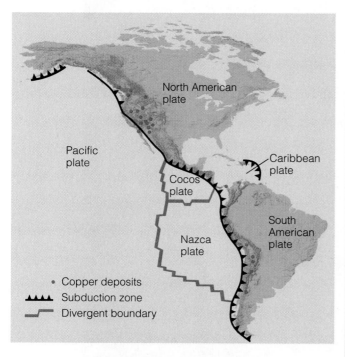

Figure 2.22 Copper Deposits and Convergent Plate Boundaries

Valuable copper deposits are located along the west coasts of North and South America in association with convergent plate boundaries. The rising magma and associated hydrothermal activity resulting from subduction carried small amounts of copper, which became trapped and concentrated in the surrounding rocks through time.

CHAPTER 3
MINERALS: THE BUILDING BLOCKS OF ROCKS

Cluster of crocoite crystals, a rare and brightly colored lead chromate mineral (Smithsonian Museum).

Introduction

Liquid water, gases, and trees are not minerals, but ice is because it meets the criteria geologists use to define the term **mineral**: a naturally occurring, inorganic, crystalline solid, with a narrowly defined chemical composition and characteristic physical properties. Of course, ice is naturally occurring and inorganic, it is composed of hydrogen and oxygen (H_2O), and it has distinguishing properties such as hardness and density. The only part of this definition that may be unfamiliar is crystalline solid, which means that the atoms in a mineral are arranged in a specific three-dimensional framework, as opposed to the atoms in glass, which have no such orderly arrangement.

We cannot overstate the importance of minerals in many human endeavors. Indeed, ores of iron, copper, and aluminum, as well as minerals and rocks used for fertilizer and animal feed supplements, are essential to our economic well-being. The United States and Canada owe much of their economic success to abundant mineral and energy resources, but both nations must import some essential commodities, including aluminum ore and manganese. As a matter of fact, the distribution of minerals, rocks, and energy resources has important implications in foreign relations and accounts for economic ties between nations. Even some rather common minerals are valuable: Quartz, for example, is used for sandpaper, and muscovite accounts for the lustrous sheen of paint used on automobiles and appliances.

One important reason to study minerals is that they are the building blocks of rocks. Granite, for instance, is made up of specific percentages of the minerals quartz, potassium feldspars, and plagioclase feldspars. Some minerals are attractive and eagerly sought out by collectors and for museum displays, whereas others are known as gemstones—that is, precious and semiprecious minerals and rocks used for decorative purposes, especially jewelry (Figure 3.1).

mineral A naturally occurring, inorganic, crystalline solid that has characteristic physical properties and a narrowly defined chemical composition.

LEARNING OUTCOMES

After reading this unit, you should be able to do the following:

LO1 Define matter

LO2 Explore the world of minerals

LO3 Identify mineral groups recognized by geologists

LO4 Identify physical properties of minerals

LO5 Recognize rock-forming minerals

LO6 Explain how minerals form

LO7 Recognize natural resources and reserves

COPYRIGHT AND PHOTOGRAPH BY DR. PARVINDER S. SETHI

Figure 3.1 Diamond Necklace
The diamond pendant in this necklace in the Smithsonian Institution is the 68-carat Victoria Transvaal diamond from South Africa. Their hardness, brilliance, beauty, durability, and scarcity make diamonds the most sought-after gemstones.

Scientists recognize a fourth state of matter known as *plasma,* an ionized gas as found in fluorescent and neon lights and matter in the sun and stars.

© ROYALTY FREE/CORBIS

four states or phases of matter: *liquids, gases, solids,* and *plasma,* an ionized gas, as in the Sun. Here our main concern is with solids because, by definition, minerals are solids.

ATOMS AND ELEMENTS

Matter is made up of chemical **elements**, which, in turn, are composed of **atoms**, the smallest units of matter that retain the characteristics of a particular element (Figure 3.2). That is, elements cannot be changed into different substances except through radioactive decay (discussed in Chapter 17). Thus, an element is made up of atoms, all of which have the same properties. Some of the 92 naturally occurring elements are listed in Figure 3.2. All naturally occurring elements and most artificial ones have a name and a symbol—for example, oxygen (O), aluminum (Al), and potassium (K).

At the center of an atom is a tiny **nucleus** made up of one or more **protons**, which have a positive electrical charge, and **neutrons**, which are electrically neutral. The nucleus is only about 1/100,000th of the diameter of an atom, yet it contains virtually all of the atom's mass. **Electrons**, particles with a negative electrical charge, orbit rapidly around the nucleus at specific distances in one or more **electron shells** (Figure 3.2). The electrons determine how an atom interacts with other atoms, but the nucleus determines how many electrons an atom has, because the positively charged protons attract and hold the negatively charged electrons in their orbits.

The number of protons in its nucleus determines an atom's identity and its **atomic number**. Hydrogen (H), for instance, has one proton in its nucleus and thus has an atomic number of 1. The nuclei of helium (He) atoms possess 2 protons, whereas those of carbon (C) have 6, and uranium (U) 92, so their atomic numbers are 2, 6, and 92, respectively. An atom's **atomic mass number** is the sum of protons and neutrons in the nucleus (electrons contribute negligible mass to atoms). However, atoms of the same chemical element might have different atomic mass numbers, because the number of neutrons can vary. All carbon (C) atoms have six protons—otherwise they would not be carbon—but the number of neutrons is 6, 7, or 8. Thus we recognize three types of carbon, or what are known as isotopes (Figure 3.3), each with a different atomic mass number.

The isotopes of carbon, or those of any other element, behave the same chemically; carbon 12 and

element A substance composed of atoms that all have the same properties; atoms of one element can change to atoms of another element by radioactive decay, but otherwise they cannot be changed by ordinary chemical means.

atom The smallest unit of matter that retains the characteristics of an element.

nucleus The central part of an atom consisting of protons and neutrons.

proton A positively charged particle found in the nucleus of an atom.

neutron An electrically neutral particle found in the nucleus of an atom.

From our discussion so far, we have a formal definition of the term *mineral,* and we know that minerals are the basic constituents of rocks. Now let's delve deeper into what minerals are made of by considering matter, atoms, elements, and bonding.

LO1 Matter: What Is It?

Matter is anything that has mass and occupies space, and accordingly includes water, plants, animals, the atmosphere, and minerals and rocks. Physicists generally recognize

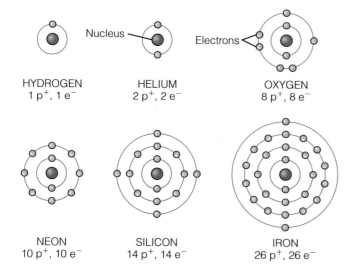

Element	Symbol	Atomic Number	First Shell	Second Shell	Third Shell	Fourth Shell
Hydrogen	H	1	1	—	—	—
Helium	He	2	2	—	—	—
Carbon	C	6	2	4	—	—
Oxygen	O	8	2	6	—	—
Neon	Ne	10	2	8	—	—
Sodium	Na	11	2	8	1	—
Magnesium	Mg	12	2	8	2	—
Aluminum	Al	13	2	8	3	—
Silicon	Si	14	2	8	4	—
Phosphorus	P	15	2	8	5	—
Sulfur	S	16	2	8	6	—
Chlorine	Cl	17	2	8	7	—
Potassium	K	19	2	8	8	1
Calcium	Ca	20	2	8	8	2
Iron	Fe	26	2	8	14	2

(Table header: Distribution of Electrons)

Figure 3.2 Shell Models for Common Atoms

The shell model for several atoms and their electron configurations. A blue circle represents the nucleus of each atom, but remember that atomic nuclei are made up of protons and neutrons, as shown in Figure 3.3.

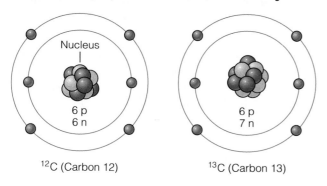

^{12}C (Carbon 12) ^{13}C (Carbon 13) ^{14}C (Carbon 14)

Figure 3.3 Carbon Isotopes

Schematic representation of the isotopes of carbon. Carbon has an atomic number of 6 and an atomic mass number of 12, 13, or 14, depending on the number of neutrons in its nucleus.

carbon 14 are both present in carbon dioxide (CO_2), for example. However, some isotopes are radioactive, meaning that they spontaneously decay or change to other elements. Carbon 14 is radioactive, whereas both carbon 12 and carbon 13 are not. Radioactive isotopes are important for determining the absolute ages of rocks (see Chapter 17).

BONDING AND COMPOUNDS

Interactions among electrons around atoms can result in two or more atoms joining together, a process known as **bonding**. If atoms of two or more elements bond, the resulting substance is a **compound**. Gaseous oxygen consists of only oxygen atoms and is thus an element, whereas the mineral quartz, consisting of silicon and oxygen atoms, is a compound. Most minerals are compounds, although gold, platinum, and several others are important exceptions.

To understand bonding, it is necessary to delve deeper into the structure of atoms. Recall that negatively charged electrons orbit the nuclei of atoms in electron shells. With the exception of hydrogen, which has only one proton and one electron, the innermost electron shell of an atom contains only two electrons. The other shells contain various numbers of electrons, but the outermost shell never has more than eight (Figure 3.2). The electrons in the outermost shell are usually involved in chemical bonding.

Ionic and *covalent bonding* are important in minerals, and many minerals contain both types

electron A negatively charged particle of very little mass that encircles the nucleus of an atom.

electron shell Electrons orbit an atom's nucleus at specific distances in electron shells.

atomic number The number of protons in the nucleus of an atom.

atomic mass number The number of protons plus neutrons in the nucleus of an atom.

bonding The process whereby atoms join to other atoms.

compound Any substance resulting from the bonding of two or more different elements (e.g., water, H_2O, and quartz, SiO_2).

of bonds. Two other types of bonds, *metallic* and *van der Waals*, are much less common, but are important in determining the properties of some useful minerals.

Ionic Bonding

Notice in Figure 3.2 that most atoms have fewer than eight electrons in their outermost electron shell. However, some elements, including neon and argon, have complete outer shells with eight electrons; because of this electron configuration, these elements, known as the *noble gases*, do not react readily with other elements to form compounds. Interactions among atoms tend to produce electron configurations similar to those of the noble gases. That is, atoms interact so that their outermost electron shell is filled with eight electrons, unless the first shell (with two electrons) is also the outermost electron shell, as in helium.

One way that the noble gas configuration is attained is by the transfer of one or more electrons from one atom to another. Common salt is composed of the elements sodium (Na) and chlorine (Cl); each element is poisonous, but when combined chemically, they form the mineral halite, a compound of sodium chloride (NaCl). Notice in Figure 3.4a that sodium has 11 protons and 11 electrons; thus the positive electrical charges of the protons are exactly balanced by the negative charges of the electrons, and the atom is electrically neutral. Likewise, chlorine with 17 protons and 17 electrons is electrically neutral (Figure 3.4a). However, neither sodium nor chlorine has eight electrons in its outermost electron shell; sodium has only one, whereas chlorine has seven. To attain a stable configuration, sodium loses the electron in its outermost electron shell, leaving its next shell with eight electrons as the outermost one (Figure 3.4a). Sodium now has one fewer electron (negative charge) than it has protons (positive charge), so it is an electrically charged **ion** and is symbolized Na^+.

The electron lost by sodium is transferred to the outermost electron shell of chlorine, which had seven electrons to begin with. The addition of one more electron gives chlorine an outermost electron shell of eight electrons, the configuration of a noble gas. But its total number of electrons is now 18, which exceeds by 1 the number of protons. Accordingly, chlorine also becomes an ion, but it is negatively charged (Cl^-). An **ionic bond** forms between sodium and chlorine because of the attractive force between the positively charged sodium ion and the negatively charged chlorine ion (Figure 3.4a).

ion An electrically charged atom produced by adding or removing electrons from the outermost electron shell.

ionic bond A chemical bond resulting from the attraction between positively and negatively charged ions.

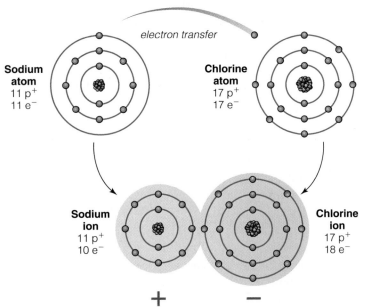

Sodium atom
11 p$^+$
11 e$^-$

electron transfer

Chlorine atom
17 p$^+$
17 e$^-$

Sodium ion
11 p$^+$
10 e$^-$

Chlorine ion
17 p$^+$
18 e$^-$

+   −

a. Transfer of the electron in the outermost shell of sodium to the outermost shell of chlorine. After electron transfer, the sodium and chlorine atoms are positively and negatively charged ions, respectively.

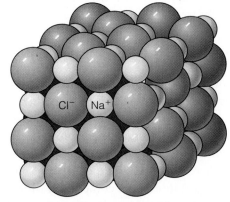

Cl$^-$ Na$^+$

b. This diagram shows the relative sizes of the sodium and chlorine atoms and their locations in a crystal of halite.

Figure 3.4 Ionic Bond to Form the Mineral Halite (NaCl)

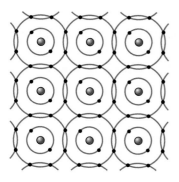 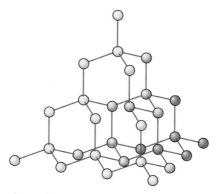 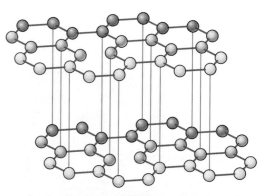

a. The orbits in the outermost electron shell overlap, so electrons are shared in diamond.

b. Covalent bonding of carbon atoms in diamond forms a three-dimensional framework.

c. Covalent bonds in graphite form strong sheets, but the van der Waals bonds between sheets are weak.

Figure 3.5 Covalent Bonds

In ionic compounds, such as sodium chloride (the mineral halite), the ions are arranged in a three-dimensional framework that results in overall electrical neutrality. In halite, sodium ions are bonded to chlorine ions on all sides, and chlorine ions are surrounded by sodium ions (Figure 3.4b).

Covalent Bonding

Covalent bonds form between atoms when their electron shells overlap and they share electrons. For example, atoms of the same element, such as carbon, cannot bond by transferring electrons from one atom to another. Carbon (C), which forms the minerals graphite and diamond, has four electrons in its outermost electron shell (Figure 3.5a). If these four electrons were transferred to another carbon atom, the atom receiving the electrons would have the noble gas configuration of eight electrons in its outermost electron shell, but the atom contributing the electrons would not.

In such situations, adjacent atoms share electrons by overlapping their electron shells. A carbon atom in diamond, for instance, shares all four of its outermost electrons with a neighbor to produce a stable noble gas configuration (Figure 3.5a).

Among the most common minerals, the silicates, the element silicon forms partly covalent and partly ionic bonds with oxygen.

Metallic and van der Waals Bonds

Metallic bonding results from an extreme type of electron sharing. The electrons of the outermost electron shell of metals such as gold, silver, and copper readily move from one atom to another. This electron mobility accounts for the fact that metals (1) have a metallic luster (their appearance in reflected light), (2) provide good electrical and thermal conductivity, and (3) can be easily reshaped. Only a few minerals possess metallic bonds, but those that do are very useful; copper, for example, is used for electrical wiring because of its high electrical conductivity.

> **covalent bond** A chemical bond formed by the sharing of electrons between atoms.

Some electrically neutral atoms and molecules* have no electrons available for ionic, covalent, or metallic bonding. They nevertheless have a weak attractive force between them, called a *van der Waals* or *residual bond*, when in proximity. The carbon atoms in the mineral graphite are covalently bonded to form sheets, but the sheets are weakly held together by van der Waals bonds (Figure 3.5c). This type of bonding makes graphite useful for pencil lead; when a pencil point is moved across a piece of paper, small pieces of graphite flake off along the planes held together by van der Waals bonds and adhere to the paper.

LO2 Explore the World of Minerals

We defined a mineral as an inorganic, naturally occurring, crystalline solid with a narrowly defined chemical composition and characteristic physical properties.

---

*A molecule is the smallest unit of a substance that has the properties of that substance. A water molecule (H_2O), for example, possesses two hydrogen atoms and one oxygen atom.

© TETRA IMAGES/JUPITERIMAGES

The distinction between minerals and rocks is not easy for beginning students to understand. Minerals are made up of chemical elements, and rocks consist of one or more minerals, but students commonly mistake one for the other. Can you think of analogies that might help you to remember the difference between minerals and rocks?

Furthermore, we know from the preceding section that most minerals are compounds of two or more chemically bonded elements, as in quartz (SiO_2).

NATURALLY OCCURRING INORGANIC SUBSTANCES

crystalline solid A solid in which the constituent atoms are arranged in a regular, three-dimensional framework.

crystal A naturally occurring solid of an element or compound with a specific internal structure that is manifested externally by planar faces, sharp corners, and straight edges.

The criterion *naturally occurring* excludes from minerals all substances manufactured by humans, such as synthetic diamonds and rubies. Some geologists think the term *inorganic* in the mineral definition is unnecessary, but it does remind us that animal matter and vegetable matter are not minerals. Nevertheless, some organisms, including corals, clams, and several other animals and plants, construct their shells of the compound calcium carbonate ($CaCO_3$), which is either the mineral aragonite or calcite, or their shells are made of silicon dioxide (SiO_2) as in quartz.

MINERAL CRYSTALS

By definition, minerals are **crystalline solids**, in which the constituent atoms are arranged in a regular, three-dimensional framework (Figure 3.5b). Under ideal conditions, crystalline solids grow and form perfect **crystals** that possess planar surfaces (crystal faces), sharp corners, and straight edges (Figure 3.6). In other words, the regular geometric shape of a mineral crystal is the exterior manifestation of an ordered internal atomic arrangement. Not all rigid substances are crystalline solids; natural and manufactured glass lack the ordered arrangement of atoms and is said to be *amorphous*, meaning "without form." Minerals are, by definition, crystalline solids; however, crystalline solids do not always yield well-formed crystals. The reason is that when crystals form, they may grow in proximity and form an interlocking mosaic in which individual crystals are not apparent or easily discerned.

So how do we know that minerals with no obvious crystals are actually crystalline? X-ray beams and light transmitted through mineral crystals or crystalline solids behave in a predictable manner, which provides compelling evidence for an internal orderly structure. Another way we can determine that minerals are actually crystalline is by their *cleavage*, the property of breaking or splitting repeatedly along smooth, closely spaced planes. Not all minerals have cleavage planes, but many do, and such regularity certainly indicates that cleavage are controlled by internal structure.

As early as 1669, the Danish scientist Nicholas Steno determined that the angles of intersection of equivalent crystal faces on different specimens of quartz are identical. Since then, this *constancy of interfacial angles* has been demonstrated for many other minerals, regardless of their size, shape, age, or geographic occurrence. Steno postulated that mineral crystals are made up of very small, identical building blocks, and that the arrangement of these building blocks determines the external form of mineral crystals, a proposal that has since been verified.

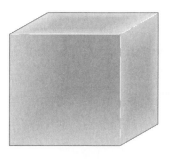

a. Cubic crystals are typical of the minerals halite and galena.

b. Pyritohedron crystals such as those of pyrite have 12 sides.

c. Diamond has octahedral, or eight-sided, crystals.

d. A prism terminated by a pyramid is found in quartz.

Figure 3.6 A Variety of Mineral Crystal Shapes

CHEMICAL COMPOSITION OF MINERALS

Mineral composition is shown by a chemical formula, which is a shorthand way of indicating the numbers of atoms of different elements that make up a mineral. The mineral quartz consists of one silicon (Si) atom for every two oxygen (O) atoms and thus has the formula SiO_2; the subscript number indicates the number of atoms. Orthoclase is composed of one potassium, one aluminum, three silicon, and eight oxygen atoms, so its formula is $KAlSi_3O_8$. Some minerals known as native elements consist of a single element and include silver (Ag), platinum (Pt), gold (Au), and graphite and diamond, both of which are composed of carbon (C).

For many minerals, the chemical composition does not vary, but some minerals have a range of compositions, because one element can substitute for another if the atoms of two or more elements are nearly the same size and the same charge. Notice in Figure 3.7 that iron

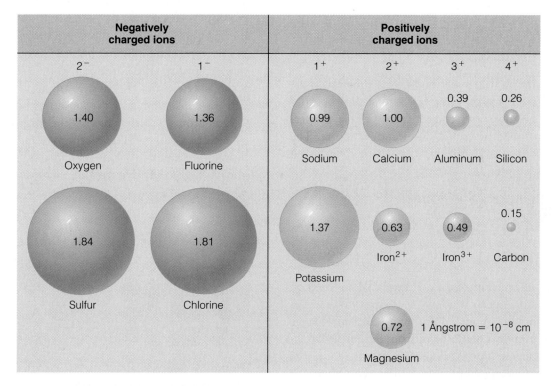

Figure 3.7 Sizes and Charges of Ions
Electrical charges and relative sizes of ions common in minerals. The numbers within the ions are the radii shown in Ångstrom units.

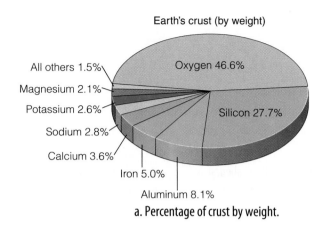

Earth's crust (by weight)

All others 1.5%
Magnesium 2.1%
Potassium 2.6%
Sodium 2.8%
Calcium 3.6%
Iron 5.0%
Aluminum 8.1%
Oxygen 46.6%
Silicon 27.7%

a. Percentage of crust by weight.

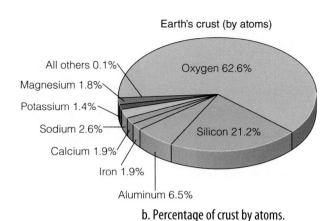

Earth's crust (by atoms)

All others 0.1%
Magnesium 1.8%
Potassium 1.4%
Sodium 2.6%
Calcium 1.9%
Iron 1.9%
Aluminum 6.5%
Oxygen 62.6%
Silicon 21.2%

b. Percentage of crust by atoms.

Figure 3.8 Common Elements in Earth's Crust

SOURCE: FROM G. T. MILLER, *LIVING IN THE ENVIRONMENT: PRINCIPLES, CONCEPTS, AND SOLUTIONS* 9/E (BELMONT, CA: WADSWORTH PUBLISHING, 1996), FIGURE 8.3.

and magnesium atoms are about the same size; therefore, they can substitute for each other. The chemical formula for the mineral olivine is $(Mg,Fe)_2SiO_4$, meaning that, in addition to silicon and oxygen, it may contain only magnesium, only iron, or a combination of both. Several other minerals also have ranges of compositions, so these are mineral groups with several members.

PHYSICAL PROPERTIES OF MINERALS

The last criterion in our definition of a mineral, *characteristic physical properties*, refers to such properties as hardness, color, and crystal form. These properties are controlled by composition and structure. The physical properties of minerals are further discussed later in this chapter.

LO3 Mineral Groups Recognized by Geologists

Geologists have identified and described more than 3,500 minerals, but only a few—perhaps two dozen— are common. One might think that an extremely large number of minerals could form from 92 naturally occurring elements; however, several factors limit the number possible. For one thing, many combinations of elements simply do not occur; no compounds are composed of only potassium and sodium or of silicon and iron, for example. Another important factor is that the bulk of Earth's crust is made up of only eight

chemical elements, and even among these eight, silicon and oxygen are by far the most common. In fact, the most common minerals in Earth's crust consist of silicon, oxygen, and one or more of the elements in Figure 3.8.

Geologists recognize mineral classes or groups, each with members that share the same negatively charged ion or ion group (Table 3.1). We have mentioned that ions are atoms that have either a positive or negative electrical charge resulting from the loss or gain of electrons in their outermost shell. In addition to ions, some minerals contain tightly bonded, complex groups of different atoms known as *radicals* that act as single units. A good example is the carbonate radical, consisting of a carbon atom bonded to three oxygen atoms and thus having the formula CO_3 and a -2 electrical charge. Other common radicals and their charges are sulfate (SO_4, -2), hydroxyl (OH, -1), and silicate (SiO_4, -4) (Figure 3.9).

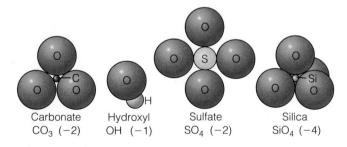

Carbonate
CO_3 (-2)

Hydroxyl
OH (-1)

Sulfate
SO_4 (-2)

Silica
SiO_4 (-4)

Figure 3.9 Radicals

Many minerals contain radicals, which are complex groups of atoms tightly bonded together. The silica and carbonate radicals are particularly common in many minerals, such as quartz (SiO_2) and calcite ($CaCO_3$).

TABLE 3.1

MINERAL GROUPS RECOGNIZED BY GEOLOGISTS

MINERAL GROUP	NEGATIVELY CHARGED ION OR RADICAL	EXAMPLES	COMPOSITION
Carbonate	$(CO_3)^{-2}$	Calcite	$CaCO_3$
		Dolomite	$CaMg(CO_3)_2$
Halide	Cl^{-1}, F^{-1}	Halite	$NaCl$
		Fluorite	CaF_2
Hydroxide	$(OH)^{-1}$	Brucite	$Mg(OH)_2$
Native element	—	Gold	Au
		Silver	Ag^*
		Diamond	C
Phosphate	$(PO_4)^{-3}$	Apatite	$Ca_5(PO_4)_3(F,Cl)$
Oxide	O^{-2}	Hematite	Fe_2O_3
		Magnetite	Fe_3O_4
Silicate	$(SiO_4)^{-4}$	Quartz	SiO_2
		Potassium feldspar	$KAlSi_3O_8$
		Olivine	$(Mg,Fe)_2SiO_4$
Sulfate	$(SO_4)^{-2}$	Anhydrite	$CaSO_4$
		Gypsum	$CaSO_4 \cdot 2H_2O$
Sulfide	S^{-2}	Galena	PbS
		Pyrite	FeS_2
		Argentite	Ag_2S^*

*Note that silver is found as both a native element and a sulfide mineral.

SILICATE MINERALS

Because silicon and oxygen are the two most abundant elements in Earth's crust, it is not surprising that many minerals contain these elements. A combination of silicon and oxygen is known as **silica**, and minerals that contain silica are **silicates**. Quartz (SiO_2) is pure silica, because it is composed entirely of silicon and oxygen. But most silicates have one or more additional elements, as in orthoclase ($KAlSi_3O_8$) and olivine [$(Mg,Fe)_2SiO_4$]. Silicate minerals include about one-third of all known minerals, but their abundance is even more impressive when one considers that they make up perhaps 95% of Earth's crust.

The basic building block of all silicate minerals is the **silica tetrahedron**, consisting of one silicon atom and four oxygen atoms (Figure 3.10a). These atoms are arranged so that the four oxygen atoms surround a silicon atom, which occupies the space between the oxygen atoms, thus forming a four-faced pyramidal structure (Figure 3.10b). The silicon atom has a positive charge of 4, and each of the four oxygen atoms has a negative charge of 2, resulting in a radical with a total negative charge of 4 $(SiO_4)^{-4}$.

Because the silica tetrahedron has a negative charge, it does not exist in nature as an isolated ion group; rather, it combines with positively charged ions or shares its oxygen atoms with other silica tetrahedra. In the simplest silicate minerals, the silica tetrahedra exist as single units bonded to positively charged ions. In minerals that contain isolated tetrahedra, the silicon-to-oxygen ratio is 1:4, and the negative charge of the silica ion is balanced by positive ions (Figure 3.10c). Olivine [$(Mg,Fe)_2SiO_4$],

silicate A mineral that contains silica, such as quartz (SiO_2).

silica A compound of silicon and oxygen.

silica tetrahedron The basic building block of all silicate minerals; consists of one silicon atom and four oxygen atoms.

for example, has either two magnesium (Mg^{+2}) ions, two iron (Fe^{+2}) ions, or one of each to offset the -4 charge of the silica ion.

Silica tetrahedra may also join together to form chains of indefinite length (Figure 3.10d). Single chains, as in the pyroxene minerals, form when each tetrahedron shares two of its oxygens with an adjacent tetrahedron, resulting in a silicon-to-oxygen ratio of 1:3. Enstatite, a pyroxene-group mineral, reflects

this ratio in its chemical formula $MgSiO_3$. Individual chains, however, possess a net -2 electrical charge, so they are balanced by positive ions, such as Mg^{+2}, that link parallel chains together (Figure 3.10d).

The amphibole group of minerals is characterized by a double-chain structure in which alternate tetrahedra in two parallel rows are cross-linked (Figure 3.10d). The formation of double chains results in a silicon-to-oxygen ratio of 4:11, so each double

Figure 3.10 The Silica Tetrahedron and Silicate Materials

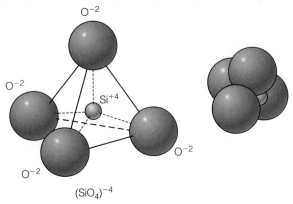

O^{-2}

O^{-2}

Si^{+4}

O^{-2}

O^{-2}

$(SiO_4)^{-4}$

a. Expanded view of the silica tetrahedron (*left*) and how it actually exists with its oxygen atoms touching.

b. View of the silica tetrahedron from above. Only the oxygen atoms are visible.

			Formula of negatively charged ion group		Example
c.	Isolated tetrahedra	△	$(SiO_4)^{-4}$	No oxygen atoms shared	Olivine
d.	Continuous chains of tetrahedra	Single chain	$(SiO_3)^{-2}$	Each tetrahedra shares two oxygen atoms with adjacent tetrahedra	Pyroxene group (augite)
		Double chain	$(Si_4O_{11})^{-6}$	Single chains linked by sharing oxygen atoms	Amphibole group (hornblende)
e.	Continuous sheets		$(Si_4O_{10})^{-4}$	Three oxygen atoms shared with adjacent tetrahedra	Micas (muscovite)
f.	Three-dimensional networks	SiO_2	$(SiO_2)^0$	All four oxygen atoms in tetrahedra shared	Quartz Potassium feldspars Plagioclase feldspars

c–f. Structures of the common silicate minerals shown by various arrangements of the silica tetrahedra.

chain possesses a -6 electrical charge. Mg^{+2}, Fe^{+2}, and Al^{+2} are usually involved in linking the double chains together.

In sheet-structure silicates, three oxygens of each tetrahedron are shared by adjacent tetrahedra (Figure 3.10e). Such structures result in continuous sheets of silica tetrahedra with silicon-to-oxygen ratios of 2:5. Continuous sheets also possess a negative electrical charge satisfied by positive ions located between the sheets. This particular structure accounts for the characteristic sheet structure of the *micas*, such as biotite and muscovite, and the *clay minerals*.

Three-dimensional networks of silica tetrahedra form when all four oxygens of the silica tetrahedra are shared by adjacent tetrahedra (Figure 3.10f). Such sharing of oxygen atoms results in a silicon-to-oxygen ratio of 1:2, which is electrically neutral. Quartz is a common framework silicate.

Geologists recognize two subgroups of silicates: ferromagnesian and nonferromagnesian silicates. The **ferromagnesian silicates** are those that contain iron (Fe), magnesium (Mg), or both. These minerals are commonly dark and more dense than nonferromagnesian silicates. Some of the common ferromagnesian silicate minerals are olivine, the pyroxenes, the amphiboles, and biotite (Figure 3.11a).

The **nonferromagnesian silicates** lack iron and magnesium, are generally light colored, and are less dense than ferromagnesian silicates (Figure 3.11b). The most common minerals in Earth's crust are nonferromagnesian silicates known as *feldspars*. Feldspar is a general name, however, and there are two distinct groups, each of which includes several species.

The *potassium feldspars* are represented by microcline and orthoclase ($KAlSi_3O_8$). The second group of feldspars, the *plagioclase feldspars*, range from calcium-rich ($CaAl_2Si_2O_8$) to sodium-rich ($NaAlSi_3O_8$) varieties.

Quartz (SiO_2) is another common nonferromagnesian silicate. It is a framework silicate that can usually be recognized by its glassy appearance and hardness. Another fairly common nonferromagnesian silicate is muscovite, which is a mica (Figure 3.11b).

CARBONATE MINERALS

Carbonate minerals, those that contain the negatively charged carbonate radical $(CO_3)^{-2}$, include calcium carbonate ($CaCO_3$) as the minerals *aragonite* or *calcite* (Figure 3.12a). Aragonite is unstable and commonly changes to calcite, the main constituent of the sedimentary rock *limestone*. Several other carbonate minerals are known, but only one need concern us: *Dolomite* [$CaMg(CO_3)_2$] forms by the chemical alteration of calcite by the addition of magnesium. Sedimentary rock composed of the mineral dolomite is *dolostone*.

> **ferromagnesian silicate** Any silicate mineral that contains iron, magnesium, or both.
>
> **nonferromagnesian silicate** A silicate mineral that has no iron or magnesium.
>
> **carbonate mineral** A mineral with the carbonate radical $(CO_3)^{-2}$, as in calcite ($CaCO_3$) and dolomite [$CaMg(CO_3)_2$].

a. Ferromagnesian silicates.

b. Nonferromagnesian silicates.

Figure 3.11 Common Silicate Materials

OTHER MINERAL GROUPS

Even though minerals from these groups are less common than silicates and carbonates, many are found in rocks in small quantities and others are important resources (Table 3.1). In the oxides, an element combines with oxygen as in hematite (Fe_2O_3) and magnetite (Fe_3O_4). Rocks with high concentrations of these minerals are sources of iron ores for the manufacture of steel. The related hydroxides form mostly by the chemical alteration of other minerals.

We have noted that the *native elements* are minerals composed of a single element, such as diamond and graphite (C) and the precious metals gold (Au), silver (Ag), and platinum (Pt). Some elements, such as silver and copper, are found both as native elements and as compounds and are thus also included in other mineral groups; argentite (Ag_2S), a silver sulfide, is an example.

Several minerals and rocks that contain the phosphate radical $(PO_4)^{-3}$ are important sources of phosphorus for fertilizers. The sulfides, such as galena (PbS), the ore of lead, have a positively charged ion combined with sulfur (S^{-2}) (Figure 3.12b), whereas the sulfates have an element combined with the complex radical $(SO_4)^{-2}$, as in gypsum ($CaSO_4·2H_2O$) (Figure 3.12c). The halides contain the halogen elements, fluorine (F^{-1}) and chlorine (Cl^{-1}); examples are halite (NaCl) (Figure 3.12d) and fluorite (CaF_2).

LO4 Physical Properties of Minerals

Many physical properties of minerals are remarkably constant for a given mineral species, but some, especially color, may vary. Most common minerals can be identified by using the physical properties described next.

LUSTER AND COLOR

Luster (not to be confused with *color*) is the quality and intensity of light reflected from a mineral's surface. Geologists define luster as *metallic*, having the appearance of a metal, and *nonmetallic*. Of the four minerals shown in Figure 3.12, only galena has a metallic luster. Among the several types of nonmetallic luster are glassy or vitreous (as in quartz), dull or earthy, waxy, greasy, and brilliant (as in diamond) (Figure 3.1).

> **luster** The appearance of a mineral in reflected light. Luster is metallic or nonmetallic, although the latter has several subcategories.

a. Calcite ($CaCO_3$) is the most common carbonate mineral.

SUE MONROE

b. The sulfide mineral galena (PbS) is the ore of lead.

SUE MONROE

c. Gypsum ($CaSO_4·2H_2O$) is a common sulfate mineral.

SUE MONROE

d. Halite (NaCl) is a good example of a halide mineral.

SUE MONROE

Figure 3.12 Representative Specimens from Four Mineral Groups

Beginning geology students are distressed by the fact that the color of some minerals varies considerably, making the most obvious physical property of little use for identifying some minerals. In any case, we can make some helpful generalizations about color. Ferromagnesian silicates are typically black, brown, or dark green, although olivine is olive green (Figure 3.11a). Nonferromagnesian silicates vary considerably in color but are rarely very dark. White, cream, colorless, and shades of pink and pale green are more typical (Figure 3.11b).

Another helpful generalization is that the color of minerals with a metallic luster is more consistent than is the color of nonmetallic minerals. For example, galena is always lead-gray (Figure 3.12b), and pyrite is invariably brassy yellow. In contrast, quartz, a nonmetallic mineral, may be colorless, smoky brown to almost black, rose, yellow-brown, milky white, blue, or violet to purple.

CRYSTAL FORM

As we noted, many mineral specimens do not show the perfect crystal form typical of that mineral species. Nevertheless, some minerals do typically occur as crystals (Figure 3.13a). For example, 12-sided crystals of garnet are common, as are 6- and 12-sided crystals of pyrite. Minerals that grow in cavities or are

precipitated from circulating hot water (hydrothermal solutions) in cracks and crevices in rocks also commonly occur as crystals.

Crystal form is useful for mineral identification, but some minerals have the same crystal form. Pyrite (FeS_2), galena (PbS), and halite (NaCl) all occur as cubic crystals, but they can be easily identified by other properties, such as color, luster, hardness, and density.

CLEAVAGE AND FRACTURE

Not all minerals possess **cleavage**, but those that do have this property, break or split along a smooth plane or planes of weakness determined by the strength of their chemical bonds. Cleavage is characterized in terms of quality (perfect, good, poor), direction, and angles of intersection of cleavage planes. Biotite, a common ferromagnesian silicate, has perfect cleavage in one direction (Figure 3.14a). Biotite is a sheet silicate, with the sheets of silica tetrahedra weakly bonded to one another by iron and magnesium ions.

Feldspars possess two directions of cleavage that intersect at right angles (Figure 3.14b), and the mineral halite has three directions of cleavage, all of which

> **cleavage** Breakage along internal planes of weakness in mineral crystals.

Figure 3.13 Mineral Crystals

a. Cubic crystals of fluorite (CaF_2).

SUE MONROE

b. Calcite ($CaCO_3$) crystals.

SUE MONROE

c. Blade-shaped crystals of barite ($BaSO_4$).

SUE MONROE

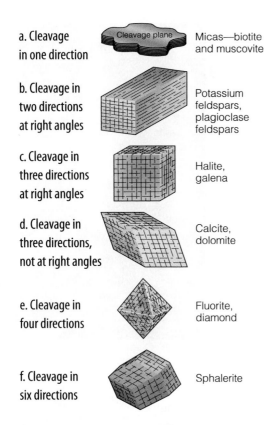

a. Cleavage in one direction — Micas—biotite and muscovite

b. Cleavage in two directions at right angles — Potassium feldspars, plagioclase feldspars

c. Cleavage in three directions at right angles — Halite, galena

d. Cleavage in three directions, not at right angles — Calcite, dolomite

e. Cleavage in four directions — Fluorite, diamond

f. Cleavage in six directions — Sphalerite

Figure 3.14 Several Types of Mineral Cleavage

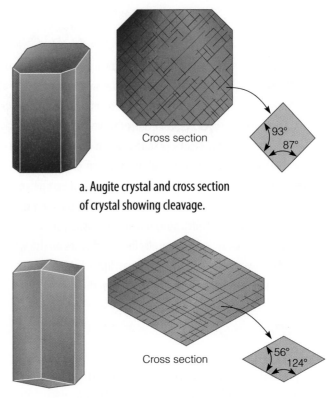

Cross section

93°
87°

a. Augite crystal and cross section of crystal showing cleavage.

Cross section

56°
124°

b. Hornblende crystal and cross section of crystal showing cleavage.

Figure 3.15 Cleavage in Augite and Hornblende

intersect at right angles (Figure 3.14c). Calcite also possesses three directions of cleavage, but none of the intersection angles is a right angle, so cleavage fragments of calcite are rhombohedrons (Figure 3.14d). Minerals with four directions of cleavage include fluorite and diamond (Figure 3.14e). Ironically, diamond, the hardest mineral, can be easily cleaved. A few minerals, such as sphalerite, an ore of zinc, have six directions of cleavage (Figure 3.14f).

Cleavage is an important diagnostic property of minerals, and recognizing it is essential in distinguishing between some minerals. The pyroxene mineral augite and the amphibole mineral hornblende, for example, look much alike: Both are dark green to black, have the same hardness, and possess two directions of cleavage. But the cleavage planes of augite intersect at about 90 degrees, whereas the cleavage planes of hornblende intersect at angles of 56 degrees and 124 degrees (Figure 3.15).

In contrast to cleavage, *fracture* is mineral breakage along irregular

hardness A term used to express the resistance of a mineral to abrasion.

specific gravity The ratio of a substance's weight, especially a mineral, to an equal volume of water at 4°C.

surfaces. Any mineral can be fractured if enough force is applied, but the fracture surfaces are commonly uneven or conchoidal (curved) rather than smooth.

HARDNESS

An Austrian geologist, Friedrich Mohs, arbitrarily assigned a hardness value of 10 to diamond, the hardest mineral known, and lesser values to the other minerals. Relative hardness is easily determined by the use of Mohs hardness scale (Table 3.2). Quartz will scratch fluorite but cannot be scratched by fluorite, gypsum can be scratched by a fingernail, and so on. So **hardness** is defined as a mineral's resistance to abrasion and is controlled mostly by internal structure. Both graphite and diamond are composed of carbon, but the former has a hardness of 1 to 2, whereas the latter has a hardness of 10.

SPECIFIC GRAVITY (DENSITY)

A mineral's **specific gravity** is the ratio of its weight to the weight of an equal volume of pure water at 4°C.

TABLE 3.2

MOHS HARDNESS SCALE

HARDNESS	MINERAL	HARDNESS OF SOME COMMON OBJECTS
10	Diamond	
9	Corundum	
8	Topaz	
7	Quartz	
		Steel file ($6\frac{1}{2}$)
6	Orthoclase	
		Glass ($5\frac{1}{2}$–6)
5	Apatite	
4	Fluorite	
3	Calcite	Copper penny (3)
		Fingernail ($2\frac{1}{2}$)
2	Gypsum	
1	Talc	

Thus, a mineral with a specific gravity of 3.0 is three times as heavy as water. **Density,** in contrast, is a mineral's mass (weight) per unit of volume expressed in grams per cubic centimeter. So the specific gravity of galena (Figure 3.12b) is 7.58, and its density is 7.58 g/cm^3. In most instances, we will refer to a mineral's density, and in some of the following chapters, we will mention the density of rocks.

Because ferromagnesian silicates contain iron, magnesium, or both, they tend to be denser than nonferromagnesian silicates. In general, the metallic minerals, such as galena and hematite, are denser than nonmetals. Pure gold, with a density of 19.3 g/cm^3, is about 2.5 times as dense as lead. Diamond and graphite, both of which are composed of carbon (C), illustrate how structure controls specific gravity or density. The specific gravity of diamond is 3.5, whereas that of graphite varies from 2.09 to 2.33.

OTHER USEFUL MINERAL PROPERTIES

Talc has a distinctive soapy feel, graphite writes on paper, halite tastes salty, and magnetite is magnetic. Calcite possesses the property of *double refraction,* meaning that an object when viewed through a transparent piece of calcite will have a double image. Some sheet silicates are plastic and, when bent into a new shape, will retain that shape; others are flexible and, if bent, will return to their original position when the forces that bent them are removed.

A simple chemical test to identify the minerals calcite and dolomite involves applying a drop of dilute hydrochloric acid to the mineral specimen. If the mineral is calcite, it will react vigorously with the acid and release carbon dioxide, which causes the acid to bubble or effervesce. Dolomite, in contrast, will not react with hydrochloric acid unless it is powdered.

LO5 Rock-Forming Minerals

Geologists use the term **rock** for a solid aggregate of one or more minerals, but the term also refers to masses of mineral-like matter, as in the natural glass obsidian, and masses of solid organic matter, as in coal. And even though some rocks may contain many minerals, only a few, designated **rock-forming minerals,** are sufficiently common for rock identification and classification (Table 3.3 and Figure 3.16). Others, known as *accessory minerals,* are present in such small quantities that they can be disregarded.

Given that silicate minerals are by far the most common ones in Earth's crust, it follows that most rocks are composed of these minerals. Indeed, feldspar minerals (plagioclase feldspars and potassium feldspars) and quartz make up more than 60% of Earth's crust. So, even though there are hundreds of silicates, only a few are particularly common in rocks.

The most common nonsilicate rock-forming minerals are the carbonates calcite ($CaCO_3$) and dolomite [$CaMg(CO_3)_2$], the main constituents of the sedimentary rocks limestone and dolostone, respectively. Among the sulfates and halides, gypsum ($CaSO_4 \cdot 2H_2O$) in rock gypsum and halite (NaCl) in rock salt, are common enough to qualify as rock-forming minerals. Even though these minerals and their corresponding rocks might be common in some areas, however, their overall abundance is limited compared to the silicate and carbonate rock-forming minerals.

density The mass of an object per unit volume; usually expressed in grams per cubic centimeter (g/cm^3).

rock A solid aggregate of one or more minerals, as in limestone and granite, or a consolidated aggregate of rock fragments, as in conglomerate, or masses of rocklike materials, such as coal and obsidian.

rock-forming mineral Any mineral common in rocks that is important in their identification and classification.

TABLE 3.3

IMPORTANT ROCK-FORMING MINERALS

MINERAL	PRIMARY OCCURRENCE
Ferromagnesian silicates	
Olivine	Igneous and metamorphic rocks
Pyroxene group Augite most common	Igneous and metamorphic rocks
Amphibole group Hornblende most common	Igneous and metamorphic rocks
Biotite	All rock types
Nonferromagnesian silicates	
Quartz	All rock types
Potassium feldspar group Orthoclase, microcline	All rock types
Plagioclase feldspar group	All rock types
Muscovite	All rock types
Clay mineral group	Soils, sedimentary rocks, and some metamorphic rocks
Carbonates	
Calcite	Sedimentary rocks
Dolomite	Sedimentary rocks
Sulfates	
Anhydrite	Sedimentary rocks
Gypsum	Sedimentary rocks
Halides	
Halite	Sedimentary rocks

Plagioclase feldspar

Quartz

Potassium feldspar

Biotite

COPYRIGHT AND PHOTOGRAPH BY DR. PARVINDER S. SETHI

Figure 3.16 Minerals in Granite

The igneous rock granite (see Chapter 4) is made up of mostly three minerals—quartz, potassium feldspar, and plagioclase feldspar—but it may also contain small amounts of biotite, muscovite, and hornblende.

LO6 How Do Minerals Form?

Thus far, we have discussed the composition, structure, and physical properties of minerals, but we have not addressed how they originate. One phenomenon that accounts for minerals is the cooling of molten rock material known as *magma* (magma that reaches the surface is called *lava*).

As magma or lava cools, minerals crystallize and grow, thereby determining the mineral composition of igneous rocks such as basalt (dominated by ferromagnesian silicates) and granite (dominated by nonferromagnesian silicates). Hot-water solutions derived from magma commonly invade cracks and crevasses in adjacent rocks, and from these solutions, a variety of minerals crystallize, some of which have economic importance. Minerals also originate when water in hot springs cools, and when hot, mineral-rich water discharges onto the seafloor at hot springs known as hydrothermal vents.

Dissolved materials in seawater, more rarely in lake water, combine to form minerals such as halite (NaCl), gypsum ($CaSO_4 \cdot 2H_2O$), and several others when the water evaporates. Aragonite and/or calcite, both varieties of calcium carbonate ($CaCO_3$), might also form from evaporating water, but most originate when organisms such as clams, oysters, corals, and floating microorganisms use this compound to construct their shells. A few plants and animals use silicon dioxide (SiO_2) for their skeletons, which accumulate as mineral matter on the seafloor when the organisms die.

Some clay minerals form when chemical processes compositionally and structurally alter other minerals, and others originate when rocks are changed during metamorphism. In fact, the agents that cause metamorphism—heat, pressure, and chemically active fluids—are responsible for the origin of many minerals.

LO7 Natural Resources and Reserves

Geologists at the U.S. Geological Survey define a **resource** as a concentration of naturally occurring solid, liquid, or gaseous material in or on Earth's crust in such form and amount that economic extraction of a commodity from the concentration is currently or potentially feasible.

Natural resources are mostly concentrations of minerals, rocks, or both, but liquid petroleum and natural gas are also included. In fact, some resources are referred to as *metallic resources* (copper, tin, iron ore, etc.), *nonmetallic resources* (sand and gravel, crushed stone, salt, sulfur, etc.), and energy resources (petroleum, natural gas, coal, and uranium). All of these are indeed resources, but we must distinguish between a resource, the total amount of a commodity, whether discovered or undiscovered, and a **reserve**, which is only that part of the resource base that is known and can be economically recovered.

The distinction between a resource and a reserve is simple enough in principle, but in practice it depends on several factors, not all of which remain constant. Geographic location may be important. For instance, a resource in a remote region might not be mined because transportation costs are too high, and what might be deemed a resource rather than a reserve in the United States and Canada may be mined in a developing country, where labor costs are low. The commodity in question is also important. Gold or diamonds in sufficient quantity can be mined profitably just about anywhere, whereas sand and gravel deposits must be close to their market areas.

Obviously, the market price is important in evaluating any resource. From 1935 until 1968, the U.S. government maintained the price of gold at $35 per troy ounce (1 troy ounce = 31.1 g). When this restriction was removed, demand determined the market price, gold prices rose, and many marginal deposits became reserves, and several abandoned mines were reopened.

The status of a resource is also affected by changes in technology. By the time of World War II (1939–1945), the richest iron ore deposits of the Great Lakes region in the United States and Canada had been mostly depleted. But the development of a method for separating the iron from unusable rock and shaping it into pellets ideal for use in blast furnaces made it profitable to mine rocks that contained less iron.

Most people know that industrialized societies depend on natural resources, but they know little about their occurrence, methods of recovery, and economics. Geologists are, of course, essential in finding and evaluating deposits; however, extraction involves engineers, chemists, miners, and many people in support. Ultimately, though, the decision about whether a deposit should be mined or not is made by people trained in business and economics. In short, extraction must yield a profit.

In addition to resources such as petroleum, gold, and ores of iron and copper, some quite common minerals are also essential. For example, pure quartz sand is used to manufacture glass and optical instruments, as well as sandpaper and steel alloys. Clay minerals are needed to make ceramics and paper; feldspars are used for porcelain, ceramics, enamel, and glass; and phosphate-bearing rock is used for fertilizers. Muscovite is used in a variety of products, such as lipstick, glitter, eye shadow, and the lustrous paints on appliances and automobiles.

Many resources are *nonrenewable*, which means that there is a limited supply and they cannot be replenished by natural processes as fast as they are depleted. Accordingly, once a resource is depleted, suitable substitutes, if available, must be found. For some essential resources, the United States is totally dependent on imports; for example, no cobalt is mined in this country. Yet the United States, the world's largest consumer of cobalt, uses this essential metal in gas-turbine aircraft engines and magnets. Obviously, all cobalt is imported, as is all manganese, an element essential for making steel. The United States also imports all of the aluminum ore it uses, as well as all or some of many other resources.

To ensure continued supplies of essential minerals and energy resources, geologists and other scientists, government agencies, and leaders in business and industry continually assess the status of resources in view of changing economic and political conditions and changes in science and technology. The U.S. Geological Survey, for instance, keeps detailed statistical records of mine production, imports, and exports, and regularly publishes reports on the status of numerous commodities. Similar reports appear regularly in the *Canadian Minerals Yearbook.*

resource A concentration of naturally occurring solid, liquid, or gaseous material in or on Earth's crust in such form and amount that economic extraction of a commodity from the concentration is currently or potentially feasible.

reserve The part of the resource base that can be extracted economically.

CHAPTER 4
IGNEOUS ROCKS AND INTRUSIVE IGNEOUS ACTIVITY

SUE MONROE

View of the Sierra Nevada taken west of Lone Pine, California. The rocks in this view are part of the Sierra Nevada batholith—a huge mass of granite and related rocks made up of intrusive bodies. The high peak toward the right is Mount Whitney, which at 4,421 m is the highest peak in the continental United States.

Introduction

You already know that the term *rock* applies to solid aggregates of minerals, mineral-like matter (obsidian), and masses of altered organic matter (coal). In Chapter 1, we also briefly defined the three main families of rocks—igneous, sedimentary, and metamorphic (see Figure 1.15). Our concern in this chapter is with igneous rocks, that is, rocks that formed from molten rock matter (magma) that cooled and crystallized to form minerals, and rocks made up of particulate matter ejected during explosive volcanic eruptions. We are most familiar with eruptions of lava and particulate matter from volcanoes, because they can be observed. And yet, most magma never reaches Earth's surface, but rather cools and crystallizes underground, thus forming several types of igneous rock bodies that geologists call *plutons*, named for Pluto, the Roman god of the underworld.

LEARNING OUTCOMES

After reading this unit, you should be able to do the following:

LO1 Describe the properties and behavior of magma and lava

LO2 Explain how magma originates and changes

LO3 Identify and classify igneous rocks by their characteristics

LO4 Recognize intrusive igneous bodies, or plutons

LO5 Explain how batholiths intrude into Earth's crust

Granite and related rocks are attractive, especially when sawed and polished. They are used for tombstones, mantelpieces, kitchen counters, facing stones on buildings, monuments, pedestals for statues, and for statuary itself.

Granite, composed mostly of feldspars and quartz, and similar-appearing rocks are the most common in the large plutons, such as the one in the Sierra Nevada in California. These exposures of granite formed far below the surface and were revealed in their present form only after uplift and deep erosion.

Although we discuss intrusive igneous activity and the origin of plutons in this chapter, whereas volcanism is considered in Chapter 5, both processes are related. In fact, the same kinds of magmas are involved; however, magma varies in its mobility, so only some of it reaches the surface. Furthermore, plutons that lie beneath areas of volcanism are the source of the overlying lava flows and particulate matter erupted by volcanoes.

One important reason to study igneous rocks and intrusive igneous processes is that igneous rocks make up large parts of the continents and all of the oceanic crust, which forms continuously at divergent plate boundaries. Another reason to study this topic is that most plutons and volcanoes are found at or near divergent and convergent plate boundaries, so the presence of igneous rocks is one criterion for recognizing ancient plate boundaries.

magma Molten rock material generated within Earth.

lava Magma that reaches Earth's surface.

lava flow A stream of magma flowing over Earth's surface.

pyroclastic materials Fragmental substances, such as ash, that are explosively ejected from a volcano.

igneous rock Any rock formed by cooling and crystallization of magma or lava or the consolidation of pyroclastic materials.

volcanic (extrusive igneous) rock An igneous rock formed when magma is extruded onto Earth's surface where it cools and crystallizes, or when pyroclastic materials become consolidated.

plutonic (intrusive igneous) rock Igneous rock that formed from magma intruded into or formed in place within the crust.

felsic magma Magma with more than 65% silica and considerable sodium, potassium, and aluminum, but little calcium, iron, and magnesium.

mafic magma Magma with between 45% to 52% silica and proportionately more calcium, iron, and magnesium than intermediate and felsic magma.

LO1 The Properties and Behavior of Magma and Lava

Geologists define **magma** as any mass of molten rock material below Earth's surface, whereas **lava** is simply magma that reaches the surface. Thus, it is the same material in both locations, although when lava reaches the surface, it tends to lose some of its gas content because of decreased pressure. Any magma is less dense than the rock from which it formed, so it tends to rise toward the surface, but much of it solidifies deep underground, thereby accounting for the origin of plutons. The magma that does reach the surface issues forth as **lava flows**, and some of it is forcefully ejected into the atmosphere as particles known as **pyroclastic materials** (from the Greek *pyro* for "fire" and *klastos* for "broken").

All **igneous rocks** derive ultimately from magma; however, two separate processes account for them. They form when (1) magma or lava cools and crystallizes to form aggregates of minerals, or (2) pyroclastic materials are consolidated. Those igneous rocks derived from lava flows and pyroclastic materials, both of which are extruded onto the surface, are known as **volcanic rocks** or **extrusive igneous rocks**. **Plutonic rocks** or **intrusive igneous rocks** are formed when magma cooled below the surface.

COMPOSITION OF MAGMA

By far the most abundant minerals in Earth's crust are silicates such as quartz, feldspars, and several ferromagnesian silicates, all made up of silicon and oxygen, and other elements shown in Figure 3.8. As a result, melting of the crust yields mostly silica-rich magmas that also contain considerable aluminum, calcium, sodium, iron, magnesium, and potassium, and several other elements in lesser quantities. Another source of magma is Earth's upper mantle, which is composed of rocks that contain mostly ferromagnesian silicates. Thus, magma from this source contains comparatively less silicon and oxygen (silica) and more iron and magnesium.

Although there are a few exceptions, the primary constituent of magma is silica, which varies enough to distinguish magmas classified as felsic, intermediate, and mafic.* **Felsic magma**, with more than 65% silica, is silica rich and contains considerable sodium, potassium, and aluminum, but little calcium, iron, and magnesium. In contrast, **mafic magma**, with less than 52% silica, is silica poor and contains proportionately more calcium, iron, and magnesium. And as you would expect, **intermediate magma** has a composition between felsic and mafic magma (Table 4.1).

How Hot Are Magma and Lava?

Erupting lava generally has a temperature in the range of 700° to 1,200°C, although a temperature of 1,350°C was recorded above a lava lake in Hawaii where volcanic gases reacted with the atmosphere. Magma must be even hotter than lava; however, no direct measurements of magma temperatures have ever been made.

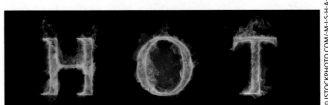

© ISTOCKPHOTO.COM/-M-I-S-H-A-

*Lava from some volcanoes in Africa cools to form carbonitite, an igneous rock with at least 50% carbonate minerals, mostly calcite and dolomite.

TABLE 4.1

THE MOST COMMON TYPES OF MAGMAS AND THEIR CHARACTERISTICS

TYPE OF MAGMA	SILICA CONTENT (%)	SODIUM, POTASSIUM, AND ALUMINUM	CALCIUM, IRON, AND MAGNESIUM
Ultramafic	< 45		Increase
Mafic	45–52		↑
Intermediate	53–65		
Felsic	> 65	Increase	

Most lava temperatures are taken at volcanoes that show little or no explosive activity, so our best information comes from mafic lava flows such as those in Hawaii (Figure 4.1). In contrast, eruptions of felsic lava are not as common, and the volcanoes that these flows issue from tend to be explosive and thus cannot be approached safely. Nevertheless, the temperatures of

COPYRIGHT AND PHOTOGRAPH BY DR. PARVINDER S. SETHI

Figure 4.1 How Hot Is Lava?
A geologist uses a remotely operated, handheld device for measuring temperature of a lava flow visible through an opening in the rocks in Hawaiian Volcanoes National Park, Hawaii.

some bulbous masses of felsic lava in lava domes have been measured at a distance with an optical pyrometer. The surfaces of these lava domes are as hot as 900°C, but their interiors must surely be even hotter.

The reason that lava and magma retain heat so well is that rock conducts heat so poorly. Accordingly, the interiors of thick lava flows and pyroclastic flow deposits may remain hot for months or years, whereas plutons, depending on their size and depth, may not cool completely for thousands to millions of years.

VISCOSITY: RESISTANCE TO FLOW

All liquids have the property of **viscosity,** or resistance to flow. Water's viscosity is very low, so it is highly fluid and flows readily. For other liquids, such as cold motor oil and syrup, their viscosity is so high that they flow much more slowly. But when these liquids are heated, their viscosity is much lower and they flow more easily. Accordingly, you might suspect that temperature controls the viscosity of magma and lava, and this inference is partly correct. We can generalize and say that hot magma or lava moves more readily than cooler magma or lava, but we must qualify this statement by noting that temperature is not the only control of viscosity.

Silica content strongly controls magma and lava viscosity. With increasing silica content, numerous networks of silica tetrahedra form and retard flow, because for flow to take place, the strong bonds of the networks must be ruptured. Mafic magma and lava with 45% to 52% silica have fewer silica tetrahedra networks and, as a result, are more mobile than felsic magma and lava flows (Figure 4.2). One mafic flow in 1783 in Iceland flowed nearly 80 km. Felsic magma, in contrast, because of its higher viscosity, does not reach the surface as often as mafic magma. And when felsic lava flows do occur, they tend to be slow moving and thick, and move only short distances.

Temperature and silica content are important controls on the viscosity of magma and lava, but other factors include gases, mostly CO_2, as well as the presence of mineral crystals and friction from the surface over which lava flows. Lava with a high content of dissolved

intermediate magma
Magma with a silica content between 53% to 65% and an overall composition intermediate between mafic and felsic magma.

viscosity A fluid's resistance to flow.

Figure 4.2 Viscosity of Magma and Lava

Temperature is an important control on viscosity, but so is composition. Mafic lava tends to be fluid, whereas felsic lava is much more viscous.

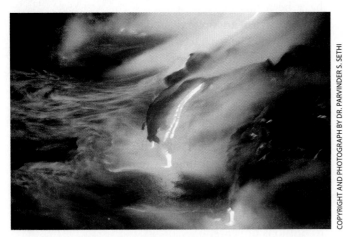

a. A mafic lava flow in 1984 on Mauna Loa Volcano in Hawaii. These flows move rapidly and form thin layers.

b. The Novarupta lava dome in Katmai National Park in Alaska. The lava is felsic and viscous, so it was extruded as a bulbous mass. This image was taken in 1987.

gases flows more readily than one with a lesser amount of gases, whereas lava with many crystals or that flows over a rough surface tends to be more viscous.

LO2 How Does Magma Originate and Change?

Even if you have not witnessed a volcanic eruption, you are probably familiar with lava flows and pyroclastic eruptions. In any case, we know about some aspects of igneous activity, but most people are unaware of how and where magma originates, how it rises, and how it might change. Indeed, there is a misconception that lava comes from a continuous layer of molten rock beneath the crust or that it comes from Earth's molten outer core.

First, let us address how magma originates. We know that the atoms in a solid are in constant motion and that, when a solid is heated, the energy of motion exceeds the binding forces and the solid melts. We are all familiar with this phenomenon, and we are also aware that not all solids melt at the same temperature. Likewise, if

magma chamber A reservoir of magma within Earth's upper mantle or lower crust.

Bowen's reaction series A series of minerals that form in a specific sequence in cooling magma or lava; originally proposed to explain the origin of intermediate and felsic magma from mafic magma.

heated, the minerals in rocks begin to melt but not all at the same time. And once magma forms, it tends to rise, because it is less dense than the rock that melted.

Magma may come from 100 to 300 km deep, but most forms at much shallower depths in the upper mantle or lower crust and accumulates in reservoirs known as **magma chambers**. Beneath spreading ridges, where the crust is thin, magma chambers exist at a depth of only a few kilometers; however, along convergent plate boundaries, magma chambers are commonly a few tens of kilometers deep. The volume of a magma chamber ranges from a few to many hundreds of cubic kilometers of molten rock within the otherwise solid lithosphere. Some simply cools and crystallizes within Earth's crust, thus accounting for the origin of plutons, whereas some rises to the surface and is erupted as lava flows or pyroclastic materials.

BOWEN'S REACTION SERIES

During the early 1900s, N. L. Bowen knew that minerals crystallize in a predictable sequence from cooling magma. Based on his observations and laboratory experiments, Bowen proposed a mechanism, now called **Bowen's reaction series**, to account for the derivation of intermediate and felsic magmas from mafic magma. Bowen's reaction series consists of two branches: a *discontinuous branch* and a *continuous branch* (Figure 4.3). As the temperature of magma decreases, minerals crystallize along both branches simultaneously, but for convenience, we will discuss them separately.

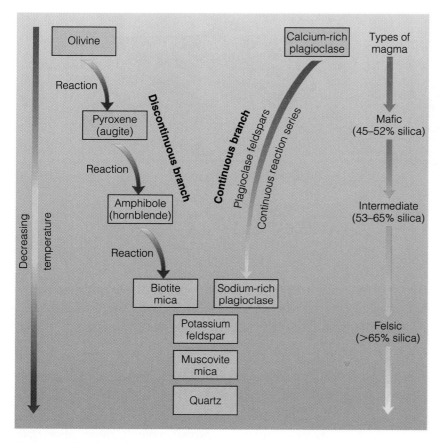

Figure 4.3 Bowen's Reaction Series
Bowen's reaction series consists of a discontinuous branch, along which a succession of ferromagnesian silicates crystallize as the magma's temperature decreases, and a continuous branch, along which plagioclase feldspars with increasing amounts of sodium crystallize. Notice also that the composition of the initial mafic magma changes as crystallization takes place along the two branches.

In the discontinuous branch, which contains only ferromagnesian silicates, one mineral changes to another over specific temperature ranges (Figure 4.3). As the temperature decreases, a temperature range is reached in which a given mineral begins to crystallize. A previously formed mineral reacts with the remaining liquid magma (the melt) so that it forms the next mineral in the sequence. For instance, olivine $[(Mg,Fe)_2SiO_4]$ is the first ferromagnesian silicate to crystallize. As the magma continues to cool, it reaches the temperature range at which pyroxene is stable; a reaction occurs between the olivine and the remaining melt, and pyroxene forms.

With continued cooling, a reaction takes place between pyroxene and the melt, and the pyroxene structure is rearranged to form amphibole. Further cooling causes a reaction between the amphibole and the melt, and its structure is rearranged so that the sheet structure of biotite mica forms. Although the

reactions just described tend to convert one mineral to the next in the series, the reactions are not always complete. Olivine, for example, might have a rim of pyroxene, indicating an incomplete reaction. If magma cools rapidly enough, the early-formed minerals do not have time to react with the melt, and thus all of the ferromagnesian silicates in the discontinuous branch can be in one rock. In any case, by the time biotite has crystallized, essentially all of the magnesium and iron present in the original magma have been used up.

Plagioclase feldspars, which are nonferromagnesian silicates, are the only minerals in the continuous branch of Bowen's reaction series (Figure 4.3). Calcium-rich plagioclase crystallizes first. As the magma continues to cool, calcium-rich plagioclase reacts with the melt, and plagioclase containing proportionately more sodium crystallizes until all of the calcium and sodium are used up. In many cases, cooling is too rapid for a complete transformation from calcium-rich to sodium-rich plagioclase to take place. Plagioclase forming under these conditions is *zoned*, meaning that it has a calcium-rich core surrounded by zones progressively richer in sodium.

As minerals crystallize simultaneously along the two branches of Bowen's reaction series, iron and magnesium are depleted, because they are used in ferromagnesian silicates, whereas calcium and sodium are used up in plagioclase feldspars. At this point, any leftover magma is enriched in potassium, aluminum, and silicon, which combine to form orthoclase $(KAlSi_3O_8)$, a potassium feldspar, and if water pressure is high, the sheet silicate muscovite forms. Any remaining magma is enriched in silicon and oxygen (silica) and forms the mineral quartz (SiO_2). The crystallization of orthoclase and quartz is not a true reaction series, because they form independently rather than by a reaction of orthoclase with the melt.

THE ORIGIN OF MAGMA AT SPREADING RIDGES

One fundamental observation regarding the origin of magma is that Earth's temperature, or *geothermal*

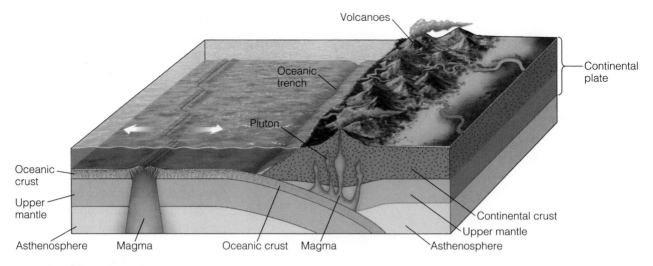

Figure 4.4 The Origin of Magma
Magma forms beneath spreading ridges, because as plates separate, pressure is reduced on the hot rocks and partial melting of the upper mantle begins. Invariably, the magma formed is mafic. Magma also forms at subduction zones, where water from the subducted plate causes partial melting of the upper mantle. This magma is also mafic, but as it rises, melting of the lower crust makes it more felsic.

gradient increases with depth and averages about 25°C/km. Accordingly, rocks at depth are hot but remain solid, because their melting temperature rises with increasing pressure. However, beneath spreading ridges, the temperature locally exceeds the melting temperature, at least in part because pressure decreases. That is, plate separation at ridges probably causes a decrease in pressure on the already hot rocks at depth, thus initiating melting (Figure 4.4). In addition, the presence of water decreases the melting temperature beneath spreading ridges, because water aids thermal energy in breaking the chemical bonds in minerals.

Magma formed beneath spreading ridges is invariably mafic (45% to 52% silica). However, the upper mantle rocks from which this magma is derived are characterized as ultramafic (<45% silica), consisting mostly of ferromagnesian silicates and lesser amounts of nonferromagnesian silicates. To explain how mafic magma originates from ultramafic rock, geologists propose that the magma forms from source rock that only partially melts. This phenomenon of partial melting takes place because not all of the minerals in rocks melt at the same temperature.

Recall the sequence of minerals in Bowen's reaction series (Figure 4.3). The order in which these minerals melt is the opposite of their order of crystallization. Accordingly, rocks made up of quartz, potassium feldspar, and sodium-rich plagioclase begin melting at lower temperatures than those composed of ferromagnesian silicates and the calcic varieties of plagioclase.

So when ultramafic rock starts to melt, the minerals richest in silica melt first, followed by those containing less silica. Therefore, if melting is not complete, mafic magma containing proportionately more silica than the source rock results.

Subduction Zones and the Origin of Magma

Another fundamental observation regarding magma is that, where an oceanic plate is subducted beneath either a continental plate or another oceanic plate, a belt of volcanoes and plutons is found near the leading edge of the overriding plate (Figure 4.4). It would seem, then, that subduction and the origin of magma must be related in some way, and indeed they are. Furthermore, magma at these convergent plate boundaries is mostly intermediate (53% to 65% silica) or felsic (>65% silica).

Once again, geologists invoke the phenomenon of partial melting to explain the origin and composition of magma at subduction zones. As a subducted plate descends toward the asthenosphere, it eventually reaches the depth where the temperature is high enough to initiate partial melting. In addition, the oceanic crust descends to a depth at which dewatering of hydrous minerals takes place, and as the water rises into the overlying mantle, it enhances melting and magma forms.

Recall that partial melting of ultramafic rock at spreading ridges yields mafic magma. Similarly, partial

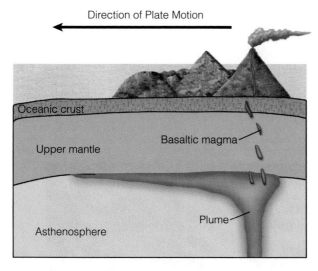

Direction of Plate Motion

Oceanic crust

Upper mantle

Basaltic magma

Plume

Asthenosphere

Figure 4.5 Mantle Plume and Hot Spot
A mantle plume beneath oceanic crust with a hot spot. Rising magma forms a series of volcanoes that become younger in the direction of plate movement.

melting of mafic rocks of the oceanic crust yields intermediate (53% to 65% silica) and felsic (>65% silica) magmas, both of which are richer in silica than the source rock. Moreover, some of the silica-rich sediments and sedimentary rocks of continental margins are probably carried downward with the subducted plate and contribute their silica to the magma. Also, mafic magma rising through the lower continental crust must be contaminated with silica-rich materials, which changes its composition.

HOT SPOTS AND THE ORIGIN OF MAGMA

Most volcanism occurs at divergent and convergent plate boundaries; however, there are some chains of volcanic outpourings in the ocean basins and on continents that are not near either of these boundaries. The Emperor Seamount–Hawaiian Islands, for instance, form a chain of volcanic islands 6,000 km long, and the volcanic rocks become progressively older toward the northwest (see Figure 2.18). In 1963, Canadian geologist J. Tuzo Wilson proposed that the Hawaiian Islands and other areas showing similar trends lay above a **hot spot** over which a plate moves, thereby yielding a succession of volcanoes (Figure 4.5).

Many geologists now think that hot-spot volcanism results from a rising **mantle plume**, a cylindrical plume of hot mantle rock that rises from perhaps near the core-mantle boundary. As it rises toward the surface, the pressure decreases on the hot rock and melting begins, thus yielding magma. Hot-spot volcanism may also account for vast flat-lying areas of overlapping lava flows or what geologists call *flood basalts*. Figure 2.11 shows the locations of many hot spots.

PROCESSES THAT CAUSE COMPOSITIONAL CHANGES IN MAGMA

Once magma forms, its composition may change by **crystal settling**, which involves the physical separation of minerals by crystallization and gravitational settling (Figure 4.6a). Olivine, the first ferromagnesian silicate to form in the discontinuous branch of Bowen's reaction series, has a density greater than the remaining magma and tends to sink. Accordingly, the remaining magma becomes richer in silica, sodium, and potassium, because much of the iron and magnesium were removed as olivine and perhaps pyroxene minerals crystallized and settled.

Although crystal settling does take place, it does not do so on a scale that would yield very much felsic magma from mafic magma. In some thick, sheetlike plutons called *sills*, the first-formed ferromagnesian silicates are indeed concentrated in their lower parts, thus making their upper parts less mafic. But even in these plutons, crystal settling has yielded very little felsic magma.

To yield a particular volume of granite (a felsic igneous rock), approximately 10 times as much mafic magma would have to be present initially for crystal settling to yield the volume of granite in question. If this were so, then mafic intrusive igneous rocks would be much more common than felsic ones. However, just the opposite is the case, so it appears that mechanisms other than crystal settling must account for the large volume of felsic magma. Partial melting of mafic oceanic crust and

hot spot Localized zone of melting below the lithosphere that probably overlies a mantle plume.

mantle plume A cylindrical mass of magma rising from the mantle toward the surface; recognized at the surface by a hot spot, an area such as the Hawaiian Islands where volcanism takes place.

crystal settling The physical separation and concentration of minerals in the lower part of a magma chamber or pluton by crystallization and gravitational settling.

Figure 4.6 Crystal Settling and Assimilation

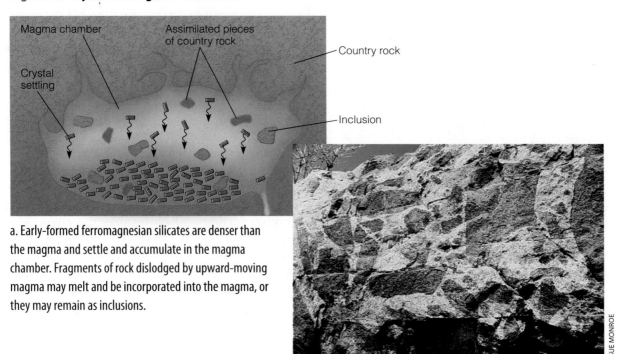

a. Early-formed ferromagnesian silicates are denser than the magma and settle and accumulate in the magma chamber. Fragments of rock dislodged by upward-moving magma may melt and be incorporated into the magma, or they may remain as inclusions.

b. Dark inclusions in granitic rock.

melting silica-rich sediments of continental margins during subduction yields magma richer in silica than the source rock. Furthermore, magma rising through the continental crust absorbs some felsic materials and becomes more enriched in silica.

The composition of magma also changes by **assimilation**, a process by which magma reacts with preexisting rock, called **country rock**, with which it comes in contact (Figure 4.6a). The walls of a volcanic conduit or magma chamber are heated by the adjacent magma, which may reach temperatures of 1,300°C. Some of these rocks partially or completely melt, provided that their melting temperature is lower than that of the magma. Because the assimilated rocks seldom have the same composition as the magma, the composition of the magma changes.

The fact that assimilation occurs is indicated by *inclusions*, incompletely melted pieces of rock that are fairly common in igneous rocks. Many inclusions were simply wedged loose from the country rock as magma forced its way into preexisting fractures (Figure 4.6b). No one doubts that assimilation takes place, but its effect on the bulk composition of magma must be slight, because the heat for melting comes from the magma itself, which then causes cooling of the magma. Only a limited amount of rock can be assimilated by magma, and that amount is insufficient to bring about a major compositional change.

Neither crystal settling nor assimilation can produce a significant amount of felsic magma from a mafic one. But both processes, if operating concurrently, can bring about greater changes than either process acting alone. Some geologists think that this is one way that intermediate magma (53% to 65% silica) forms where oceanic lithosphere is subducted beneath continental lithosphere.

A single volcano can erupt lavas of different composition, indicating that magmas of differing composition are present. It seems likely that some of these magmas would come into contact and mix with one another. If this is the case, we would expect that the composition of the magma resulting from **magma mixing** would be a modified version of the parent magmas.

assimilation A process whereby magma changes composition as it reacts with country rock.

country rock Any preexisting rock that has been intruded by a pluton or altered by metamorphism.

magma mixing The process whereby magmas of different composition mix together to yield a modified version of the parent magmas.

LO3 Igneous Rocks: Their Characteristics and Classification

We have already defined *plutonic* or *intrusive igneous rocks* and *volcanic* or *extrusive igneous rocks*. Here we will have considerably more to say about the texture, composition, and classification of these rocks, which constitute one of the three major rock families depicted in the rock cycle (see Figure 1.14).

IGNEOUS ROCK TEXTURES

The term *texture* refers to the size, shape, and arrangement of the minerals that make up igneous rocks. Size is the most important characteristic, because mineral crystal size is related to the cooling history of magma or lava and generally indicates whether an igneous rock is volcanic or plutonic. The atoms in magma and lava are in constant motion, but when cooling begins,

some atoms bond to form small nuclei. As other atoms in the liquid chemically bond to these nuclei, they do so in an orderly geometric arrangement, and the nuclei grow into crystalline *mineral grains*, the individual particles that make up igneous rocks.

During rapid cooling, as takes place in lava flows, the rate at which mineral nuclei form exceeds the rate of growth, and an aggregate of many small mineral grains forms. The result is a fine-grained or **aphanitic texture**, in which individual minerals are too small to be seen without magnification (Figure 4.7a). With slow cooling, the rate of growth exceeds the rate of nuclei formation, and large mineral grains form, thus yielding a coarse-grained or **phaneritic texture**, in which minerals are

> **aphanitic texture** A texture in igneous rocks in which individual mineral grains are too small to be seen without magnification; results from rapid cooling of magma and generally indicates an extrusive origin.
>
> **phaneritic texture** Igneous rock texture in which minerals are easily visible without magnification.

Phenocrysts

a

b

c

Figure 4.7 Textures of Igneous Rocks
a. Rapid cooling as in lava flows results in many small minerals and an aphanitic (fine-grained) texture.
b. Slower cooling in plutons yields a phaneritic texture.
c. These porphyritic textures indicate a complex cooling history.
d. Obsidian has a glassy texture because magma cooled too quickly for mineral crystals to form.
e. Gases expand in lava to yield a vesicular texture.
f. Microscopic view of a rock with a fragmental texture. The colorless, angular particles of volcanic glass measure up to 2 mm.

d

e

f

SUE MONROE (ALL PHOTOS)

clearly visible (Figure 4.7b). Aphanitic textures usually indicate an extrusive origin, whereas rocks with phaneritic textures are usually intrusive. However, shallow plutons might have an aphanitic texture, and the rocks that form in the interiors of thick lava flows might be phaneritic.

Another common texture in igneous rocks is one termed **porphyritic**, in which minerals of markedly different size are present in the same rock. The larger minerals are *phenocrysts* and the smaller ones collectively make up the *groundmass*, which is simply the grains between phenocrysts (Figure 4.7c).

The groundmass can be either aphanitic or phaneritic; the only requirement for a porphyritic texture is that the phenocrysts be considerably larger than the minerals in the groundmass. Igneous rocks with porphyritic textures are designated *porphyry*, as in basalt porphyry. These rocks have more complex cooling histories than those with aphanitic or phaneritic textures and might involve, for example, magma partly cooling beneath the surface followed by its eruption and rapid cooling at the surface.

Lava may cool so rapidly that its constituent atoms do not have time to become arranged in the ordered, three-dimensional frameworks of minerals. As a consequence, a *natural glass*, such as *obsidian* forms (Figure 4.7d). Some magmas contain large amounts of water vapor and other gases. These gases may be trapped in cooling lava, where they form numerous small holes or cavities known as **vesicles;** rocks with many vesicles are termed *vesicular*, as in vesicular basalt (Figure 4.7e).

A **pyroclastic** or **fragmental texture** characterizes igneous rocks formed by explosive volcanic activity (Figure 4.7f). For example, ash discharged high into the atmosphere eventually settles to the surface where it accumulates; if consolidated, it forms pyroclastic igneous rock.

COMPOSITION OF IGNEOUS ROCKS

Most igneous rocks, like the magma from which they originate, are mafic (45% to 52% silica), intermediate (53% to 65% silica), or felsic (<65% silica). A few

are called *ultramafic* (<45% silica); however, these are probably derived from mafic magma by a process to be discussed later. The parent magma plays an important role in determining the mineral composition of igneous rocks, yet it is possible for the same magma to yield a variety of igneous rocks, because its composition can change as a result of the sequence in which minerals crystallize, or by crystal settling, assimilation, and magma mixing (Figures 4.3 and 4.6).

CLASSIFYING IGNEOUS ROCKS

Geologists use texture and composition to classify most igneous rocks. Notice in Figure 4.8 that all rocks except peridotite are in pairs; the members of a pair have the same composition but different textures. Basalt and gabbro, andesite and diorite, and rhyolite and granite are compositional (mineralogical) pairs, but basalt, andesite, and rhyolite are aphanitic and most commonly extrusive (volcanic), whereas gabbro, diorite, and granite are phaneritic and mostly intrusive (plutonic). The extrusive and intrusive members of each pair can usually be distinguished by texture, but remember that rocks in some shallow plutons may be aphanitic, and rocks that formed in thick lava flows may be phaneritic. In other words, all of these rocks exist in a textural continuum.

The igneous rocks in Figure 4.8 are also differentiated by their mineral content. Reading across the chart from rhyolite to andesite to basalt, for example, we see that the proportions of nonferromagnesian and ferromagnesian silicates change. The differences in composition, however, are gradual along a compositional continuum. In other words, there are rocks with compositions that correspond to the lines between granite and diorite, basalt and andesite, and so on.

Ultramafic Rocks

Ultramafic rocks (<45% silica) are composed mostly of ferromagnesian silicates. *Peridotite* contains mostly olivine, lesser amounts of pyroxene, and usually a little plagioclase feldspar (Figure 4.9), and *pyroxenite* is composed predominantly of pyroxene. Because these minerals are dark, the rocks are black or green. Ultramafic rocks in Earth's crust probably originate by concentration of the early-formed ferromagnesian minerals that separated from mafic magmas.

Ultramafic lava flows, called *komatiites*, are known in rocks older than 2.5 billion years but are

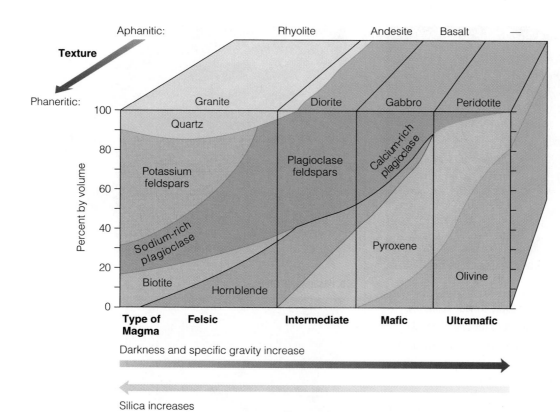

rare or absent in younger ones. The reason is that in order to erupt, ultramafic lava must have a near-surface temperature of approximately 1,600°C; the surface temperatures of present-day mafic lava flows are rarely more than 1,200°C. During early Earth history, though, more radioactive decay heated the mantle to as much as 300°C hotter than now, and ultramafic lavas formed. Because the amount of radiogenic heat has decreased over time, Earth has cooled, and eruptions of ultramafic lava flows have ceased.

Basalt-Gabbro

Basalt and *gabbro* are the aphanitic and phaneritic rocks that crystallize from mafic magma (45% to 52% silica) (Figure 4.10). Thus, both have the same composition—mostly calcium-rich plagioclase and pyroxene, with smaller amounts of olivine and amphibole (Figure 4.8). Because they contain a large proportion of ferromagnesian silicates, basalt and gabbro are dark; those that are porphyritic typically contain calcium plagioclase or olivine phenocrysts.

Extensive basalt lava flows cover vast areas in Washington, Oregon, Idaho, and northern California (see Chapter 5). Oceanic islands such as Iceland, the Galápagos, the Azores, and the Hawaiian Islands are composed mostly of basalt, which makes up the upper part of the oceanic crust.

Gabbro is much less common than basalt, at least in the continental crust or where it can be easily observed. Small intrusive bodies of gabbro are present in the continental crust, but intermediate to felsic intrusive rocks are much more common. However, the lower part of the oceanic crust is composed of gabbro.

Andesite-Diorite

Intermediate-composition magma (53% to 65% silica) crystallizes to form *andesite* and *diorite*, which are compositionally equivalent fine- and coarse-grained igneous rocks (Figure 4.11). Andesite and diorite are composed predominantly of plagioclase feldspar, with the typical ferromagnesian component being amphibole or biotite (Figure 4.8). Andesite is generally medium to dark gray, but diorite has a salt-and-pepper appearance because of its white to light-gray plagioclase and dark ferromagnesian silicates.

Andesite is a common extrusive igneous rock formed from lava erupted in volcanic chains at convergent plate boundaries. The volcanoes of the Andes

Figure 4.9 Peridotite
This specimen of the ultramafic rock peridotite is made up mostly of olivine. Notice in Figure 4.8 that peridotite is the only phaneritic rock that does not have an aphanitic counterpart. Peridotite is rare at Earth's surface, but is very likely the rock that makes up the mantle.

SUE MONROE

a. Basalt is aphanitic.

COPYRIGHT AND PHOTOGRAPH BY DR. PARVINDER S. SETHI

b. Gabbro is phaneritic. Notice the light reflected from the crystal faces.

COPYRIGHT AND PHOTOGRAPH BY DR. PARVINDER S. SETHI

Figure 4.10 Mafic Igneous Rocks

Mountains of South America and the Cascade Range in western North America are composed, in part, of andesite. Intrusive bodies of diorite are fairly common in the continental crust.

Rhyolite-Granite

Rhyolite and *granite* crystallize from felsic magma (>65% silica) and are therefore silica-rich rocks (Figure 4.12). They consist mostly of potassium feldspar, sodium-rich plagioclase, and quartz, with perhaps some biotite and rarely hornblende (Figure 4.8). Because nonferromagnesian silicates predominate, rhyolite and granite are typically light colored. Rhyolite is fine grained, although most often it contains phenocrysts of potassium feldspar or quartz, and granite is coarse grained.

Rhyolite lava flows are much less common than andesite and basalt flows. Recall that one control of magma viscosity is silica content. Thus, if felsic magma rises to the surface, it begins to cool, the pressure on it decreases, and gases are released explosively, usually yielding rhyolitic pyroclastic materials. The rhyolitic lava flows that do occur are thick and highly viscous and move only short distances.

Granite is a coarsely crystalline igneous rock with a composition corresponding to that of the field shown in Figure 4.8. Strictly speaking, not all rocks in this field are granites. For example, a rock with a composition close to the line separating granite and diorite is

called *granodiorite*. To avoid the confusion that might result from introducing more rock names, we will follow the practice of referring to rocks to the left of the granite-diorite line in Figure 4.8 as *granitic*.

Granitic rocks are by far the most common intrusive igneous rocks, and they are restricted to the continents. Most granitic rocks were intruded at or near convergent plate margins during mountain-building episodes. When these mountainous regions are uplifted and eroded, the vast bodies of granitic rocks forming their cores are exposed.

Pegmatite

The term *pegmatite* refers to a particular texture rather than a specific composition, but most pegmatites are

a. This specimen of andesite has hornblende phenocrysts that are so numerous the rock may be classified as an andesite hornblende porphyry.

b. Diorite has a salt-and-pepper appearance because it contains light-colored nonferromagnesian silicates and dark-colored ferromagnesian silicates.

Figure 4.11 Intermediate Igneous Rocks

composed mostly of quartz, potassium feldspar, and sodium-rich plagioclase, thus corresponding closely to granite. The most remarkable feature of pegmatites is the size of their minerals, which measure at least 1 cm across, and in some pegmatites they measure tens of centimeters or meters (Figure 4.13). Many pegmatites are adjacent to large granite plutons and are composed of minerals that formed from the water-rich magma that remained after most of the granite crystallized.

When felsic magma cools and forms granite, the remaining water-rich magma has a lower density and viscosity and commonly invades cracks in the nearby rocks where minerals crystallize. This water-rich

magma also contains elements that rarely enter into the common minerals that form granite. Pegmatites that are essentially very coarsely crystalline granite are simple pegmatites, whereas those with minerals containing elements such as lithium, beryllium, cesium, boron, and several others are complex pegmatites.

The formation and growth of mineral-crystal nuclei in pegmatites are similar to those processes in other magmas, but with one critical difference: The water-rich magma from which pegmatites crystallize inhibits the formation of nuclei. However, some nuclei do form, and because the appropriate ions in the liquid can move easily and attach themselves to a growing

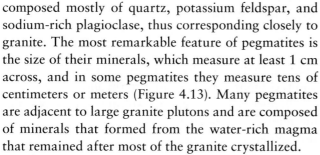

a. Rhyolite with small phenocrysts of orthoclase (K) feldspar.

b. Granite.

Figure 4.12 Felsic Igneous Rocks
These rocks are typically light colored, because they contain mostly nonferromagnesian silicate minerals. The dark spots in the granite specimen are biotite mica. The white and pinkish minerals are feldspars, whereas the glassy-appearing minerals are quartz.

STEVE STAHL

a. This pegmatite, the light-colored rock, is exposed in the Black Hills of South Dakota.

Figure 4.13 **Pegmatite**

COPYRIGHT AND PHOTOGRAPH BY DR. PARVINDER S. SETHI

b. Closeup view of a specimen from a pegmatite with minerals measuring 8 to 10 cm across.

crystal, individual minerals have the opportunity to grow very large.

Other Igneous Rocks

Geologists classify the igneous rocks in Figure 4.8 by texture and composition, but a few others are identified primarily by their textures. Much of the fragmental material erupted by volcanoes is *ash*, a designation for pyroclastic materials measuring less than 2.0 mm, most of which consists of pieces of minerals or shards of volcanic glass (Figure 4.7f). The consolidation of ash forms the pyroclastic rock *tuff* (Figure 4.14a). Most tuff is silica rich and light colored and is appropriately called *rhyolite tuff*. Some ash flows are so hot that as they come to rest, the ash particles fuse together and form a *welded tuff*. Consolidated deposits of larger pyroclastic materials, such as cinders, blocks, and bombs, are *volcanic breccia*.

Both *obsidian* and *pumice* are varieties of volcanic glass (Figure 4.14b, c). Obsidian may be black, dark gray, red, or brown, depending on the presence of iron. Obsidian breaks with the conchoidal (smoothly curved) fracture typical of glass. Analyses of many samples indicate that most obsidian has a high silica content and is compositionally similar to rhyolite.

Pumice is a variety of volcanic glass containing numerous vesicles that develop when gas escapes through lava and forms a froth (Figure 4.14c). If pumice falls into water, it can be carried great distances, because it is so porous and light that it floats. Another vesicular rock is *scoria*. It is more crystalline and denser than pumice, but it has more vesicles than solid rock (Figure 4.14d).

LO4 Intrusive Igneous Bodies: Plutons

Unlike volcanism and the origin of volcanic rocks, we can study intrusive igneous bodies, collectively called **plutons**, only indirectly, because intrusive rocks form when magma cools and crystallizes within Earth's crust (Figure 4.15a). We can observe these rock bodies only when uplift and deep erosion have taken place, thereby exposing them at the surface. Furthermore, geologists cannot duplicate the conditions under which intrusive rocks form except in small laboratory experiments.

Geologists recognize several types of plutons based on their geometry (three-dimensional shape) and relationships to the country rocks. In terms of their

pluton An intrusive igneous body that forms when magma cools and crystallizes within the crust, such as a batholith or sill.

a. Tuff is made up of pyroclastic materials, such as those shown in Figure 4.7f.

b. The natural glass obsidian.

c. Pumice is glassy and extremely vesicular.

d. Scoria is also vesicular, but it is darker, denser, and more crystalline than pumice.

Figure 4.14 Igneous Rocks Classified Primarily by Their Texture

geometry, plutons are tabular, cylindrical, or irregular (massive). Furthermore, they may be **concordant**, meaning they have boundaries that parallel the layering in the country rock, or **discordant**, with boundaries that cut across the country rock's layering (Figure 4.15a).

DIKES, SILLS, AND LACCOLITHS

Dikes and **sills** are tabular or sheetlike igneous bodies that differ only in that dikes are discordant and sills are concordant (Figure 4.15a). Dikes are quite common and range from a few centimeters to more than 100 m thick (Figure 4.15b). Invariably, they are intruded into preexisting fractures or where fluid pressure is great enough for them to form their own fractures as they move upward into county rock.

Sills are tabular just as dikes are, but they are concordant. Many sills are a meter or less thick, although some are much thicker. Most sills were intruded into sedimentary rocks, but eroded volcanoes also reveal that sills are injected into piles of volcanic rocks. In fact, some of the inflation of a volcano that precedes

an eruption may be caused by the injection of sills (see Chapter 5). In contrast to dikes, which follow zones of weakness, sills are intruded between layers in country rock when the fluid pressure is great enough for the magma to actually lift the overlying rocks.

Under some circumstances, a sill inflates and causes the overlying rocks to bow upward, forming an igneous body called a **laccolith** (Figure 4.15a). A laccolith has a flat floor and is domed up in its central part, giving it a mushroom-like geometry. Like sills, laccoliths are rather shallow intrusions that lift the overlying rocks.

concordant pluton
Intrusive igneous body whose boundaries parallel the layering in the country rock.

discordant pluton Pluton with boundaries that cut across the layering in the country rock.

dike A tabular or sheetlike discordant pluton.

sill A tabular or sheetlike concordant pluton.

laccolith A concordant pluton with a mushroom-like geometry.

a. Block diagram showing various plutons. Some plutons cut across the layering in country rock and are discordant, whereas others parallel the layering and are concordant.

b. The dark material in this image is a dike that cuts through Cenozoic-age sedimentary rocks near Dulce, New Mexico.

c. A volcanic neck in Monument Valley Navajo Tribal Park, Arizona. This landform is 457 m high. Most of the original volcano was eroded, leaving only this remnant.

Figure 4.15 Plutons

Volcanic Pipes and Necks

Volcanoes have a cylindrical conduit known as a **volcanic pipe** that connects to an underlying magma chamber. Magma rises through this structure; however, when a volcano ceases to erupt, its slopes are attacked by weathering and erosion, but the magma in the pipe is commonly more resistant to erosion and is left as a remnant called a **volcanic neck** (Figure 4.15c). Several volcanic necks are found in the southwestern United States, especially in Arizona and New Mexico, and others are recognized elsewhere.

Batholiths and Stocks

By definition, a **batholith**, the largest of all plutons, must have at least 100 km² of surface area, and most are far larger (Figure 4.15a). A **stock**, in contrast, is similar but smaller. Some stocks are simply parts of large plutons that, once exposed by erosion, are parts of batholiths. Both batholiths and stocks are mostly discordant, although locally they may be concordant, and batholiths, especially, consist of multiple intrusions. In other words, a batholith is a large composite body produced by repeated, voluminous intrusions of magma in the same region.

The igneous rocks that make up batholiths are mostly granitic, although diorite may also be present. Batholiths and stocks are emplaced mostly near convergent plate boundaries during episodes of mountain building. One example is the Sierra Nevada batholith of California, which formed over millions of years. Other large batholiths in North America include the Idaho batholith, the Boulder batholith in Montana, and the Coast Range batholith in British Columbia, Canada.

Figure 4.16 **Emplacement of a Batholith by Forceful Injection and Stoping**

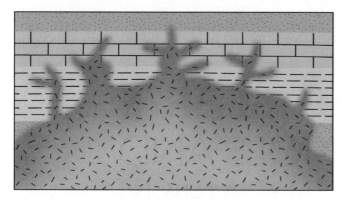

a. Forceful injection is when magma rises and forces its way into fractures and planes between layers in the country rock.

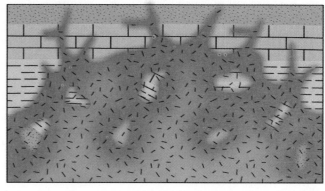

b. Stopping is the process in which blocks of country rock are detached and engulfed in the magma, thereby making room for the magma to rise farther. Some of the engulfed blocks might be assimilated, and some may remain as inclusions (Figure 4.6b).

Mineral resources are found in rocks of batholiths and stocks, and in the adjacent country rocks. The copper deposits at Butte, Montana, are in rocks near the margins of the granitic rocks of the Boulder batholith. Near Salt Lake City, Utah, copper is mined from the mineralized rocks of the Bingham stock, a composite pluton composed of granite and granite porphyry. Granitic rocks also are the primary source of gold, which forms from mineral-rich solutions moving through cracks and fractures of the igneous body.

LO5 How Are Batholiths Intruded into Earth's Crust?

Geologists realized long ago that the origin of batholiths posed a space problem. What happened to the rock that was once in the space now occupied by a batholith? One solution is that these large igneous bodies melted their way into the crust. In other words, they simply assimilated the country rock as they moved upward (Figure 4.6). The presence of inclusions, especially near the tops of some plutons, indicates that assimilation does occur. Nevertheless, as we noted, assimilation is a limited process, because magma cools as country rock is assimilated. Calculations indicate that far too little heat is available in magma to assimilate the huge quantities of country rock necessary to make room for a batholith.

Geologists now generally agree that batholiths were emplaced by *forceful injection* as magma moved upward (Figure 4.16a). Recall that granite is derived from viscous felsic magma and therefore rises slowly. It appears that the magma deforms and shoulders aside the country rock, and as it rises farther, some of the country rock fills the space beneath the magma.

Some batholiths do show evidence of having been emplaced by forceful injection, but this mechanism probably occurs in the deeper parts of the crust where temperature and pressure are high and the country rocks are easily deformed in the manner described. At shallower depths, the crust is more rigid and tends to deform by fracturing. In this environment, batholiths may move upward by **stoping**, a process in which rising magma detaches and engulfs pieces of country rock (Figure 4.16b). According to this concept, magma moves up along fractures and the planes separating layers of country rock. Eventually, pieces of country rock detach and settle into the magma. No new room is created during stoping; the magma simply fills the space formerly occupied by country rock.

volcanic pipe The conduit connecting the crater of a volcano with an underlying magma chamber.

volcanic neck An erosional remnant of the material that solidified in a volcanic pipe.

batholith An irregularly shaped, discordant pluton with at least 100 km^2 of surface area.

stock An irregularly shaped discordant pluton with a surface area smaller than 100 km^2.

stoping A process in which rising magma detaches and engulfs pieces of the country rock.

CHAPTER 5
VOLCANOES AND VOLCANISM

Lava from Kilauea flows into the sea, forming new land and making Hawaii grow larger.

Introduction

No other geologic phenomenon has captured the public imagination more than erupting volcanoes, especially lava flows issuing forth as fiery streams or particulate matter blasted into the atmosphere in sensational pyrotechnic displays. What better subject for a disaster movie? Several such movies of varying quality and scientific accuracy have been released in recent years, but one of the best was *Dante's Peak* in 1997. Certainly the writers and the director exaggerated some aspects of volcanism, but the movie rather accurately depicted the phenomenal power of an explosive eruption. Incidentally, the volcano called Dante's Peak was a 10-m-high model built of wood and steel.

Incandescent streams of molten rock are often portrayed in movies as a great danger to humans, and, in fact, on a few occasions lava flows have caused fatalities. In 2002, lava flows in Goma, Zaire (Democratic Republic of the Congo), caused gasoline storage tanks to explode and kill 147 people. Furthermore, lava flows destroy buildings and cover otherwise productive land, but actually they are the least dangerous manifestation of volcanic eruptions. Explosive eruptions, in contrast, are quite dangerous, especially if they occur near populated areas.

Ironically, when considered in the context of Earth history, volcanism is actually a constructive process. The atmosphere and surface waters most likely resulted from the emission of gases during Earth's early history, and oceanic crust is continuously produced by volcanism at spreading ridges. Oceanic islands such as the Hawaiian Islands, Iceland, and the Azores owe their existence to volcanism, and weathering of lava flows, pyroclastic materials, and volcanic mudflows in tropical areas such as Indonesia converts them to productive soils.

One very good reason to study volcanic eruptions is that they illustrate the complex interactions among

LEARNING OUTCOMES

After reading this unit, you should be able to do the following:

LO1 Understand volcanism and volcanoes

LO2 Identify the types of volcanoes

LO3 Identify other volcanic landforms

LO4 Identify the distribution of volcanoes

LO5 Understand the relationship between plate tectonics, volcanoes, and plutons

LO6 Understand volcanic hazards, volcano monitoring, and forecasting eruptions

ERUPTIONS IN THE CONTINENTAL U.S.

Eruptions in the continental United States have occurred only three times since 1914, all in the Cascade Range, which stretches from northern California through Oregon and Washington and into southern British Columbia, Canada.

volcanism The process whereby magma and its associated gases rise through the crust and are extruded onto the surface or into the atmosphere.

Earth's systems. Volcanism, especially the emission of gases and pyroclastic materials, has an immediate and profound impact on the atmosphere, hydrosphere, and biosphere, at least in the vicinity of an eruption. And in some cases, the effects are worldwide, as they were following the eruptions of Tambora in 1815, Krakatau in 1883, and Pinatubo in 1991. Furthermore, the fact that lava flows and explosive eruptions cause property damage, injuries, fatalities, and at least short-term atmospheric changes indicates that volcanic eruptions are catastrophic events, at least from the human perspective.

LO1 Volcanism and Volcanoes

What do we mean by the terms *volcanism* and *volcano*? The latter is a landform—that is, a feature on Earth's surface—whereas **volcanism** is the process by which magma rises through Earth's crust and issues forth at the surface as lava flows and/or pyroclastic materials and gases. Volcanism is responsible for all extrusive igneous rocks, such as basalt, tuff, and obsidian.

About 550 volcanoes are *active;* that is, they have erupted during historic time, but only about a dozen are erupting at any one time. Most eruptions are minor and go unreported in the popular press unless they occur near populated areas or have tragic consequences. However, large eruptions that cause extensive property damage, injuries, and fatalities are not uncommon, so a great amount of effort is devoted to understanding and more effectively anticipating large eruptions.

In addition to active volcanoes, Earth has numerous *dormant* volcanoes that have not erupted during historic time but may do so in the future. The largest volcanic outburst in nearly a century was when Mount

Vog

Residents of the island of Hawaii have coined the term *vog* for volcanic smog. Kilauea volcano has been erupting continuously since 1983, releasing small amounts of lava and copious quantities of carbon dioxide and sulfur dioxide. Carbon dioxide is no problem, because it dissipates quickly in the atmosphere, but sulfur dioxide produces a haze and the unpleasant odor of sulfur. Vog probably poses little or no health risk for tourists, but a long-term threat exists for residents of the west side of the island, where vog is most common.

Pinatubo in the Philippines erupted in 1991 after lying dormant for 600 years. Chaitén Volcano in Chile was dormant for more than 9,000 years until it erupted huge clouds of gases and volcanic ash in May 2008. Some volcanoes have not erupted in historic time and show no signs of doing so again; thousands of these *extinct* or *inactive* volcanoes are known.

All terrestrial planets and Earth's Moon were volcanically active during their early histories, but now active volcanoes are known only on Earth and on one or two other bodies in the solar system. Triton, one of Neptune's moons, probably has active volcanoes, and Jupiter's moon Io is by far the most volcanically active body in the solar system. Many of its hundred or so volcanoes are erupting at any given time.

VOLCANIC GASES

Volcanic gases from present-day volcanoes are 50% to 80% water vapor, with lesser amounts of carbon dioxide, nitrogen, sulfur gases—especially sulfur dioxide and hydrogen sulfide—and very small amounts of carbon monoxide, hydrogen, and chlorine. In areas of recent volcanism, emission of gases from fumeroles (volcanic vents) continues, and one cannot help noticing the rotten-egg odor of hydrogen sulfide gas (Figure 5.1).

When magma rises toward the surface, the pressure is reduced and the contained gases begin to expand. In highly viscous, felsic magma, expansion is inhibited and gas pressure increases. Eventually, the pressure may become great enough to cause an explosion and produce pyroclastic materials such as volcanic ash. In contrast, low-viscosity mafic magma allows gases to expand and escape easily. Accordingly, mafic magma usually erupts rather quietly.

Most volcanic gases quickly dissipate in the atmosphere and pose little danger to humans, but on several occasions, they have caused fatalities. In 1783, toxic gases, probably fluorine and sulfur dioxide, erupting from Laki fissure in Iceland had devastating effects. About 75% of the nation's livestock died, and the haze

Figure 5.1 Fumerole in Lassen Volcanic National Park, California
Gases emitted from a vent (fumerole) at Lassen Volcanic National Park, California.

resulting from the gas caused lower temperatures and crop failures; about 24% of Iceland's population died as a result of the ensuing Blue Haze Famine. The country suffered its coldest winter in 225 years in 1783–1784, with temperatures 4.8°C below the long-term average.

In 1986, in the African nation of Cameroon, 1,746 people died when a cloud of carbon dioxide engulfed them. The gas accumulated in the waters of Lake Nyos, which occupies a volcanic caldera. Scientists disagree about what caused the gas to suddenly burst forth from the lake, but once it did, it flowed downhill along the surface because it was denser than air. In fact, the density and velocity of the gas cloud were great enough to flatten vegetation, including trees, a few kilometers from the lake. Unfortunately, thousands of animals and many people, some as far as 23 km from the lake, were asphyxiated.

LAVA FLOWS

Although lava flows are portrayed in movies and on television as a great danger to humans, they only rarely cause fatalities. The reason is that most lava flows do not move very fast, and because they are fluid, they follow existing low areas. Even low-viscosity lava flows do not move very rapidly. Flows can move much faster, though, when their margins cool to form a channel, and especially when insulated on all sides as in a **lava tube**, where a speed of more than 50 km per hour has been recorded. A conduit known as a lava tube within a lava flow forms when the margins and upper surface of the flow solidify. Thus confined and insulated, the flow moves rapidly and over great distances. As an eruption ceases, the tube drains, leaving an empty tunnel-like structure (Figure 5.2a). Part of the roof of a lava tube may collapse to form a *skylight* through which an active flow can be observed (Figure 5.2b), or access can be gained to an inactive lava tube.

Geologists define two types of lava flows, both

lava tube A tunnel beneath the solidified surface of a lava flow through which lava moves; also, the hollow space left when the lava within a tube drains away.

SUE MONROE

a. Entrance to a lava tube on Medicine Lake Volcano, a huge shield volcano in northern California. It has a circumference of 240 km, and it last erupted about 1,000 years ago.

J.B. JUDD/USGS

Figure 5.2 Lava Tubes
Lava tubes consisting of hollow spaces beneath the surfaces of lava flows are common in many areas.

b. An active lava tube in Hawaii. Part of the tube's roof has collapsed, forming a skylight.

a. An excellent example of the taffy-like appearance of pahoehoe.

COPYRIGHT AND PHOTOGRAPH BY DR. PARVINDER S. SETHI

ROBERT TILLING/USGS

b. An aa lava flow advances over an older pahoehoe flow. Notice the rubbly nature of the aa flow.

Figure 5.3 Pahoehoe and aa Lava Flows
Pahoehoe and aa were named for lava flows in Hawaii, but the same kinds of flows are found in many other areas.

named for Hawaiian flows. A **pahoehoe** (pronounced *pah-hoy-hoy*) flow has a ropy surface much like taffy (Figure 5.3a). The surface of an **aa** (pronounced *ah-ah*) flow is characterized by rough, jagged, angular blocks and fragments (Figure 5.3b). Pahoehoe flows are less viscous than aa flows; indeed, the latter are viscous

enough to break up into blocks and move forward as a wall of rubble.

Pressure on the partly solidified crust of a still-moving lava flow causes the surface to buckle into *pressure ridges* (Figure 5.4a). Gases escaping from a flow hurl globs of lava into the air, which fall back to the surface and adhere to one another, thus forming small, steep-sided *spatter cones*, or spatter ramparts if they are elongated (Figure 5.4b). Spatter cones a few meters high are common on lava flows in Hawaii, and you can see ancient ones in many areas.

Many lava flows, especially mafic ones, have a distinctive pattern of columns bounded by fractures, or what geologists call **columnar joints**. Once a lava flow stops moving, it contracts as it cools and produces forces that cause fractures called *joints* to open. On the surface of a lava flow, the joints are commonly polygonal (often six-sided) cracks that extend downward, thus forming parallel columns with their long axes perpendicular to the cooling surface (Figure 5.5).

Much of the igneous rock in the upper part of the oceanic crust is a distinctive type consisting of bulbous masses of basalt that resemble pillows, hence the name **pillow lava**. It was long recognized that pillow lava forms when lava is rapidly chilled beneath water, but its formation was not observed until 1971. Divers near Hawaii saw pillows form when a blob of lava broke through the crust of an underwater lava flow and cooled almost instantly, forming a pillow-shaped structure with a glassy exterior. The remaining fluid inside then broke through the crust of the pillow, repeating the process and resulting in an accumulation of interconnected pillows (Figure 5.6).

pahoehoe A type of lava flow with a smooth, ropy surface.

aa Lava flow with a surface of rough, angular blocks and fragments.

columnar jointing The phenomenon of forming columns bounded by fractures in some igneous rocks as they cooled and contracted.

pillow lava Bulbous masses of basalt, resembling pillows, formed when lava is rapidly chilled under water.

Figure 5.4 Pressure Ridges and Spatter Cones

SUE MONROE

a. The buckled surface of this lava flow on Medicine Lake Volcano in California is a pressure ridge. It formed when the solidified surface of a lava flow was bent upward by a still-moving flow beneath.

SUE MONROE

b. These two chimney-like pillars of rocks are spatter cones on the surface of a lava flow in the Coso volcanic field of California.

Figure 5.5 Columnar Jointing

Columnar jointing is seen mostly in mafic lava flows and related intrusive rocks.

a. As lava cools and contracts, three-pronged cracks form that grow and intersect to form four- to seven-sided columns, most of which are six-sided.

b. Columnar joints in a basalt lava flow at Devil's Postpile National Monument in California. The rubble in the foreground is collapsed columns.

c. Surface view of the columns from (b). The straight lines and polish resulted from abrasion by a glacier that moved over this surface.

PYROCLASTIC MATERIALS

In addition to lava flows, erupting volcanoes eject pyroclastic materials, especially **volcanic ash**, a designation for pyroclastic particles that measure less than 2.0 mm (Figure 5.7). In some cases, ash is ejected into the atmosphere and settles to the surface as an *ash fall*. In contrast to an ash fall, an *ash flow* is a cloud of ash and gas that flows along or close to the land surface. Ash flows can move faster than 100 km per hour, and some cover vast areas.

In populated areas adjacent to volcanoes, ash falls and ash flows pose serious problems, and volcanic ash in the atmosphere is a hazard to aviation. In 1989, ash from Redoubt volcano in Alaska caused all four jet engines to fail on KLM Flight 867. The plane, carrying 231 passengers,

volcanic ash Pyroclastic materials that measure less than 2 mm.

Figure 5.6 Pillow Lava

Much of the upper part of the oceanic crust is made up of pillow lava that formed when lava erupted underwater.

NOAA

a. Pillow lava on the seafloor in the Pacific Ocean about 150 miles west of Oregon that formed about five years before the photo was taken.

JAMES. S. MONROE

b. Ancient pillow lava that formed on the seafloor is now on land in Marin County, California. The largest pillow measures about 0.6 m across.

SUE MONROE

Figure 5.7 Pyroclastic Materials

Pyroclastic materials are all particles ejected from volcanoes, especially during explosive eruptions. The volcanic bomb is elongate, because it was molten when it descended through the air. The lapilli was collected at a small volcano in Oregon, whereas the ash came from the 1980 eruption of Mount St. Helens in Washington.

LO2 What Are the Types of Volcanoes?

Simply put, a **volcano** is a hill or mountain that forms around a vent where lava, pyroclastic materials, and gases erupt. Although volcanoes vary in size and shape, all have a conduit or conduits leading to a magma chamber beneath the surface. Vulcan, the Roman deity of fire, was the inspiration for calling these mountains volcanoes, and because of their danger and obvious connection to Earth's interior, they have been held in awe by many cultures.

© RADIUS IMAGES/JUPITERIMAGES

nearly crashed when it fell more than 3 km before the crew could restart the engines. The plane landed safely in Anchorage, Alaska, but it required $80 million in repairs.

In addition to volcanic ash, volcanoes erupt *lapilli*, consisting of pyroclastic materials that measure from 2 mm to 64 mm, and *blocks* and *bombs*, both larger than 64 mm (Figure 5.7). Bombs have a twisted, streamlined shape, which indicates that they were erupted as globs of magma that cooled and solidified during their flight through the air. Blocks, in contrast, are angular pieces of rock ripped from a volcanic conduit or pieces of a solidified crust of a lava flow. Because of their size, lapilli, bombs, and blocks are confined to the immediate area of an eruption.

Most volcanoes have a circular depression known as a **crater** at their summit, or on their flanks, that forms by explosions or collapse. Craters are generally less than 1 km across, whereas much larger rimmed depressions on volcanoes are **calderas**. In fact, some volcanoes have a summit crater within a caldera. Calderas are huge structures that form following voluminous eruptions, during which part of a magma chamber drains and the mountain's summit collapses into the vacated space below. An excellent example is misnamed Crater Lake in Oregon (Figure 5.8). Crater Lake is actually a steep-rimmed caldera that formed

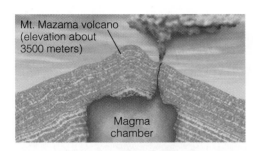

Mt. Mazama volcano (elevation about 3500 meters)

Magma chamber

a. Eruption begins as huge quantities of ash are ejected from the volcano.

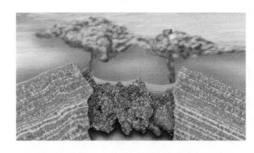

c. The eruption continues as more ash and pumice are ejected into the air and pyroclastic flows move down the flanks of the mountain.

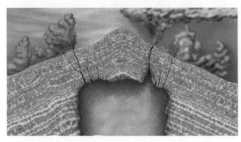

b. The collapse of the summit into the partially drained magma chamber forms a huge caldera.

Wizard island

Rim of Crater Lake caldera (elevation 2000 meters)

d. Postcaldera eruptions partly cover the caldera floor, and the small cinder cone called Wizard Island forms.

FROM HOWELL WILLIAMS, *CRATER LAKE: THE STORY OF ITS ORIGIN* (BERKELEY, CALIF. UNIVERSITY OF CALIFORNIA PRESS): ILLUSTRATIONS FROM P. 84 © 1941 REGENTS OF THE UNIVERSITY OF CALIFORNIA, © RENEWED 1969, HOWELL WILLIAMS.

JAMES S. MONROE

e. View from the rim of Crater Lake showing Wizard Island. The lake is 594 m deep, making it the second deepest in North America.

Figure 5.8 The Origin of Crater Lake, Oregon
Remember, Crater Lake is actually a caldera that formed by partial draining of a magma chamber.

about 7,700 years ago in the manner just described. As impressive as Crater Lake is, it is not nearly as large as some other calderas, such as the Toba caldera in Sumatra, which is 100 km long and 30 km wide.

Geologists recognize several major types of volcanoes, but one must realize that each volcano is unique in its history of eruptions and development. For instance, the frequency of eruptions varies considerably; the Hawaiian volcanoes and Mount Etna on Sicily have erupted repeatedly, whereas Pinatubo in the Philippines erupted in 1991 for the first time in 600 years. Some volcanoes are complex mountains that have the characteristics of more than one type of volcano.

SHIELD VOLCANOES

Shield volcanoes look like the outer surface of a shield lying on the ground with the convex side up (Figure 5.9). They have low, rounded profiles with gentle slopes ranging from about 2 to 10 degrees and are composed mostly of mafic

volcano A hill or mountain formed around a vent as a result of the eruption of lava and pyroclastic materials.

crater An oval to circular depression at the summit of a volcano resulting from the eruption of lava, pyroclastic materials, and gases.

caldera A large, steep-sided, oval to circular depression usually formed when a volcano's summit collapses into a partially drained underlying magma chamber.

shield volcano A dome-shaped volcano with a low, rounded profile built up mostly by overlapping basalt lava flows.

Figure 5.9 Shield Volcanoes

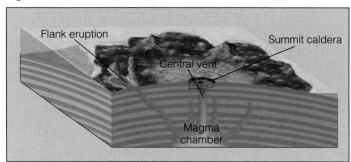

a. Shield volcanoes consist of numerous thin basalt lava flows that build up mountains with slopes rarely exceeding 10 degrees.

b. Sierra Grande is an extinct shield volcano in the Raton-Clayton volcanic field of northeastern New Mexico. It stands about 670 m above the surrounding plain, and it is 15 km in diameter. This 2.6- to 4.0-million-years-old shield volcano is unusual in that it is made up of andesite rather than basalt lava flows.

flows that had low viscosity, so the flows spread out and formed thin, gently sloping layers. Eruptions from shield volcanoes, sometimes called *Hawaiian-type eruptions*, are quiet compared to those of volcanoes such as Mount St. Helens. Lava most commonly rises to the surface with little explosive activity and poses little danger to humans. Lava fountains, some up to 400 m high, contribute some pyroclastic materials to shield volcanoes, but otherwise they are composed largely of basalt lava flows.

Although eruptions of shield volcanoes tend to be rather quiet, some of the Hawaiian volcanoes have, on occasion, produced sizable explosions when ground-water instantly vaporized as it came into contact with magma. One such explosion in 1790 killed about 80 warriors in a party headed by Chief Keoua, who was leading them across the summit of Kilauea volcano.

Kilauea volcano is impressive, because it has been erupting continuously since January 3, 1983, making it the longest recorded eruption. During these 26 years, more than 2.5 km³ of molten rock has flowed out at the surface, much of it reaching the sea and forming 2.2 km² of new land on the island of Hawaii. Unfortunately, lava flows from Kilauea have also destroyed about 200 homes and caused some $61 million in damages.

Shield volcanoes are most common in the ocean basins, but some are also present on the continents—in East Africa, for instance. The island of Hawaii is made up of five huge shield volcanoes, two of which, Kilauea and Mauna Loa, are active much of the time. Mauna Loa is nearly 100 km across its base and stands more than 9.5 km above the surrounding seafloor; it has a volume estimated at 50,000 km³, making it the world's largest volcano.

CINDER CONES

Small, steep-sided **cinder cones** made up of particles resembling cinders form when pyroclastic materials accumulate around a vent from which they erupted (Figure 5.10). Cinder cones are small, rarely exceeding 400 m high, with slope angles up to 33 degrees, depending on the angle that can be maintained by the angular pyroclastic materials. Many of these small volcanoes have a large, bowl-shaped crater, and if they issue any lava flows, they usually break through the base or lower flanks of the mountains. Although all cinder cones are conical, their symmetry varies from those that are almost perfectly symmetrical to those that formed when prevailing winds caused pyroclastic materials to build up higher on the downwind side of the vent.

Many cinder cones form on the flanks or within the calderas of larger volcanoes and represent the final stages of activity, particularly in areas of basaltic volcanism. Wizard Island in Crater Lake, Oregon, is a small cinder cone that formed after the summit of Mount Mazama collapsed to form a caldera (Figure

cinder cone A small, steep-sided volcano made up of pyroclastic materials resembling cinders that accumulate around a vent.

Figure 5.10 Cinder Cones
This 400-m-high cinder cone named Paricutín formed in a short time in Mexico in 1943, when pyroclastic materials began to erupt in a farmer's field. Lava flows from the volcano covered two nearby villages, but all activity ceased by 1952.

5.8). Cinder cones are common in the southern Rocky Mountain states, particularly New Mexico and Arizona, and many others are in California, Oregon, and Washington.

COMPOSITE VOLCANOES (STRATOVOLCANOES)

Pyroclastic layers, as well as lava flows, both of intermediate composition, are found in **composite volcanoes**, which are also called **stratovolcanoes**. As the lava flows cool, they typically form andesite; recall that intermediate lava flows are more viscous than mafic ones. Geologists use the term **lahar** for volcanic mudflows, which are also common on composite volcanoes. A lahar may form when rain falls on unconsolidated pyroclastic materials and creates a muddy slurry that moves downslope. On November 13, 1985, a minor eruption of Nevado del Ruiz in Colombia melted snow and ice on the volcano, causing lahars that killed 23,000 people.

Composite volcanoes differ from shield volcanoes and cinder cones in composition and shape. Remember that shield volcanoes have very low slopes, whereas cinder cones are small, steep-sided, conical mountains. In contrast, composite volcanoes are steep-sided near their summits, perhaps as much as 30 degrees, but the slope decreases toward the base, where it may be no more than 5 degrees. Mayon volcano in the Philippines is one of the most nearly symmetrical composite volcanoes anywhere. It erupted in 1999 for the 13th time during the 1900s.

When most people think of volcanoes, they picture the graceful profiles of composite volcanoes, which are the typical large volcanoes found on the continents and island arcs. And some of these volcanoes are indeed large: Mount Shasta in northern California is made up of about 350 km³ of material and measures 20 km across its base (Figure 5.11). Mount Pinatubo in the Philippines erupted violently on June 15, 1991. Huge quantities of gas and an estimated 3 to 5 km³ of ash were discharged into the atmosphere, making this the world's largest eruption since 1912. Fortunately, warnings of an impending eruption were heeded, and 200,000 people were evacuated from around the volcano. Nevertheless, the eruption was responsible for 722 deaths.

LAVA DOMES

Less common volcanoes are **lava domes**, also known as *volcanic domes* and *plug domes*. These are steep-sided, bulbous mountains form when viscous felsic magma, and occasionally intermediate magma, is forced toward the surface (Figure 5.12). Because felsic magma is so viscous, it moves upward very slowly and only when the pressure from below is great.

composite volcano (stratovolcano) A volcano composed of lava flows and pyroclastic layers, typically of intermediate composition, and mudflows.

lahar A mudflow composed of pyroclastic materials such as ash.

lava dome A bulbous, steep-sided mountain formed by viscous magma moving upward through a volcanic conduit.

Figure 5.11 Composite Volcanoes

A view of Mt. Shasta, a huge composite volcano in northern California. Mount Shasta is about 24 km across its base and rises more than 3,400 m above its surroundings.

Figure 5.12 Lava Dome

Chaos Crags in the distance are made up of at least four lava domes that formed less than 1,200 years ago in Lassen Volcanic National Park in California. The debris in the foreground, called Chaos Jumbles, formed when parts of the domes collapsed.

Lava dome eruptions are some of the most violent and destructive. In 1902, viscous magma accumulated beneath the summit of Mount Pelée on the island of Martinique. Eventually, the pressure increased until the side of the mountain blew out in a tremendous explosion, ejecting a mobile, dense cloud of pyroclastic materials and a glowing cloud of gases and dust called a **nuée ardente** (French for "glowing cloud"). The pyroclastic flow followed a valley to the sea, but the nuée ardente jumped a ridge and engulfed the city of St. Pierre.

A tremendous blast hit St. Pierre, leveling buildings, and hurling boulders, trees, and pieces of masonry down the streets. Accompanying the blast was a swirling cloud of incandescent ash and gases with an internal temperature of 700°C that incinerated everything in its path. The nuée ardente passed through St. Pierre in two or three minutes, only to be followed by a firestorm as combustible materials burned and casks of rum exploded. But by then most of the 28,000 residents of the city were already dead. In fact, in the area covered by the nuée ardente, only two people survived!* One survivor was on the outer edge of the nuée ardente, but even there, he was terribly burned and his family and neighbors were all killed. The other survivor, a stevedore incarcerated the night before for disorderly conduct, was in a windowless cell partly below ground level. He remained in his cell badly burned for four days after the eruption until rescue workers heard his cries for help.

LO3 Other Volcanic Landforms

In some areas of volcanism, volcanoes fail to develop at all. For instance, during *fissure eruptions*, fluid lava pours out and simply builds up rather flat-lying areas, whereas huge explosive eruptions might yield *pyroclastic sheet deposits*, which, as their name implies, have a sheetlike geometry.

FISSURE ERUPTIONS AND BASALT PLATEAUS

Rather then erupting from central vents, the lava flows making up **basalt plateaus** issue from long cracks or

---

*Although reports commonly claim that only two people survived the eruption, at least 69 and possibly as many as 111 people survived beyond the extreme margins of the nuée ardente and on ships in the harbor. Many, however, were badly injured.

a. About 20 lava flows of the Columbia River basalt are exposed in the canyon of the Grand Ronde River in Washington.

b. Basalt lava flows of the Snake River Plain near Twin Falls, Idaho.

Figure 5.13 Basalt Plateaus
Basalt plateaus are vast areas of overlapping lava flows that issued from long fissures. Fissure eruptions take place today in Iceland; however, in the past, they formed basalt plateaus in various areas.

fissures during **fissure eruptions**. The lava has such low viscosity that it spreads out and covers vast areas. A good example is the Columbia River basalt in eastern Washington and parts of Oregon and Idaho. This huge accumulation of 17- to 6-million-year-old overlapping lava flows covers about 164,000 km² (Figure 5.13a) and has an aggregate thickness of more than 1,000 m.

nuée ardente A fast-moving, dense cloud of hot pyroclastic materials and gases ejected from a volcano.

basalt plateau A plateau built up by horizontal or nearly horizontal overlapping lava flows that erupted from fissures.

fissure eruption A volcanic eruption in which lava or pyroclastic materials issue from a long, narrow fissure (crack) or group of fissures.

Similar accumulations of vast, overlapping lava flows are also found in the Snake River Plain in Idaho (Figure 5.13b). These flows are 5.0 to 1.6 million years old, and they represent a style of eruption between fissure eruptions and those of shield volcanoes. In fact, there are small, low shields, as well as fissure flows, in the Snake River Plain.

Currently, fissure eruptions occur only in Iceland. Iceland has several volcanoes, but the bulk of the island is composed of basalt lava flows that issued from fissures. In fact, about half of the lava that erupted during historic time in Iceland came from two fissure eruptions, one in A.D. 930 and the other in 1783. The 1783 eruption from Laki fissure, which is more than 30 km long, accounted for lava that covered 560 km² and, in one place, filled a valley to a depth of about 200 m.

PYROCLASTIC SHEET DEPOSITS

Geologists have long been aware of vast areas covered by felsic volcanic rocks a few meters to hundreds of meters thick. Based on observations of historic pyroclastic flows, such as the nuée ardente erupted by Mount Pelée in 1902, it seems that these ancient rocks originated as pyroclastic flows—hence the name **pyroclastic sheet deposits**. They cover far greater areas than any observed during historic time, however, and apparently erupted from long fissures rather than from a central vent. The pyroclastic materials of many of these flows were so hot that they fused together to form *welded tuff*.

Geologists now think that major pyroclastic flows issue from fissures formed during the origin of calderas. Similarly, the Bishop Tuff of eastern

pyroclastic sheet deposit Vast, sheetlike deposit of felsic pyroclastic materials erupted from fissures.

circum-Pacific belt A zone of seismic and volcanic activity and mountain building that nearly encircles the Pacific Ocean basin.

Mediterranean belt A zone of seismic and volcanic activity extending through the Mediterranean region of southern Europe and eastward to Indonesia.

Cascade Range A mountain range stretching from Lassen Peak in northern California north through Oregon and Washington to Meager Mountain in British Columbia, Canada.

California erupted shortly before the formation of the Long Valley caldera. Interestingly, earthquake activity in the Long Valley caldera and nearby areas beginning in 1978 may indicate that magma is moving upward beneath part of the caldera. Thus, the possibility of future eruptions in that area cannot be discounted.

LO4 Distribution of Volcanoes

Most of the world's active volcanoes are in well-defined zones or belts rather than being randomly distributed. The **circum-Pacific belt**, popularly called the Ring of Fire, has more than 60% of all active volcanoes. It includes volcanoes in the Andes of South America; the volcanoes of Central America, Mexico, and the Cascade Range of North America; as well as the Alaskan volcanoes and those in Japan, the Philippines, Indonesia, and New Zealand (Figure 5.14).

The second area of active volcanism is the **Mediterranean belt** (Figure 5.14). About 20% of all active volcanism takes place in this belt, where the famous Italian volcanoes such as Mounts Etna and Vesuvius and the Greek volcano Santorini are found.

The **Cascade Range** (Figure 5.15) stretches from Lassen Peak in northern California north through Oregon and Washington to Meager Mountain in British Columbia, Canada. Most of the large volcanoes in the range are composite volcanoes, but Lassen Peak in California is the world's largest lava dome. It erupted from 1914 to 1917, but has since been quiet except for ongoing hydrothermal activity (Figure 5.15b).

What was once a nearly symmetrical composite volcano changed markedly on May 6, 1980, when Mount St. Helens in Washington erupted explosively, killing 57 people and leveling some 600 km² of forest. A huge lateral blast caused much of the damage and fatalities, but snow and ice on the volcano melted, and pyroclastic materials displaced water in lakes and rivers, causing lahars and extensive flooding.

Mount St. Helens's renewed activity, beginning in late September 2004, has resulted in dome growth and small steam and ash explosions. Scientists at the Cascades Volcano Observatory in Vancouver, Washington, issued a low-level alert for an eruption and continue to monitor the volcano. However, in January 2008, after 40 months of activity, it became dormant.

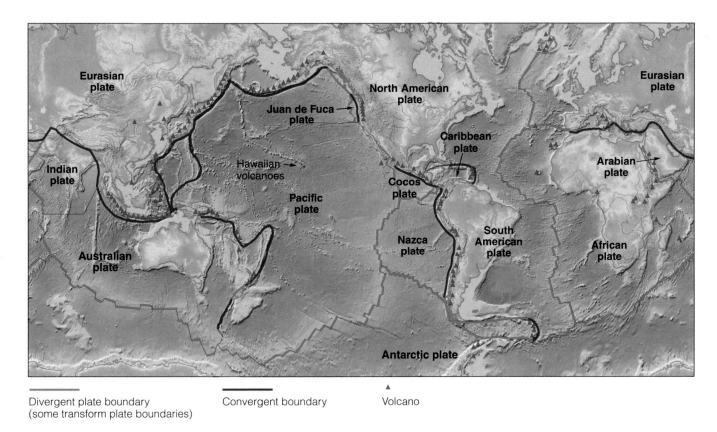

Divergent plate boundary
(some transform plate boundaries)

Convergent boundary

Volcano

Figure 5.14 Volcanoes of the World

Most volcanoes are at or near convergent and divergent plate boundaries. The two major volcano belts are the circum-Pacific belt, commonly known as the Ring of Fire, with about 60% of all active volcanoes, and the Mediterranean belt, with 20% of active volcanoes. Most of the rest lie near the mid-oceanic ridges.

LO5 Plate Tectonics, Volcanoes, and Plutons

In Chapter 4, we discussed the origin and evolution of magma and concluded that (1) mafic magma is generated beneath spreading ridges, and (2) intermediate and felsic magma form where an oceanic plate is subducted beneath another oceanic plate or a continental plate. Accordingly, most volcanism and emplacement of plutons take place at or near divergent and convergent plate boundaries.

IGNEOUS ACTIVITY AT DIVERGENT PLATE BOUNDARIES

Much of the mafic magma that originates at spreading ridges is emplaced as vertical dikes and gabbro plutons, thus composing the lower part of the oceanic crust. However, some rises to the surface and issues forth as submarine lava flows and pillow lava (Figure 5.6), which constitutes the upper part of the oceanic crust. Much of this volcanism goes undetected, but researchers in submersibles have seen the results of recent eruptions.

Mafic lava is very fluid, allowing gases to escape easily, and at great depth in the oceans, the water pressure is so great that explosive volcanism is prevented. In short, pyroclastic materials are rare to absent unless a volcanic center builds up above sea level. Even if this occurs, however, the mafic magma is so fluid that it forms the gently sloping layers found on shield volcanoes.

Excellent examples of divergent plate boundary volcanism are found along the Mid-Atlantic Ridge, particularly where it rises above sea level as in Iceland.

The East Pacific Rise and the Indian Ridge are areas of similar volcanism. A divergent plate boundary is also present in Africa as the East African Rift system, which is well known for its volcanoes.

IGNEOUS ACTIVITY AT CONVERGENT PLATE BOUNDARIES

Nearly all of the large active volcanoes in both the circum-Pacific and Mediterranean belts are composite volcanoes near the leading edges of overriding plates at convergent plate boundaries (Figure 5.14). The overriding plate, with its chain of volcanoes, may be oceanic, as in the case of the Aleutian Islands, or it may be continental as is, for instance, the South American plate with its chain of volcanoes along its western edge.

As we noted, these volcanoes at convergent plate boundaries consist mostly of lava flows and pyroclastic materials of intermediate to felsic composition. Remember that when mafic oceanic crust partially melts, some of the magma generated is emplaced near plate boundaries as plutons and some is erupted to build up composite volcanoes. More viscous magmas, usually of felsic composition, are emplaced as lava domes, thus accounting for the explosive eruptions that typically occur at convergent plate boundaries.

Good examples of volcanism at convergent plate boundaries are the explosive eruptions of Mount Pinatubo and Mayon volcano in the Philippines; both are near a plate boundary beneath which an oceanic plate is subducted. Mount St. Helens, Washington, is similarly situated, but it is on a continental rather than an oceanic plate.

Figure 5.15 The Cascade Range of the Pacific Northwest

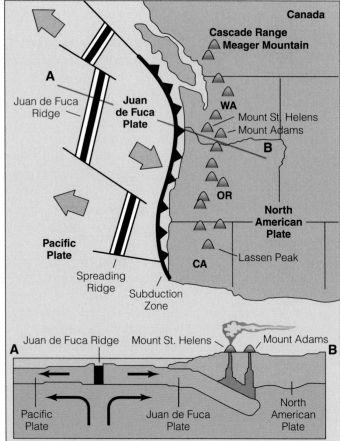

a. Plate tectonic setting for the Pacific Northwest. Subduction of the Juan de Fuca plate accounts for ongoing volcanism in the region.

b. Lassen Peak in California erupted from 1914 to 1917. This eruption took place in 1915.

INTRAPLATE VOLCANISM

Mauna Loa and Kilauea on the island of Hawaii and Loihi just 32 km to the south are within the interior of a rigid plate located far from any divergent or convergent plate boundary (Figure 5.14). The magma is derived from the upper mantle, as it is at spreading ridges, and accordingly is mafic, so it builds up shield volcanoes. Loihi is particularly interesting, because it represents an early stage in the origin of a new Hawaiian island. It is a submarine volcano that rises more than 3,000 m above the adjacent seafloor, but its summit is still about 940 m below sea level.

Even though the Hawaiian volcanoes are not at or near a spreading ridge or a subduction zone, their evolution is nevertheless related to plate movements. Notice in Figure 2.18 that the ages of the rocks that make up the Hawaiian Islands increase toward the northwest. Kauai formed 5.6 to 3.8 million years ago, whereas Hawaii began forming less than 1 million years ago, and Loihi began to form even more recently. The islands have formed in succession as the Pacific plate moves continuously over a hot spot that is now beneath Hawaii and just to the south at Loihi.

LO6 Volcanic Hazards, Volcano Monitoring, and Forecasting Eruptions

Undoubtedly you suspect that living near an active volcano poses some risk, and of course this assessment is correct, but further questions remain: What exactly are volcanic hazards? Is there any way to anticipate eruptions? What can we do to minimize the dangers of eruptions? We have already mentioned that lava flows, with few exceptions, pose little threat to humans, although they may destroy property. Lava flows, nuée ardentes, and volcanic gases are threats during an eruption, but lahars and landslides may take place even when no eruption has occurred for a long time.

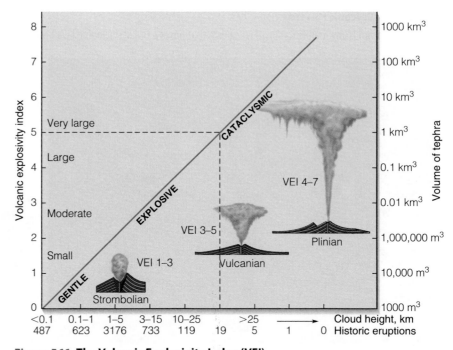

Figure 5.16 The Volcanic Explosivity Index (VEI)

In this example, an eruption with a VEI of 5 has an eruption cloud up to 25 km high and ejects at least 1 km³ of *tephra*, a collective term for all pyroclastic materials. Geologists characterize eruptions as Hawaiian (nonexplosive), Strombolian, Vulcanian, and Plinian.

HOW LARGE IS AN ERUPTION, AND HOW LONG DO ERUPTIONS LAST?

The most widely used indication of the size of a volcanic eruption is the **volcanic explosivity index (VEI)** (Figure 5.16). The VEI ranges from 0 (gentle) to 8 (cataclysmic) and is based on several aspects of an eruption, such as the volume of material explosively ejected and the height of the eruption plume. However, the volume of lava, fatalities, and property damage are not considered. For instance, the 1985 eruption of Nevado del Ruiz in Colombia killed 23,000 people, yet has a VEI value of only 3. In contrast, the huge eruption (VEI = 6) of Novarupta in Alaska in 1912 caused no fatalities or injuries. Since A.D. 1500, only the 1815 eruption of Tambora had a value of 7; it was both large and deadly. Nearly 5,700 eruptions during the last 10,000 years have been assigned VEI numbers, but none has exceeded 7, and most (62%) were assigned a value of 2.

volcanic explosivity index (VEI) A semi-quantitative scale for determining the size of a volcanic eruption based on evaluation of criteria, such as volume of material explosively erupted and height of eruption cloud.

Figure 5.17 Volcanic Hazards

A volcanic hazard is any manifestation of volcanism that poses a threat, including lava flows and, more importantly, volcanic gas, ash, and lahars.

a. This 2002 lava flow in Goma, Democratic Republic of Congo, killed 147 people, mostly by causing gasoline storage tanks to explode.

b. This sign at Mammoth Mountain volcano in California warns of the potential danger of CO_2 gas, which has killed 170 acres of trees.

The duration of eruptions varies considerably. Fully 42% of about 3,300 historic eruptions lasted less than one month. About 33% erupted for one to six months, but some 16 volcanoes have been active more or less continuously for more than 20 years. Stromboli and Mount Etna in Italy and Erta Ale in Ethiopia are good examples. For some explosive volcanoes, the time from the onset of their eruptions to the climactic event is weeks or months.

A case in point is the colossal explosive eruption of Mount St. Helens on May 18, 1980, which occurred two months after eruptive activity began. Unfortunately, many volcanoes give little or no warning of such large-scale events. Of 252 explosive eruptions, 42% erupted most violently during their first day of activity. As one might imagine, predicting eruptions is complicated by those volcanoes that give so little warning of impending activity.

volcanic tremor Ground motion lasting from minutes to hours, resulting from magma moving beneath the surface, as opposed to the sudden jolts produced by most earthquakes.

IS IT POSSIBLE TO FORECAST ERUPTIONS?

Only a few of Earth's potentially dangerous volcanoes are monitored, including some in Japan, Italy, Russia, New Zealand, and the United States. Volcano monitoring involves recording and analyzing physical and chemical changes at volcanoes (Figure 5.18). Tiltmeters detect changes in the slopes of a volcano as it inflates when magma rises beneath it, and a geodimeter uses a laser beam to measure horizontal distances, which change as a volcano inflates. Geologists also monitor gas emissions, changes in groundwater level and temperature, hot springs activity, and changes in the local magnetic and electrical fields. Even the accumulating snow and ice, if any, are evaluated to anticipate hazards from floods if an eruption takes place.

Of critical importance in volcano monitoring and warning of an imminent eruption is the detection of **volcanic tremor**, continuous ground motion that lasts for minutes to hours as opposed to the sudden, sharp jolts produced by most earthquakes. Volcanic tremor,

Figure 5.18 Volcanic Monitoring
Some important techniques used to monitor volcanoes.

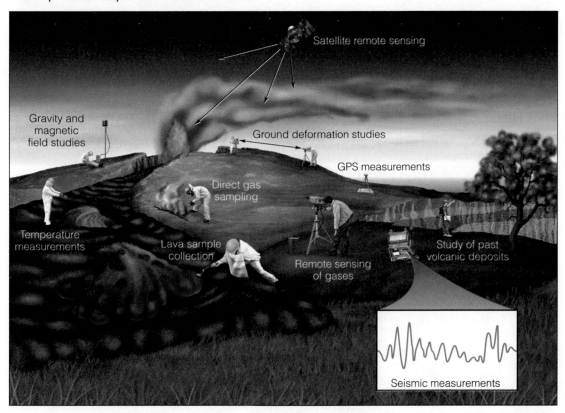

also known as *harmonic tremor*, indicates that magma is moving beneath the surface.

To more fully anticipate the future activity of a volcano, its eruptive history must be known. Accordingly, geologists study the record of past eruptions preserved in rocks. Detailed studies before 1980 indicated that Mount St. Helens, Washington, had erupted explosively 14 or 15 times during the last 4,500 years, so geologists concluded that it was one of the most likely Cascade Range volcanoes to erupt again.

Geologists successfully gave timely warnings of impending eruptions of Mount St. Helens in Washington and Mount Pinatubo in the Philippines, but in both cases, the climactic eruptions were preceded by eruptive activity of lesser intensity. In some cases, however, the warning signs are much more subtle and difficult to interpret. Numerous small earthquakes and other warning signs indicated to United States Geological Survey (USGS) geologists that magma was moving beneath the surface of the Long Valley caldera in eastern California, so in 1987, they issued a low-level warning, and then nothing happened.

Volcanic activity in the Long Valley caldera occurred as recently as 250 years ago, and there is every reason to think that it will occur again. Unfortunately, the local populace was largely unaware of the geologic history of the region, the USGS did a poor job of communicating its concerns, and premature news releases caused more concern than was justified. In any case, local residents were outraged, because the warnings caused a decrease in tourism (Mammoth Mountain on the margins of the caldera is the second largest ski area in the country) and property values plummeted. Monitoring continues in the Long Valley caldera, and the signs of renewed volcanism, including earthquake swarms, trees being killed by carbon dioxide gas apparently emanating from magma (Figure 5.17b), and hot spring activity, cannot be ignored. In April 2006, three members of a ski patrol were killed by carbon dioxide gas that accumulated in a low area.

CHAPTER 6
WEATHERING, SOIL, AND SEDIMENTARY ROCKS

An arch in the Entrada Sandstone in Arches National Park in Utah. The park is famous for its pillars, spires, and arches, all of which formed by weathering and erosion.

COPYRIGHT AND PHOTOGRAPH BY DR. PARVINDER S. SETHI

Introduction

"All rocks are important in deciphering Earth history, but sedimentary rocks have a special place in this endeavor."

All rocks at or near Earth's surface—as well as rocklike substances such as pavement and concrete in sidewalks, bridges, and foundations—decay and crumble with age. In short, they experience **weathering**, defined as the physical breakdown and chemical alteration of Earth materials as they are exposed to the atmosphere, hydrosphere, and biosphere. Actually, weathering is a group of physical and chemical processes that alter Earth materials so that they are more nearly in equilibrium with a new set of environmental conditions. Many rocks form within the crust where little or no oxygen or water is present, but at or near the surface they are exposed to both, as well as to lower temperature and pressure and the activities of organisms.

The rock acted on by weathering, or **parent material**, is disaggregated to form smaller pieces, and some of its constituent minerals are altered or dissolved. Some of this weathered material may accumulate and be further modified to form **soil**. Much of it, however, is removed by **erosion**, which is the wearing away of soil and rock by geologic agents such as running water (Figure 6.1). This eroded material is transported elsewhere by running water, wind, glaciers, and marine currents and is eventually deposited as *sediment*, the raw material for *sedimentary rocks*.

Earth's crust is composed mostly of *crystalline rock*, a term that refers loosely to metamorphic and igneous rocks, except those made up of pyroclastic materials. Nevertheless, sediment and sedimentary rocks are

LEARNING OUTCOMES

After reading this unit, you should be able to do the following:

LO1 Explain how Earth materials are altered

LO2 Explain how soil forms and deteriorates

LO3 Know how weathering and resources are related

LO4 Identify sediment and sedimentary rocks

LO5 Explain how sedimentary rocks are classified

LO6 Understand sedimentary facies

LO7 Read the story preserved in sedimentary rocks

LO8 Recognize important resources in sedimentary rocks

weathering The physical breakdown and chemical alteration of rocks and minerals at or near Earth's surface.

parent material The material that is chemically and mechanically weathered to yield sediment and soil.

soil Regolith consisting of weathered materials, water, air, and humus that can support vegetation.

erosion The removal of weathered materials from their source area by running water, wind, glaciers, and waves.

EROSION AND NATURAL RESOURCES

Weathering and erosion have yielded many areas of exceptional scenery, including Bryce Canyon and Arches National Parks, both in Utah. In addition to forming interesting landscapes, weathering is also responsible for the origin of some natural resources (e.g., aluminum ore), and it enriches other resources by removing soluble materials. In fact, some sediments and sedimentary rocks are resources in their own right, or they are the host rocks for petroleum and natural gas.

Figure 6.1 Weathering
One of several spectacular waterfalls in Yosemite National Park, California. Rocks in Yosemite are weathered or disaggregated into smaller pieces, some of which are subsequently removed through erosion.

differential weathering
Weathering that occurs at different rates on rocks, thereby yielding an uneven surface.

mechanical weathering
Disaggregation of rocks by physical processes that yields smaller pieces that retain the composition of the parent material.

frost action The disaggregation of rocks by repeated freezing and thawing of water in cracks and crevasses.

by far the most common materials in surface exposures and in the shallow subsurface, even though they make up perhaps only 50% of the crust. They cover approximately two-thirds of the continents and most of the seafloor, except spreading ridges. All rocks are important in deciphering Earth history, but sedimentary rocks have a special place in this endeavor, because they preserve evidence of surface processes responsible for them, as well as most fossils, which are evidence of prehistoric life.

LO1 How Are Earth Materials Altered?

Weathering takes place at or near the surface, but the rocks it acts upon are not structurally and composi-

tionally homogeneous throughout, which accounts for **differential weathering**. That is, weathering that takes place at different rates even in the same area, so uneven surfaces develop. Differential weathering and *differential erosion*—that is, variable rates of erosion—combine to yield some unusual and even bizarre features, such as *hoodoos*, *spires*, and *arches* (see chapter opening photo).

The two recognized types of weathering, *mechanical* and *chemical*, both proceed simultaneously on parent material, as well as on materials in transport and those deposited as sediment. In short, all surface or near-surface materials weather, although one type of weathering may predominate depending on such variables as climate and rock type.

Mechanical Weathering

Mechanical weathering takes place when physical forces break earth materials into smaller pieces that retain the composition of the parent material. Granite, for instance, might yield smaller pieces of granite or individual grains of quartz, potassium feldspars, plagioclase feldspars, and biotite.

Frost action involving water repeatedly freezing and thawing in cracks and pores in rocks is particularly

a. When water seeps into cracks and expands as it freezes, it causes frost wedging. Repeated freezing and thawing pry angular pieces of rock loose.

b. Talus has accumulated at the base of this slope. Notice that the source rocks are highly fractured.

Figure 6.2 Frost Wedging and the Accumulation of Talus

effective where temperatures commonly fluctuate above and below freezing. Frost action is so effective because water expands by about 9% when it freezes, thus exerting great force on the walls of a crack, widening and extending it by *frost wedging* (Figure 6.2a). Repeated freezing and thawing dislodge angular pieces of rock from the parent material that tumble downslope and accumulate as **talus** (Figure 6.2b).

Some rocks form at depth and are stable under tremendous pressure. Granite crystallizes far below the surface, so when it is uplifted and eroded, its contained energy is released by outward expansion, a phenomenon known as **pressure release**. The outward expansion results in the origin of fractures called *sheet joints* that more or less parallel the exposed rock surface. Sheet-joint-bounded slabs of rock slip or slide off the parent rock, leaving large, rounded masses known as **exfoliation domes** (Figure 6.3). That solid rock expands and produces fractures might be counterintuitive but is nevertheless a well-known phenomenon. In deep mines, masses of rock detach from the sides of the excavation,

often explosively. These *rock bursts* and less violent *popping* pose a danger to mine workers.

During **thermal expansion and contraction**, the volume of rocks changes as they heat up and then cool down. The temperature may vary as much as 30°C a day in a desert, and rock, being a poor conductor of heat, heats and expands on its outside more than its inside. Even dark minerals absorb heat faster than light-colored ones, so differential expansion takes place between minerals. Surface expansion

talus Accumulation of coarse, angular rock fragments at the base of a slope.

pressure release A mechanical weathering process in which rocks that formed under pressure expand on being exposed at the surface.

exfoliation dome A large, rounded dome of rock resulting when concentric layers of rock are stripped from the surface of a rock mass.

thermal expansion and contraction A type of mechanical weathering in which the volume of rocks changes in response to heating and cooling.

might generate enough stress to cause fracturing, but experiments in which rocks are heated and cooled repeatedly to simulate years of such activity indicate that thermal expansion and contraction are of minor importance in mechanical weathering.

The formation of salt crystals exerts enough force to widen cracks and dislodge particles in porous, granular rocks such as sandstone. And even in rocks with an interlocking mosaic of crystals, such as granite, **salt crystal growth** pries loose individual minerals. It takes place mostly in hot, arid regions, but also probably affects rocks in some coastal areas.

salt crystal growth A mechanical weathering process in which salt crystals growing in cracks and pores disaggregate rocks.

Animals, plants, lichens, and bacteria all participate in the mechanical and chemical alteration of rocks (Figure 6.4a). Burrowing animals, such as worms,

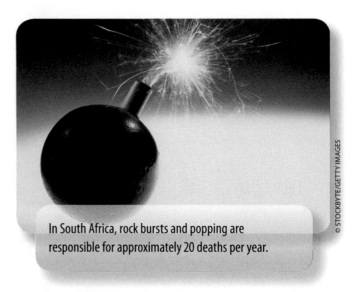

In South Africa, rock bursts and popping are responsible for approximately 20 deaths per year.

reptiles, rodents, termites, and ants, constantly mix soil and sediment particles and bring material from depth to the surface, where further weathering occurs.

Figure 6.3 Sheet Joints and Exfoliation Domes
Sheet joints in this body of granite in the Sierra Nevada of California parallel the surface of exposed rock. The rounded mass formed by this process is an exfoliation dome.

LO2 How Does Soil Form and Deteriorate?

Most of Earth's land surface is covered by a layer of **regolith**, a collective term for sediment, layers of pyroclastic materials, and the residue formed in place by weathering. Some regolith consisting of weathered materials, air, water, and organic matter supports vegetation and is called **soil**. Almost all land-dwelling organisms depend directly or indirectly on soil for their existence. Plants grow in soil, from which they derive their nutrients and most of their water, whereas many land-dwelling animals depend on plants for nutrients.

About 45% of good soil for farming and gardening is composed of weathered particles, with much of the remaining volume simply void spaces filled with air and/or water. In addition, a small but important amount of *humus* is usually present. Humus is carbon derived by bacterial decay of organic matter and is highly resistant to further decay. Even a fertile soil might have as little as 5% humus, but it is nevertheless important as

© C SQUARED STUDIOS/PHOTODISC/ GETTY IMAGES

The Scent of Soil

Threadlike soil bacteria give freshly plowed soil its earthy aroma.

a source of plant nutrients, and it enhances a soil's capacity to retain moisture.

Some weathered materials in soils are sand- and silt-sized minerals, especially quartz, which hold soil particles apart, allowing oxygen and water to circulate more freely. Clay minerals are also important in soils and aid in the retention of water, as well as supplying nutrients to plants.

Residual soils form when parent material weathers in place. For example, if a body of granite weathers and the weathering residue accumulates over the granite and is converted to residual soil. In contrast, *transported soils* develop on weathered material that was eroded and carried to a new location, where it is altered to soil.

THE SOIL PROFILE

In vertical cross section, soil consists of layers or **soil horizons** that differ in texture, structure, composition, and color (Figure 6.8). Starting from the top, the soil horizons are designated O, A, B, and C, but the boundaries between horizons are transitional. Because soil formation begins at the surface and works downward, horizon A is more altered from the parent material than the layers below.

Horizon O, which is only a few centimeters thick, consists of organic matter. The remains of plant materials are clearly recognizable in the upper part of horizon O, but its lower part consists of humus.

Horizon A, called *topsoil*, contains more organic matter than horizons B and C below, and is characterized by intense biological activity, because plant roots, bacteria, fungi, and animals such as worms are abundant. In soils developed over a long period of time, horizon A is mostly clays and chemically stable minerals such as quartz. Water percolating through horizon A dissolves soluble minerals and carries them away or downward to lower levels in the soil, a process called *leaching*.

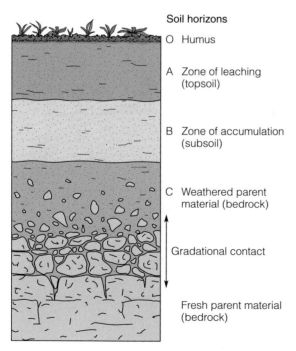

Soil horizons

O   Humus

A   Zone of leaching (topsoil)

B   Zone of accumulation (subsoil)

C   Weathered parent material (bedrock)

Gradational contact

Fresh parent material (bedrock)

Figure 6.8 The Composition of Soil Horizons
The soil horizons of a fully developed soil.

regolith The layer of unconsolidated rock and mineral fragments and soil that covers most of the land surface.

soil horizon A distinct soil layer that differs from other soil layers in texture, structure, composition, and color.

pedalfer Soil formed in humid regions with an organic-rich A horizon and aluminum-rich clays and iron oxides in horizon B.

Accordingly, horizon A is also referred to as the *zone of leaching*.

Horizon B, or *subsoil*, contains fewer organisms and less organic matter than horizon A (Figure 6.8). Horizon B is also known as the *zone of accumulation*, because soluble minerals leached from horizon A accumulate as irregular masses. If horizon A is eroded, leaving horizon B exposed, plants do not grow as well, and if it is clayey, it is harder when dry and stickier when wet than other soil horizons.

Horizon C has little organic matter and consists of partially altered parent material grading down into unaltered parent material (Figure 6.8). In horizons A and B, the parent material has been so thoroughly altered that it is no longer recognizable. In contrast, rock fragments and minerals of parent material retain their identity in horizon C.

FACTORS THAT CONTROL SOIL FORMATION

Climate is the single most important factor in soil origins, but complex interactions among several factors account for soil type, thickness, and fertility (Figure 6.9). A very general classification recognizes three major soil types characteristic of different climatic settings: (1) pedalfer, (2) pedocal, and (3) laterite.

Soils that develop in humid regions such as the eastern United States and much of Canada are **pedalfers**, a name derived from the Greek word *pedon*, meaning "soil," and from the chemical symbols for aluminum (Al) and iron (Fe). Because these soils form where abundant moisture is present, most of the soluble minerals have been leached from horizon A. Horizon A is commonly dark because of abundant organic matter, and aluminum-rich clays and iron oxides tend to accumulate in horizon B.

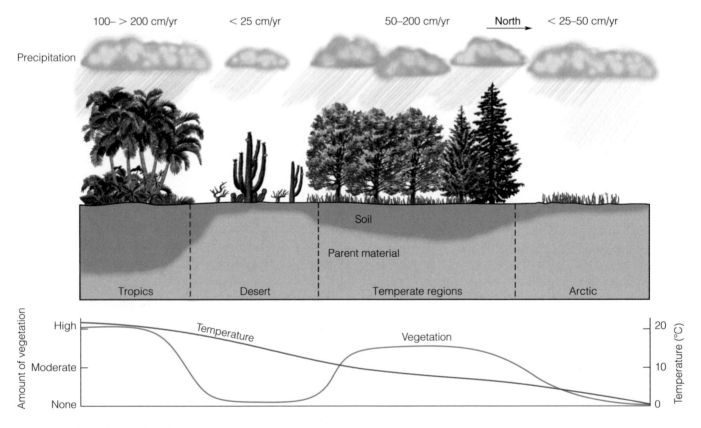

Figure 6.9 Climate and Soil Formation
Generalized diagram showing soil formation as a function of the relationships between climate and vegetation, which alter parent material over time. Soil formation operates most vigorously where precipitation and temperatures are high, as in the tropics.

Soils found in much of the arid and semiarid western United States, especially the Southwest, are **pedocals**. Pedocal derives its name, in part, from the first three letters of "calcite." These soils contain less organic matter than pedalfers, so horizon A is lighter in color and contains more unstable minerals because of less intense chemical weathering. As soil water evaporates, calcium carbonate leached from above precipitates in horizon B, where it forms irregular masses of *caliche*. Precipitation of sodium salts in some desert areas where soil water evaporation is intense yields *alkali soils* that are so alkaline that they cannot support plants.

Laterite forms in the tropics, where chemical weathering is intense and leaching of soluble minerals is complete. These soils are red, extend to depths of several tens of meters, and are composed mostly of aluminum hydroxides, iron oxides, and clay minerals; even quartz, a chemically stable mineral, is leached out.

The same rock type can yield different soils in different climatic regimes and, in the same climatic regime, the same soils can develop on different rock types. Thus, climate is more important than parent material in determining the type of soil. Nevertheless, rock type does exert some control. For example, the metamorphic rock quartzite will have a thin soil over it, because it is chemically stable, whereas an adjacent body of granite will have a much deeper soil.

Soils depend on organisms for their fertility, and in return they provide a suitable habitat for many organisms. Earthworms, ants, sowbugs, termites, centipedes, millipedes, and nematodes, along with fungi, algae, and single-celled organisms, make their homes in soil. All contribute to soil formation and provide humus when they die and decompose.

Much of the humus in soils comes from grasses or leaf litter that microorganisms decompose to obtain food. In so doing, they break down organic compounds in plants and release nutrients back into the soil. In addition, organic acids from decaying soil organisms are important in further weathering of parent materials and soil particles. Burrowing animals constantly churn and mix soils, and their burrows provide avenues for gases and water. Soil organisms, especially some types of bacteria, are extremely important in changing atmospheric nitrogen into a form of soil nitrogen suitable for use by plants.

The difference in elevation between high and low points in a region is called *relief*. Because climate is such an important factor in soil formation and climate changes with elevation, areas with considerable relief have different soils in mountains and adjacent lowlands. *Slope*, another important control, influences soil formation in two ways. One is *slope angle*; steep slopes have little or no soil, because weathered materials erode faster than soil-forming processes operate. The other factor is *slope direction*. In the Northern Hemisphere, north-facing slopes receive less sunlight than do south-facing slopes and have cooler internal temperatures, support different vegetation, and if in a cold climate, remain snow covered or frozen longer.

How much time is needed to develop a centimeter of soil or a fully developed soil a meter or so deep? We can give no definite answer, because weathering proceeds at vastly different rates depending on climate and parent material, but an overall average might be about 2.5 cm per century. Nevertheless, a lava flow a few centuries old in Hawaii may have a well-developed soil on it, whereas a flow the same age in Iceland will have considerably less soil.

SOIL DEGRADATION

From the human perspective, soil forms so slowly that it is a nonrenewable resource. So any soil losses that exceed the rate of soil formation are viewed with alarm. Likewise, any reduction in soil fertility or production is cause for concern. Any process that removes soil or makes it less productive is defined as **soil degradation**, a serious problem in many parts of the world and includes erosion, chemical deterioration, and physical changes.

Erosion is a natural process, but it is usually slow enough for soil formation to keep pace. Unfortunately, some human practices add to the problem. Removing natural vegetation by plowing, overgrazing, overexploitation for firewood, and deforestation all contribute to erosion by wind and running water. The Dust Bowl that developed in several U.S. Great Plains states during the 1930s is a poignant example of just how effective wind erosion is on soil that has been pulverized and exposed by plowing.

pedocal Soil characteristic of arid and semiarid regions with a thin A horizon and a calcium carbonate–rich B horizon.

laterite A red soil, rich in iron or aluminum, or both, resulting from intense chemical weathering in the tropics.

soil degradation Any process leading to a loss of soil productivity; may involve erosion, chemical pollution, or compaction.

Figure 6.10 Soil Degradation Resulting from Erosion

a. Rill erosion in a field in Michigan during a rainstorm. The rill was later plowed over.

b. A large gully in the upper basin of the Rio Reventado in Costa Rica. Notice the man at the right for scale.

Wind has caused considerable soil erosion in some areas, but running water is much more powerful. Some soil is removed by *sheet erosion*, which involves the removal of thin layers of soil more or less evenly over a broad, sloping surface. *Rill erosion*, in contrast, takes place when running water scours small, troughlike channels. Channels shallow enough to be eliminated by plowing are *rills*, but those too deep (about 30 cm) to be plowed over are *gullies* (Figure 6.10). Where gullying is extensive, croplands can no longer be tilled and must be abandoned.

Soil undergoes chemical deterioration when its nutrients are depleted and its productivity decreases. Loss of soil nutrients is most notable in many of the populous developing nations, where soils are overused to maintain high levels of agricultural productivity. Chemical deterioration is also caused by insufficient use of fertilizers and by clearing soils of their natural vegetation.

Other types of chemical deterioration are pollution and *salinization*, which occurs when the concentration of salts increases in a soil, making it unfit for agriculture. Improper disposal of domestic and industrial wastes, oil and chemical spills, and the concentration of insecticides and pesticides in soils all cause pollution. Soil deteriorates physically when it is compacted by the weight of heavy machinery and livestock, especially cattle. Compacted soils are more costly to plow, and plants have a more difficult time emerging from them. Furthermore, water does not readily infiltrate, so more runoff occurs; this in turn accelerates the rate of water erosion.

Problems experienced in the past have stimulated the development of methods to minimize soil erosion on agricultural lands. Crop rotation, contour plowing, and terracing have all proved helpful. So has no-till planting, in which the residue from the harvested crop is left on the ground to protect the surface from the ravages of wind and water.

LO3 Weathering and Resources

Soils are certainly one of our most precious natural resources. Indeed, if it were not for soils, food production on Earth would be vastly different and capable of supporting far fewer people. In addition, other aspects of soils are important economically. We discussed the origin of laterite in response to intense chemical weathering in the tropics. Laterite is not very productive for agriculture, but if the parent material is rich in aluminum, the ore of aluminum called *bauxite* accumulates in horizon B. Some bauxite is found in Arkansas, Alabama, and Georgia, but at present, it is cheaper to import rather than mine these deposits, so both the United States and Canada depend on foreign sources of aluminum ore.

Bauxite and other accumulations of valuable minerals form by the selective removal of soluble substances during chemical weathering and are known as *residual concentrations*. Bauxite is a good example, but other deposits that formed in a similar manner are those rich in iron, manganese, clays, nickel, phosphate, tin, diamonds, and gold. Some of the sedimentary iron deposits in the Lake Superior region of the United States and Canada were enriched by chemical weathering when soluble parts of the deposits were carried away. Some kaolinite deposits in the southern United States formed when chemical weathering altered feldspars in pegmatites or as residual concentrations of clay-rich limestones and dolostones.

LO4 Sediment and Sedimentary Rocks

All **sediment** is derived by mechanical and chemical weathering, but the term encompasses two categories of particles: (1) *detrital sediment* consists of the solid particles from preexisting rocks; and (2) *chemical sediment* is minerals that precipitate from solutions and minerals derived from water by organisms to build their shells. **Sedimentary rock** is simply any rock made up of sediment.

One criterion for classifying detrital sediment is particle size. Particles described as *gravel* measure more than 2 mm, whereas sand measures 1/16 mm to 2 mm, and silt is any particle between 1/256 mm and 1/16 mm. None of these designations implies anything about composition; most gravel is made up of rock fragments—that is, small pieces of granite, basalt, or any other rock type—but sand and silt grains are usually single minerals, especially quartz. Particles smaller than 1/256 mm are termed *clay*, but clay has two meanings. One is a size designation, but the term also refers to certain types of sheet silicates known as *clay minerals*. However, most clay minerals are also clay sized.

Sediment Transport and Deposition

Weathering is fundamental to the origin of sediment and sedimentary rocks, and so are erosion and *deposition*—that is, the movement of sediment by natural processes and its accumulation in some area. Because glaciers are moving solids, they can carry sediment of any size, whereas wind transports only sand and smaller sediment. Waves and marine currents transport sediment along shorelines, but running water is

Figure 6.11 Rounding and Sorting in Sediments

a. Deposit of rounded, moderately sorted gravel. The largest particle is about 5 cm across.

b. Angular, poorly sorted gravel. Note the quarter for scale.

by far the most common way to transport sediment from its source to other locations.

During transport, *abrasion* reduces the size of particles, and the sharp corners and edges are worn smooth, a process known as *rounding*, as pieces of sand and gravel collide with one another (Figure 6.11a). Transport and processes that operate where sediment accumulates also result in *sorting*, which refers to the particle-size distribution in a sedimentary deposit. Sediment is characterized as well sorted if all particles are about the same size, and poorly sorted if a wide range of particle sizes is present (Figure 6.11b). Both rounding and sorting have important implications for other aspects of sediment and sedimentary rocks, such as how readily fluids move through them, and they also help geologists decipher the history of a deposit.

sediment Loose aggregate of solids derived by weathering from preexisting rocks, or solids precipitated from solution by inorganic chemical processes or extracted from solution by organisms.

sedimentary rock Any rock composed of sediment, such as limestone and sandstone.

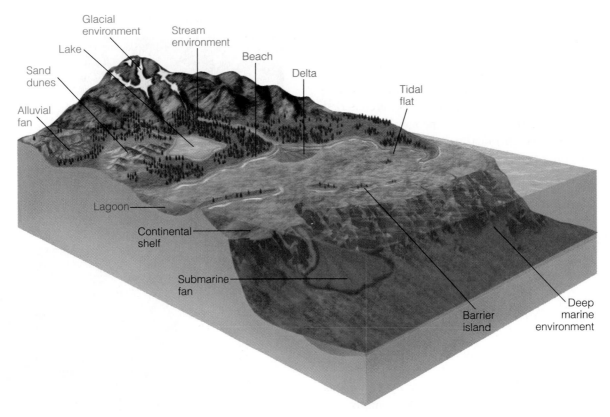

Figure 6.12 Depositional Environment
Continental environments are shown in red type. The environments along the shoreline, shown in blue type, are transitional between continental and marine. The others, shown in black type, are marine environments.

Regardless of how sediment is transported, it is eventually deposited in some geographic area known as a **depositional environment**, such as a beach, or on the seafloor, where physical, chemical, and biological processes impart various characteristics to the accumulating sediment. Geologists recognize three major depositional settings: (1) continental (on the land), (2) transitional (on or near seashores), and (3) marine, each with several specific depositional environments (Figure 6.12).

HOW DOES SEDIMENT BECOME SEDIMENTARY ROCK?

Mud in lakes and sand and gravel in stream channels or on beaches are good examples of sediment. To convert these aggregates of particles into sedimentary rocks requires **lithification** by compaction, cementation, or both (Figure 6.13).

To illustrate the relative importance of compaction and cementation, consider detrital deposits of mud and another composed of sand. In both cases, the sediment consists of solid particles and *pore spaces*, the voids between particles. These deposits are subjected to **compaction** from their own weight and the weight of any additional sediment deposited on top of them, thereby reducing the amount of pore space and the volume of the deposit. Our hypothetical mud deposit may have 80% water-filled pore space, but after compaction, its volume is reduced by as much as 40% (Figure

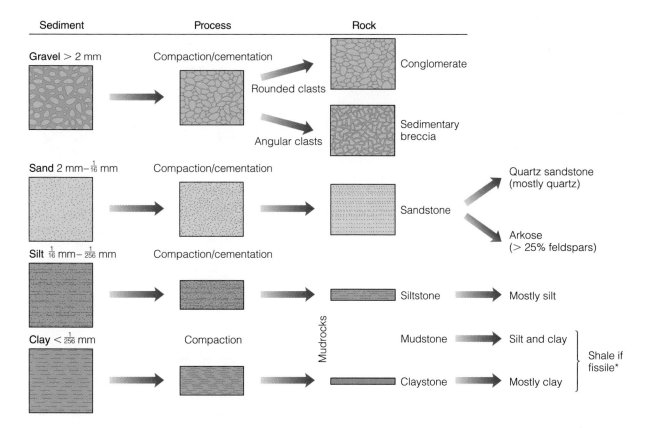

Sediment	Process	Rock

Gravel > 2 mm → Compaction/cementation → Rounded clasts → Conglomerate

→ Angular clasts → Sedimentary breccia

Sand 2 mm–$\frac{1}{16}$ mm → Compaction/cementation → Sandstone → Quartz sandstone (mostly quartz) / Arkose (> 25% feldspars)

Silt $\frac{1}{16}$ mm–$\frac{1}{256}$ mm → Compaction/cementation → Siltstone → Mostly silt

Clay < $\frac{1}{256}$ mm → Compaction → Mudrocks

Mudstone → Silt and clay

Claystone → Mostly clay

Shale if fissile*

*Fissile refers to rocks capable of splitting along closely spaced planes.

Figure 6.13 Lithification and Classification of Detrital Sedimentary Rocks
Notice that little compaction takes place in gravel and sand.

6.13). The sand deposit, with as much as 50% pore space, is also compacted, so that the grains fit more tightly together (Figure 6.13).

Compaction alone is sufficient for lithification of mud, but for sand and gravel, **cementation** involving the precipitation of minerals in pore spaces is also necessary. The two most common chemical cements are calcium carbonate ($CaCO_3$) and silicon dioxide (SiO_2), but iron oxide and hydroxide cement, such as hematite (Fe_2O_3) and limonite [$FeO(OH)\cdot nH_2O$], are found in some sedimentary rocks. Recall that calcium carbonate readily dissolves in water that contains a small amount of carbonic acid, and chemical weathering of feldspars and other minerals yields silica in solution. Cementation takes place when minerals precipitate in the pore spaces of sediment from circulating water, thereby binding the loose particles together. Iron oxide and hydroxide cements account for the red, yellow, and brown sedimentary rocks found in many areas (see the chapter opening photograph).

By far the most common chemical sediments are calcium carbonate mud and sand- and gravel-sized accumulations of calcium carbonate grains, such as shells and shell fragments. Compaction and cementation also take place in these sediments, converting them into limestone, but compaction is generally less effective, because cementation takes place soon after deposition. In any case, the cement is calcium carbonate derived by partial solution of some of the particles in the deposit.

LO5 Types of Sedimentary Rocks

Thus far, we have considered the origin of sediment, its transport, deposition, and lithification. We now turn to the types of sedimentary rocks and how they are classified. The two broad classes or types of sedimentary rocks are *detrital* and *chemical*, although the latter has a subcategory known as *biochemical* (Table 6.1).

cementation The process whereby minerals crystallize in the pore spaces of sediment and bind the loose particles together.

TABLE 6.1

CLASSIFICATION OF CHEMICAL AND BIOCHEMICAL SEDIMENTARY ROCKS

CHEMICAL SEDIMENTARY ROCKS		
TEXTURE	COMPOSITION	ROCK NAME
Varies	Calcite (CaCO₃)	Limestone
Varies	Dolomite [CaMg(CO₃)₂]	Dolostone
Crystalline	Gypsum (CaSO₄·2H₂O)	Rock gypsum
Crystalline	Halite (NaCl)	Rock salt
BIOCHEMICAL SEDIMENTARY ROCKS		
Clastic	Calcite (CaCO₃) shells	Limestone (various types, such as chalk and coquina)
Usually crystalline	Altered microscopic shells of SiO₂	Chert (various color varieties)
Amorphous	Carbon from altered land plants	Coal (lignite, bituminous, anthracite)

Carbonate rocks (Limestone, Dolostone)
Evaporites (Rock gypsum, Rock salt)

DETRITAL SEDIMENTARY ROCKS

Detrital sedimentary rocks are made up of solid particles such as sand and gravel, and all of these have a *clastic texture*, meaning they are composed of particles or fragments known as *clasts*. The several varieties of detrital rocks are classified by the size of their constituent particles, although composition is used to modify some rock names.

Both *conglomerate* and *sedimentary breccia* are composed of gravel-sized particles (Figure 6.13 and Figure 6.14a, b), but conglomerate has rounded gravel, whereas sedimentary breccia has angular gravel. Conglomerate is common, but sedimentary breccia is rare, because gravel becomes rounded very quickly during transport. Considerable energy is needed to transport gravel, so conglomerate is usually found in environments such as stream channels and beaches.

Sand is a size designation for particles between 1/16 mm and 2 mm, so any mineral or rock fragment can be in *sandstone*. Geologists recognize varieties of sandstone based on mineral content (Figure 6.13, 6.14a). *Quartz sandstone* is the most common and, as the name implies, is made up mostly of quartz

sand. Another variety of sandstone called *arkose* contains at least 25% feldspar minerals. Sandstone is found in many depositional environments, including stream channels, sand dunes, beaches, barrier islands, deltas, and the continental shelf.

Mudrock is a general term that includes all detrital sedimentary rocks composed of silt- and clay-sized particles (Figure 6.13). Varieties include *siltstone* (mostly silt-sized particles), *mudstone* (a mixture of silt and clay), and *claystone* (primarily clay-sized particles). Some mudstones and claystones are designated *shale* if they are fissile, meaning that they break along closely spaced parallel planes (Figure 6.14c). Even weak currents transport silt- and clay-sized particles, and deposition takes place only where currents and fluid turbulence are minimal, as in the quiet offshore waters of lakes or in lagoons.

CHEMICAL AND BIOCHEMICAL SEDIMENTARY ROCKS

Several compounds and ions taken into solution during chemical weathering are the raw materials for **chemical sedimentary rocks** (Table 6.1). Some of these rocks have a *crystalline texture*, meaning they are composed of a mosaic of interlocking mineral crystals as in rock salt. Others, though, have a clastic texture; some limestones, for instance, are composed of fragmented seashells. Organisms play an important role in the origin of chemical sedimentary rocks designated as **biochemical sedimentary rocks**.

detrital sedimentary rock Sedimentary rock made up of the solid particles (detritus) of preexisting rocks.

chemical sedimentary rock Sedimentary rock made up of minerals that were dissolved during chemical weathering and later precipitated from seawater, more rarely lake water, or extracted from solution by organisms.

Figure 6.14 Detrital and Sedimentary Rocks

a. A layer of conglomerate overlying sandstone. The largest clast measures about 30 cm across.

b. Sedimentary breccia in Death Valley, California. Notice the angular gravel-sized particles. The largest clast is about 12 cm across.

c. Exposure of shale in Tennessee.

Limestone, composed of calcite ($CaCO_3$), and dolostone, made up of dolomite [$CaMg(CO_3)_2$], the most abundant chemical sedimentary rocks, are known as **carbonate rocks** because each is made up of minerals that contain the carbonate radical (CO_3^{-2}). Recall that calcite rapidly dissolves in acidic water, but the chemical reaction leading to dissolution is reversible, so calcite can precipitate from solution in some circumstances. However, most limestone is biochemical, because organisms are so important in its origin—the rock in coral reefs and limestone composed of seashells, for instance (Figure 6.15).

A type of limestone composed almost entirely of fragmented seashells is known as *coquina* (Figure 6.15b). Chalk is a soft variety of limestone made up mostly of microscopic shells. One distinctive variety of limestone contains small spherical grains called *ooids* that have a small nucleus around which concentric layers of calcite precipitated (Figure 6.15c). Lithified deposits of ooids form *oolitic limestones*.

Dolostone is similar to limestone, but most or all of it formed secondarily by the alteration of limestone. The consensus among geologists is that dolostone originates when magnesium replaces some of the calcium in calcite, thereby converting calcite to dolomite.

Some of the dissolved substances derived by chemical weathering precipitate from evaporating water and thus form

biochemical sedimentary rock Any sedimentary rock produced by the chemical activities of organisms.

carbonate rock Any rock, such as limestone and dolostone, made up mostly of carbonate minerals.

Figure 6.15 Varieties of Limestone

a. Limestone with numerous fossil shells.

SUE MONROE

b. Coquina is limestone composed of broken shells.

SUE MONROE

STAN CELESTIAN/GLENDALE COMMUNITY COLLEGE

SUE MONROE

c. This limestone is made up partly of ooids (see inset), which are rather spherical grains of calcium carbonate.

chemical sedimentary rocks known as **evaporites** (Table 6.1). *Rock salt*, composed of halite (NaCl), and *rock gypsum* ($CaSO_4 \cdot 2H_2O$) are the most common (Figure 6.16), although several others are known and some are important resources. Compared with mudrocks, sandstone, and limestone, evaporites are not very common, but nevertheless are significant deposits in areas.

Chert is a hard rock composed of microscopic crystals of quartz (Table 6.1 and Figure 6.16c). Some of the color varieties of chert are *flint*, which is black because of inclusions of organic matter, and *jasper*, which is colored red or brown by iron oxides. Because chert is hard and lacks cleavage, it can be shaped to form sharp cutting edges, so it has been used to manufacture tools, spear points, and arrowheads. Chert is found as irregular masses or *nodules* in other rocks, especially limestone, and as distinct layers of *bedded chert* made up of tiny shells of silica-secreting organisms.

Coal consists of compressed, altered remains of land plants, so it is a biochemical sedimentary rock (Figure 6.16d). It forms in swamps and bogs where the water is oxygen deficient or where organic matter accumulates faster than it decomposes. In oxygen-deficient swamps and bogs, the bacteria that decompose vegetation can live without oxygen, but their wastes must be oxidized, and because little or no oxygen is present, wastes accumulate and kill the bacteria. Bacterial decay ceases, and the vegetation is not completely decomposed and forms organic muck. When buried

evaporite Any sedimentary rock, such as rock salt, formed by inorganic chemical precipitation of minerals from evaporating water.

Figure 6.16 Evaporites, Chert, and Coal

a. This cylindrical core of rock salt was taken from an oil well in Michigan.

SUE MONROE

c. Bedded chert exposed in Marin County, California. Most of the layers are about 5 cm thick.

JAMES S. MONROE

b. Rock gypsum. When deeply buried, gypsum ($CaSO_4 \cdot 2H_2O$) loses its water and is converted to anhydrite ($CaSO_4$).

SUE MONROE

d. Bituminous coal is the most common type of coal used for fuel.

SUE MONROE

and compressed, the muck becomes *peat*, which looks somewhat like coarse pipe tobacco. Where peat is abundant, as in Ireland and Scotland, it is used for fuel.

Peat represents the first step in forming coal. If peat is more deeply buried and compressed, and especially if it is heated, too, it is converted to dull black coal called *lignite*. During this change, the easily vaporized or volatile elements are driven off, enriching the residue in carbon; lignite has about 70% carbon, whereas only about 50% is present in peat. *Bituminous coal*, with about 80% carbon, is dense, black, and so thoroughly altered that plant remains are rarely seen. It burns more efficiently than lignite, but the highest-grade coal is *anthracite*, a metamorphic type of coal (see Chapter 7), with up to 98% carbon.

LO6 Sedimentary Facies

Long ago, geologists realized that when they traced a layer of sediment or sedimentary rock laterally, it generally changed in composition, texture, or both. They concluded that these changes resulted from the simultaneous operation of different processes in adjacent depositional environments. For example, sand may be deposited in a high-energy nearshore marine environment, whereas mud and carbonate sediments accumulate simultaneously in the laterally adjacent low-energy offshore environments (Figure 6.17). Deposition in each environment produces **sedimentary facies**, bodies of sediment that possess distinctive physical, chemical, and biological attributes.

Three Stages of Marine Transgression

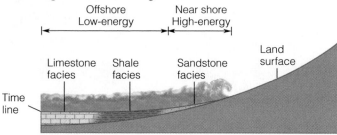

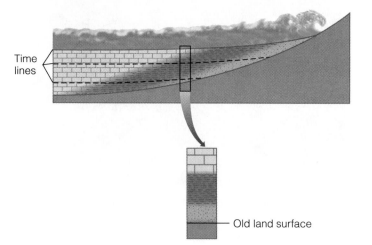

Three Stages of Marine Regression

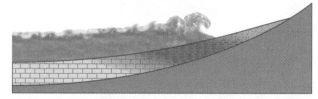

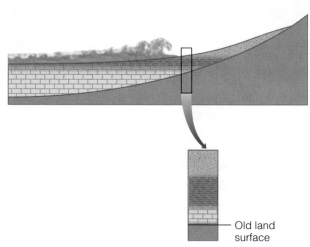

Figure 6.17 Marine Transgressions and Regressions
Notice that the sandstone, shale, and limestone facies are deposited simultaneously in adjacent environments. Also note the vertical succession of facies that result from marine transgressions and regressions.

Many sedimentary rocks in the interiors of continents show clear evidence of deposition in marine environments. The rock layers in Figure 6.17 (*left*), for example, consist of a sandstone facies that was deposited in a nearshore marine environment overlain by shale and limestone facies deposited in offshore environments. Geologists explain this vertical sequence of facies by deposition occurring during a time when sea level rose with respect to the continents. As sea level rises, the shoreline moves inland, giving rise to a **marine transgression** (Figure 6.17), and the depositional environments parallel to the shoreline migrate landward. As a result, offshore facies are superimposed over nearshore facies, thus accounting for the vertical succession of sedimentary facies. Even though the nearshore environment is long and narrow at any particular time, deposition takes place continuously as the environment migrates landward. The sand deposit may be tens to hundreds of meters thick but has horizontal dimensions of length and width measured in hundreds of kilometers.

The opposite of a marine transgression is a **marine regression** (Figure 6.17). If sea level falls with respect to a continent, the shoreline and environments that parallel the shoreline move seaward. The vertical sequence produced by a marine regression has facies of the nearshore environment superposed over facies of offshore environments.

LO7 Reading the Story Preserved in Sedimentary Rocks

No one was present when ancient sediments were deposited, so geologists must evaluate those aspects of sedimentary rocks that allow them to make inferences about the original depositional environment. Sedimentary textures such as sorting and rounding can give clues to depositional processes. Windblown dune sands tend to be well sorted and well rounded, but poor sorting is typical of glacial deposits. The geometry or three-dimensional shape is another important aspect of sedimentary rock bodies.

Marine transgressions and regressions yield sediment bodies with a blanket or sheetlike geometry, but sand deposits in stream channels are long and narrow

and are described as having a shoestring geometry. Sedimentary textures and geometry alone are usually insufficient to determine depositional environment, but when considered with other sedimentary rock properties, especially *sedimentary structures* and *fossils*, geologists can reliably determine the history of a deposit.

SEDIMENTARY STRUCTURES

Physical and biological processes operating in depositional environments are responsible for features known as **sedimentary structures**. One of the most common is distinct layers known as **strata** or **beds** (Figure 6.18a), with individual layers of less than a millimeter up to many meters thick. These strata or beds are separated from one another by surfaces above and below in which the rocks differ in composition, texture, color, or a combination of features.

Many sedimentary rocks have **cross-bedding**, in which layers are arranged at an angle to the surface on which they were deposited (Figure 6.18b and c). Cross-beds are found in many depositional environments, such as sand dunes in deserts and along shorelines, as well as in stream-channel deposits. Invariably, cross-beds result from transport and deposition by wind or water currents, and the cross-beds are inclined downward in the same direction that the current flowed. Thus, ancient deposits with cross-beds inclined down toward the south, for example, indicate that the currents responsible for them flowed from north to south.

marine transgression
The invasion of a coastal area or a continent by the sea, resulting from a rise in sea level or subsidence of the land.

marine regression The withdrawal of the sea from a continent or coastal area, resulting in the emergence of the land as sea level falls or the land rises with respect to sea level.

sedimentary structure Any feature in sedimentary rock that formed at or shortly after the time of deposition, such as cross-bedding, animal burrows, and mud cracks.

strata (singular, **stratum**) Refers to layering in sedimentary rocks.

bed An individual layer of rock, especially sediment or sedimentary rock.

cross-bedding A type of bedding in which layers are deposited at an angle to the surface on which they accumulate, as in sand dunes.

VIRTUAL FIELD TRIP

Sedimentary Rocks

Ready to Go!

Now that you've read some of the key facts about sedimentary rocks, we are going to go "into the field" and see some up close. During this field trip, we will be visiting two of Utah's five national parks. Arches National Park and Capitol Reef National Park are geologically rich, making them ideal settings for learning about sedimentary rocks, including examining the effects of weathering, different types of sedimentary structures, dating strata by sight, and much more.

This field trip is designed to help you think about these fascinating places from a geologic perspective. Several types of sedimentary rocks are present in Arches National Park and Capitol Reef National Park, but most of them are sandstones. Here we will learn about the rocks' composition, textures, color, structures, and how they were deposited—all topics we talk about in this chapter, under the section titled "Sediment and Sedimentary Rocks." In addition, the parks have a fantastic variety of landforms such as windows, columns, pinnacles, and balanced rocks (which we learned about in the earlier pages of this chapter), all of which resulted from differential weathering and erosion. For an example of this phenomenon, revisit Figure 6.5 in the text, which shows differential weathering and erosion in Joshua Tree National Park, California.

In addition to being great examples of sedimentary rocks, the rock exposures in these parks help you visualize some of the principles of deciphering geologic history, a topic that is discussed in greater depth in Chapter 17.

Onward!

FOLLOW-UP QUESTIONS

1. How do windows form by differential weathering and erosion? Are there examples in the park of windows that are in their earliest stages of formation? If so, which one(s)?

2. The rocks in Arches and Capitol Reef have been only minimally deformed. Suppose that some of them were crumpled and fractured so that now some of the layers were upside down. What kinds of sedimentary structures would help you figure out if some layers had in fact been flipped over?

3. How can you use the Principles of Superposition and Original Horizontality to help you decipher the geologic history of Arches and Capitol Reef National Parks?

What to See When You Go

There is a lot to see when you visit these parks in Virtual Field Trips, but here are some of the highlights...

The photo above shows great examples of pillars, columns, and arches, all of which formed by differential weathering and erosion of the Entrada Sandstone. Also notice in the above photo that in addition to the Entrada Sandstone, the Navajo Sandstone is also present. Consider how it differs from the Entrada Sandstone, and in what environment it may have been deposited.

Sedimentary rocks possess several types of sedimentary structures that formed when the sediment was deposited or shortly thereafter and provide evidence about the environment of deposition. These are wave-formed ripple marks much like those shown in Figure 6.20 in this chapter.

Perhaps the most important observations you can make on a trip to these parks are those that illustrate that Earth is a complex system of interacting rocks, water, air, and life. Also, it is important to realize that rocks of any kind record events that happened in the far distant past. We discuss the methods geologists use to interpret geologic history in Chapter 17, but even now you can determine which of the rock layers in the above photos is oldest and youngest. You simply have to note where the rock layers are in a sequence of layers, because the oldest is at the bottom while the youngest is at the top. Also, note in the above photo that one layer is gray rather than red. The gray layer is made up of bentonite (a clay mineral), which indicates ancient volcanoes erupted large amounts of volcanic ash.

For the rocks in the photo above, the caption in the Virtual Field Trip notes that the rocks from the bottom upward were deposited in tidal flat environments, slightly deeper water, and sand dunes. Geologists made these interpretations based on the features of the rocks, especially rock textures (sorting and rounding) and especially sedimentary structures. (See the section titled "Determining Environment of Deposition" in this chapter.) Note that these rocks are records of events that took place in this area long before any humans were present to make records of their own.

Figure 6.18 Bedding (Stratification) and Cross Bedding

a. These rocks seen from Dead Horse State Park, Utah, clearly show bedding or stratification. The deep canyon was eroded by the Colorado River.

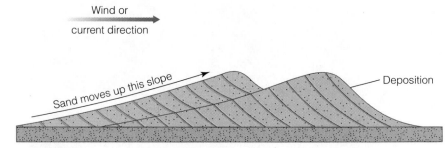

b. Origin of cross-bedding by deposition on the sloping surface of a desert dune. Cross-bedding is also common in dunelike structures in stream and river channels.

c. Cross-bedding in sandstone at Natural Bridges National Monument in Utah. The current moved from left to right.

Some individual sedimentary rock layers show an upward decrease in grain size, termed **graded bedding**, mostly formed by turbidity current deposition. A *turbidity current* is an underwater flow of sediment and water with a greater density than sediment-free water. Because of its greater density, a turbidity current flows downslope until it reaches the relatively flat seafloor, or lake floor, where it slows and begins depositing large particles followed by progressively smaller ones (Figure 6.19).

The surfaces that separate layers in sand deposits commonly have **ripple marks**, small ridges with intervening troughs, giving them a corrugated appearance. Some ripple marks are asymmetrical in cross section, with a gentle slope on one side and a steep slope on the other. Currents that flow in one direction, as in stream channels, generate these so-called *current ripple marks* (Figure 6.20a, b). Because the steep slope of these ripples is on the downstream side, they are good indications of ancient current directions. In contrast, *wave-formed ripple marks* tend to be symmetrical in cross section and, as their name implies, are generated by the to-and-fro motion of waves (Figure 6.20c, d).

graded bedding
Sedimentary layer in which a single bed shows a decrease in grain size from bottom to top.

ripple mark Wavelike (undulating) structure produced in granular sediment, especially sand, by unidirectional wind and water currents or by oscillating wave currents.

Figure 6.19 Turbidity Currents and the Origin of Graded Bedding

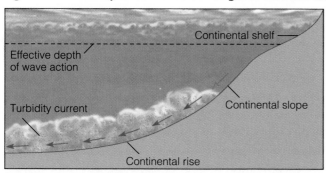

a. A turbidity current flows downslope along the seafloor (or a lake bottom) because it is denser than sediment-free water.

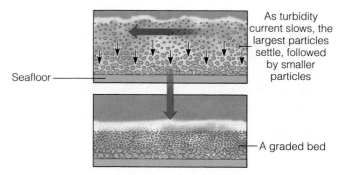

b. The flow slows and deposits progressively smaller particles, thus forming a graded bed.

Figure 6.20 Current and Wave-Formed Ripple Marks

a. Current ripple marks form in response to currents that flow in one direction, as in a stream. The enlargement shows the cross-beds in an individual ripple.

c. The to-and-fro motion of waves in shallow water yield wave-formed ripple marks.

b. Current ripple marks that formed in a stream channel. Flow was from right to left.

d. These symmetric wave-formed ripple marks are preserved in Triassic-age rocks in Capitol Reef National Park, Utah.

When clay-rich sediment dries, it shrinks and develops intersecting fractures called **mud cracks** (Figure 6.21). Mud cracks in ancient sedimentary rocks indicate that the sediment was deposited in an environment where periodic drying took place, such as on a river floodplain, near a lakeshore, or where muddy deposits are exposed along sea-coasts at low tide.

mud crack A crack in clay-rich sediment that forms in response to drying and shrinkage.

Figure 6.21 Mud Cracks
Mud cracks form in clay-rich sediments when they dry and contract.

COURTESY OF R. V. DIETRICH

a. Mud cracks in a present-day environment.

JAMES S. MONROE

b. Ancient mud cracks in Glacier National Park in Montana. Note that the cracks have been filled in with sediment.

Fossils: Remains and Traces of Ancient Life

Fossils, the remains or traces of once-living organisms, are interesting as evidence of prehistoric life (Figure 6.22) and are also important for determining depositional environments. The actual remains of organisms are known as *body fossils*, whereas any indication of organic activity, such as tracks and trails, are *trace fossils*. Most people are familiar with fossils of dinosaurs and some other land-dwelling animals but are unaware that fossils of invertebrates—animals lacking a segmented vertebral column, such as corals, clams, oysters, and a variety of microorganisms—are much more useful because they are so common.

It is true that the remains of land-dwelling creatures and plants can be washed into marine environments, but most are preserved in rocks deposited on land or perhaps transitional environments such as deltas. In contrast, fossils of corals tell us that the rocks in which they are preserved were deposited in the ocean.

Microfossils are particularly useful for environmental studies, because hundreds or even thousands can be recovered from small rock samples. In oil-drilling operations, small rock chips known as *well cuttings* are brought to the surface. These samples may contain numerous microfossils but rarely

fossil The remains or traces of once-living organisms.

have entire fossils of larger organisms. These fossils are routinely used to determine depositional environments and to match up rocks of the same relative age (see Chapter 17).

Determining the Environment of Deposition

Geologists rely on textures, sedimentary structures, and fossils to interpret how a particular sedimentary rock body was deposited. Furthermore, they compare the features seen in ancient rocks with those in deposits forming today. But are we justified in using present-day processes and environments to make inferences about what happened when no human observers were present? Perhaps some examples will help answer this question.

The Navajo Sandstone of the southwestern United States is an ancient dune deposit that formed when the prevailing winds blew from the northeast. What evidence justifies this conclusion? This 300-m-thick sandstone is made up of well-sorted, well-rounded sand grains measuring 0.2 mm to 0.5 mm in diameter. Furthermore, it has cross-beds up to 30 m high (Figure 6.23a) and current ripple marks, both typical of dunes. Some of the sand layers have preserved dinosaur tracks and tracks of other land-dwelling animals, ruling out the possibility of a marine origin. In

Figure 6.22 **Fossils**

a. These body fossils of trilobites are on display in the Mesalands Community College's Dinosaur Museum in Tucumcari, New Mexico. Although distantly related to today's insects, spiders, crabs, and lobsters, trilobites went extinct at the end of the Paleozoic Era, 251 million years ago.

JAMES S. MONROE

b. These circular depressions in sandstone at Copper Ridge in Utah are trace fossils. They are tracks, which measure about 35 cm across, and were made by a large dinosaur.

JAMES S. MONROE

short, the Navajo Sandstone possesses features that point to a dune depositional environment. Finally, the cross-beds are inclined downward toward the southwest, indicating that the prevailing winds were from the northeast.

In the Grand Canyon of Arizona, several formations are well exposed; a *formation* is a widespread unit of rock, especially sedimentary rock, that is recognizably different from the rocks above and below it. A vertical sequence consisting of the Tapeats Sandstone, Bright Angel Shale, and Muav Limestone is present in the lower part of the canyon (Figure 6.23b), all of which contain features, including fossils, clearly indicating that they were deposited in transitional and marine environments. As a matter of fact, all three were forming simultaneously in different adjacent environments, and during a marine transgression they were deposited in the vertical sequence now seen. They conform closely to the sequence shown in Figure 6.17.

LO8 Important Resources in Sedimentary Rocks

Sand and gravel are essential to the construction industry; pure clay deposits are used for ceramics, and limestone is used in the manufacture of cement and in blast furnaces, where iron ore is refined to make steel. Evaporites are the source of table salt, as well as chemical compounds, and rock gypsum is used to manufacture wallboard. Phosphate-bearing sedimentary rock is used in fertilizers and animal-feed supplements.

Placer deposits are surface accumulations resulting from the separation and concentration of materials of greater density from those of lesser density in streams or on beaches. Much of the gold recovered during the initial stages of the California gold rush (1849–1853) was mined from placer deposits.

Most coal mined in the United States has been bituminous coal from the Appalachian region that

Figure 6.23 Ancient Sedimentary Rocks and Their Interpretation

a. The Jurassic-aged Navajo Sandstone in Zion National Park in Utah is a wind-blown dune deposit. Vertical fractures intersect cross-beds, hence the name Checkerboard Mesa for this rock exposure.

b. View of three formations in the Grand Canyon of Arizona. These rocks were deposited during a marine transgression. Compare with the vertical sequence of rocks in Figure 6.17.

formed in coastal swamps during the Pennsylvanian Period. Huge lignite and subbituminous coal deposits in the western United States are becoming increasingly important. During 2007, about a billion tons of coal were mined in this country, more than 60% of it from mines in Wyoming, West Virginia, and Kentucky.

Coke, a hard, gray substance consisting of the fused ash of bituminous coal, is used in blast furnaces where steel is produced. Synthetic oil and gas and several other products are also made from bituminous coal and lignite.

PETROLEUM AND NATURAL GAS

Petroleum and natural gas are *hydrocarbons*, meaning that they are composed of hydrogen and carbon. The remains of microscopic organisms settle to the seafloor, or lake floor in some cases, where little oxygen is present to decompose them. If buried beneath layers of sediment, they are heated and transformed into petroleum and natural gas. The rock in which hydrocarbons form is *source rock*, but for them to accumulate in economic quantities, they must migrate from the source rock into some kind of *reservoir rock*. Finally, the reservoir rock must have an overlying, nearly impervious *cap rock*; otherwise, the hydrocarbons would eventually

reach the surface and escape (Figure 6.24a, b). Effective reservoir rocks must have appreciable pore space and good *permeability*, the capacity to transmit fluids; otherwise, hydrocarbons cannot be extracted from them in reasonable quantities.

Many hydrocarbon reservoirs consist of nearshore marine sandstones with nearby fine-grained, organic-rich source rocks. These are called *stratigraphic traps*, because they owe their existence to variations in the strata (Figure 6.24a). Indeed, some of the oil in the Persian Gulf region and Michigan is trapped in ancient reefs that are also good stratigraphic traps. *Structural traps* result when rocks are deformed by folding, fracturing, or both. In sedimentary rocks that have been deformed into a series of folds, hydrocarbons migrate to the high parts of these structures (Figure 6.24b).

URANIUM

Most of the uranium used in nuclear reactors in North America comes from the complex potassium-, uranium-, vanadium-bearing mineral *carnotite*, which is found in some sedimentary rocks. Some uranium is also derived from *uraninite* (UO_2), a uranium oxide in granitic rocks and hydrothermal veins. Uraninite is

Figure 6.24 Oil and Natural Gas

The arrows in (a) and (b) indicate the direction of migration of hydrocarbons.

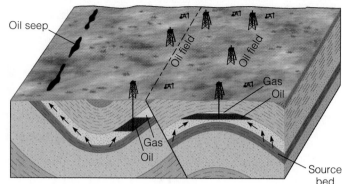

a. Two examples of stratigraphic traps: one in sand within shale and the other in a buried reef.

b. Two examples of structural traps: one formed by folding and the other by faulting.

easily oxidized and dissolved in groundwater, transported elsewhere, and chemically reduced and precipitated in the presence of organic matter.

The richest uranium ores in the United States are widespread in the Colorado Plateau area of Colorado and adjoining parts of Wyoming, Utah, Arizona, and New Mexico. These ores, consisting of fairly pure masses and encrustations of carnotite, are associated with plant remains in sandstones that formed in ancient stream channels. Although most of these ores are associated with fragmentary plant remains, some petrified trees also contain large quantities of uranium.

Large reserves of low-grade uranium ore are also found in the Chattanooga Shale. The uranium is finely disseminated in this black, organic-rich mudrock that underlies large parts of several states, including Illinois, Indiana, Ohio, Kentucky, and Tennessee. Canada is the world's largest producer and exporter of uranium.

BANDED IRON FORMATION

The chemical sedimentary rock known as *banded iron formation* consists of alternating thin layers of chert and iron minerals, mostly the iron oxides hematite and magnetite. Banded iron formations are present on all continents and account for most of the iron ore mined in the world today. Vast banded iron formations are present in the Lake Superior region of the United States and Canada and in the Labrador Trough of eastern Canada.

CHAPTER 7
METAMORPHISM AND METAMORPHIC ROCKS

This slate exposed in the Wasatch Mountains just east of Salt Lake City, Utah, demonstrates "slaty cleavage," wherein the rock tends to split into flat, plate-like pieces along cleavage planes.

Introduction

"A useful analogy for metamorphism is baking a cake: the outcome of each depends on the ingredients, their proportions, how they are mixed together, the temperature, and time."

Its homogeneity, softness, and various textures have made marble, a metamorphic rock formed from limestone or dolostone, a favorite rock of sculptors throughout history. As the value of authentic marble sculptures has increased over the years, the number of forgeries has also increased. With the price of some marble sculptures in the millions of dollars, private collectors and museums need some means of ensuring the authenticity of the work that they are buying. Aside from the monetary considerations, it is important that forgeries not become part of the historical and artistic legacy of human endeavor.

Experts have traditionally relied on artistic style and weathering characteristics to determine whether a marble sculpture is authentic or a forgery. Because marble is not very resistant to weathering, however, forgers have been able to produce the weathered appearance of an authentic work.

Using newly developed techniques, geologists can now distinguish a naturally weathered marble surface from one that has been artificially altered. Yet, there are examples in which expert opinion is still divided on whether a sculpture is authentic. One of the best examples is the Greek *kouros* (a sculptured figure of a Greek youth) that the J. Paul Getty Museum in Malibu, California, purchased for a reputed price of $7 million in 1984. Because some of its stylistic features caused some experts to question its authenticity, the museum had a variety of geochemical and mineralogical tests performed in an effort to determine the authenticity of the kouros.

Although numerous scientific tests have not unequivocally proved authenticity, they have shown that the weathered surface layer of the kouros bears more similarities to naturally occurring weathered surfaces of dolomitic marble than to known artificially produced surfaces. Furthermore, no evidence indicates that the surface alteration of the kouros is of modern origin.

Unfortunately, despite intensive study by scientists, archaeologists, and art historians, opinion is still divided as to the authenticity of the Getty kouros. Most scientists accept that the kouros was carved sometime around 530 B.C. Pointing to inconsistencies in its style of sculpture for that period, other art historians think that it is a modern forgery.

Regardless of whether the Getty kouros is proven to be authentic or a forgery, geologic testing to authenticate marble sculptures is now an important part of many museums' curatorial functions. To help geologists in the authentication of marble sculptures, a large body of data about the characteristics and origin of marble is being amassed as more sculptures and their quarries are analyzed.

Having examined igneous and sedimentary rocks, we now turn our attention to the third major rock group, the **metamorphic rocks** (from the Greek *meta*, "change," and *morphe*, "shape"), which result from the transformation of other rocks by processes that typically occur beneath Earth's surface (see Figure 1.14). During

metamorphic rock Any rock that has been changed from its original condition by heat, pressure, and the chemical activity of fluids, as in marble and slate.

LEARNING OUTCOMES

After reading this unit, you should be able to do the following:

LO1 Identity the agents of metamorphism

LO2 Identify the three types of metamorphism

LO3 Explain how metamorphic rocks are classified

LO4 Recognize the difference between metamorphic zones and facies

LO5 Understand how plate tectonics affects metamorphism

LO6 Understand the relationship between metamorphism and natural resources

Figure 7.1 Gneiss

This gneiss, found in Canada, is estimated to be about 4.0 billion years old. Gneiss is a foliated metamorphic rock.

metamorphism, rocks are subjected to sufficient heat, pressure, and fluid activity to change their mineral composition, texture, or both, thus forming new rocks that usually do not look anything like the original rock before it was metamorphosed. These transformations take place below the melting temperature of the rock; otherwise, an igneous rock would result.

A useful analogy for metamorphism is baking a cake. Just like a metamorphic rock, the resulting cake depends on the ingredients, their proportions, how they are mixed together, how much water or milk is added, and the temperature and length of time used for baking the cake.

Except for marble and slate, most people are not familiar with metamorphic rocks. We are frequently asked by students why it is important to study metamorphic rocks and processes. Our answer is always, "Just look around you." A large portion of Earth's continental crust is composed of metamorphic and igneous rocks. Together, they form the crystalline basement rocks underlying the sedimentary rocks of a continent's surface. Some of the oldest known rocks, dated at about 4.0 billion years (Figure 7.1), are metamorphic, which means that they formed from even older rocks! Many metamorphic minerals and rocks, such as garnets, talc, asbestos, marble, and slate, are also economically important and useful.

metamorphism The phenomenon of changing rocks subjected to heat, pressure, and fluids so that they are in equilibrium with a new set of environmental conditions.

heat An agent of metamorphism.

LO1 The Agents of Metamorphism

The three principal agents of metamorphism are *heat*, *pressure*, and *fluid activity*. Time is also important to the metamorphic process, because chemical reactions proceed at different rates and thus require different amounts of time to complete. Reactions involving silicate compounds are particularly slow, and because most metamorphic rocks are composed of silicate minerals, it is thought that metamorphism is a very slow geologic process.

During metamorphism, the original rock, which was in equilibrium with its environment, meaning that it was chemically and physically stable under those conditions, undergoes changes to achieve equilibrium with its new environment. These changes may result in the formation of new minerals, a change in the texture of the rock, or both. In some instances, the change is minor, and features of the original rock can still be recognized. In other cases, the rock changes so much that the identity of the original rock can be determined only with great difficulty, if at all.

HEAT

Heat is an important agent of metamorphism, because it increases the rate of chemical reactions that may produce minerals different from those in the original rock. Heat may come from lava, magma, or as a result of deep burial in the crust due to subduction along a convergent plate boundary.

When rocks are intruded by bodies of magma, they are subjected to intense heat that affects the surrounding rock; the most intense heating usually occurs adjacent to the magma body and gradually decreases with distance from the intrusion. The zone of metamorphosed rocks that forms in the country rock adjacent to an intrusive igneous body is usually distinct and easy to recognize.

Recall that temperature increases with depth. Rocks that form at the surface may be transported to great depths by subduction along a convergent plate boundary and thus subjected to increasing temperature and pressure. During subduction, some minerals may be transformed into other minerals that are more stable under the higher temperature and pressure conditions.

Figure 7.2 Lithostatic Pressure

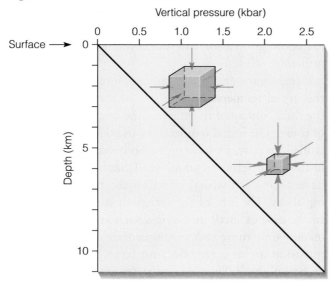

1 kilobar (kbar) = 1,000 bars
Atmospheric pressure at sea level = 1 bar

a. Lithostatic pressure is applied equally in all directions in Earth's crust due to the weight of overlying rocks. Thus, pressure increases with depth, as indicated by the sloping black line. From C. Gillen, *Metamorphic Geology*, Figure 4.4, p. 73. Copyright 1982 Kluwer Academic Publishers. Reprinted by permission of the author.

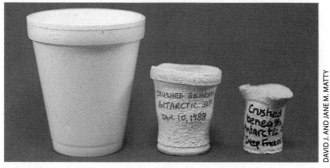

b. A similar situation occurs when 200-ml cups composed of Styrofoam™ are lowered to ocean depths of approximately 750 m and 1,500 m. Increased water pressure is exerted equally in all directions on the cups, and they consequently decrease in volume while maintaining their general shape.

PRESSURE

During burial, rocks are subjected to increasingly greater pressure, just as you feel greater pressure the deeper you dive into a body of water. Whereas the pressure you feel is known as *hydrostatic pressure*, because it comes from the water surrounding you, rocks undergo **lithostatic pressure**, which means that the stress (force per unit area) on a rock in Earth's crust is the same in all directions (Figure 7.2a). A similar situation occurs when an object is immersed in water. For

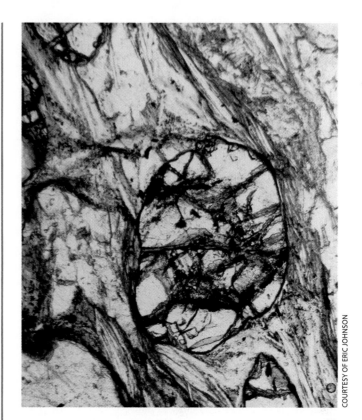

Figure 7.3 Differential Pressure

Differential pressure results from stress that is unequally applied to an object. Rotated garnets are a good example of the effects of differential pressure applied to a rock during metamorphism. In this example from a schist in northeast Sardinia, stress was applied in opposite directions on the left and right side of the garnet (*center*), causing it to rotate.

example, the deeper a cup composed of Styrofoam™ is submerged in the ocean, the smaller it gets because pressure increases with depth and is exerted on the cup equally in all directions, thereby compressing the cup (Figure 7.2b).

Along with lithostatic pressure resulting from burial, rocks may also experience **differential pressure** (Figure 7.3). In this case, the stresses are not equal in all directions, but are stronger from some directions than from others. Differential pressures typically occur when two plates collide, thus producing distinctive metamorphic textures and features.

FLUID ACTIVITY

In almost every region of metamorphism, water and carbon dioxide (CO_2) are present in varying amounts along mineral

lithostatic pressure
Pressure exerted on rocks by the weight of overlying rocks.

differential pressure
Pressure that is not applied equally to all sides of a rock body.

grain boundaries or in the pore spaces of rocks. These fluids, which may contain ions in solution, enhance metamorphism by increasing the rate of chemical reactions. Under dry conditions, most minerals react very slowly, but when even small amounts of fluid are introduced, reaction rates increase, mainly because ions can move readily through the fluid and thus enhance chemical reactions and the formation of new minerals.

The following reaction provides a good example of how new minerals can be formed by **fluid activity**. Seawater moving through hot basaltic rock in the oceanic crust transforms olivine into the metamorphic mineral serpentine:

$$2Mg_2SiO_4 + 2H_2O \longrightarrow Mg_3Si_2O_5(OH)_4 + MgO$$

olivine   water   serpentine   carried away in solution

The chemically active fluids important in the metamorphic process come primarily from three sources: (1) water trapped in the pore spaces of sedimentary rocks as they form, (2) the volatile fluid within magma, and (3) the dehydration of water-bearing minerals such as gypsum ($CaSO_4 \cdot 2H_2O$) and some clays.

LO2 The Three Types of Metamorphism

Geologists recognize three major types of metamorphism: (1) *contact (thermal) metamorphism*, in which magmatic heat and fluids act to produce change; (2) *dynamic metamorphism*, which is principally the result of high differential pressures associated with intense deformation; and (3) *regional metamorphism*, which occurs within a large area and is associated with major mountain-building episodes. Even though we will discuss each type of metamorphism separately, the boundary between them is not always distinct and depends largely on which of the three metamorphic agents was dominant.

fluid activity An agent of metamorphism in which water and carbon dioxide promote metamorphism by increasing the rate of chemical reactions.

contact (thermal) metamorphism Metamorphism of country rock adjacent to a pluton.

aureole A zone surrounding a pluton in which contact metamorphism took place.

CONTACT METAMORPHISM

Contact (thermal) metamorphism takes place when a body of magma alters the surrounding country rock. At shallow depths, intruding magma raises the temperature of the surrounding rock, causing thermal alteration. Furthermore, the release of hot fluids into the country rock by the cooling intrusion can aid in the formation of new minerals.

Important factors in contact metamorphism are the initial temperature, the size of the intrusion, and the fluid content of the magma, the country rock, or all of these. The initial temperature of an intrusion is controlled, in part, by its composition; mafic magmas are hotter than felsic magmas (see Chapter 4) and hence have a greater thermal effect on the rocks surrounding them. The size of the intrusion is also important. In the case of small intrusions, such as dikes and sills, usually only those rocks in immediate contact with the intrusion are affected. Because large intrusions, such as batholiths, take a long time to cool, the increased temperature in the surrounding rock may last long enough for a larger area to be affected.

The area of metamorphism surrounding an intrusion is an **aureole**, and the boundary between an intrusion and its aureole may be either sharp or transitional (Figure 7.4). Metamorphic aureoles vary in width depending on the size, temperature, and composition of the intruding magma, as well as the mineralogy of the surrounding country rock. Aureoles range from a few centimeters wide bordering small dikes and sills, to several hundred meters or even several kilometers wide around large plutons.

The degree of metamorphic change within an aureole generally decreases with distance from the intrusion, reflecting the decrease in temperature from the original heat source. The region or zone closest to the intrusion, and hence subject to the highest temperatures, commonly contains high-temperature metamorphic minerals (that is, minerals in equilibrium with the higher-temperature environment) such as sillimanite. The outer zones, that is, those farthest from the intrusion, are typically characterized by lower-temperature metamorphic minerals such as chlorite, talc, and epidote.

Contact metamorphism can also result from lava flows, either along mid-ocean ridges or from lava flowing over land and thermally altering the underlying rocks (Figure 7.5). Whereas recognizing a recent lava flow and the resulting contact metamorphism of the rocks below is easy, it is less obvious whether an igneous body is intrusive or extrusive in a rock outcrop where sedimentary rocks occur above and below the igneous body. Recognizing which sedimentary rock units have been metamorphosed enables geologists to determine whether the igneous body is intrusive (such as a sill or dike) or extrusive (lava flow).

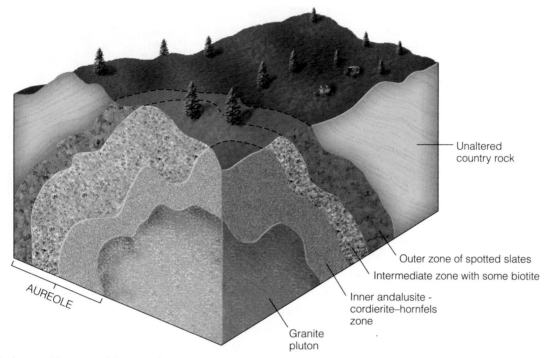

Figure 7.4 Metamorphic Aureole
A metamorphic aureole, the area surrounding an intrusion, consists of zones that reflect the degree of metamorphism. The metamorphic aureole associated with this idealized granite pluton contains three zones of mineral assemblages reflecting the decrease in temperature with distance from the intrusion. An inner andalusite–cordierite hornfels zone forms adjacent to the pluton and is reflective of the high temperatures near the intrusion. This is followed by an intermediate zone of extensive recrystallization in which some biotite develops, and farthest from the intrusion is an outer zone characterized by spotted slates.

Figure 7.5 Contact Metamorphism from a Lava Flow
A highly weathered basaltic lava flow near Susanville, California, has altered an underlying rhyolitic volcanic ash by contact metamorphism. The red zone below the lava flow has been baked by the heat of the lava when it flowed over the ash layer. The lava flow displays spheroidal weathering, a type of weathering common in fractured rocks (see Chapter 6).

Fluids also play an important role in contact metamorphism. Magma is usually wet and contains hot, chemically active fluids that may emanate into the surrounding rock. These fluids can react with the rock and aid in the formation of new minerals. In addition, the country rock may contain pore fluids that, when heated by magma, also increase reaction rates.

Because heat and fluids are the primary agents of contact metamorphism, two types of contact metamorphic rocks are generally recognized: those resulting from baking of country rock and those altered by hot solutions. Many of the rocks resulting from contact metamorphism have the texture of porcelain; that is, they are hard and fine grained. This is particularly true for rocks with a high clay content, such as shale. Such texture results because the clay minerals in the rock are baked, just as a clay pot is baked when fired in a kiln.

During the final stages of cooling, when an intruding magma begins to crystallize, large amounts of hot, watery solutions are often released. These solutions may react with the country rock and produce new metamorphic minerals. This process, which usually occurs near Earth's surface, is called *hydrothermal alteration* (from the Greek *hydro*, "water," and *therme*, "heat") and may result in valuable mineral deposits.

DYNAMIC METAMORPHISM

Most **dynamic metamorphism** is associated with fault (fractures along which movement has occurred) zones where rocks are subjected to high differential pressure. The metamorphic rocks that result from pure dynamic metamorphism are called *mylonites* and are typically restricted to narrow zones adjacent to faults. Mylonites are hard, dense, fine-grained rocks, many of which are characterized by thin laminations (Figure 7.6). Tectonic settings where mylonites occur include the Moine Thrust Zone in northwest Scotland, the Adirondack highlands in New York, and portions of the San Andreas fault in California (see Chapter 2).

dynamic metamorphism
Metamorphism in fault zones where rocks are subjected to high differential pressure.

regional metamorphism
Metamorphism that occurs over a large area, resulting from high temperatures, tremendous pressure, and the chemical activity of fluids within the crust.

REGIONAL METAMORPHISM

Most metamorphic rocks result from **regional meta-**

COURTESY OF ERIC JOHNSON

Figure 7.6 Mylonite
An outcrop of mylonite from the Adirondack Highlands, New York. Mylonites result from dynamic metamorphism, where rocks are subjected to high levels of differential pressure. Note the thin laminations (closely spaced layers), which are characteristic of many mylonites.

morphism, which occurs over a large area and is usually caused by tremendous temperatures, pressures, and deformation all occurring together within the deeper portions of the crust. Regional metamorphism is most obvious along convergent plate boundaries, where rocks are intensely deformed and recrystallized during convergence and subduction. Within these metamorphic rocks there is usually a gradation of metamorphic intensity from areas that were subjected to the most intense pressures and/or highest temperatures to areas of lower pressures and temperatures. Such a gradation in metamorphism can be recognized by the metamorphic minerals that are present.

Regional metamorphism is not, however, confined only to convergent margins. It can also occur in areas where plates diverge, although usually at much shallower depths because of the high geothermal gradient associated with these areas.

INDEX MINERALS AND METAMORPHIC GRADE

From field studies and laboratory experiments, certain minerals are known to form only within specific temperature and pressure ranges. Such minerals are known

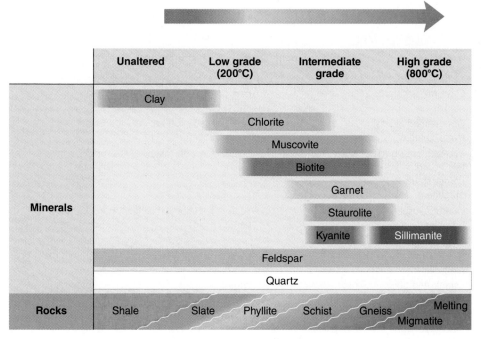

Figure 7.7 Metamorphic Grade Change in mineral assemblage and rock type with increasing metamorphism of shale. When a clay-rich rock such as shale is subjected to increasing metamorphism, new minerals form, as shown by the colored bars. The progressive appearance of certain minerals, known as index minerals, allows geologists to recognize low-, intermediate-, and high-grade metamorphism.

as **index minerals**, because their presence allows geologists to recognize low-, intermediate-, and high-grade metamorphism (Figure 7.7).

Metamorphic grade is a term that generally characterizes the degree to which a rock has undergone metamorphic change (Figure 7.7). Although the boundaries between the different metamorphic grades are not sharp, the distinction is nonetheless useful for communicating in a general way the degree to which rocks have been metamorphosed. The presence of index minerals thus helps determine metamorphic grade. For example, when a clay-rich rock such as shale is metamorphosed, it undergoes regional metamorphism, the mineral chlorite first begins to crystallize under relatively low temperatures of about 200°C, and so its presence thus indicates low-grade metamorphism. If temperatures and pressures continue to increase, new minerals form to replace chlorite, because they are more stable under those new conditions. Thus, there is a progression in the appearance of new minerals from chlorite, whose presence indicates low-grade metamorphism, to biotite and garnet, which are good index minerals for intermediate-grade metamorphism, to sillimanite, whose presence indicates high-grade metamorphism and temperatures exceeding 500°C (Figure 7.7).

Different rock compositions develop different sets of index minerals. For example, clay-rich rocks such as shale will develop the index minerals shown in Figure 7.7, whereas a sandy dolomite will produce a different set

of index minerals as metamorphism progresses, because it has a different mineral composition than shale. Thus, a particular set of index minerals will form based on the original composition of the parent rock undergoing metamorphism.

index mineral A mineral that forms within specific temperature and pressure ranges during metamorphism.

foliated texture A texture in metamorphic rocks in which platy and elongate minerals are aligned in a parallel fashion.

LO3 How Are Metamorphic Rocks Classified?

For purposes of classification, metamorphic rocks are commonly divided into two groups: those exhibiting a *foliated texture* (from the Latin *folium*, "leaf") and those with a *nonfoliated texture* (Table 7.1).

FOLIATED METAMORPHIC ROCKS

Rocks subjected to heat and differential pressure during metamorphism typically have minerals arranged in a parallel fashion, giving them a **foliated texture** (Figure 7.8). The size and shape of the mineral grains determine whether the foliation is fine or coarse. Low-grade metamorphic rocks, such as slate, have a finely

TABLE 7.1

CLASSIFICATION OF COMMON METAMORPHIC ROCKS

TEXTURE	METAMORPHIC ROCK	TYPICAL MINERAL	METAMORPHIC GRADE	CHARACTERISTICS OF ROCKS	PARENT ROCK
Foliated	Slate	Clays, micas, chlorite	Low	Fine-grained, splits easily into flat pieces	Mudrocks, volcanic ash
	Phyllite	Fine-grained quartz, micas, chlorite	Low to medium	Fine-grained, glossy or lustrous sheen	Mudrocks
	Schist	Micas, chlorite, quartz, talc, hornblende, garnet, staurolite, graphite	Low to high	Distinct foliation, minerals visible	Mudrocks, carbonates, mafic igneous rocks
	Gneiss	Quartz, feldspars, hornblende, micas	High	Segregated light and dark bands visible	Mudrocks, sandstones, felsic igneous rocks
	Amphibolite	Hornblende, plagioclase	Medium to high	Dark, weakly foliated	Mafic igneous rocks
	Migmatite	Quartz, feldspars, hornblende, micas	High	Streaks or lenses of granite intermixed with gneiss	Felsic igneous rocks mixed with sedimentary rocks
Nonfoliated	Marble	Calcite, dolomite	Low to high	Interlocking grains of calcite or dolomite, reacts with HCl	Limestone or dolostone
	Quartzite	Quartz	Medium to high	Interlocking quartz grains, hard, dense	Quartz sandstone
	Greenstone	Chlorite, epidote, hornblende	Low to high	Fine-grained, green	Mafic igneous rocks
	Hornfels	Micas, garnets, andalusite, cordierite, quartz	Low to medium	Fine-grained, equidimensional grains, hard, dense	Mudrocks
	Anthracite	Carbon	High	Black, lustrous, subconcoidal fracture	Coal

Figure 7.8 Foliated Texture

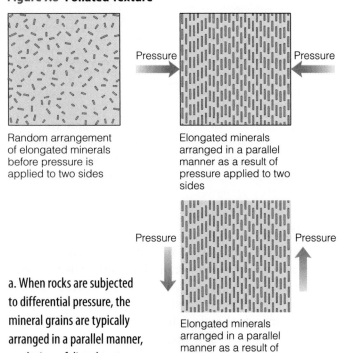

Random arrangement of elongated minerals before pressure is applied to two sides

Pressure → ← Pressure

Elongated minerals arranged in a parallel manner as a result of pressure applied to two sides

Pressure ↓ ↑ Pressure

Elongated minerals arranged in a parallel manner as a result of shear

a. When rocks are subjected to differential pressure, the mineral grains are typically arranged in a parallel manner, producing a foliated texture.

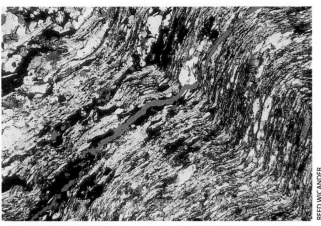

REED WICANDER

b. Photomicrograph of a metamorphic rock with a foliated texture showing the parallel arrangement of mineral grains.

a. Hand specimen of red slate.

COPYRIGHT AND PHOTOGRAPH
BY DR. PARVINDER S. SETHI

JAMES S. MONROE

Figure 7.9 Slate

b. Slate roof of Chalet Enzian, Switzerland.

foliated texture in which the mineral grains are so small that they cannot be recognized without magnification. High-grade foliated rocks, such as gneiss, are coarse-grained, such that the individual grains can easily be seen without magnification. Foliated metamorphic rocks can be arranged in order of increasingly coarse grain size and perfection of foliation.

Slate is a very fine-grained metamorphic rock that commonly exhibits *slaty cleavage* (Figure 7.9). It results from regional metamorphism of shale or, more rarely, volcanic ash. Because it can easily be split along cleavage planes into flat pieces, slate is an excellent rock for roofing and floor tiles, billiard and pool table tops, and blackboards. The different colors of slate are caused by minute amounts of graphite (black), iron oxide (red and purple), and chlorite (green).

Phyllite is similar in composition to slate but is coarser grained. The minerals, however, are still too small to be identified without magnification. Phyllite can be distinguished from slate by its glossy or

lustrous sheen (Figure 7.10) and represents an intermediate grain size between slate and schist.

Schist is most commonly produced by regional metamorphism. The type of schist formed depends on the intensity of metamorphism and the character of the original rock (Figure 7.11). Metamorphism of many rock types can yield schist, but most schist appears to have formed from clay-rich sedimentary rocks (Table 7.1).

All schists contain more than 50% platy and elongated minerals, all of which are large enough to be clearly visible. Their mineral composition imparts a *schistosity* or *schistose foliation* to the rock that usually produces a wavy type of parting when split. Schistosity is common in low- to high-grade metamorphic environments, and each type of schist is known by its most conspicuous mineral or minerals, such as mica schist, chlorite schist, or garnet-mica schist (Figure 7.11).

Gneiss is a high-grade metamorphic rock that is streaked or has segregated bands of light and dark minerals. Gneisses are composed of granular minerals such as quartz, feldspar, or both, with lesser percentages of platy or elongated minerals such as micas or amphiboles (Figure 7.12). Quartz and feldspar are the principal minerals composing the light-colored bands of minerals, whereas biotite and hornblende comprise the dark mineral bands. Gneiss typically breaks in an irregular manner, much like coarsely crystalline nonfoliated rocks.

Most gneiss probably results from recrystallization of clay-rich sedimentary rocks during regional metamorphism (Table 7.1). Gneiss also can form from

REED WICANDER

Figure 7.10 Phyllite
Hand specimen of phyllite. Note the lustrous sheen, as well as the bedding (*upper left to lower right*) at an angle to the cleavage of the specimen.

Figure 7.11 **Schist**

a. Almandine garnet crystals in a mica schist.

b. Hornblende–mica schist.

igneous rocks such as granite or older metamorphic rocks.

Another fairly common foliated metamorphic rock is *amphibolite*. A dark rock, it is composed mainly of hornblende and plagioclase. The alignment of the hornblende crystals produces a slightly foliated texture. Many amphibolites result from intermediate- to high-grade metamorphism of basalt and ferromagnesian-rich mafic rocks.

In some areas of regional metamorphism, exposures of "mixed rocks" called *migmatites*, having both igneous and high-grade metamorphic characteristics, are present (Figure 7.13). Migmatites are thought to result from the extremely high temperatures produced

nonfoliated texture
A metamorphic texture in which there is no discernable preferred orientation of minerals.

Figure 7.12 **Gneiss**
Gneiss is characterized by segregated bands of light and dark minerals. This folded gneiss is exposed at Wawa, Ontario, Canada.

during metamorphism. However, part of the problem in determining the origin of migmatites is explaining how the granitic component formed. According to one model, the granitic magma formed in place by the partial melting of rock during intense metamorphism. Such an origin is possible provided that the host rocks contained quartz and feldspars and that water was present.

Others argue that the characteristic layering or wavy appearance of migmatites arises by the redistribution of minerals during recrystallization in the solid state—that is, through purely metamorphic processes.

NONFOLIATED METAMORPHIC ROCKS

In some metamorphic rocks, the mineral grains do not show a discernable preferred orientation. Instead, these rocks consist of a mosaic of roughly equidimensional minerals and are characterized as having a **nonfoliated texture** (Figure 7.14). Most nonfoliated metamorphic rocks result from contact or regional metamorphism of rocks with no platy or elongate minerals. Frequently, the only indication that a granular rock has been metamorphosed is the large grain size resulting from recrystallization.

Nonfoliated metamorphic rocks are generally of two types: those composed of mainly one mineral—for example, marble or quartzite—and those in which the different mineral grains are too small to be seen without magnification, such as greenstone and hornfels.

Marble is a well-known metamorphic rock composed predominantly of calcite or dolomite; its grain

Figure 7.13 Migmatites
A migmatite boulder in the Rocky Mountain National Park, near Estes Park, Colorado. Migmatities consist of high-grade metamorphic rock intermixed with streaks or lenses of granite.

COPYRIGHT AND PHOTOGRAPH BY DR. PARVINDER S. SETHI

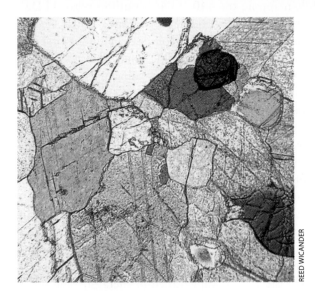

REED WICANDER

Figure 7.14 Nonfoliated Texture
Nonfoliated textures are characterized by a mosaic of roughly equidimensional minerals, as in this photomicrograph of marble.

size ranges from fine to coarsely granular. Marble results from either contact or regional metamorphism of limestones or dolostones (Figure 7.15 and Table 7.1). Pure marble is snowy white or bluish; however, many color varieties exist because of the presence of mineral impurities in the original sedimentary rock.

Quartzite is a hard, compact rock typically formed from quartz sandstone under intermediate- to high-grade metamorphic conditions during contact or regional metamorphism (Figure 7.16). Because recrystallization is so complete, metamorphic quartzite is of uniform strength and therefore usually breaks across the component quartz grains rather than around them when it is struck. Pure quartzite is white; however, iron and other impurities commonly impart a pinkish-red or other color to it. Quartzite is commonly used as foundation material for road and railway beds.

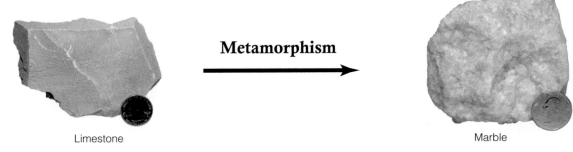

Metamorphism

Limestone

Marble

COPYRIGHT AND PHOTOGRAPHS BY DR. PARVINDER S. SETHI

Figure 7.15 Marble Results from the Metamorphism of the Sedimentary Rock Limestone or Dolostone

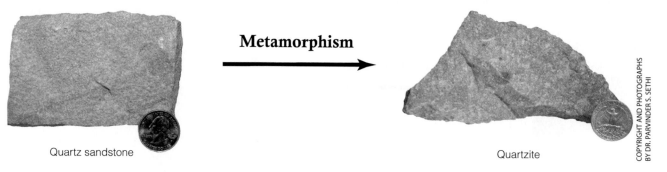

The softness of marble, its uniform texture, and its varying colors have made it the favorite rock of builders and sculptors throughout history.

REED WICANDER

The name *greenstone* is applied to any compact, dark-green, altered, mafic igneous rock that formed under low- to high-grade metamorphic conditions. The green color results from the presence of chlorite, epidote, and hornblende.

Hornfels is a common, fine-grained, nonfoliated metamorphic rock resulting from contact metamorphism and composed of various equidimensional mineral grains. The composition of hornfels depends directly on the composition of the original rock, and many compositional varieties are known. The majority of hornfels, however, are apparently derived from contact metamorphism of clay-rich sedimentary rocks or impure dolostones.

Anthracite is a black, lustrous, hard coal that contains a high percentage of fixed carbon and a low percentage of volatile matter. It usually forms from the metamorphism of lower-grade coals by heat and pressure, and many geologists consider it to be a metamorphic rock.

LO4 Metamorphic Zones and Facies

While mapping the 440- to 400-million-year-old Dalradian schists of Scotland in the late 1800s, George Barrow and other British geologists made the first systematic study of metamorphic zones. Here, clay-rich sedimentary rocks have been subjected to regional metamorphism, and the resulting metamorphic rocks can be divided into different zones based on the presence of distinctive silicate mineral assemblages. These mineral assemblages, each recognized by the presence of one or more index minerals, indicate different degrees of metamorphism. The index minerals that Barrow and his associates chose to represent increasing metamorphic intensity were chlorite, biotite, garnet, staurolite, kyanite, and sillimanite (Figure 7.7), which we now know all result from the recrystallization of clay-rich sedimentary rocks.

The successive appearance of metamorphic index minerals indicates gradually increasing or decreasing intensity of metamorphism. Going from lower- to

Metamorphism

Quartz sandstone

Quartzite

COPYRIGHT AND PHOTOGRAPHS BY DR. PARVINDER S. SETHI

Figure 7.16 Quartzite Results from the Metamorphism of the Sedimentary Rock Quartz Sandstone

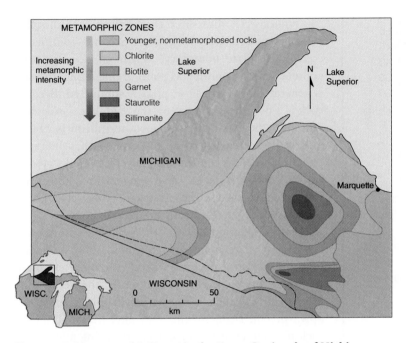

METAMORPHIC ZONES

Increasing metamorphic intensity

- Younger, nonmetamorphosed rocks
- Chlorite
- Biotite
- Garnet
- Staurolite
- Sillimanite

Lake Superior

MICHIGAN

Marquette

Lake Superior

WISCONSIN

0   50
km

WISC.
MICH.

Figure 7.17 Metamorphic Zones in the Upper Peninsula of Michigan
The zones in this region are based on the presence of distinctive silicate mineral assemblages resulting from the metamorphism of sedimentary rocks during an interval of mountain building and minor granitic intrusion during the Proterozoic Eon, approximately 1.5 billion years ago. The lines separating the different metamorphic zones are isograds

SOURCE: FROM H. L. JAMES, *G. S. A. BULLETIN,* VOL. 66 (1955), PLATE 1, PAGE 1454, WITH PERMISSION OF THE PUBLISHER, THE GEOLOGICAL SOCIETY OF AMERICA, BOULDER, COLORADO, USA. COPYRIGHT © 1955 GEOLOGICAL SOCIETY OF AMERICA.

same general degree of metamorphism. By mapping adjoining metamorphic zones, geologists can reconstruct metamorphic conditions throughout an entire area (Figure 7.17).

Not long after Barrow and his coworkers completed their work, geologists in Norway and Finland came up with a different method of mapping metamorphism that was more useful than the metamorphic zone approach. A **metamorphic facies** is defined as a group of metamorphic rocks characterized by particular mineral assemblages formed under broadly similar temperature and pressure conditions (Figure 7.18). Each facies is named after its most characteristic rock or mineral. For example, the green metamorphic mineral chlorite, which forms under relatively low temperatures and pressures, yields rocks belonging to the *greenschist facies.* Under increasingly higher temperatures and pressures, mineral assemblages indicative of the *amphibolite* and *granulite facies* develop.

Although usually applied to areas where the original rocks were clay rich, the concept of metamorphic facies can be used with modification in other situations. It cannot, however, be used in areas where the original rocks were pure quartz sandstones or pure limestones or dolostones. Such rocks would yield only quartzites and marbles, respectively.

higher-grade metamorphic zones, the first appearance of a particular index mineral indicates the location of the minimum temperature and pressure conditions needed for the formation of that mineral. When the locations of the first appearances of that index mineral are connected on a map, the result is a line of equal metamorphic intensity or an *isograd.* The region between two adjacent isograds makes up a single **metamorphic zone**—a belt of rocks displaying the

metamorphic zone
The region between lines of equal metamorphic intensity known as isograds.

metamorphic facies A group of metamorphic rocks characterized by particular minerals that formed under the same broad temperature and pressure conditions.

Remember that metamorphic zones are identified by the appearance of a single index mineral within rocks of the same general composition that occur throughout an area. However, rocks of greatly different composition within an area can belong to the same metamorphic facies, because each facies has its own characteristic assemblage of minerals whose presence indicates metamorphism with the same broad temperature–pressure range unique to that facies.

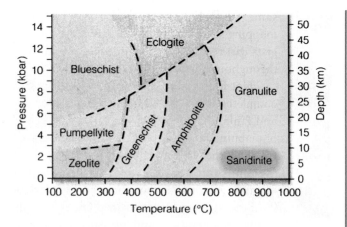

Figure 7.18 Metamorphic Facies and Their Associated Temperature–Pressure Conditions
A temperature–pressure diagram showing under what conditions various metamorphic facies occur. A metamorphic facies is characterized by a particular mineral assemblage that formed under the same broad temperature–pressure conditions. Each facies is named after its most characteristic rock or mineral.

LO5 Plate Tectonics and Metamorphism

Although metamorphism is associated with all three types of plate boundaries, it is most common along convergent plate margins. Metamorphic rocks form at convergent plate boundaries, because temperature and pressure increase as a result of plate collisions.

Figure 7.19 illustrates the various metamorphic facies produced along a typical oceanic–continental convergent plate boundary. When an oceanic plate collides with a continental plate, tremendous pressure is generated as the oceanic plate is subducted. Because rock is a poor heat conductor, the cold descending oceanic plate heats slowly, and metamorphism is caused mostly by increasing pressure with depth. Metamorphism in such an environment produces rocks typical of the *blueschist facies* (low temperature, high pressure). Geologists use the presence of blueschist facies rocks as evidence of ancient subduction zones.

As subduction along the oceanic–continental convergent plate boundary continues, both temperature and pressure increase with depth and yield high-grade metamorphic rocks. Eventually, the descending plate begins to melt and generate magma that moves upward. This rising magma may alter the surrounding rock by contact metamorphism, producing migmatites in the deeper portions of the crust and hornfels at shallower depths. High temperatures and low to medium pressures characterize such an environment.

Although metamorphism is most common along convergent plate margins, many divergent plate boundaries are characterized by contact metamorphism. Rising magma at mid-oceanic ridges heats the adjacent rocks, producing contact metamorphic minerals and textures. In addition, fluids emanating from the rising magma—and its reaction with seawater—very commonly produce metal-bearing hydrothermal solutions that may precipitate minerals of economic value, such as the copper ores of Cyprus.

Figure 7.19 Relationship of Facies to Major Tectonic Features at an Oceanic–Continental Convergent Plate Boundary

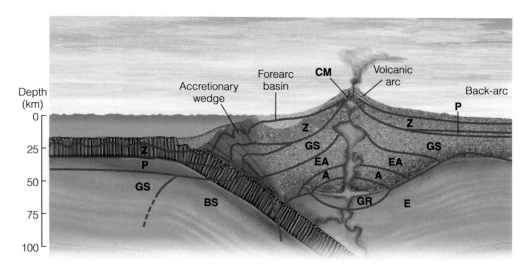

A = amphibolite facies
BS = blueschist facies
CM = contact metamorphic zone shown in green
E = eclogite facies

EA = epidote-amphibolite facies
GR = granulite facies
GS = greenschist facies
P = prehnite-pumpellyite facies
Z = zeolite facies

LO6 Metamorphism and Natural Resources

Many metamorphic minerals and rocks are valuable natural resources. Although these resources include various types of ore deposits, the two most familiar and widely used metamorphic rocks are marble and slate (Figure 7.9b), which have been used for centuries in a variety of ways.

Many ore deposits result from contact metamorphism during which hot, ion-rich fluids migrate from igneous intrusions into the surrounding rock, thereby producing rich ore deposits. The most common sulfide ore minerals associated with contact metamorphism are bornite (copper), chalcopyrite (copper), galena (lead), pyrite, and sphalerite (zinc); two common iron oxide ore minerals are hematite and magnetite. Tin and tungsten are also important ores associated with contact metamorphism.

Other economically important metamorphic minerals include asbestos, used for insulation and fireproofing in buildings and building materials, talc for talcum powder, graphite for pencils and dry lubricants, and garnets and corundum, which are used as abrasives or gemstones, depending on their quality.

CHAPTER 8
EARTHQUAKES AND EARTH'S INTERIOR

This shop in Olema, California, is rather whimsically called the Epicenter, alluding to the fact that it is in the San Andreas fault zone.

JAMES S. MONROE

Introduction

Rocks subjected to intense forces bend until they break and then return to their original position, releasing energy in the process.

At 5:54 a.m., on May 27, 2006, a 6.3-magnitude earthquake struck Java, Indonesia. When the earthquake was over, more than 6,200 people were dead, at least 38,600 were injured, more than 127,000 houses were destroyed, and approximately 451,000 structures were damaged. The amount of economic destruction inflicted by this earthquake is estimated to be a staggering $3.1 billion. All in all, this was a disaster of epic proportions. Yet it was not the first, nor will it be the last, major devastating earthquake in this region or other parts of the world.

Earthquakes, along with volcanic eruptions, are manifestations of Earth's dynamic and active makeup. As one of nature's most frightening and destructive phenomena, earthquakes have always aroused feelings of fear and have been the subject of myths and legends. What makes an earthquake so frightening is that when it begins, there is no way to tell how long it will last or how violent it will be. Approximately 13 million people have died in earthquakes during the past 4,000 years, with about 2.7 million of these deaths occurring during the last century (Table 8.1).

Geologists define an **earthquake** as the shaking or trembling of the ground caused by the sudden release of energy, usually as a result of faulting, which involves the displacement of rocks along fractures (we discuss the different types of faults in Chapter 10). After an earthquake, continuing adjustments along a fault may generate a series of earthquakes known as *aftershocks*. Most aftershocks are smaller than the main shock, but they can still cause considerable damage to already weakened structures.

Why should you study earthquakes? The obvious reason is that they are destructive and cause many deaths and injuries to the people living in earthquake-prone areas. Earthquakes also affect the economies of many countries in terms of cleanup costs, lost jobs, and lost business revenues. From a purely personal standpoint, you someday may be caught in an earthquake. Even if you don't live in an area that is subject to earthquakes, you probably will sooner or later travel where there is the threat of earthquakes, and you should know what to do if you experience one. Such knowledge may help you avoid serious injury or even death.

LEARNING OUTCOMES

After reading this unit, you should be able to do the following:

LO1 Explain Elastic Rebound Theory

LO2 Describe seismology

LO3 Identify where earthquakes occur, and how often

LO4 Identify different seismic waves

LO5 Discuss how earthquakes are located

LO6 Explain how the strength of an earthquake is measured

LO7 Describe the destructive effects of earthquakes

LO8 Discuss earthquake prediction methods

LO9 Discuss earthquake control methods

LO10 Describe Earth's interior

LO11 Examine Earth's core

LO12 Examine Earth's mantle

LO13 Describe Earth's internal heat

LO14 Examine Earth's crust

LO1 Elastic Rebound Theory

Based on studies conducted after the 1906 San Francisco earthquake, H. F. Reid of The Johns Hopkins University proposed the **elastic rebound theory** to explain how

earthquake Vibrations caused by the sudden release of energy, usually as a result of displacement of rocks along faults.

elastic rebound theory An explanation for the sudden release of energy that causes earthquakes when deformed rocks fracture and rebound to their original undeformed condition.

TABLE 8.1

SOME SIGNIFICANT EARTHQUAKES

YEAR	LOCATION	MAGNITUDE (ESTIMATED BEFORE 1935)	DEATHS (ESTIMATED)
1556	China (Shanxi Province)	8.0	1,000,000
1755	Portugal (Lisbon)	8.6	70,000
1906	USA (San Francisco, California)	8.3	3,000
1923	Japan (Tokyo)	8.3	143,000
1960	Chile	9.5	5,700
1964	USA (Anchorage, Alaska)	8.6	131
1976	China (Tangshan)	8.0	242,000
1985	Mexico (Mexico City)	8.1	9,500
1988	Armenia	6.9	25,000
1990	Iran	7.3	50,000
1993	India	6.4	30,000
1995	Japan (Kobe)	7.2	6,000+
1998	Afghanistan	6.9	5,000+
1999	Turkey	7.4	17,000
2001	India	7.9	14,000+
2003	Iran	6.6	43,000
2004	Indonesia	9.0	>220,000
2005	Pakistan	7.6	>86,000

energy is released during earthquakes. Reid studied three sets of measurements taken across a portion of the San Andreas fault that had broken during the 1906 earthquake. The measurements revealed that points on opposite sides of the fault had moved 3.2 m during

The energy stored in rocks undergoing deformation is analogous to the energy stored in a tightly wound watch spring. The tighter the spring is wound, the more energy is stored, thus making more energy available for release. If the spring is wound so tightly that it breaks, then the stored energy is released as the spring rapidly unwinds and partially regains its original shape.

the 50-year period prior to breakage in 1906, with the west side moving northward (Figure 8.1).

According to Reid, rocks on opposite sides of the San Andreas fault had been storing energy and bending slightly for at least 50 years before the 1906 earthquake. Any straight line, such as a fence or road that crossed the San Andreas fault, was gradually bent, because rocks on one side of the fault moved relative to rocks on the other side (Figure 8.1). Eventually, the strength of the rocks was exceeded and then fractured, the rocks on opposite sides of the fault rebounded or "snapped back" to their former undeformed shape, and the energy stored was released as earthquake waves radiating outward from the break (Figure 8.1). Additional field and laboratory studies conducted by Reid and others have confirmed that elastic rebound is the mechanism by which energy is released during earthquakes.

A useful analogy is that of bending a long, straight stick over your knee. As the stick bends, it deforms and eventually reaches the point at which it breaks. When this happens, the two pieces of the original stick snap back into their original straight position. Likewise,

Figure 8.1 The Elastic Rebound Theory

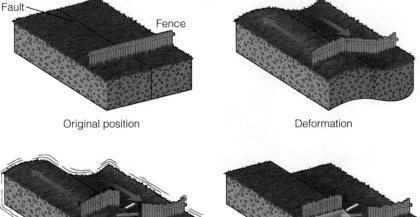

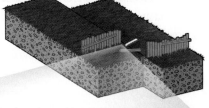

Original position

Deformation

Rupture and release of energy

Rocks rebound to original undeformed shape

a. According to the elastic rebound theory, rocks experiencing deformation store energy and bend. When the initial strength of the rocks is exceeded, they rupture, releasing their accumulated energy, and "snap back" or rebound to their former undeformed shape. This sudden release of energy is what causes an earthquake.

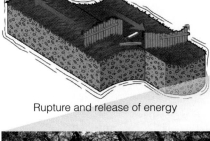

COPYRIGHT AND PHOTOGRAPH BY DR. PARVINDER S. SETHI

b. During the 1906 San Francisco earthquake, this fence in Marin County was displaced by almost 5 m. Whereas many people would see a broken fence, a geologist sees that the fence has moved or been displaced and would look for evidence of a fault. A geologist would also notice that the ground has been displaced toward the right side, relative to his or her view. Regardless of what side of the fence you stand on, you must look to the right to see the other part of the fence. Try it!

rocks subjected to intense forces bend until they break and then return to their original position, releasing energy in the process.

LO2 Seismology

Seismology, the study of earthquakes, emerged as a true science during the 1880s with the development of **seismographs**, instruments that detect, record, and measure the vibrations produced by an earthquake (Figure 8.2). The record made by a seismograph is called a *seismogram*. Modern seismographs have electronic sensors and record movements precisely using computers, rather than simply relying on the drum strip charts commonly used on older seismographs.

When an earthquake occurs, energy in the form of *seismic waves* radiates out from the point of release (Figure 8.3). These waves are somewhat analogous to the ripples that move out concentrically from the point where a stone is thrown into a pond. Unlike waves on a pond, however, seismic waves move outward in all directions from their source.

Earthquakes take place because rocks are capable of storing energy; however, their strength is limited, so if enough force is present, they rupture and thus release their stored energy. In other words, most earthquakes

seismology The study of earthquakes.

seismograph An instrument that detects, records, and measures the various waves produced by earthquakes.

Figure 8.2 Seismographs

REED WICANDER

a. Seismographs record ground motion during an earthquake. The record produced is a seismogram. This seismograph records earthquakes on a strip of paper attached to a rotating drum.

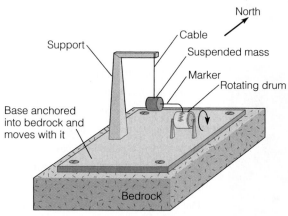

b. A horizontal-motion seismograph. Because of its inertia, the heavy mass that contains the marker remains stationary, while the rest of the structure moves along with the ground during an earthquake. As long as the length of the arm is not parallel to the direction of ground movement, the marker will record the earthquake waves on the rotating drum. This seismograph would record waves from west or east, but to record waves from the north or south, another seismograph at right angles to this one is needed.

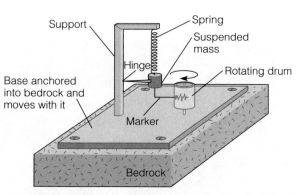

c. A vertical-motion seismograph. This seismograph operates on the same principle as the horizontal-motion instrument and records vertical ground movement.

result when movement occurs along faults, most of which are related, at least indirectly, to plate movements. Once rupturing begins, it moves along the fault at several kilometers per second for as long as conditions for failure exist. The longer the fault along which movement occurs, the more time it takes for the stored energy to be released, and therefore the longer the ground will shake. During some very large earthquakes, the ground might shake for 3 minutes, a seemingly brief, but interminable, time if you are experiencing the earthquake firsthand.

focus The site within Earth where an earthquake originates and energy is released.

epicenter The point on Earth's surface directly above an earthquake's focus.

THE FOCUS AND EPICENTER OF AN EARTHQUAKE

The location within Earth's lithosphere where rupturing begins—that is, the point at which energy is first released—is an earthquake's **focus**, or *hypocenter*. What we usually hear in news reports, however, is the location of the **epicenter**, the point on Earth's surface directly above the focus (Figure 8.3a).

Figure 8.3 The Focus and Epicenter of an Earthquake

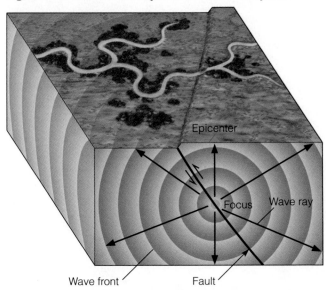

a. The focus of an earthquake is the location where the rupture begins and energy is released. The place on the surface vertically above the focus is the epicenter. Seismic wavefronts move out in all directions from their source, the focus of an earthquake.

b. The epicenter of the December 26, 2004, earthquake that caused the devastating tsunami in the Indian Ocean. The epicenter was located 160 km off the west coast of northern Sumatra and had a focal depth of 30 km.

Seismologists recognize three categories of earthquakes based on focal depth. *Shallow-focus* earthquakes have focal depths of less than 70 km from the surface, whereas those with foci between 70 and 300 km are *intermediate-focus* earthquakes, and the foci of those characterized as *deep-focus* earthquakes are more than 300 km deep. However, earthquakes are not evenly distributed among these three categories. Approximately 90% of all earthquake foci are at depths of less than 100 km, whereas only about 3% of all earthquakes are deep. Shallow-focus earthquakes are, with few exceptions, the most destructive, because the energy they release has little time to dissipate before reaching the surface.

A definite relationship exists between earthquake foci and plate boundaries. Earthquakes generated along divergent or transform plate boundaries are invariably shallow focus, whereas many shallow-focus earthquakes and nearly all intermediate- and deep-focus earthquakes occur along convergent margins (Figure 8.4). Furthermore, a pattern emerges when the focal depths of earthquakes near island arcs and

their adjacent ocean trenches are plotted. Notice in Figure 8.5 that the focal depth increases beneath the Tonga Trench in a narrow, well-defined zone that dips approximately 45 degrees. Dipping seismic zones, called *Benioff* or *Benioff–Wadati zones*, are common to convergent plate boundaries where one plate is subducted beneath another. Such dipping seismic zones indicate the angle of plate descent along a convergent plate boundary.

LO3 Where Do Earthquakes Occur, and How Often?

No place on Earth is immune to earthquakes, but almost 95% take place in seismic belts corresponding to plate boundaries where plates converge, diverge, and slide past each other. The relationship between plate margins and the distribution of earthquakes is readily apparent when the locations of earthquake epicenters are superimposed on a map showing the boundaries of Earth's plates.

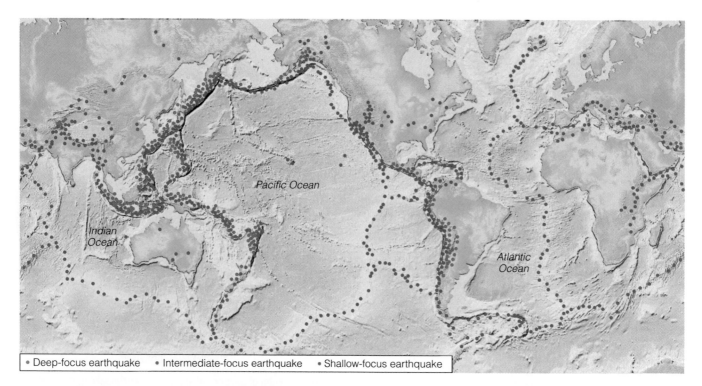

Figure 8.4 Earthquake Epicenters and Plate Boundaries

This map of earthquake epicenters shows that most earthquakes occur within seismic zones that correspond closely to plate boundaries. Approximately 80% of earthquakes occur within the circum-Pacific belt, 15% within the Mediterranean–Asiatic belt, and the remaining 5% within plate interiors and along oceanic spreading ridges. The dots represent earthquake epicenters and are divided into shallow-, intermediate-, and deep-focus earthquakes. Along with many shallow-focus earthquakes, nearly all intermediate- and deep-focus earthquakes occur along convergent plate boundaries.

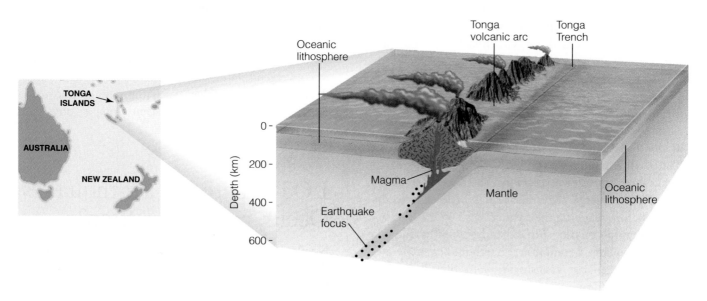

Figure 8.5 Benioff Zones

Focal depth increases in a well-defined zone that dips approximately 45 degrees beneath the Tonga volcanic arc in the South Pacific. Dipping seismic zones are called *Benioff* or *Benioff–Wadati zones*.

The majority of all earthquakes (approximately 80%) occur in the *circum-Pacific belt*, a zone of seismic activity nearly encircling the Pacific Ocean basin. Most of these earthquakes result from convergence along plate margins, as in the case of the 1995 Kobe, Japan, earthquake. The earthquakes along the North American Pacific Coast, especially in California, are also in this belt, but here plates slide past one another rather than converge. The October 17, 1989, Loma Prieta earthquake (Figure 8.6) occurred along this plate boundary.

The second major seismic belt, accounting for 15% of all earthquakes, is the *Mediterranean–Asiatic belt*. This belt extends westward from Indonesia through the Himalayas, across Iran and Turkey, and westward through the Mediterranean region of Europe. The 2005 earthquake in Pakistan that killed more than 86,000 people is a recent example of the destructive earthquakes that strike this region (Table 8.1).

The remaining 5% of earthquakes occur mostly in the interiors of plates and along oceanic spreading-ridge systems. Most of these earthquakes are not strong, although several major intraplate earthquakes are worthy of mention. For example, the 1811 and 1812 earthquakes near New Madrid, Missouri, killed approximately 20 people and nearly destroyed the town. So strong were these earthquakes that they were felt from the Rocky Mountains to the Atlantic Ocean and from the Canadian border to the Gulf of Mexico. Within the immediate area, numerous buildings were destroyed and forests were flattened. The land sank several meters in some areas, causing flooding, and

Figure 8.6 Earthquake Damage in the Circum-Pacific Belt Damage in Oakland, California, resulting from the October 1989 Loma Prieta earthquake. The columns supporting the upper deck of Interstate 880 failed, causing the upper deck to collapse onto the lower one.

reportedly the Mississippi River reversed its flow during the shaking and changed its course slightly.

The cause of intraplate earthquakes is not well understood, but geologists think that they arise from localized stresses caused by the compression that most plates experience along their margins. A useful analogy is moving a house. Regardless of how careful the movers are, moving something so large without its internal parts shifting slightly is impossible. Similarly, plates are not likely to move without some internal stresses that occasionally cause earthquakes.

More than 900,000 earthquakes are recorded annually by the worldwide network of seismograph stations. Many of these are too small to be felt but are nonetheless recorded. These small earthquakes result from the energy released as continual adjustments take place between the various plates. However, on average, more than 31,000 earthquakes per year are strong enough to be felt and can cause various amounts of damage, depending on how strong they are and where they occur.

LO4 Seismic Waves

Many people have experienced an earthquake, but most are probably unaware that the shaking they feel and the damage to structures are caused by the arrival of *seismic waves*, a general term encompassing all waves generated by an earthquake. When movement on a fault takes place, energy is released in the form of two kinds of seismic waves that radiate outward in all directions from an earthquake's focus. *Body waves*, so called because they travel through the solid body of Earth, are somewhat like sound waves, and *surface waves*, which travel along the ground surface, are analogous to undulations or waves on water surfaces.

Body Waves

An earthquake generates two types of body waves: P-waves and S-waves (Figure 8.7). **P-waves** or *primary waves* are the fastest seismic waves and can travel through solids, liquids, and gases. P-waves are *compressional*, or *push-pull*, *waves* and are similar to sound waves in that they move material forward and backward along a line in the same

P-wave A compressional, or push-pull, wave; the fastest seismic wave and one that can travel through solids, liquids, and gases; also called a *primary wave*.

Figure 8.7 Primary and Secondary Seismic Body Waves
Body waves travel through Earth.

a. Undisturbed material for reference.

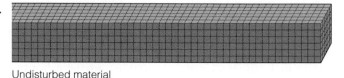

Undisturbed material

b. Primary waves (P-waves) compress and expand material in the same direction that they travel.

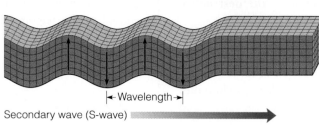

Primary wave (P-wave)   Direction of wave movement

c. Secondary waves (S-waves) move material perpendicular to the direction of wave movement.

←Wavelength→

Secondary wave (S-wave)

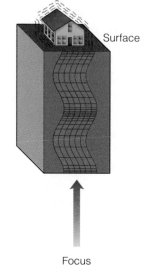

Surface

Focus

d. The effect of P- and S-waves on a surface structure.

direction that the waves themselves are moving (Figure 8.7b). Thus, the material through which P-waves travel is expanded and compressed as the waves move through it and returns to its original size and shape after the wave passes by.

S-waves or *secondary waves* are somewhat slower than P-waves and can travel only through solids. S-waves are *shear waves* because they move the material perpendicular to the direction of travel, thereby producing shear stresses in the material they move through (Figure 8.7c). Because liquids (as well as gases) are not rigid, they have no shear strength, and S-waves cannot be transmitted through them.

The velocities of P- and S-waves are determined by the density and elasticity of the materials through which they travel. For example, seismic waves travel more slowly through rocks of greater density, but more rapidly through rocks with greater elasticity. *Elasticity* is a property of solids, such as rocks, and means that once they have been deformed by an applied force, they return to their original shape when the force is no longer present. Because P-wave velocity is greater than S-wave velocity in all materials, P-waves always arrive at seismic stations first.

SURFACE WAVES

Surface waves travel along the surface of the ground, or just below it, and are slower than body waves. Unlike the sharp jolting and shaking that body waves cause, surface waves generally produce a rolling or swaying motion, much like the experience of being on a boat.

Several types of surface waves are recognized. The two most important are Rayleigh waves and Love waves, named after the British scientists who discovered them, Lord Rayleigh and A. E. H. Love. **Rayleigh waves (R-waves)** are generally the slower of the two and behave like water waves in that they move forward while the individual particles of material move

S-wave A shear wave that moves material perpendicular to the direction of travel, thereby producing shear stresses in the material it moves through; also known as a *secondary wave*; S-waves travel only through solids.

Rayleigh wave (R-wave) A surface wave in which individual particles of material move in an elliptical path within a vertical plane oriented in the direction of wave movement.

Figure 8.8 Rayleigh and Love Seismic Surface Waves
Surface waves travel along Earth's surface or just below it.

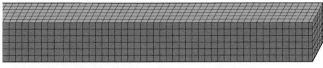

Undisturbed material

a. Undisturbed material for reference.

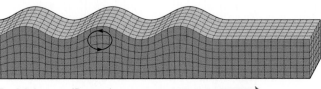

Rayleigh wave (R-wave)

b. Rayleigh waves (R-waves) move material in an elliptical path in a plane oriented parallel to the direction of wave movement.

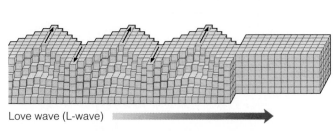

Love wave (L-wave)

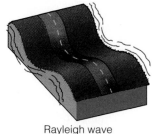

Rayleigh wave   Love wave

c. Love waves (L-waves) move material back and forth in a horizontal plane perpendicular to the direction of wave movement.

d. The arrival of R- and L-waves causes the surface to undulate and shake from side to side.

in an elliptical path within a vertical plane oriented in the direction of wave movement (Figure 8.8b).

The motion of a **Love wave** (L-wave) is similar to that of an S-wave, but the individual particles of the material move only back and forth in a horizontal plane perpendicular to the direction of wave travel (Figure 8.8c).

LO5 Locating an Earthquake

We mentioned that news articles commonly report an earthquake's epicenter, but just how is the location of an epicenter determined? Once again, geologists rely on the study of seismic waves. We know that P-waves travel faster than S-waves, nearly twice as fast in all substances, so P-waves arrive at a seismograph station first, followed some time later by S-waves. Both P- and S-waves travel directly from the focus to the seismograph station through Earth's interior, but L- and R-waves arrive last, because they are the slowest, and they also travel the longest route along the surface (Figure 8.9a). However, only the P- and S-waves need

concern us here, because they are the ones important in finding an epicenter.

Seismologists, geologists who study seismology, have accumulated a tremendous amount of data over the years and now know the average speeds of P- and S-waves for any specific distance from their source. These P- and S-wave travel times are published in *time–distance graphs* that illustrate the difference between the arrival times of the two waves as a function of the distance between a seismograph and an earthquake's focus (Figure 8.9b). That is, the farther the waves travel, the greater the *P–S time interval*, which is simply the time difference between the arrivals of P- and S-waves (Figure 8.9a, b).

If the P–S time intervals are known from at least three seismograph stations, then the epicenter of any earthquake can be determined (Figure 8.10). Here is how it works: Subtracting the arrival time of the first P-wave from the arrival time of the first S-wave gives the P–S time interval for each seismic station. Each of these

Love wave (L-wave)
A surface wave in which the individual particles of material move only back and forth in a horizontal plane perpendicular to the direction of wave travel.

Figure 8.9 Determining the Distance from an Earthquake

a. A schematic seismogram showing the arrival order and pattern produced by P-, S-, and L-waves. When an earthquake occurs, body and surface waves radiate out from the focus at the same time. Because P-waves are the fastest, they arrive at a seismograph first, followed by S-waves and then by surface waves, which are the slowest waves. The difference between the arrival times of the P- and S-waves is the P–S time interval; it is a function of the distance the seismograph station is from the focus.

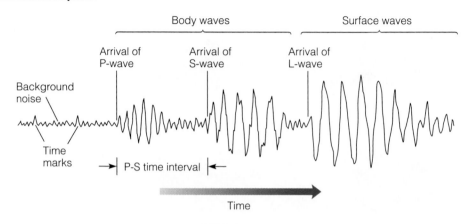

b. A time–distance graph showing the average travel times for P- and S-waves. The farther away a seismograph station is from the focus of an earthquake, the longer the interval between the arrival of the P- and S-waves, and hence the greater the distance between the P- and S-wave curves on the time–distance graph as indicated by the P–S time interval. For example, let's assume the difference in arrival times between the P- and S-waves is 10 minutes (P–S time interval). Using the Travel time (minutes) scale, measure how long 10 minutes is (P–S time interval), and move that distance between the S-wave curve and the P-wave curve until the line touches both curves as shown. Then draw a line straight down to the Distance from focus (km) scale. That number is the distance the seismograph is from the earthquake's focus. In this example, the distance is almost 9,000 km.

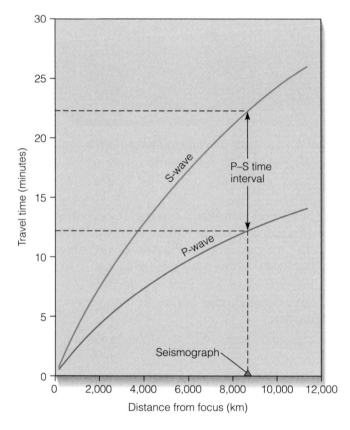

time intervals is plotted on a time–distance graph, and a line is drawn straight down to the distance axis of the graph, thus giving the distance from the focus to each seismic station (Figure 8.9b). Next, a circle whose radius equals the distance shown on the time–distance graph from each of the seismic stations is drawn on a map (Figure 8.10). The intersection of the three circles is the location of the earthquake's epicenter. It should be obvious from Figure 8.10 that P–S time intervals from at least three seismic stations are needed. If only one were used, the epicenter could be at any location on the circle drawn around that station, and using two stations would give two possible locations for the epicenter.

Determining the focal depth of an earthquake is much more difficult and considerably less precise than finding its epicenter. The focal depth is usually found by making computations based on several assumptions, comparing the results with those obtained at other seismic stations, and then recalculating and approximating the depth as closely as possible.

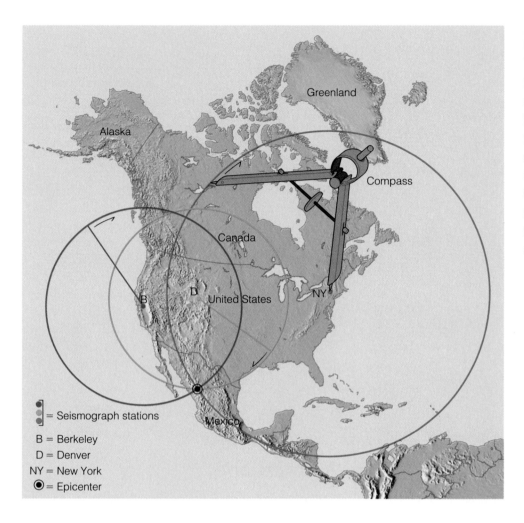

Figure 8.10 Determining the Epicenter of an Earthquake
Three seismograph stations are needed to locate the epicenter of an earthquake. The P–S time interval is plotted on a time–distance graph for each seismograph station to determine the distance that station is from the epicenter. A circle with that radius is drawn from each station, and the intersection of the three circles is the epicenter of the earthquake.

= Seismograph stations
B = Berkeley
D = Denver
NY = New York
◉ = Epicenter

LO6 Measuring the Strength of an Earthquake

Following any earthquake that causes extensive damage, fatalities, and injuries, graphic reports of the quake's violence and human suffering are common. Although descriptions of fatalities and damage give some indication of the size of an earthquake, geologists are interested in more reliable methods for determining an earthquake's size.

Two measures of an earthquake's strength are commonly used. One is *intensity*, a qualitative assessment of the kinds of damage done by an earthquake. The other, *magnitude*, is a quantitative measure of the amount of energy released by an earthquake. Each method provides important information that can be used to prepare for future earthquakes.

INTENSITY

Intensity is a subjective or qualitative measure of the kind of damage done by an earthquake as well as people's reaction to it. Since the mid-19th century, geologists have used intensity as a rough approximation of the size and strength of an earthquake. The most common intensity scale used in the United States is the **Modified Mercalli Intensity Scale**, which has values ranging from I to XII (Table 8.2).

Intensity maps can be constructed for regions hit by earthquakes by dividing the affected

intensity The subjective measure of the kind of damage done by an earthquake as well as people's reaction to it.

Modified Mercalli Intensity Scale A scale with values from I to XII used to characterize earthquakes based on damage.

TABLE 8.2

MODIFIED MERCALLI INTENSITY SCALE

I   Not felt except by a very few under especially favorable circumstances.

II   Felt by only a few people at rest, especially on upper floors of buildings.

III   Felt quite noticeably indoors, especially on upper floors of buildings, but many people do not recognize it as an earthquake. Standing automobiles may rock slightly.

IV   During the day felt indoors by many, outdoors by few. At night some awakened. Sensation like heavy truck striking building, standing automobiles rocked noticeably.

V   Felt by nearly everyone, many awakened. Some dishes, windows, etc. broken, a few instances of cracked plaster. Disturbance of trees, poles, and other tall objects sometimes noticed.

VI   Felt by all, many frightened and run outdoors. Some heavy furniture moved. A few instances of fallen plaster or damaged chimneys. Damage slight.

VII   Everybody runs outdoors. Damage negligible in buildings of good design and construction; slight to moderate in well-built ordinary structures; considerable in poorly built or badly designed structures; some chimneys broken. Noticed by people driving automobiles.

VIII   Damage slight in specially designed structures; considerable in normally constructed buildings with possible partial collapse; great in poorly built structures. Fall of chimneys, monuments, walls. Heavy furniture overturned. Sand and mud ejected in small amounts.

IX   Damage considerable in specially designed structures. Buildings shifted off foundations. Ground noticeably cracked. Underground pipes broken.

X   Some well-built wooden structures destroyed; most masonry and frame structures with foundations destroyed; ground badly cracked. Rails bent. Landsides considerable from river banks and steep slopes. Water splashed over river banks.

XI   Few, if any (masonry) structures remain standing. Bridges destroyed. Broad fissures in ground. Underground pipelines completely out of service.

XII   Damage total. Waves seen on ground surface. Objects thrown upward into the air.

Source: U.S. Geological Survey.

Richter Magnitude Scale   An open-ended scale that measures the amount of energy released during an earthquake.

magnitude   The total amount of energy released by an earthquake at its source.

region into various intensity zones. The intensity value given for each zone is the maximum intensity that the earthquake produced for that zone. Even though intensity maps are not precise, because of the subjective nature of the measurements, they do provide geologists with a rough approximation of the location of the earthquake, the kind and extent of the damage done, and the effects of local geology on different types of building construction. Because intensity is a measure of the kind of damage done by an earthquake, insurance companies still classify earthquakes on the basis of intensity.

Generally, a large earthquake will produce higher intensity values than a small earthquake, but many other factors besides the amount of energy released by an earthquake also affect its intensity. These include distance from the epicenter, focal depth of the earthquake, population density and geology of the area, type of building construction employed, and duration of shaking.

MAGNITUDE

If earthquakes are to be compared quantitatively, we must use a scale that measures the amount of energy released and is independent of intensity. Charles F. Richter, a seismologist at the California Institute of Technology, developed such a scale in 1935. The **Richter Magnitude Scale** measures earthquake **magnitude**, which is the total amount of energy released by an earthquake at its source. It is an open-ended scale with values beginning at zero. The largest magnitude recorded was a magnitude-9.5 earthquake in Chile on May 22, 1960 (Table 8.1).

Scientists determine the magnitude of an earthquake by measuring the amplitude of the largest seismic wave as recorded on a seismogram (Figure 8.11). To avoid large numbers, Richter used a conventional base-10 logarithmic scale to convert the amplitude of the largest recorded seismic wave to a numerical magnitude value. Therefore, each whole-number increase in magnitude represents a 10-fold increase in wave amplitude. For example, the amplitude of the largest seismic wave for an earthquake of magnitude 6 is 10 times that produced by an earthquake of magnitude 5, 100 times as large as a magnitude-4 earthquake, and 1,000 times that of an earthquake of magnitude 3 ($10 \times 10 \times 10 = 1000$).

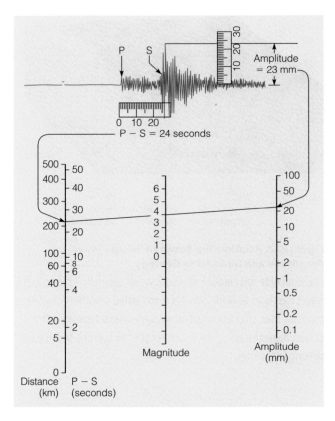

Figure 8.11 Richter Magnitude Scale
The Richter Magnitude Scale measures the total amount of energy released by an earthquake at its source. The magnitude is determined by measuring the maximum amplitude of the largest seismic wave and marking it on the right-hand scale. The difference between the arrival times of the P- and S-waves (recorded in seconds) is marked on the left-hand scale. When a line is drawn between the two points, the magnitude of the earthquake is the point at which the line crosses the center scale.

A common misconception about the size of earthquakes is that an increase of one unit on the Richter Magnitude Scale—a 7 versus a 6, for instance—means a 10-fold increase in size. It is true that each whole-number increase in magnitude represents a 10-fold increase in the wave amplitude, but each magnitude increase of one unit corresponds to a roughly 30-fold increase in the amount of energy released (actually, it is 31.5, but 30 is close enough for our purposes). This means that it would take approximately 30 earthquakes of magnitude 6 to equal the energy released in one earthquake of magnitude 7.

The Richter Magnitude Scale was devised to measure earthquake waves on a particular seismograph and at a specific distance from an earthquake. One of its limitations is that it underestimates the energy of very large earthquakes, because it measures the highest peak on a seismogram, which represents only an instant during an earthquake. For large earthquakes, though, the energy might be released over several minutes and along hundreds of kilometers of a fault. For example, during the 1857 Fort Tejon, California, earthquake, the ground shook for longer than 2 minutes and energy was released for 360 km along the fault. Seismologists now commonly use a somewhat different scale to measure magnitude. Known as the *seismic-moment magnitude scale*, this scale takes into account the strength of the rocks, the area of a fault along which rupture occurs, and the amount of movement of rocks adjacent to the fault. Because larger earthquakes rupture more rocks than smaller earthquakes, and rupture usually occurs along a longer segment of a fault and therefore for a longer duration, these very large earthquakes release more energy. For example, the December 26, 2004, Sumatra, Indonesia, earthquake that generated the devastating tsunami created the longest fault rupture and had the longest duration ever recorded.

Thus, magnitude is now frequently given in terms of both Richter magnitude and seismic-moment magnitude. For example, the 1964 Alaska earthquake is given a Richter magnitude of 8.6 and a seismic-moment magnitude of 9.2. Because the Richter Magnitude Scale is most commonly used in the news, we will use that scale here.

LO7 What Are the Destructive Effects of Earthquakes?

The number of deaths and injuries, as well as the amount of property damage, depends on several factors. Generally speaking, earthquakes that occur during working hours and school hours in densely populated urban areas are the most destructive and cause the most fatalities and injuries. However, magnitude, duration of shaking, distance from the epicenter, geology

Great earthquakes (those with a magnitude greater than 8.0) occur, on average, only once every five years.

of the affected region, and type of structures are also important considerations. Given these variables, it should not be surprising that a comparatively small earthquake can have disastrous effects, whereas a much larger one might go largely unnoticed, except perhaps by seismologists.

The destructive effects of earthquakes include ground shaking, fire, seismic sea waves, and landslides, as well as panic, disruption of vital services, and psychological shock.

GROUND SHAKING

Ground shaking, the most obvious and immediate effect of an earthquake, varies depending on the earthquake's magnitude, distance from the epicenter, and type of underlying materials in the area—unconsolidated sediment or fill versus bedrock, for instance. Certainly, ground shaking is terrifying, and it may be violent enough for fissures to open in the ground. Nevertheless, contrary to popular myth, fissures do not swallow up people and buildings and then close on them. And although California will no doubt have big earthquakes in the future, rocks cannot store enough energy to displace a landmass as large as California into the Pacific Ocean, as some alarmists claim.

The effects of ground shaking, such as collapsing buildings, falling building facades and window glass, and toppling monuments and statues, cause more damage and result in more loss of life and injuries than any other earthquake hazard.

Structures built on poorly consolidated or water-saturated material are subjected to ground shaking of longer duration and greater S-wave amplitude than structures built on bedrock (Figure 8.12). In addition, fill and water-saturated sediments tend to liquefy, or behave as a fluid, a process known as *liquefaction*. When shaken, the individual grains lose cohesion and the ground flows. Two dramatic examples of damage resulting from liquefaction are Niigata, Japan, and Turnagain Heights, Alaska. In Niigata, Japan, large apartment buildings were tipped to their sides after the water-saturated soil of the hillside collapsed (Figure

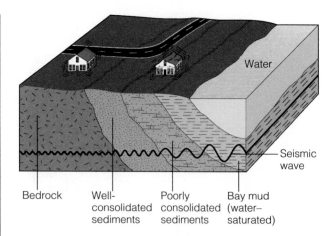

Figure 8.12 Relationship between Seismic Wave Amplitude and Underlying Geology
The amplitude and duration of seismic waves generally increase as the waves pass from bedrock to poorly consolidated or water-saturated material. Thus, structures built on weaker material typically suffer greater damage than similar structures built on bedrock, because the shaking lasts longer.

8.13). In Turnagain Heights, Alaska, many homes were destroyed when the Bootlegger Cover Clay lost all of its strength when it was shaken by the 1964 earthquake.

NATIONAL GEOPHYSICAL DATA CENTER

Figure 8.13 Liquefaction
The effects of ground shaking on water-saturated soil are dramatically illustrated by the collapse of these buildings in Niigata, Japan, during a 1964 earthquake. The buildings, which were designed to be earthquake resistant, fell over on their sides intact when the ground below them underwent liquefaction.

Besides the magnitude of an earthquake and the underlying geology, the material used and the type of construction also affect the amount of damage done. Adobe and mud-walled structures are the weakest and almost always collapse during an earthquake. Unreinforced brick structures and poorly built concrete structures are also particularly susceptible to collapse, such as was the case in the 1999 Turkey earthquake in which an estimated 17,000 people died. The 1976 earthquake in Tangshan, China, completely leveled the city, because hardly any structures were built to resist seismic forces. In fact, most had unreinforced brick walls, which have no flexibility, and consequently they collapsed during the shaking.

FIRE

In many earthquakes, particularly in urban areas, fire is a major hazard. Nearly 90% of the damage done in the 1906 San Francisco earthquake was caused by fire. The shaking severed many of the electrical and gas lines, which touched off flames and started fires throughout the city. Because the earthquake ruptured water mains, there was no effective way to fight the fires that raged out of control for three days, destroying much of the city.

Eighty-three years later, during the 1989 Loma Prieta earthquake, a fire broke out in the Marina district of San Francisco. This time, however, the fire was contained within a small area, because San Francisco had a system of valves throughout its water and gas pipeline system so that lines could be isolated from breaks.

During the September 1, 1923, earthquake in Japan, fires destroyed 71% of the houses in Tokyo and practically all of the houses in Yokohama. In all, 576,262 houses were destroyed by fire, and 143,000 people died, many as a result of fire.

TSUNAMI: KILLER WAVES

On December 26, 2004, a magnitude-9.0 earthquake struck 160 km off the west coast of northern Sumatra, Indonesia, generating the deadliest tsunami in history. Within hours, walls of water as high as 10.5 m pounded the coasts of Indonesia, Sri Lanka, India, Thailand, Somalia, Myanmar, Malaysia, and the Maldives, killing more than 220,000 people and causing billions of dollars in damage.

© ISTOCKPHOTO.COM/ CHRISTINE BALDERAS

This earthquake generated what is popularly called a "tidal wave," but is more correctly termed a *seismic sea wave* or **tsunami**, a Japanese term meaning "harbor wave." The term *tidal wave* nevertheless persists in popular literature and some news accounts, but these waves are not caused by or related to tides. Indeed, tsunami are destructive sea waves generated when the sea floor undergoes sudden, vertical movements. Many result from submarine earthquakes, but volcanoes at sea or submarine landslides can also cause them. For example, the 1883 eruption of Krakatau between Java and Sumatra generated a large sea wave that killed 36,000 on nearby islands.

tsunami A large sea wave that is usually produced by an earthquake, but can also result from submarine landslides and volcanic eruptions.

Tsunami can barrel through the ocean at 600 miles per hour. That's as fast as the cruising speed of a jet plane.

© PHOTOLINK/PHOTODISC/ GETTY IMAGES

Once a tsunami is generated, it can travel across an entire ocean and cause devastation far from its source. In the open sea, tsunami travel at several hundred kilometers per hour and commonly go unnoticed as they pass beneath ships, because they are usually less than 1 m high, and the distance between wave crests is typically hundreds of kilometers. When they enter shallow water, however, the wave slows down and water piles up to heights anywhere from a meter or two to many meters high. The 1946 tsunami that struck Hilo, Hawaii, was 16.5 m high! In any case, the tremendous energy possessed by a tsunami is concentrated on a shoreline when it hits either as a large breaking wave or, in some cases, as what appears to be a very rapidly rising tide.

Figure 8.14 Ground Failure

On August 17, 1959, an earthquake with a Richter magnitude of 7.3 shook southwestern Montana and a large area in adjacent states.

a. The fault scarp in this image was produced when the block in the background moved up several meters compared to the one in the foreground.

b. The earthquake also triggered a landslide (visible in the distance) that blocked the Madison River in Montana and created Earthquake Lake (foreground). The slide entombed approximately 26 people in a campground at the very bottom.

One of nature's warning signs of an approaching tsunami is a sudden withdrawal of the sea from a coastal region. In fact, the sea might withdraw so far that it cannot be seen and the seafloor is laid bare over a huge area. On more than one occasion, people have rushed out to inspect exposed reefs or to collect fish and shells only to be swept away when the tsunami arrived.

Following the tragic 1946 tsunami that hit Hilo, Hawaii, the U.S. Coast and Geodetic Survey established a Pacific Tsunami Early Warning System in Ewa Beach, Hawaii. This system combines seismographs and instruments that detect earthquake-generated sea waves. Whenever a strong earthquake takes place anywhere within the Pacific basin, its location is determined, and instruments are checked to see whether a tsunami has been generated. If it has, a warning is sent out to evacuate people from low-lying areas that may be affected.

Unfortunately, no such warning system exists for the Indian Ocean. If one had been in place, it is possible that the death toll from the December 26, 2004, tsunami would have been significantly lower.

GROUND FAILURE

Earthquake-triggered landslides are particularly dangerous in mountainous regions and have been responsible for tremendous amounts of damage and many deaths. The 1959 Hebgen Lake earthquake in Madison Canyon, Montana, for example, caused a huge rock slide (Figure 8.14), and the 1970 Peru earthquake caused an avalanche that destroyed the town of Yungay and killed an estimated 66,000 people.

LO8 Earthquake Prediction

A successful prediction must include a time frame for the occurrence of an earthquake, its location, and its strength. Despite the tremendous amount of information geologists have gathered about the cause of earthquakes, successful predictions are still rare. Nevertheless, if reliable predictions can be made, they can greatly reduce the number of deaths and injuries.

From an analysis of historic records and the distribution of known faults, geologists construct *seismic risk maps* that indicate the likelihood and potential severity of future earthquakes based on the intensity of past earthquakes. An international effort by scientists from several countries resulted in the publication of the first Global Seismic Hazard Assessment Map in December 1999 (Figure 8.15). Although such maps cannot be used to predict when an earthquake

will take place in any particular area, they are useful in anticipating future earthquakes and helping people plan and prepare for them.

EARTHQUAKE PRECURSORS

Studies conducted during the past several decades indicate that most earthquakes are preceded by both short-term and long-term changes within Earth. Such changes are called *precursors* and may be useful in earthquake prediction.

One long-range prediction technique used in seismically active areas involves plotting the location of major earthquakes and their aftershocks to detect areas that have had major earthquakes in the past but are currently inactive. Such regions are said to be locked and not releasing energy. Nevertheless, pressure is continuing to accumulate in these regions because of plate motions, making these *seismic gaps* prime locations for future earthquakes. Several seismic gaps along the San Andreas fault have the potential for future major earthquakes (Figure 8.16). A major earthquake that damaged Mexico City in 1985 occurred along a seismic gap in the convergence zone along the west coast of Mexico.

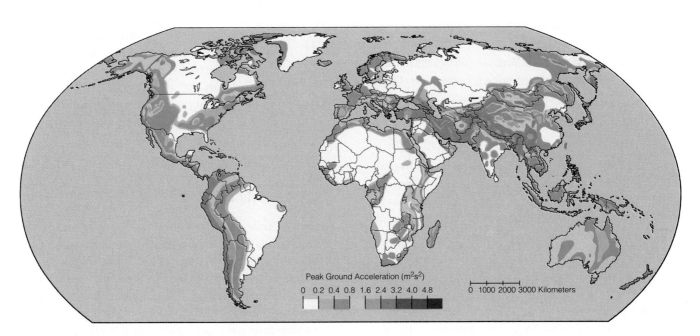

Figure 8.15 Global Seismic Hazard Assessment Map
The Global Seismic Hazard Assessment Program published this seismic hazard map showing peak ground accelerations. The values are based on a 90% probability that the indicated horizontal ground acceleration during an earthquake is not likely to be exceeded in 50 years. The higher the number, the greater the hazard. As expected, the greatest seismic risks are in the circum-Pacific belt and the Mediterranean–Asiatic belt.

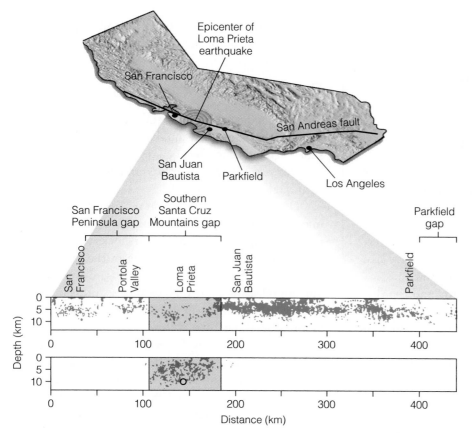

Figure 8.16 Earthquake Precursors
Seismic gaps are one type of earthquake precursor that can indicate a potential earthquake in the future. Seismic gaps are regions along a fault that are locked; that is, they are not moving and releasing energy. Three seismic gaps are evident in this cross section along the San Andreas fault from north of San Francisco to south of Parkfield. The first is between San Francisco and Portola Valley, the second near Loma Prieta Mountain, and the third is southeast of Parkfield. The top section shows the epicenters of earthquakes between January 1969 and July 1989. The bottom section shows the southern Santa Cruz Mountains gap after it was filled by the October 17, 1989, Loma Prieta earthquake (*open circle*) and its aftershocks.

Earthquake precursors that may be useful in making short-term predictions include slight changes in elevation and tilting of the land surface, fluctuations in the water level in wells, changes in Earth's magnetic field, and the electrical resistance of the ground.

EARTHQUAKE PREDICTION PROGRAMS

Currently, only four nations—the United States, Japan, Russia, and China—have government-sponsored earthquake prediction programs. These programs include laboratory and field studies of rock behavior before, during, and after large earthquakes, as well as monitoring activity along major active faults. Most earthquake prediction work in the United States is done by the U.S. Geological Survey (USGS) and involves research into all aspects of earthquake-related phenomena.

The Chinese have perhaps the most ambitious earthquake prediction program in the world, which is understandable considering their long history of destructive earthquakes. Their earthquake prediction program was initiated soon after two large earthquakes occurred at Xingtai (300 km southwest of Beijing) in 1966. This program includes extensive study and monitoring of all possible earthquake precursors. Chinese seismologists successfully predicted the 1975 Haicheng earthquake, but unfortunately, they failed to predict the devastating 1976 Tangshan earthquake that killed at least 242,000 people.

Progress is being made toward dependable, accurate earthquake predictions, and studies are underway to assess public reactions to long-, medium-, and short-term earthquake warnings. However, unless short-term warnings are actually followed by an earthquake, most people will probably ignore the warnings, as they frequently do now for hurricanes, tornadoes, and tsunami.

LO9 Earthquake Control

Reliable earthquake prediction is still in the future, but can anything be done to control or at least partly control earthquakes? Because of the tremendous energy involved, it seems unlikely that humans will ever be able to prevent earthquakes. However, it may be possible to gradually release the energy stored in rocks, thus decreasing the probability of a large earthquake occurring and extensive damage resulting from it.

During the early- to mid-1960s, Denver, Colorado, experienced numerous small earthquakes, which was surprising, because Denver had not been prone

Figure 8.17 Population Density of San Francisco
Downtown San Francisco sits atop the active plate boundary between the Pacific plate and the North American plate. The high density of population poses a risk to residents in the event of an earthquake. Although pumping seismic fluids may relieve the pressure of a fault and prevent major earthquakes from occurring, people in populated areas are reluctant to risk earthquake control, lest a major earthquake be initiated by the process.

to earthquakes in the past. Geologist David M. Evans suggested that the earthquakes were directly related to the injection of contaminated wastewater into a 3,674-m-deep disposal well at the Rocky Mountain Arsenal, northeast of Denver. A USGS study concluded that the pumping of waste fluids into the fractured rocks beneath the disposal well decreased the friction on opposites sides of the fractures, in effect lubricating them so that movement occurred, thus causing the earthquakes that Denver experienced.

Interestingly, a high degree of correlation was found when comparing the number of earthquakes in Denver against the average amount of contaminated fluids injected into the disposal well per month. Furthermore, when no waste fluids were injected, earthquake activity decreased dramatically.

Experiments conducted in 1969 at an abandoned oil field near Rangely, Colorado, confirmed the hypothesis that fluids injected into the fractured bedrock at the Rocky Mountain Arsenal was causing Denver's earthquakes. When geologists pumped water into abandoned oil wells, seismographs recorded small earthquakes in the area. When fluids were pumped out, earthquake activity declined. Geologists were thus starting and stopping earthquakes at will, and validating the relationship between pore-water pressures and earthquakes.

In a similar situation to Denver, the number of earthquakes in the Dallas–Ft. Worth area of Texas has dramatically increased such that there have been more earthquakes during 2008 and 2009 than in the previous 30 years combined. Beginning in 2004, thousands of gas wells have been drilled in the area, and almost all of them have undergone hydraulic fracturing, in which large amounts of a high-pressure water mix are injected into the wells to open up the preexisting fractures to allow gas to flow into the wells. Although no one is claiming there is a link between the hydraulic fracturing associated with the drilling and the onset of small earthquakes, geologists are looking closely at the situation to see if there is a cause-and-effect relationship between the two events.

Based on these results, some geologists have proposed that fluids be pumped into the locked segments or seismic gaps of active faults to cause small- to moderate-sized earthquakes. They think that this would relieve the pressure on the fault and prevent a major earthquake from occurring. Although this plan is intriguing, it also has many potential problems. For instance, there is no guarantee that only a small earthquake might result. Instead, a major earthquake might occur, causing tremendous property damage and loss of life, especially in a densely populated area (Figure 8.17). Who would be responsible? Certainly, a great deal more research is needed before such an experiment is performed, even in an area of low population density.

LO10 What Is Earth's Interior Like?

During most of historic time, Earth's interior was perceived as an underground world of vast caverns, heat, and sulfur gases, populated by demons. By the 1860s, scientists knew what the average density of Earth was and that pressure and temperature increased with depth. And even though Earth's interior is hidden from direct observation, scientists now have a reasonably good idea of its internal structure and composition.

Earth is generally depicted as consisting of concentric layers that differ in composition and density separated from adjacent layers by rather distinct boundaries (Figure 8.18). Recall that the outermost layer, or *crust*, is Earth's thin skin. Below the crust and extending about halfway to Earth's center is the *mantle*, which comprises more than 80% of Earth's volume. The central part of Earth consists of a *core*, which is divided into a solid inner portion and a liquid outer part (Figure 8.18).

The behavior and travel times of P- and S-waves provide geologists with information about Earth's internal structure. Seismic waves travel outward as wavefronts from their source areas, although it is most convenient to depict them as *wave rays*, which are lines showing the direction of movement of small parts of wavefronts (Figure 8.3).

As we noted earlier, P- and S-wave velocity is determined by the density and elasticity of the materials they travel through, both of which increase with depth. Wave velocity is slowed by increasing density but increases in materials with greater elasticity. Because elasticity increases with depth faster than density, a general increase in seismic wave velocity takes place as the waves penetrate to greater depths. P-waves travel faster than S-waves under all circumstances, but unlike P-waves, S-waves are not transmitted through liquids, because liquids have no shear strength (rigidity); liquids simply flow in response to shear stress.

Because Earth is not a homogeneous body, seismic waves travel from one material into another of different density and elasticity, and thus their velocity and direction of travel change. That is, the waves are bent, a phenomenon known as *refraction*, in much the same

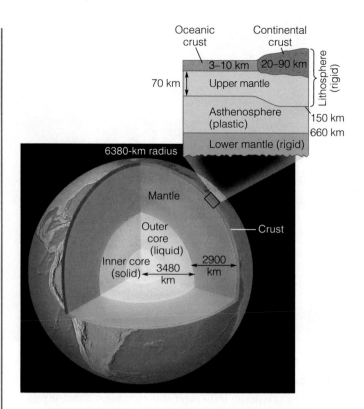

EARTH'S COMPOSITION AND DENSITY

	COMPOSITION	DENSITY (g/cm^3)
Continental crust	Average composition of granodiorite	~2.7
Oceanic crust	Upper part basalt, lower part gabbro	~3.0
Mantle	Peridotite (made up of ferromagnesian silicates)	3.3–5.7
Outer core	Iron with perhaps 12% sulfur silicon, oxygen, nickel, and potassium	9.9–12.2
Inner core	Iron with 10–20% nickel	12.6–13.0

Figure 8.18 Earth's Internal Structure
The inset shows Earth's outer part in more detail. The asthenosphere is solid but behaves plastically and flows.

way as light waves are refracted as they pass from air into more-dense water. Because seismic waves pass through materials of differing density and elasticity, they are continually refracted, so that their paths are curved; wave rays travel in a straight line only when their direction of travel is perpendicular to a boundary (Figure 8.19).

In addition to refraction, seismic rays are *reflected*, much as light is reflected from a mirror. When seismic rays encounter a boundary separating materials of

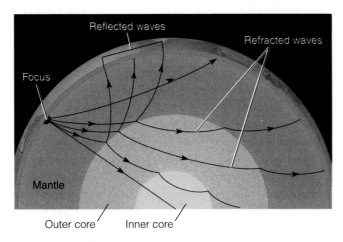

Figure 8.19 **Refraction and Reflection of Seismic Waves**
Refraction and reflection of P-waves as they encounter boundaries separating materials of different density or elasticity. Notice that the only wave ray not refracted is the one perpendicular to boundaries.

different density or elasticity, some of a wave's energy is reflected back to the surface (Figure 8.19). If we know the wave velocity and the time required for the wave to travel from its source to the boundary and back to the surface, we can calculate the depth of the reflecting boundary. Such information is useful in determining not only Earth's internal structure, but also the depths of sedimentary rocks that may contain petroleum. Seismic reflection is a common tool used in petroleum exploration.

Although changes in seismic wave velocity occur continuously with depth, P-wave velocity increases suddenly at the base of the crust and decreases abruptly at a depth of approximately 2,900 km (Figure 8.20). These marked changes in seismic wave velocity indicate a boundary called a **discontinuity**, across which a significant change in Earth materials or their properties occurs. Discontinuities are the basis for subdividing Earth's interior into concentric layers.

LO11 The Core

In 1906, R. D. Oldham of the Geological Survey of India realized that seismic waves arrived later than expected at seismic stations more than 130 degrees from an earthquake focus. He postulated that Earth has a core that transmits seismic waves more slowly than shallower Earth materials. We now know that P-wave velocity decreases markedly at a depth of 2,900 km, which indicates an important discontinuity now recognized as the core–mantle boundary (Figure 8.20).

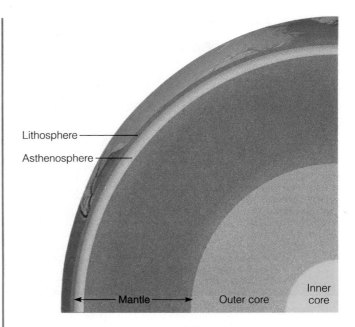

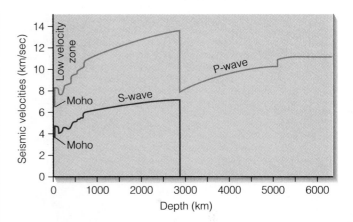

Figure 8.20 **Seismic Wave Velocities**
Profiles showing seismic wave velocities versus depth. Several discontinuities are shown, across which seismic wave velocities change rapidly.
SOURCE: FROM G. C. BROWN AND A. E. MUSSET, *THE INACCESSIBLE EARTH* (KLUWER ACADEMIC PUBLISHERS, 1981), FIGURE 12.7A. REPRINTED WITH PERMISSION OF THE AUTHOR.

Because of the sudden decrease in P-wave velocity at the core–mantle boundary, P-waves are refracted in the core, so that little P-wave energy reaches the surface in the area between 103 degrees and 143 degrees from an earthquake focus (Figure 8.21a). This **P-wave shadow zone**, as it is called, is an area in which little P-wave energy is recorded by seismographs.

The P-wave shadow zone is not a perfect

discontinuity A boundary across which seismic wave velocity or direction of travel changes abruptly, such as the mantle–core boundary.

P-wave shadow zone An area between 103 and 143 degrees from an earthquake focus where little P-wave energy is recorded by seismographs.

Figure 8.21 P-Wave and S-Wave Shadow Zones

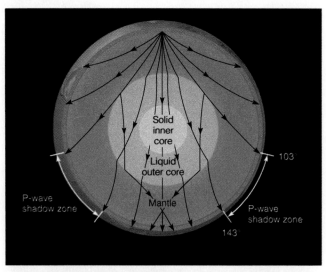

a. P-waves are refracted so that no direct P-wave energy reaches the surface in the P-wave shadow zone.

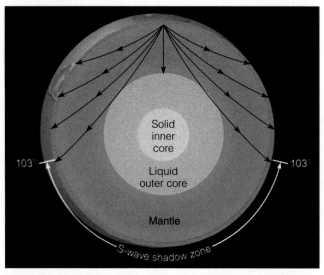

b. The presence of an S-wave shadow zone indicates that S-waves are being blocked within Earth.

shadow zone, because some weak P-wave energy is recorded within it. Scientists proposed several hypotheses to account for this observation, but all were rejected by the Danish seismologist Inge Lehmann, who in 1936 postulated that the core is not entirely liquid as previously thought. She proposed that seismic wave reflection from a solid inner core accounts for the arrival of weak

S-wave shadow zone
Those areas more than 103 degrees from an earthquake focus where no S-waves are recorded.

P-wave energy in the P-wave shadow zone, a proposal that was quickly accepted by seismologists.

In 1926, the British physicist Harold Jeffreys realized that S-waves were not simply slowed by the core, but were completely blocked by it. So, besides a P-wave shadow zone, a much larger and more complete **S-wave shadow zone** also exists (Figure 8.21b). At locations greater than 103 degrees from an earthquake focus, no S-waves are recorded, which indicates that S-waves cannot be transmitted through the core. S-waves will not pass through a liquid, so it seems that the outer core must be liquid or behave as a liquid. The inner core, however, is thought to be solid, because P-wave velocity increases at the base of the outer core.

DENSITY AND COMPOSITION OF THE CORE

The core constitutes 16.4% of Earth's volume and nearly one-third of its mass. Geologists can estimate the core's density and composition by using seismic evidence and laboratory experiments. Furthermore, meteorites, which are thought to represent remnants of the material from which the solar system formed, are used to make estimates of density and composition. From these studies, the density of the outer core has been determined to vary from 9.9 to 12.2 g/cm^3. At Earth's center, the pressure is equivalent to approximately 3.5 million times normal atmospheric pressure.

The core cannot be composed of minerals common at the surface, because, even under the tremendous pressures at great depth, they would still not be dense enough to yield an average density of 5.5 g/cm^3 for Earth. Both the outer and inner cores are thought to be composed largely of iron, but pure iron is too dense to be the sole constituent of the outer core. It must be "diluted" with elements of lesser density. Laboratory experiments and comparisons with iron meteorites indicate that perhaps 12% of the outer core consists of sulfur and possibly some silicon, oxygen, nickel, and potassium (Figure 8.18).

In contrast, pure iron is not dense enough to account for the estimated density of the inner core, so perhaps 10% to 20% of the inner core consists of nickel. These metals form an iron–nickel alloy thought to be sufficiently dense under the pressure at that depth to account for the density of the inner core.

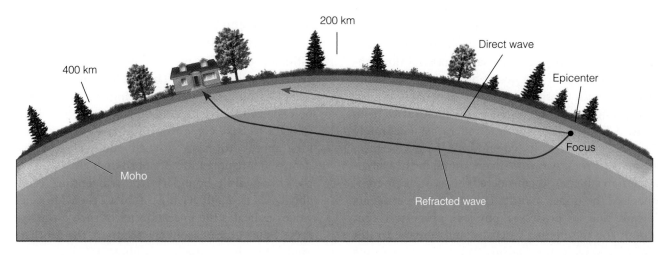

Figure 8.22 Seismic Discontinuity
Andrija Mohorovičić studied seismic waves and detected a seismic discontinuity at a depth of about 30 km. The deeper, faster seismic waves arrive at seismic stations first, even though they travel farther. This discontinuity, now known as the Moho, is between the crust and mantle.

LO12 Earth's Mantle

Another significant discovery about Earth's interior was made in 1909 when the Yugoslavian seismologist Andrija Mohorovičić detected a seismic discontinuity at a depth of about 30 km. While studying the arrival times of seismic waves from Balkan (part of southeastern Europe) earthquakes, Mohorovičić noticed that seismic stations a few hundred kilometers from an earthquake's epicenter were recording two distinct sets of P- and S-waves.

From his observations, Mohorovičić concluded that a sharp boundary separates rocks with different properties at a depth of about 30 km. He postulated that P-waves below this boundary travel at 8 km per second, whereas those above the boundary travel at 6.75 km per second. When an earthquake occurs, some waves travel directly from the focus to a seismic station, whereas others travel through the deeper layer, and some of their energy is refracted back to the surface (Figure 8.22). The waves traveling through the deeper layer (the mantle) travel farther to a seismic station, but they do so more rapidly and arrive before those that travel more slowly in the shallower layer.

The boundary identified by Mohorovičić separates the crust from the mantle and is now called the **Mohorovičić discontinuity,** or simply the **Moho.** It is present everywhere except beneath spreading ridges.

However, its depth varies: Beneath continents, it ranges from 20 to 90 km, with an average of 35 km; beneath the sea floor, it is 5 to 10 km deep.

THE MANTLE'S STRUCTURE, DENSITY, AND COMPOSITION

Although seismic wave velocity in the mantle increases with depth, several discontinuities exist. Between depths of 100 and 250 km, both P- and S-wave velocities decrease markedly. This 100- to 250-km-deep layer is the *low-velocity zone,* which corresponds closely to the *asthenosphere* (Figure 8.18), a layer in which the rocks are close to their melting point and are less elastic, accounting for the observed decrease in seismic wave velocity. The asthenosphere is an important zone, because it is where most magma is generated, especially under the ocean basins. Furthermore, it lacks strength, flows plastically, and is thought to be the layer over which the plates of the outer, rigid *lithosphere* move.

Other discontinuities are also present at deeper levels within the mantle. But unlike those between the crust and mantle or

> **Mohorovičić discontinuity (Moho)**
> The boundary between Earth's crust and mantle.

between the mantle and core, these probably represent structural changes in minerals rather than compositional changes. In other words, geologists think that the mantle is composed of the same material throughout, but the structural states of minerals such as olivine change with depth.

Although the mantle's density, which varies from 3.3 to 5.7 g/cm³, can be inferred rather accurately from seismic waves, its composition is less certain. The igneous rock *peridotite*, containing mostly ferromagnesian silicates, is considered the most likely component of the upper mantle. Laboratory experiments indicate that it possesses physical properties that account for the mantle's density and observed rates of seismic wave transmissions. Peridotite also forms the lower parts of igneous rock sequences thought to be fragments of the oceanic crust and upper mantle emplaced on land.

LO13 Earth's Internal Heat

During the 19th century, scientists realized that the temperature in deep mines increases with depth, and this same trend has been observed in deep drill holes. This temperature increase with depth, or **geothermal gradient**, is approximately 25°C/km near the surface. In areas of active or recently active volcanism, the geothermal gradient is greater than in adjacent nonvolcanic areas, and temperature rises faster beneath spreading ridges than elsewhere beneath the seafloor.

Much of Earth's internal heat is generated by radioactive decay, especially the decay of isotopes of uranium and thorium and, to a lesser degree, potassium-40. When these isotopes decay, they emit energetic particles and gamma rays that heat surrounding rocks. Because rock is such a poor conductor of heat, it takes little radioactive decay to build up considerable heat, given enough time.

Unfortunately, the geothermal gradient is not useful for estimating temperatures at great depth. If we were simply to extrapolate from the surface downward, the temperature at 100 km would be so high that, despite the great pressure, all known rocks would melt. Current estimates of the temperature at the base of the crust are 800°C to 1,200°C. The latter figure seems to be an upper limit; if it were any higher, melting would be expected. Furthermore, fragments of mantle rock, thought to have come from depths of 100 km to 300 km, appear to have reached equilibrium at these depths at a temperature of approximately 1,200°C. At the core–mantle boundary, the temperature is probably between 2,500°C and 5,000°C; the wide range of values indicates the uncertainties of such estimates. If these figures are reasonably accurate, the geothermal gradient in the mantle is only about 1°C/km.

geothermal gradient
Earth's temperature increase with depth; it averages 25°C/km near the surface but varies from area to area.

LO14 Earth's Crust

Our main concern in the latter part of this chapter is Earth's interior; however, to be complete, we must briefly discuss the crust, which along with the upper mantle constitutes the lithosphere.

Continental crust is complex, consisting of all rock types, but it is usually described as "granitic," meaning that its overall composition is similar to that of granitic rocks. With the exception of metal-rich rocks such as iron ore deposits, most rocks of the continental crust have densities between 2.5 g/cm³ and 3.0 g/cm³, with the average density of the crust being about 2.7 g/cm³. P-wave velocity in continental crust is about 6.75 km/sec, but at the base of the crust, P-wave velocity abruptly increases to about 8 km/sec. Continental crust averages 35 km thick, but its thickness varies from 20 km to 90 km. Beneath mountain ranges such as the Rocky Mountains, the Alps in Europe, and the Himalayas in Asia, continental crust is much thicker than it is in adjacent areas. In contrast, continental crust is much thinner than average beneath the Rift Valleys of East Africa and in a large area called the Basin and Range Province in the western United States and northern Mexico. The crust in these areas has been stretched and thinned in what appear to be the initial stages of rifting (see Chapter 2).

In contrast to continental crust, oceanic crust is simpler, consisting of gabbro in its lower part and overlain by basalt. It is thinnest, about 5 km, at spreading ridges, and nowhere is it thicker than 10 km. Its average density of 3.0 g/cm^3 accounts for the fact that it transmits P-waves at about 7 km/sec. In fact, this P-wave velocity is what one would expect if oceanic crust is composed of basalt and gabbro.

CHAPTER 9
THE SEAFLOOR

Underwater view of a reef in Hawaii. Coral reefs, as they are commonly called, are actually composed of the skeletons of corals, various mollusks (such as clams), as well as encrusting organisms including sponges and algae.

COURTESY OF JEFF AND DIANNA MONROE

Introduction

"Despite the commonly held misconception that the seafloor is flat and featureless, its topography is as varied as that of the continents."

According to two dialogues written in about 350 B.C. by the Greek philosopher Plato, there was a huge continent called Atlantis in the Atlantic Ocean west of the Pillars of Hercules, or what we now call the Strait of Gibraltar. According to Plato's account, Atlantis controlled a large area extending as far east as Egypt. Yet despite its vast wealth, advanced technology, and large army and navy, Atlantis was defeated in war by Athens. Following the conquest of Atlantis,

there were violent earthquakes and floods
and one terrible day and night came when . . .
Atlantis . . . disappeared beneath the sea.
And for this reason even now the sea there
has become unnavigable and unsearchable,
blocked as it is by the mud shallows which
the island produced as it sank.*

LEARNING OUTCOMES

After reading this unit, you should be able to do the following:

LO1 Examine the history and methods of oceanic exploration

LO2 Describe the structure and composition of the oceanic crust

LO3 Identify the continental margins

LO4 Discuss the features found in the deep-ocean basins

LO5 Discuss sedimentation and sediments on the deep seafloor

LO6 Explore coral reefs

LO7 Recognize the types natural resources found in the oceans

No "mud shallows" exist in the Atlantic, as Plato asserted. In fact, no geologic evidence indicates that Atlantis ever existed, so why has the legend persisted for so long? One reason is that sensational stories of lost civilizations are popular, but another is that until recently no one had much knowledge of what lies beneath the oceans. Much of the seafloor is a hidden domain, so myths and legends were widely accepted. The most basic observation we can make about Earth is that it has vast water-covered areas and continents, which, at first glance, might seem to be nothing more than parts of the planet not covered by water. Nevertheless, the continents and the ocean basins are very different.

Oceanic crust is thinner and denser than continental crust, and it is made up of basalt and gabbro, whereas continental crust has an overall composition similar to that of granite. In addition, oceanic crust is produced continually at spreading ridges and consumed at subduction zones, so none of it is older than about 180 million years. Continental crust, on the other hand, varies in age from recent to nearly 4.06 billion years old.

One important reason to study the seafloor is that it makes up the largest part of Earth's surface (Figure 9.1). Despite the commonly held misconception that the seafloor is flat and featureless, its topography is as varied as that of the continents. Furthermore, many seafloor features, as well as several aspects of the oceanic crust, provide important evidence for plate tectonic theory. And finally, natural resources are found on the marginal parts of continents, in seawater, and on the seafloor.

Our discussion of the seafloor focuses on (1) the physical attributes and composition of the oceanic crust, (2) the composition and distribution of seafloor sediments, (3) seafloor topography, and (4) the origin and evolution of the continental margins. *Oceanographers* study these topics, too, but they also study the chemistry and physics of seawater, as well as oceanic circulation and marine biology.

*From the Timaeus, quoted in E. W. Ramage, Ed., *Atlantis: Fact or Fiction?* (Bloomington: Indiana University Press, 1978), p. 13.

LO1 Exploring the Oceans

An interconnected body of saltwater that we call oceans and seas covers 71% of Earth's surface. Nevertheless, this world ocean has areas distinct enough for us to recognize the Pacific, Atlantic, Indian, and Arctic oceans. The term *ocean* refers to these large areas of saltwater, whereas *sea* designates a smaller body of water, usually a marginal part of an ocean (Figure 9.1). We should point out that whereas the oceans and their marginal seas are largely underlain by oceanic crust, the same is not true of the Dead Sea, Salton Sea, and Caspian Sea; these are actually saline lakes on the continents.

During most of historic time, people knew little of the oceans, and until recently they thought the seafloor was a vast, featureless plain. In fact, throughout most of this time, the seafloor, in one sense, was more remote than the Moon's surface, because it could not be observed.

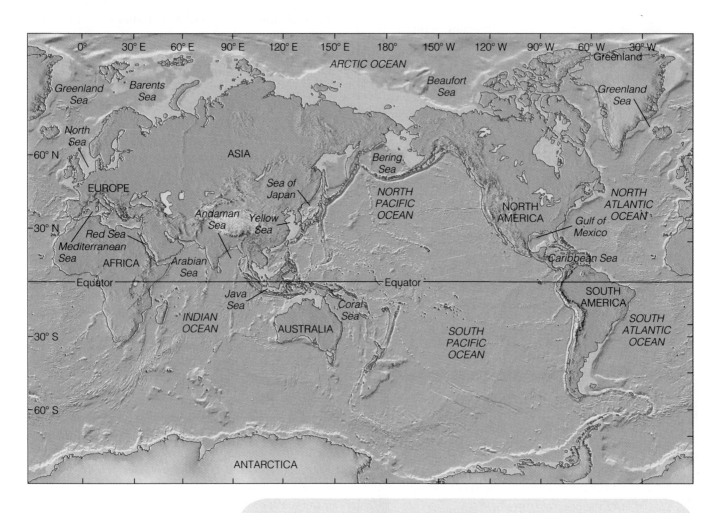

Figure 9.1 The Seafloor

This map shows the four oceans and many of the seas, which are marginal parts of oceans. The seafloor constitutes the largest part of Earth's surface.

NUMERIC DATA FOR THE OCEANS

OCEAN*	SURFACE AREA (million km²)	WATER VOLUME (million km³)	AVERAGE DEPTH (km)	MAXIMUM DEPTH (km)
Pacific	180	700	4.0	11.0
Atlantic	93	335	3.6	9.2
Indian	77	285	3.7	7.5
Arctic	15	17	1.1	5.2

Source: P. R. Pinet, 1992. Oceanography (St. Paul, MN: West, 1992).

*Excludes adjacent seas, such as the Caribbean Sea and Sea of Japan, which are marginal parts of oceans.

EARLY EXPLORATION

The ancient Greeks had determined Earth's size and shape rather accurately, but western Europeans were not aware of the vastness of the oceans until the 1400s and 1500s, when explorers sought trade routes to the Indies. Even when Christopher Columbus set sail on August 3, 1492, in an effort to find a route to the Indies, he greatly underestimated the width of the Atlantic Ocean. Contrary to popular belief, he was not attempting to demonstrate Earth's spherical shape; its shape was well accepted by then. The controversy was over Earth's circumference and the shortest route to China; on these points, Columbus's critics were correct.

These and similar voyages added considerably to the growing body of knowledge about the oceans, but truly scientific investigations did not begin until the late 1700s, when the British sought to increase their knowledge of the oceans. Scientific voyages led by Captain James Cook were launched in 1768, 1772, and 1777. From 1831 until 1836, the HMS *Beagle* sailed the seas. Aboard was Charles Darwin, who is well known for his views of organic evolution, but who also proposed a theory on the evolution of coral reefs. In 1872, the converted British warship HMS *Challenger* began a four-year voyage to sample seawater, determine oceanic depths, collect samples of seafloor sediment and rock, and name and classify thousands of species of marine organisms.

During these voyages, many oceanic islands previously unknown to Europeans were visited. Even though exploration of the oceans was limited, scientists discovered that the seafloor has varied topography just as continents do, and they recognized such features as oceanic trenches, submarine ridges, broad plateaus, hills, and vast plains.

HOW ARE OCEANS EXPLORED TODAY?

Measuring the length of a weighted line lowered to the seafloor was the first method used to determine ocean depths. Now scientists use an instrument called an *echo sounder*, which detects sound waves that travel from a ship to the seafloor and back. Depth is calculated by knowing the velocity of sound in water and the time required for the waves to reach the seafloor and return to the ship. **Seismic profiling** is similar to echo sounding but is even more useful. Strong waves from

AP PHOTO/XINHUA, MA PING

Figure 9.2 Oceanographic Research Vessels
The R/V *Chikyu* ("Earth"), a research ship in the Integrated Ocean Drilling Program.

an energy source reflect from the seafloor, and some of the waves penetrate seafloor layers and reflect from various horizons back to the surface. Seismic profiling is particularly useful for determining the structure of the oceanic crust where it is buried beneath seafloor sediments.

An international program called the Deep Sea Drilling Project began in 1968 with its research vessel the *Glomar Challenger*, but after 15 years of oceanographic research, it was replaced by the Ocean Drilling Program's research vessel JOIDES* *Resolution*. Both vessels were equipped to drill into the seafloor and retrieve samples of sediment and oceanic crust. Then, in 2003, the responsibilities passed to an even larger research vessel, the R/V *Chikyu* ("Earth"), a Japanese ship that can drill as much as 11 kilometers into the seafloor (Figure 9.2).

Submersibles are also used extensively in oceanographic research. One of the most famous of these is *Alvin*, which has carried scientists to the seafloor in many areas to make observations and collect samples. Other submersibles are

> **seismic profiling** A method in which strong waves generated at an energy source penetrate the layers beneath the seafloor. Some of the energy is reflected back from various layers to the surface, making it possible to determine the nature of the layers.

---

*JOIDES is an acronym for Joint Oceanographic Institutions for Deep Earth Sampling.

remotely controlled and towed by surface ships. In 1985, the *Argo*, equipped with sonar and radar systems, provided the first views of the British ocean liner HMS *Titanic* since it sank after hitting an iceberg in 1912. Another remotely operated submersible, the *Kaiko*, operated by the Japanese, has descended to the greatest oceanic depths.

© BANANASTOCK/JUPITER IMAGES

LO2 Oceanic Crust: Its Structure and Composition

We have mentioned that oceanic crust is composed of basalt and gabbro and is generated continuously at spreading ridges. Drilling into the oceanic crust provides some details about its composition and structure, but it has never been completely penetrated and sampled. So how do we know what it is composed of and how it varies with depth? Actually, even before it was sampled and observed, these details were known.

ophiolite A sequence of igneous rocks representing a fragment of oceanic lithosphere; composed of peridotite overlain successively by gabbro, sheeted basalt dikes, and pillow lava.

Remember that oceanic crust is consumed at subduction zones, and thus most of it is recycled, but a small amount is found in mountain ranges on land where it was emplaced by moving along large fractures called thrust faults. These preserved slivers of oceanic crust, along with part of the underlying upper mantle, are known as **ophiolites**. Detailed studies reveal that an ideal ophiolite consists of rocks of the upper oceanic crust, especially pillow lava and sheet lava flows (Figure 9.3) underlain by a sheeted dike complex consisting of vertical basaltic dikes, and then massive gabbro and layered gabbro that probably formed in the upper part of a magma chamber. And finally, the lowermost unit is peridotite from the upper mantle; this is sometimes altered by metamorphism to a greenish rock known as serpentinite. Thus, a complete ophiolite consists of oceanic crust rocks and upper mantle (Figure 9.3).

Sampling and drilling at oceanic ridges reveal that oceanic crust is indeed made up of pillow lava and sheet lava flows underlain by a sheeted dike complex, just as predicted from studies of ophiolites. But it was not until 1989 that a submersible carrying scientists descended to the walls of a seafloor fracture in the North Atlantic and verified what lay below the sheeted dike complex.

Figure 9.3 Composition of Oceanic Crust

New oceanic crust made up of the layers shown here forms as magma rises beneath oceanic ridges. The composition of oceanic crust was known from ophiolites, sequences of rock on land consisting of deep-sea sediments, oceanic crust, and upper mantle, before scientists observed it when they descended in submersibles to seafloor fractures.

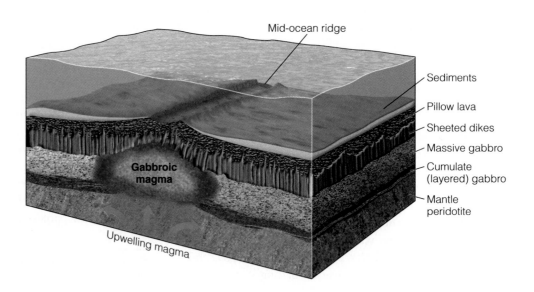

LO3 The Continental Margins

Most people think of continents as land areas outlined by the oceans, but the true geologic margin of a continent—where granitic continental crust changes to basalt and gabbro oceanic crust—is below sea level. A **continental margin** is made up of a gently sloping continental shelf, a more steeply inclined continental slope, and, in some cases, a deeper, gently sloping continental rise (Figure 9.4). Thus, the continental margins extend to increasingly greater depths until they merge with the deep seafloor. Continental crust changes to oceanic crust somewhere beneath the continental rise, so part of the continental slope and the continental rise actually rest on oceanic crust.

THE CONTINENTAL SHELF

As one proceeds seaward from the shoreline, the first area encountered is the gently sloping **continental shelf** lying between the shore and the more steeply dipping continental slope (Figure 9.4). The width of the continental shelf varies from a few tens of meters to more than 1,000 km; the shelf terminates where the inclination of the seafloor increases abruptly from 1 degree or less to several degrees. The outer margin of the continental shelf, or simply the *shelf–slope break*, is at an average depth of 135 m, so by oceanic standards, the continental shelves are covered by shallow water.

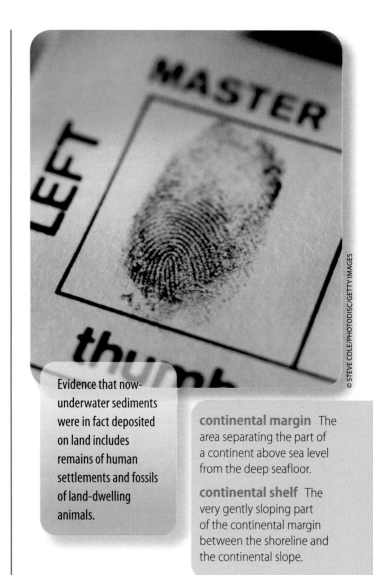

Evidence that now-underwater sediments were in fact deposited on land includes remains of human settlements and fossils of land-dwelling animals.

continental margin The area separating the part of a continent above sea level from the deep seafloor.

continental shelf The very gently sloping part of the continental margin between the shoreline and the continental slope.

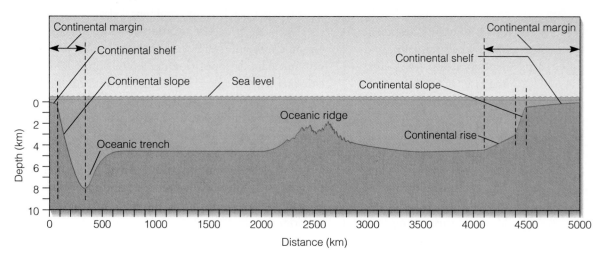

Figure 9.4 Features of Continental Margins
A generalized profile showing features of the continental margins. The vertical dimensions of the features in this profile are greatly exaggerated, because the vertical and horizontal scales differ.

At times during the Pleistocene Epoch (1.8 million to 10,000 years ago), sea level was as much as 130 m lower than it is now. As a result, the continental shelves were above sea level, where deposition in stream channels and floodplains took place. In addition, in many parts of northern Europe and North America, glaciers extended well out onto the continental shelves and deposited gravel, sand, and mud. Since the Pleistocene ended, sea level has risen, submerging these deposits, which are now being reworked by marine processes.

THE CONTINENTAL SLOPE AND RISE

The seaward margin of the continental shelf is marked by the *shelf–slope break*, where the more steeply inclined **continental slope** begins (Figure 9.4). In most areas around the Atlantic, the continental slope merges with a more gently sloping **continental rise**. This rise is absent around the margins of the Pacific, where continental slopes descend directly into an oceanic trench (Figure 9.4).

The shelf–slope break is an important feature in terms of sediment transport and deposition. Landward of the break—that is, on the shelf—sediments are affected by waves and tidal currents, but these processes have no effect on sediments seaward of the break, where gravity is responsible for their transport and deposition on the slope and rise. In fact, much of the land-derived sediment crosses the shelves and is eventually deposited on the continental slopes and rises, where more than 70% of all sediments in the oceans are found. Much of this sediment is transported through submarine canyons by turbidity currents.

continental slope The relatively steeply inclined part of the continental margin between the continental shelf and the continental rise or between the continental shelf and an oceanic trench.

continental rise The gently sloping part of the continental margin between the continental slope and the abyssal plain.

turbidity current A sediment-water mixture, denser than normal seawater, that flows downslope to the deep seafloor.

submarine fan A cone-shaped sedimentary deposit that accumulates on the continental slope and rise.

submarine canyon A steep-walled canyon best developed on the continental slope, but some extend well up onto the continental shelf.

SUBMARINE CANYONS, TURBIDITY CURRENTS, AND SUBMARINE FANS

Most graded bedding results from **turbidity currents**, underwater flows of sediment–water mixtures with densities greater than that of sediment-free water. As a turbidity current flows onto the relatively flat seafloor, it slows and begins depositing sediment, the largest particles first, followed by progressively smaller particles, thus forming a layer with graded bedding (see Figure 6.19). Deposition by turbidity currents yields a series of overlapping **submarine fans**, which constitute a large part of the continental rise (Figure 9.5a).

No one has ever observed a turbidity current in progress, so for many years, some doubted their existence. Perhaps the most compelling evidence for turbidity currents is the pattern of trans-Atlantic cable breaks that took place in the North Atlantic near Newfoundland on November 18, 1929. The breaks on the continental shelf near the epicenter occurred when an earthquake struck, but cables farther seaward were broken later and in succession (Figure 9.5b). The last cable to break was 720 km from the source of the earthquake, and it did not snap until 13 hours after the first break. In 1949, geologists realized that an earthquake-generated turbidity current had moved downslope, breaking the cables in succession. The precise time at which each cable broke was known, so calculating the velocity of the turbidity current was simple. It moved at about 80 km per hour on the continental slope, but slowed to about 27 km per hour when it reached the continental rise.

Deep, steep-sided **submarine canyons** are present on continental shelves, but they are best developed on continental slopes (Figure 9.5a). Some submarine canyons extend across the shelf to rivers on land and apparently formed as river valleys when sea level was lower during the Pleistocene. However, many have no such association, and some extend far deeper than can be accounted for by river erosion during times of lower sea level. Scientists know that turbidity currents periodically move through these canyons, and now think they are the primary agents responsible for their erosion.

Figure 9.5 Submarine Fans and Graded Bedding

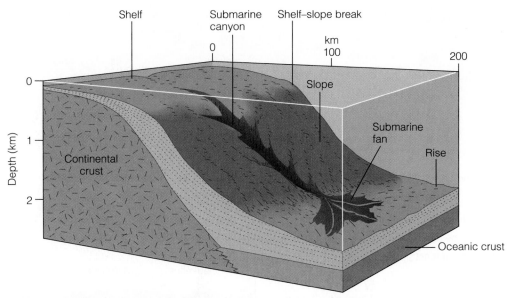

a. Much of the continental rise is made up of submarine fans that form by deposition of sediments carried down submarine canyons by turbidity currents.

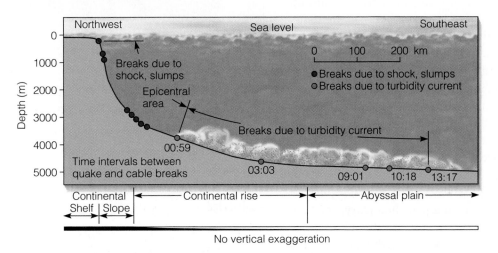

b. Submarine cable breaks caused by a turbidity current south of Newfoundland in 1929.

TYPES OF CONTINENTAL MARGINS

An **active continental margin** develops at the leading edge of a continental plate where oceanic lithosphere is subducted. The western margin of South America is a good example of where an oceanic plate is subducted beneath the continent, resulting in seismic activity, a geologically young mountain range, and active volcanism (Figure 9.6). In addition, the continental shelf is narrow, and the continental slope descends directly into an

active continental margin A continental margin with volcanism and seismicity at the leading edge of a continental plate where oceanic lithosphere is subducted.

Figure 9.6 Passive and Active Continental Margins

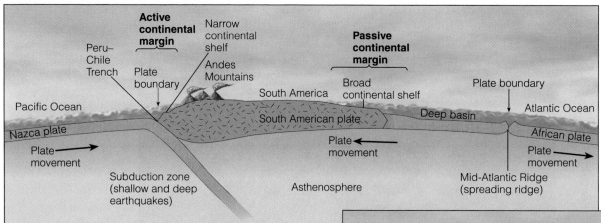

a. Active and passive continental margins along the west and east coasts of South America. Notice that the passive margins are much wider than active margins. Seafloor sediment is not shown.

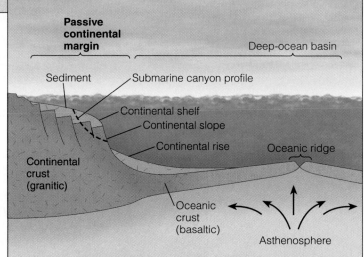

b. More detailed view of a passive continental margin showing fractures that formed during rifting and sediment accumulation. Sediment is also found on the shelves and in oceanic trenches along active continental margins. The vertical dimension in this illustration has been greatly exaggerated.

oceanic trench, so sediment is dumped into the trench and no continental rise develops. The western margin of North America is also considered an active continental margin, although much of it is now bounded by transform faults rather than a subduction zone. However, plate convergence and subduction continue in the Pacific Northwest along the continental margins of northern California, Oregon, and Washington.

The continental margins of eastern North America and South America differ considerably from their western margins. For one thing, they possess broad continental shelves, as well as a continental slope and rise, with *abyssal plains* adjacent to the rises (Figure 9.6). Furthermore, these **passive continental margins** are within a plate rather than at a plate boundary, and they lack the volcanic and seismic activity found at active continental margins. Nevertheless, earthquakes do take place occasionally.

Active and passive continental margins share some features, but they are also notably different. At both types of continental margins, turbidity currents transport sediment into deeper water. At passive margins, the sediment forms a series of overlapping submarine fans and thus develops a continental rise, whereas at an active margin, sediment is simply dumped into the trench and no rise forms. The proximity of a trench to a continent

passive continental margin A continental margin within a tectonic plate as in the East Coast of North America, where little seismic activity and no volcanism occur; characterized by a broad continental shelf and a continental slope and rise.

also explains why the continental shelves of active margins are so narrow. In contrast, land-derived sedimentary deposits at passive margins have built a broad platform extending far out into the ocean.

LO4 What Features Are Found in the Deep-Ocean Basins?

The seafloor has an average depth of 3.8 km, so most of it lies far below the depth of sunlight penetration, which is generally less than 100 m. Accordingly, the deep seafloor is completely dark, no plant life exists, the temperature is just above freezing, and the pressure varies from 200 to more than 1,000 atmospheres depending on depth. In fact, biologic productivity is low on the deep seafloor, with the exception of hydrothermal vent communities (discussed later).

ABYSSAL PLAINS

Beyond the continental rises of passive continental margins are **abyssal plains**, flat surfaces covering vast areas of the seafloor. In some areas, they are interrupted by peaks rising more than 1 km, but abyssal plains are nevertheless the flattest, most featureless areas on Earth (Figure 9.7). Their flatness is a result of sediment deposition covering the rugged topography.

Abyssal plains are invariably found adjacent to continental rises, which are composed mostly of overlapping submarine fans. Along active continental margins, sediments derived from the shelf and slope are trapped in an oceanic trench, and abyssal plains fail to develop. Accordingly, abyssal plains are common in the Atlantic Ocean basin but rare in the Pacific Ocean basin (Figure 9.7).

OCEANIC TRENCHES

Long, steep-sided depressions on the seafloor near convergent plate boundaries, called **oceanic trenches**, constitute no more than 2% of the seafloor, but they are important features, because it is here that oceanic lithosphere is consumed by subduction. Because oceanic trenches are found along active continental margins, they are common in the Pacific Ocean basin, but largely lacking in the Atlantic, those in the Caribbean being notable exceptions (Figure 9.7). On the landward sides of oceanic trenches, the continental slope descends into them at up to 25 degrees, and many have thick accumulations of sediments.

Sensitive instruments can detect the amount of heat escaping from Earth's interior by the phenomenon of *heat flow*. Heat flow is greatest in areas of active or

abyssal plain Vast flat area on the seafloor adjacent to the continental rise of a passive continental margin.

oceanic trench A long, narrow feature restricted to active continental margins and along which subduction occurs.

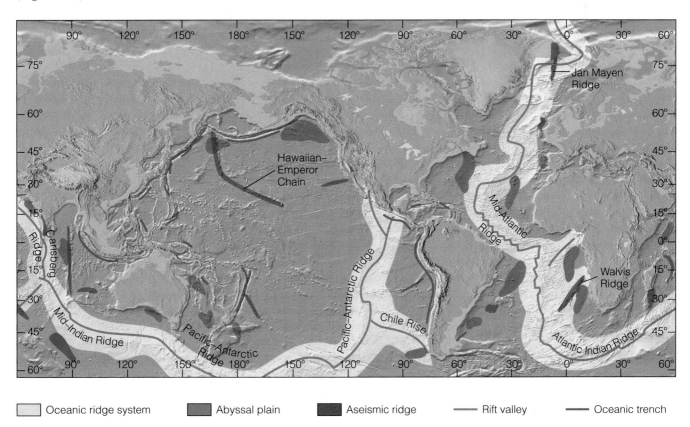

| Oceanic ridge system | Abyssal plain | Aseismic ridge | —— Rift valley | —— Oceanic trench |

Figure 9.7 Deep Seafloor Fractures
Features found on the deep seafloor include oceanic trenches (brown), abyssal plains (green), the oceanic ridge system (yellow), rift valleys (red), and some aseismic ridges (blue). Other features such as seamounts and guyots are shown in Figure 9.10.

SOURCE: FROM ALYN AND ALISON DUXBURY, *AN INTRODUCTION TO THE WORLD'S OCEANS.* COPYRIGHT © 1984 MCGRAW-HILL COMPANIES, INC. REPRINTED WITH PERMISSION.

recently active volcanism. For instance, higher-than-average heat flow takes place at spreading ridges, but at subduction zones heat flow values are less than the average for Earth as a whole. Oceanic crust at oceanic trenches is cooler and slightly denser than elsewhere.

Seismic activity also takes place at or near oceanic trenches along planes dipping at about 45 degrees. In Chapter 8, we discussed these inclined seismic zones called Benioff zones, where most of Earth's intermediate and deep earthquakes occur. Volcanism does not take place in trenches, but because these are zones where oceanic lithosphere is subducted beneath either oceanic or continental lithosphere, an arcuate chain of volcanoes is found on the overriding plate. The Aleutian Islands and the volcanoes along the western margin of South America are good examples of such chains (see Figure 2.14a and b).

OCEANIC RIDGES

When the first submarine cable was laid between North America and Europe during the late 1800s, a feature called the Telegraph Plateau was discovered in the North Atlantic. Using this data and data from a 1925–27 voyage of the German research vessel *Meteor*, scientists proposed that the plateau was actually a continuous ridge extending the length of the Atlantic Ocean basin. Subsequent investigations revealed that this conjecture was correct, and we now call this feature the Mid-Atlantic Ridge (Figure 9.7).

The Mid-Atlantic Ridge is more than 2,000 km wide and rises 2 to 2.5 km above the adjacent seafloor. Furthermore, it is part of a much larger **oceanic ridge** system of mostly submarine mountainous topography. The entire system is at least 65,000 km long, far exceeding the length of any mountain system on land. Oceanic ridges are composed almost entirely of basalt and gabbro and possess features produced by tensional forces. Mountain ranges on land, in contrast, consist of all rock types, and they formed when rocks were folded and fractured by compressive forces.

oceanic ridge A mostly submarine mountain system composed of basalt found in all ocean basins.

submarine hydrothermal vent A crack or fissure in the seafloor through which superheated water issues.

black smoker A type of submarine hydrothermal vent that emits a black plume of hot water colored by dissolved minerals.

Oceanic ridges are mostly below sea level, but they rise above the sea in Iceland, the Azores, Easter Island, and several other places. Of course, oceanic ridges are the sites where new oceanic crust is generated and plates diverge. The rate of plate divergence is important, because it determines the cross-sectional profile of a ridge. For example, the Mid-Atlantic Ridge has a comparatively steep profile because divergence is slow, allowing the new oceanic crust to cool, shrink, and subside closer to the ridge crest than it does in areas of faster divergence such as the East Pacific Rise. A ridge may also have a rift along its crest that opens in response to tension (Figure 9.8a). A rift is particularly obvious along the Mid-Atlantic Ridge, but it is absent along parts of the East Pacific Rise. These rifts are commonly 1 to 2 km deep and several kilometers wide. They open as seafloor spreading takes place and are characterized by shallow-focus earthquakes, basaltic volcanism, and high heat flow.

As part of Project FAMOUS (French-American Mid-Ocean Undersea Study), submersibles have descended to the ridges and into their rifts in several areas. Researchers have not seen any active volcanism, but they did see pillow lavas, lava tubes, and sheet lava flows, some of which formed very recently. In fact, on return visits to a site, they have seen the effects of volcanism that occurred since their previous visit. And on January 25, 1998, a submarine volcano began erupting along the Juan de Fuca Ridge west of Oregon.

SUBMARINE HYDROTHERMAL VENTS

Scientists first saw **submarine hydrothermal vents** on the seafloor in 1979 when they descended about 2,500 m to the Galapagos Rift in the eastern Pacific Ocean. Since 1979, they have seen similar vents in several other areas in the Pacific, Atlantic, and Indian Oceans and the Sea of Japan (Figure 9.8b). The vents are at or near spreading ridges where cold seawater seeps through oceanic crust, is heated by the hot rocks at depth, and then rises and discharges into the seawater as plumes of hot water with temperatures as high as 400°C. Many of the plumes are black, because dissolved minerals give them the appearance of black smoke—hence the name **black smoker** (Figure 9.8b).

Submarine hydrothermal vents are interesting from a biologic, geologic, and economic point of view. Near the vents live communities of organisms, such as bacteria, crabs, mussels, starfish, and tube worms, many

Figure 9.8 Central Rift and Submarine Hydrothermal Vents

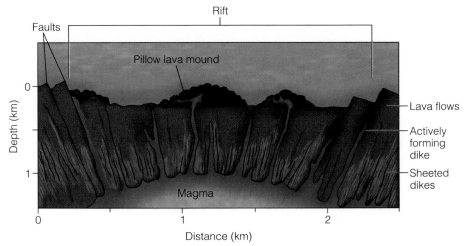

a. Cross section of the Mid-Atlantic Ridge showing its central rift where recent moundlike accumulations of volcanic rocks, mostly pillow lava, are found.

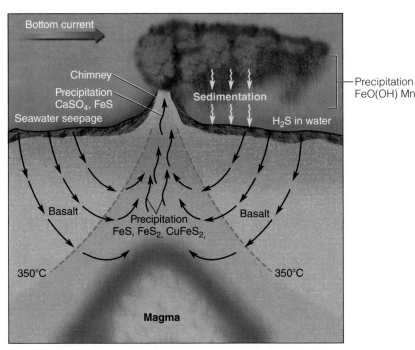

b. Cross section showing the origin of a submarine hydrothermal vent called a black smoker.

PHOTOGRAPHER: P. RONA, OAR/NATIONAL UNDERSEA RESEARCH PROGRAM/NOAA.

c. This black smoker on the East Pacific Rise is at a depth of 2,800 m. The plume of "black smoke" is heated seawater with dissolved minerals.

of which had never been seen before. No sunlight is available, so these organisms depend on bacteria that oxidize sulfur compounds for their ultimate source of nutrients, a process called *chemosynthesis*. The vents are also interesting because of their economic potential. The heated seawater reacts with oceanic crust,

transforming it into a metal-rich solution that discharges into seawater and cools, precipitating iron, copper, and zinc sulfides and other minerals. A chimney-like vent forms that eventually collapses and forms a mound of sediments rich in the elements mentioned above. Apparently, the chimneys through which black

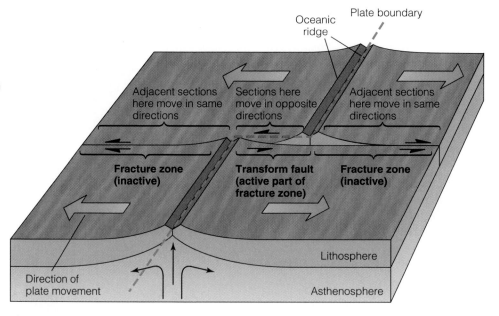

Figure 9.9 Seafloor Fractures
Diagrammatic view of an oceanic ridge offset along a fracture. That part of the fracture between displaced segments of the ridge is a transform fault. Recall from Chapter 2 that transform faults are one type of plate boundary.

smokers discharge grow rapidly. A 10-m-high chimney accidentally knocked over in 1991 by the submersible *Alvin* grew to 6 m in just three months.

In 2001, scientists announced another kind of seafloor vent in the North Atlantic responsible for massive pillars and spires as tall as 60 m. Unlike the black smokers, these vents are 14 to 15 km from spreading ridges, and they consist of light-colored minerals that were derived by chemical reaction between seawater and minerals in the oceanic crust.

SEAFLOOR FRACTURES

Oceanic ridges abruptly terminate where they are offset along fractures oriented more or less at right angles to ridge axes (Figure 9.9). These fractures are hundreds of kilometers long, although they are difficult to trace where they are buried beneath seafloor sediments. Many geologists are convinced that some geologic features on the continents are best explained by the extension of these fractures into continents.

Shallow-focus earthquakes take place along these fractures, but only between the displaced ridge segments. Furthermore, because ridges are higher than the adjacent seafloor, the offset segments yield nearly vertical escarpments 2 or 3 km high (Figure 9.9). The reason oceanic ridges have so many fractures is that plate divergence takes place irregularly on a sphere, resulting in stresses that cause fracturing.

SEAMOUNTS, GUYOTS, AND ASEISMIC RIDGES

As previously noted, the seafloor is not a flat, featureless plain except for the abyssal plains. In fact, a large number of volcanic hills, seamounts, and guyots rise above the seafloor in all ocean basins, but they are particularly abundant in the Pacific. All are of volcanic origin and differ mostly in size. **Seamounts** rise more than 1 km above the seafloor, and if flat topped, they are called guyots (Figure 9.10). **Guyots** are volcanoes that originally extended above sea level. However, as the plate upon which they were located continued to move, they were carried away from a spreading ridge, and as the oceanic crust cooled, it descended to greater depths. Thus, what was once an island slowly sank beneath the sea, and as it did, wave erosion produced the typical flat-topped appearance (Figure 9.10). Many other volcanic features smaller than seamounts exist on the seafloor, but they probably originated in the same way. These so-called *abyssal hills* average only about 250 m high.

Other common features in the ocean basins are long, narrow ridges and broad plateau-like features rising as much as 2 to 3 km above the surrounding

seamount A submarine volcanic mountain rising at least 1 km above the seafloor.

guyot A flat-topped seamount of volcanic origin rising more than 1 km above the seafloor.

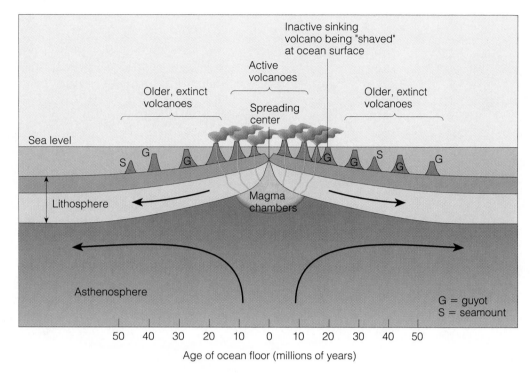

Figure 9.10 The Origin of Seamounts and Guyots
As the plate on which a volcano rests moves into greater water depths, the submerged volcanic island is called a seamount. Those that are flat-topped are called guyots.

seafloor. These **aseismic ridges** are so called because they lack seismic activity. A few of these ridges are probably small fragments separated from continents during rifting and are referred to as *microcontinents*.

Most aseismic ridges form as a linear succession of hot-spot volcanoes. These may develop at or near an oceanic ridge, but each volcano so formed is carried laterally with the plate upon which it originated. The net result is a line of seamounts/guyots extending from an oceanic ridge (Figure 9.10). Aseismic ridges also form over hot spots unrelated to ridges—the Hawaiian–Emperor chain in the Pacific, for example (see Figure 2.18).

LO5 Sedimentation and Sediments on the Deep Seafloor

Sediments on the deep seafloor consist mostly of silt- and clay-sized particles, because few processes transport sand and gravel very far from land. Certainly, icebergs carry sand and gravel, and, in fact, a broad band of glacial-marine sediment is adjacent to Antarctica and Greenland. Floating vegetation might also carry large particles far out to sea, but it contributes very little sediment to the deep seafloor.

Most of the sediment on the deep seafloor is derived from (1) windblown dust and volcanic ash from the continents and volcanic islands, and (2) the shells of microscopic plants and animals that live in the near-surface waters. Minor sources are chemical reactions in seawater that yield manganese nodules found in all ocean basins and cosmic dust. Perhaps as many as 40,000 metric tons of cosmic dust fall to Earth each year, but this is a trivial quantity compared to the volume of sediment derived from the two primary sources.

Most sediment on the deep seafloor is *pelagic*, meaning that it settled from suspension far from land (Figure 9.11). Pelagic sediment is further characterized as pelagic clay and ooze. **Pelagic clay**

aseismic ridge A ridge or broad area rising above the seafloor that lacks seismic activity.

pelagic clay Brown or red deep-sea sediment composed of clay-sized particles.

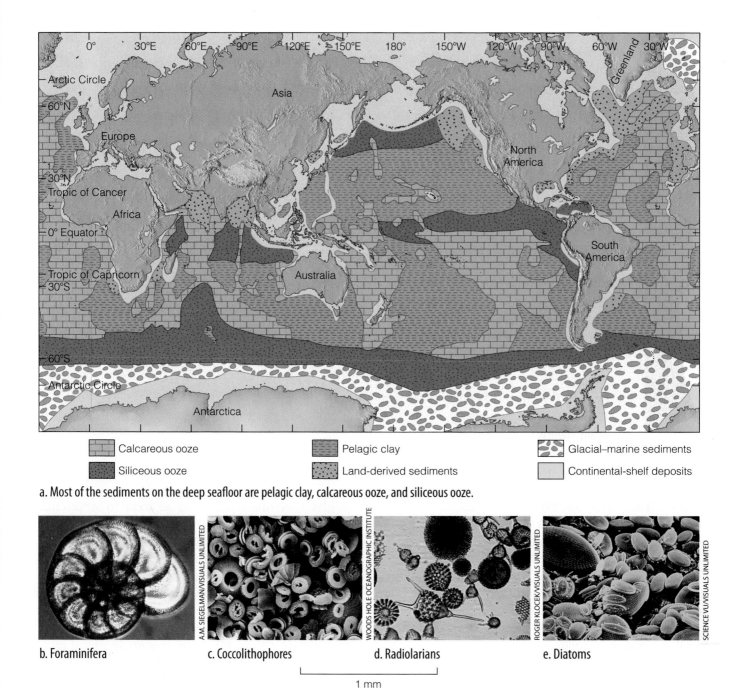

a. Most of the sediments on the deep seafloor are pelagic clay, calcareous ooze, and siliceous ooze.

Calcareous ooze	Pelagic clay	Glacial–marine sediments
Siliceous ooze	Land-derived sediments	Continental-shelf deposits

b. Foraminifera   c. Coccolithophores   d. Radiolarians   e. Diatoms

1 mm

Figure 9.11 Sediments on the Deep Seafloor

The particles making up the calcareous ooze are skeletons of (b) foraminifera (floating single-celled animals) and (c) coccolithophores (floating single-celled plants), whereas siliceous ooze is composed of skeletons of (d) radiolarians (single-celled floating animals) and (e) diatoms (single-celled floating plants).

ooze Deep-sea sediment composed mostly of shells of marine animals and plants.

is brown or red and, as its name implies, is composed of clay-sized particles from the continents or oceanic islands. **Ooze**, in contrast, is made up mostly of tiny shells of marine organisms. *Calcareous ooze* consists primarily of calcium carbonate ($CaCO_3$) skeletons of marine organisms such as foraminifera, and *siliceous ooze* is composed of the silica (SiO_2) skeletons of single-celled organisms such as radiolarians (animals) and diatoms (plants) (Figure 9.11).

LO6 Reefs

The term **reef** has a variety of meanings, such as shallowly submerged rocks that pose a hazard to navigation; however, here we restrict it to mean a moundlike, wave-resistant structure composed of the skeletons of marine organisms (Figure 9.12a). Although commonly called coral reefs, they actually have a solid framework composed of skeletons of corals and mollusks, such as clams, and encrusting organisms, including sponges and algae. Reefs are restricted to shallow, tropical seas where the water is clear and its temperature does not fall below about 20°C. The depth to which reefs grow, rarely more than 50 m, depends on sunlight penetra-

Figure 9.12 Reefs and Their Origin

COURTESY OF CARL ROESSLER

a. Underwater view of a reef in the Red Sea.

The best-known barrier reef in the world is the 2,000-km-long Great Barrier Reef of Australia.

tion, because many of the corals rely on symbiotic algae that must have sunlight for energy.

Reefs of many shapes are known, but most are one of three basic varieties: fringing, barrier, and atoll (Figure 9.12b). *Fringing reefs* are solidly attached to the margins of an island or continent. They have a rough, tablelike surface, are as much as 1 km wide, and, on their seaward side, slope steeply down to the seafloor. Barrier reefs (Figure 9.13) are similar to fringing reefs, except that a lagoon separates them from the mainland. Circular to oval reefs surrounding a lagoon are *atolls*. They form around volcanic islands that subside below sea level as the plate on which they rest is carried progressively farther from an oceanic ridge. As subsidence proceeds, the reef organisms construct the reef upward so that the living part of the reef remains in shallow water. However, the island eventually subsides below sea level, leaving a circular lagoon surrounded by a more or less continuous reef. Atolls are particularly common in the western Pacific Ocean basin. Many of them began as fringing reefs, but as the plate they were on was carried into deeper water, they evolved first to barrier reefs and finally to atolls.

> **reef** A mound-like, wave-resistant structure composed of the skeletons of organisms.

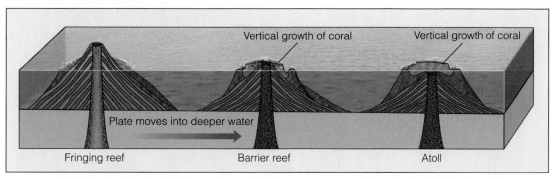

b. Three stages in the evolution of a reef. A fringing reef forms around a volcanic island, but as the island is carried into deeper water on a moving plate, the reef is separated from the island by a lagoon and becomes a barrier reef. Continued plate movement carries the island into even deeper water. The island disappears below sea level, but the reef grows upward, forming an atoll.

Figure 9.13 Barrier Reef
The white line of breaking waves marks the site of a barrier reef around Rarotonga in the Cook Islands in the Pacific Ocean. The island is only about 12 km long.

LO7 Resources from the Oceans

Seawater contains many elements in solution, some of which are extracted for industrial and domestic uses. Sodium chloride (table salt) is produced by the evaporation of seawater, and a large proportion of the world's magnesium comes from seawater. Numerous other elements and compounds can be extracted from seawater, but for many, such as gold, the cost is prohibitive.

In addition to substances in seawater, deposits on the seafloor or within seafloor sediments are becoming increasingly important. Many of these potential resources lie well beyond continental margins, so their ownership is a political and legal problem that has not yet been resolved.

Most nations bordering the ocean claim those resources within their adjacent continental margin. The United States, by a presidential proclamation issued on March 10, 1983, claims sovereign rights over an area designated the **Exclusive Economic Zone (EEZ)**. The EEZ extends seaward 200 nautical miles (371 km) from the coast and includes areas adjacent to U.S. territories such as Guam, American Samoa, Wake Island, and

Exclusive Economic Zone (EEZ) An area extending 371 km seaward from the coast of the United States and its possessions in which the United States claims rights to all resources.

Puerto Rico. In short, the United States claims rights to all resources within an area about 1.7 times larger than its land area. Many other nations make similar claims.

Numerous resources are found within the EEZ, some of which have been exploited for many years. Sand and gravel for construction are mined from the continental shelf in several areas, and approximately 34% of U.S. oil production comes from wells on the continental shelf. Ancient shelf deposits in the Persian Gulf region contain the world's largest reserves of oil.

A potential resource within the EEZ is methane hydrate, consisting of single methane molecules bound up in networks formed by frozen water. These methane hydrates are stable at water depths of more than 500 m and near-freezing temperatures. According to one estimate, the carbon in these deposits is double that in all coal, oil, and natural gas reserves. However, no one knows yet whether methane hydrates can be effectively recovered and used as an energy source. Additionally, their contribution to global warming must be assessed, because a volume of methane 3,000 times greater than in the atmosphere is present in seafloor deposits—and methane is 10 times more effective than carbon dioxide as a greenhouse gas.

The manganese nodules previously discussed are another potential seafloor resource. These spherical objects are composed mostly of manganese and iron oxides but also contain copper, nickel, and cobalt. The United States, which must import most of the manganese and cobalt it uses, is particularly interested in these nodules as a potential resource.

Other seafloor resources of interest include massive sulfide deposits that form by submarine hydrothermal activity at spreading ridges. These deposits, containing iron, copper, zinc, and other metals, have been identified within the EEZ at the Gorda Ridge off the coasts of California and Oregon; similar deposits occur at the Juan de Fuca Ridge within the Canadian EEZ.

Within the EEZ, manganese nodules are found near Johnston Island in the Pacific Ocean and on the Blake Plateau off the east coast of South Carolina and Georgia. In addition, seamounts and seamount chains within the EEZ in the Pacific are known to have metalliferous oxide crusts several centimeters thick from which cobalt and manganese could be mined.

REVIEW

HE DID

GEOL puts a multitude of study aids at your fingertips. After reading the chapters check out these resources for further help:

- **Chapter in Review cards**, found in the back of your book, include all learning outcomes, definitions, and visual summaries for each chapter.

- **Online printable flashcards** give you three additional ways to check your comprehension of key concepts.

Other great ways to help you study include:
- Virtual Field Trips in Geology
- eBook
- video exercises
- animations
- tutorial quizzes with feedback

You can find it all at **4ltrpress.cengage.com**.

CHAPTER 10
DEFORMATION, MOUNTAIN BUILDING, AND THE CONTINENTS

The Great Smoky Mountain Range rises along the border between North Carolina and Tennessee. The oldest rocks in the mountain range formed more than a billion years ago from the accumulation of marine sediments and igneous intrusions. The highest point of the Great Smoky Mountain Range rises to an elevation of 2,025 meters.

COPYRIGHT AND PHOTOGRAPH BY DR. PARVINDER S. SETHI

Introduction

The phrase "solid as a rock" implies permanence and durability, but you know from earlier chapters that rocks disaggregate and decompose, and rocks behave very differently at great depth than they do at or near Earth's surface. Indeed, under the tremendous pressures and high temperature at several kilometers below the surface, rock layers actually crumple or fold, yet remain solid, and at shallower depths, they yield by fracturing or a combination of folding and fracturing. In either case, dynamic forces within Earth cause **deformation**, a general term encompassing all changes in the shape or volume of rocks.

The action of dynamic forces within Earth is obvious from ongoing seismic activity, volcanism, plate movements, and the continuing evolution of mountains in South America, Asia, and elsewhere. In short, Earth is an active planet with processes driven by internal heat, particularly plate movements; most of Earth's seismic activity, volcanism, and rock deformation take place at divergent, convergent, and transform plate boundaries (see Chapter 2).

The origin of Earth's truly large mountain ranges on the continents involves tremendous deformation, usually accompanied by emplacement of plutons, volcanism, and metamorphism, at convergent plate boundaries. And, in some cases, this activity continues even now. Thus, deformation and mountain building are closely related topics, and accordingly we consider both in this chapter.

The past and continuing evolution of continents involves not only deformation at continental margins, but also additions of new material to existing continents, a phenomenon known as *continental accretion* (see Chapter 18). North America, for instance, has not always had its present shape and area. Indeed, it began evolving during the Archean Eon (4.6 to 2.5 billion years ago) as new material was added to the continent at deformation belts along its margins.

Much of this chapter is devoted to a review of *geologic structures*, such as folded and fractured rock layers resulting from deformation, their descriptive terminology, and the forces responsible for them. There are several practical reasons to study deformation and mountain building. For one thing, deformed rock layers provide a record of the kinds and intensities of forces that operated during the past. Thus, interpretations of these structures allow us to satisfy our curiosity about Earth history, and, in addition, such studies are essential in engineering endeavors such as choosing sites for dams, bridges, and nuclear power plants, especially if they are in areas of ongoing deformation. Also, many aspects of mining and exploration for petroleum and natural gas rely on correctly identifying geologic structures.

> **deformation** A general term for any change in shape or volume, or both, of rocks in response to stress; involves folding and fracturing.
>
> **stress** The force per unit area applied to a material such as rock.

LEARNING OUTCOMES

After reading this unit, you should be able to do the following:

LO1 Explain how rock deformation occurs

LO2 Understand strike and dip—the orientation of deformed rock layers

LO3 Identify the types of deformation and geologic structures

LO4 Understand deformation and the origin of mountains

LO5 Describe Earth's continental crust

LO1 Rock Deformation: How Does It Occur?

We defined *deformation* as a general term referring to changes in the shape or volume (or both) of rocks; that is, rocks may be crumpled into folds or fractured as a result of **stress**, which results from force applied to a given area of rock. If the intensity of the stress is greater than the rock's internal strength, the rock undergoes **strain**, which is simply deformation caused by stress. The terminology is a little confusing at first, but keep in mind that *deformation* and *strain* are synonyms, and *stress* is the force that causes deformation or strain. The following discussion and Figure 10.1 will help clarify the meaning of stress and the distinction between stress and strain.

STRESS AND STRAIN

Remember that stress is the force applied to a given area of rock, usually expressed in kilograms per square centimeter (kg/cm^2). For example, the stress, or force, exerted by a person walking on an ice-covered pond is a function of the person's weight and the area beneath her or his feet. The ice's internal strength resists the stress unless the stress is too great, in which case the ice may bend or crack as it is strained (deformed) (Figure 10.1). To avoid breaking through the ice, the person may lie down; this does not reduce the weight of the person on the ice, but it does distribute the weight over a larger area, thus reducing the stress per unit area.

Although stress is force per unit area, it comes in three varieties: *compression*, *tension*, and *shear*. In **compression**, materials are squeezed or compressed by forces directed toward one another along the same line, as when you squeeze a rubber ball in your hand. Rock layers in compression tend to be shortened in the direction of stress by either folding or fracturing (Figure 10.2a). **Tension** results from forces acting along the same line, but in opposite directions. Tension tends to lengthen rocks or pull them apart

strain Deformation caused by stress.

compression Stress resulting when materials are squeezed by external forces directed toward one another.

tension A type of stress in which forces act in opposite directions but along the same line, thus tending to stretch an object.

shear stress The result of forces acting parallel to one another but in opposite directions; results in deformation by displacement of adjacent layers along closely spaced planes.

elastic strain A type of deformation in which the material returns to its original shape when stress is relaxed.

plastic strain Permanent deformation of a solid with no failure by fracturing.

fracture A break in rock resulting from intense applied pressure.

Like people, rocks under stress and strain either bounce back, bend, or break.

© BRAND X PICTURES/JUPITERIMAGES

(Figure 10.2b). Incidentally, rocks are much stronger in compression than they are in tension. In **shear stress**, forces act parallel to one another, but in opposite directions, resulting in deformation by displacement along closely spaced planes (Figure 10.2c).

TYPES OF STRAIN

Geologists characterize strain as **elastic strain** if deformed rocks return to their original shape when the deforming forces are relaxed. In Figure 10.1, the ice on the pond may bend under a person's weight but return to its original shape once the person leaves. As you might expect, rocks are not very elastic, but Earth's crust behaves elastically when loaded by glacial ice and depressed into the mantle.

As stress is applied, rocks respond first by elastic strain, but when strained beyond their elastic limit, they undergo **plastic strain** as when they yield by folding, or they behave like brittle solids and **fracture**. In either folding or fracturing, the strain is permanent; that is, the rocks do not recover their original shape or volume even if the stress is removed.

Figure 10.1 Stress and Strain Exerted on an Ice-Covered Pond

a. The woman weighs 65 kg (6,500 g). Her weight is imparted to the ice through her feet, which have a contact area of 120 cm². The stress she exerts on the ice (6,500/120 cm²) is 54 g/cm². This is sufficient stress to cause the ice to crack.

b. To avoid plunging into the freezing water, the woman lays out flat, thereby decreasing the stress she exerts on the ice. Her weight remains the same, but her contact area with the ice is 3,150 cm², so the stress is only about 2 g/cm² (6,500 g/3,150 cm²), which is well below the threshold needed to crack the ice.

Figure 10.2 Stress and Possible Types of Resulting Deformation

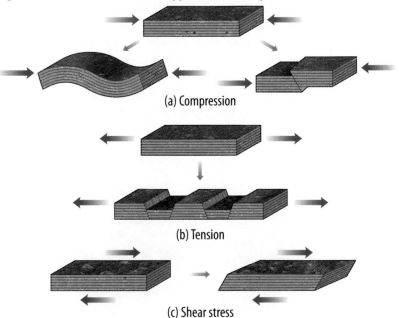

(a) Compression

(b) Tension

(c) Shear stress

a. Compression causes shortening of rock layers by folding or faulting.

b. Tension lengthens rock layers and causes faulting.

c. Shear stress causes deformation by displacement along closely spaced planes.

Whether strain is elastic, plastic, or fracture depends on the kind of stress applied, pressure and temperature, rock type, and the length of time rocks are subjected to stress. A small stress applied over a long period, as on a mantelpiece supported only at its ends, will cause the rock to sag; that is, the rock deforms plastically. By contrast, a large stress applied rapidly to the same object, as when struck by a hammer, results in fracture. Rock type is important, because not all rocks have the same internal strength and thus respond to stress differently. Some rocks are *ductile*, whereas others are *brittle*. Brittle rocks show little or no plastic strain before they fracture, but ductile rocks exhibit a great deal.

Many rocks show the effects of plastic strain that must have taken place deep within the crust. At or near the surface, rocks commonly behave like brittle solids and fracture, but at depth, they become more ductile with increasing pressure and temperature.

LO2 Strike and Dip: The Orientation of Deformed Rock Layers

One concept in geology is the *principle of original horizontality*, meaning that sediments accumulate in horizontal or nearly horizontal layers. Thus, if we observe steeply inclined sedimentary rocks, we are justified in inferring that they were deposited nearly horizontally, lithified, and then tilted into their present position. Rock layers deformed by folding, faulting, or both are no longer in their original position, so geologists use *strike* and *dip* to describe their orientation with respect to a horizontal plane.

By definition, **strike** is the direction of a line formed by the intersection of a horizontal plane and an inclined plane. The surfaces of the rock layers in Figure 10.3 are good examples of inclined planes, whereas the water surface is a horizontal plane. The direction of the line formed at the intersection of these planes is the strike of the rock layers. The strike line's orientation is determined by using a compass to measure its angle with respect to north. **Dip** is a measure of an inclined plane's deviation from horizontal, so it must be measured at right angles to strike direction (Figure 10.3).

Geologic maps showing the age, aerial distribution, and geologic structures of rocks in an area use a special symbol to indicate strike and dip. A long line oriented in the appropriate direction indicates strike, and a short line perpendicular to the strike line shows the direction of dip (Figure 10.3). Adjacent to the strike and dip symbol is a number corresponding to the dip angle. The usefulness of strike and dip symbols will become apparent in the sections on folds and faults.

LO3 Deformation and Geologic Structures

Remember that *deformation* and its synonym *strain* refer to changes in the shape or volume of rocks. During deformation, rocks might be crumpled into folds, or they might be fractured, or perhaps folded and fractured. Any of these features resulting from deformation is referred to as a **geologic structure**. Geologic structures are present almost everywhere that rock exposures can be observed, and many are detected far below the surface by drilling and several geophysical techniques.

FOLDED ROCK LAYERS

Geologic structures known as **folds**, in which planar features are crumpled and bent, are quite common. Compression is responsible for most folding, as when you place your hands on a tablecloth and move them toward one another, thereby producing a series of up- and down-arches in the fabric. Rock layers in the crust respond similarly to compression, but unlike the tablecloth, folding in rock layers is permanent. Most folding probably takes place deep in the crust, where rocks

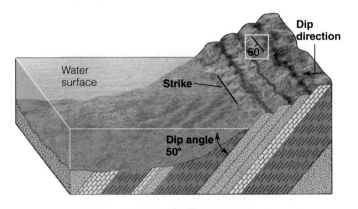

Figure 10.3 Strike and Dip of Deformed Rocks
Strike is the intersection of a horizontal plane (the water surface) with an inclined plane (the surface of the rock layer). Dip is the maximum angular deviation of the inclined layer from horizontal. Notice the strike and dip symbol.

strike The direction of a line formed by the intersection of an inclined plane and a horizontal plane.

dip A measure of the maximum angular deviation of an inclined plane from horizontal.

geologic structure Any feature in rocks that results from deformation, such as folds, joints, and faults.

fold A type of geologic structure in which planar features in rock layers such as bedding and foliation have been bent.

Figure 10.4 Monocline
A monocline in the Bighorn Mountains in Wyoming.

are more ductile than they are at or near the surface. The configuration of folds and the intensity of folding vary considerably, but there are only three basic types of folds: *monoclines*, *anticlines*, and *synclines*.

Monoclines

A simple bend or flexure in otherwise horizontal or uniformly dipping rock layers is a **monocline**. The large monocline in Figure 10.4 formed when the Bighorn Mountains in Wyoming rose vertically along a fracture. The fracture did not penetrate to the surface, so as uplift of the mountains proceeded, the near-surface rocks were bent so that they now appear to be draped over the margin of the uplifted block.

Anticlines and Synclines

An **anticline** is an up-arched or convex upward fold with the oldest rock layers in its core, whereas a **syn**cline is a down-arched or concave downward fold in which the youngest rock layers are in its core (Figure 10.5). Anticlines and synclines have an axial plane connecting the points of maximum curvature of each folded layer (Figure 10.6); the axial plane divides folds into halves, each half being a *limb*. Because folds are most often found in a series of anticlines alternating with synclines, an anticline and adjacent syncline share a limb.

Folds are commonly exposed to view in areas of deep erosion, but even where eroded, strike and dip and the relative ages of the folded rock layers

monocline A bend or flexure in otherwise horizontal or uniformly dipping rock layers.

anticline A convex upward fold in which the oldest exposed rocks coincide with the fold axis and all strata dip away from the axis.

syncline A down-arched fold in which the youngest exposed rocks coincide with the fold axis and all strata dip toward the axis.

> *It is important to remember that anticlines and synclines are simply folded rock layers and do not necessarily correspond to high and low areas at the surface.*

easily distinguish anticlines from synclines. Notice in Figure 10.7 that in the surface view of the anticline, each limb dips outward or away from the center of the fold, and the oldest exposed rocks are in the fold's core. In an eroded syncline, though, each limb dips inward toward the fold's center, where the youngest exposed rocks are found.

The folds described so far are *upright*, meaning that their axial planes are vertical, and both fold limbs

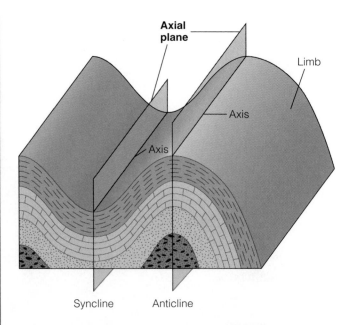

Figure 10.6 Syncline and Anticline Axial Planes
Syncline and anticline showing the axial plane, axis, and fold limbs.

Figure 10.5 Anticlines and Synclines
Folded rocks in the Calico Mountains of southeastern California. Compression was responsible for these folds, which are, from left to right, a syncline, an anticline, and a syncline.

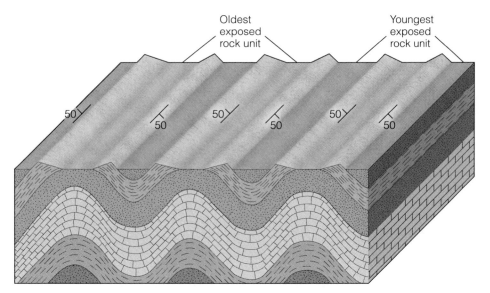

Oldest exposed rock unit

Youngest exposed rock unit

50 50 50 50 50 50

Figure 10.7 Eroded Anticlines and Synclines
Geologists identify eroded anticlines and synclines by strike and dip and the relative ages of the folded rock layers.

dip at the same angle (Figure 10.7). In many folds, the axial plane is not vertical, the limbs dip at different angles, and the folds are characterized as *inclined* (Figure 10.8a). If both limbs dip in the same direction, the fold is *overturned*. That is, one limb has been rotated more than 90 degrees from its original position so that it is now upside down (Figure 10.8b). In some areas, deformation has been so intense that axial planes of folds are now horizontal, giving rise to what geologists call *recumbent folds* (Figure 10.8c). Overturned and recumbent folds are particularly common in mountains resulting from compression at convergent plate boundaries.

Plunging Folds

In some folds, the fold axis—a line formed by the intersection of the axial plane with the folded layers—is horizontal and the folds are *nonplunging* (Figure 10.7). Much more commonly, though, fold axes are inclined so that they appear to plunge beneath adjacent rocks, and the folds are said to be *plunging* (Figure 10.9).

It might seem that with this additional complication, differentiating plunging anticlines from plunging synclines would be much more difficult. However, geologists use exactly the same criteria that they use for nonplunging folds. Therefore, all rock layers dip away from the fold axis in plunging anticlines and toward the axis in plunging synclines. The oldest exposed rocks are in the core of an eroded plunging anticline, whereas the youngest exposed rock layers

are found in the core of an eroded plunging syncline (Figure 10.9b).

In Chapter 6, we noted that anticlines form one type of structural trap in which petroleum and natural gas might accumulate. As a matter of fact, most of the world's petroleum production comes from anticlines, although other geologic structures and stratigraphic traps are also important.

Domes and Basins

Anticlines and synclines are elongate structures, meaning that their length greatly exceeds their width. In contrast, folds that are nearly equidimensional (i.e., circular) are *domes* and *basins*. In a **dome**, all of the folded strata dip outward from a central point (as opposed to outward from a line as in an anticline), and the oldest exposed rocks are at the center of the fold (Figure 10.10a). In contrast, a **basin** has all strata dipping inward toward a central point, and the youngest exposed rocks are at the fold's center (Figure 10.10b).

Unfortunately, the terms *dome* and *basin* are also used to distinguish high and low areas of Earth's surface, but domes and basins as defined here do not necessarily correspond with

> **dome** A rather circular geologic structure in which all rock layers dip away from a central point and the oldest exposed rocks are in the dome's center.
>
> **basin** A circular fold in which all strata dip inward toward a central point and the youngest exposed strata are in the center.

Figure 10.8 Inclined, Overturned, and Recumbent Folds

Inclined

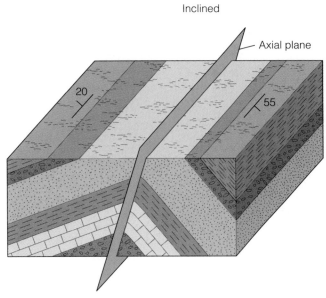

Axial plane

20

55

a. An inclined fold. The axial plane is not vertical, and the fold limbs dip at different angles.

Overturned

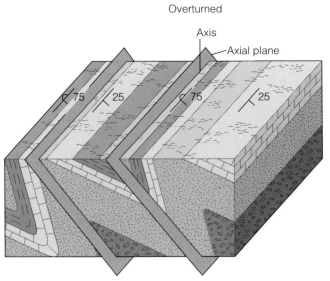

Axis

Axial plane

75   25   75   25

b. Overturned folds. Both fold limbs dip in the same direction, but one limb is inverted. Notice the special strike-and-dip symbol to indicate overturned beds.

Recumbent

Oldest rocks

Youngest rocks

Axial plane

c. Recumbent folds are folds in which the axial planes are horizontal.

SUE MONROE

d. An overturned fold in Switzerland.

Figure 10.9 Plunging Folds

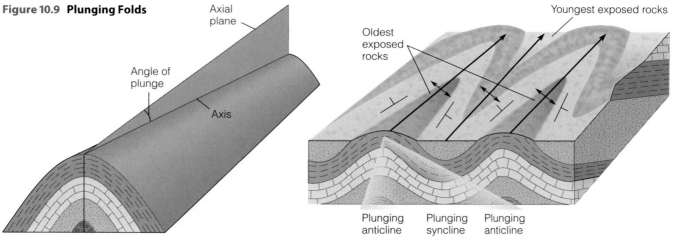

a. A plunging fold.

b. Surface and cross-sectional views of plunging folds. The long arrow is the geologic symbol for a plunging fold; it shows the direction of plunge.

Figure 10.10 Domes and Basins

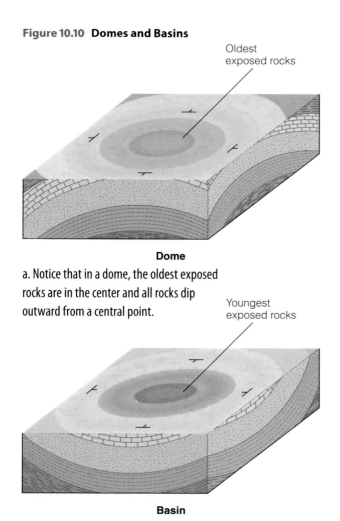

Dome

a. Notice that in a dome, the oldest exposed rocks are in the center and all rocks dip outward from a central point.

Basin

b. In a basin, the youngest exposed rocks are in the center and all rocks dip inward toward a central point.

Figure 10.11 Joints

Fractures along which no movement has taken place parallel with the fracture surface are called joints. Joints intersecting at right angles yield this rectangular pattern in Wales.

mountains or valleys. In some of the following discussions, we will use these terms in other contexts, but we will try to be clear when we refer to surface elevations as opposed to geologic structures.

JOINTS

Rocks are also permanently deformed by fracturing. **Joints** are fractures along which no movement has taken place parallel with the fracture surface (Figure 10.11), although they may open up; that is, they show movement perpendicular to the fracture. Remember that rocks

joint A fracture along which no movement has occurred or where movement is perpendicular to the fracture surface.

Figure 10.12 Faults

Fractures along which movement has occurred parallel to the fracture surface are called faults.

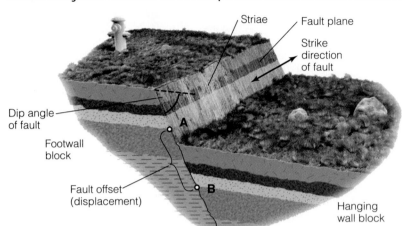

b. A series of faults at Arches National Park, Utah.

a. Terms used to describe the orientation of a fault plane. Striae are scratch marks that form when one block slides past another. You can measure offset or displacement on a fault wherever the truncated end of one feature (point A) can be related to its equivalent along the fault (B).

fault A fracture along which rocks on opposite sides of the fracture have moved parallel with the fracture surface.

fault plane A fault surface that is more or less planar.

hanging wall block The block of rock that overlies a fault plane.

footwall block The block of rock that lies beneath a fault plane.

dip-slip fault A fault on which all movement is parallel with the dip of the fault plane.

near the surface are brittle and therefore commonly fail by fracturing when subjected to stress. In fact, almost all near-surface rocks have joints that form in response to compression, tension, and shearing. They vary from minute fractures to those extending for many kilometers and are often arranged in two or perhaps three prominent sets.

We have discussed columnar joints that form when lava or magma in some shallow plutons cools and contracts (see Figure 5.5). A different type of jointing previously discussed is sheet jointing that forms in response to pressure release (see Figure 6.3).

FAULTS

Another type of fracture known as a **fault** is one along which blocks of rock on opposite sides of the fracture have moved parallel with the fracture surface, which is

a **fault plane**. Notice the designations *hanging wall block* and *footwall block* in Figure 10.12. The **hanging wall block** consists of the rock overlying the fault, whereas the **footwall block** lies beneath the fault plane. You can recognize these two blocks on any fault except a vertical one. To identify some kinds of faults, you must not only correctly identify these two blocks, but also determine which one moved relatively up or down. We use the phrase *relative movement*, because you usually cannot tell which block moved or if both moved. In Figure 10.12, the footwall block may have moved up, the hanging wall block may have moved down, or both could have moved. Nevertheless, the hanging wall block appears to have moved down relative to the footwall block.

Recall our discussion of strike and dip of rock layers. Fault planes are also inclined planes, and they, too, are characterized by strike and dip (Figure 10.12a). In fact, the two basic varieties of faults are defined by whether the blocks on opposite sides of the fault plane moved parallel to the direction of dip (dip-slip faults) or along the direction of strike (strike-slip faults).

Dip-Slip Faults

All movement on **dip-slip faults** takes place parallel with the fault's dip; that is, movement is vertical, either up or down the fault plane. In Figure 10.13a, for example, the hanging wall block moved down relative to the

Figure 10.13 Types of Faults

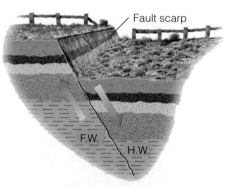

a. Normal fault—hanging wall block (HW) moves down relative to the footwall block (FW).

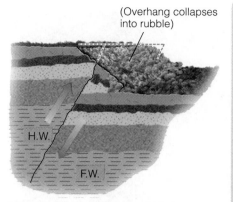

(Overhang collapses into rubble)

b. Reverse fault—hanging wall block moves up relative to the footwall block.

normal fault A dip-slip fault on which the hanging wall block has moved downward relative to the footwall block.

reverse fault A dip-slip fault on which the hanging wall block has moved upward relative to the footwall block.

thrust fault A type of reverse fault in which a fault plane dips less than 45 degrees.

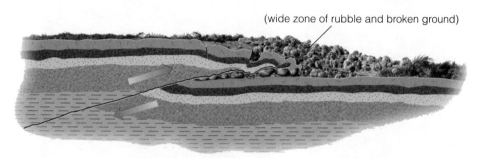

(wide zone of rubble and broken ground)

c. A thrust is a type of reverse fault with a fault plane dipping at less than 45 degrees.

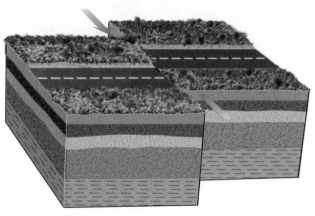

d. Left-lateral strike-slip fault.

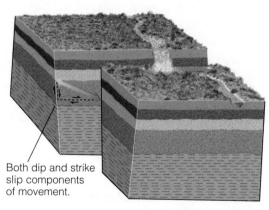

Both dip and strike slip components of movement.

e. An oblique-slip fault involves a combination of dip-slip and strike-slip movements.

footwall block, giving rise to a **normal fault** (Figure 10.14a). In contrast, in a **reverse fault**, the hanging wall block moves up relative to the footwall block (Figure 10.13b). In Figure 10.13c, the hanging wall block also moved up relative to the footwall block, but the fault has a dip of less than 45 degrees and is a special variety of reverse fault known as a **thrust fault**.

Numerous normal faults are present along one or both sides of mountain ranges in the Basin and Range Province of the western United States where the crust is being stretched and thinned. The Sierra Nevada at the western margin of the Basin and Range is bounded by normal faults, and the range has risen along these faults so that it now stands more than 3,000 m above

Figure 10.14 Faults

JAMES S. MONROE

a. A normal fault at Mt. Carmel Junction, Utah. The hanging wall block (on the right) has moved down about 2 m relative to the footwall block.

LINDIE BREWER/USGS

b. Oblique slip took place on this fault in central Nevada during a 1915 earthquake. Notice the fence that shows right-lateral displacement and the fault plane that shows dip-slip displacement.

strike-slip fault A fault involving horizontal movement of blocks of rock on opposite sides of a fault plane.

the lowlands to the east. Also, an active normal fault is found along the eastern margin of the Teton Range in Wyoming, accounting for the 2,100-m elevation difference between the valley floor and the highest peaks in the mountains.

Large-scale examples of reverse and thrust faults are both found in mountain ranges that formed at convergent plate margins, where one would expect compression (Figure 10.13b,c).

Strike-Slip Faults

Strike-slip faults, resulting from shear stresses, show horizontal movement with blocks on opposite sides of the fault sliding past one another (Figure 10.13d). In other words, all movement is in the direction of the fault plane's strike. Several large strike-slip faults are known, but the best studied is the San Andreas fault, which cuts through coastal California. Recall from Chapter 2 that the San Andreas fault is called a *transform fault* in plate tectonics terminology.

Strike-slip faults are characterized as right-lateral or left-lateral, depending on the apparent direction of offset. In Figure 10.13d, for example, observers looking at the block on the opposite side of the fault from their location notice that it appears to have moved to the left. Accordingly, this is a *left-lateral strike-slip fault*. If it had been a *right-lateral strike-slip fault*, the block across the fault from the observers would appear to have moved to the right.

Oblique-Slip Faults

The movement on most faults is primarily dip-slip or strike-slip, but on **oblique-slip faults**, both types of movement take place. Strike-slip movement might be accompanied by a component of dip-slip, giving rise to a combined movement that includes left-lateral and reverse, or right-lateral and normal (Figure 10.13e, 10.14b).

LO4 Deformation and the Origin of Mountains

Mountains form in several ways, but the truly large mountains on continents result mostly from compression-induced deformation at convergent plate boundaries. *Mountain* is a designation for any area of land that stands significantly higher, at least 300 m, than the surrounding country and has a restricted summit area. Some mountains are single, isolated peaks, but more commonly they are parts of linear associations of peaks and ridges known as *mountain ranges* that are related in age and origin. A *mountain system*, a complex linear zone of deformation and crustal thickening, on the other hand, consists of several or many mountain ranges. The Teton Range in Wyoming is one of many ranges in the Rocky Mountains.

MOUNTAIN BUILDING

Block-faulting is one way that mountains form, which is caused by movement on normal faults so that one or more blocks are elevated relative to adjacent blocks (Figure 10.15). A classic example is the Basin and Range Province, which is centered on Nevada but extends into adjacent areas. Differential movement on faults has produced uplifted blocks called *horsts* and down-dropped blocks called *grabens*. Erosion of the horsts has yielded mountainous topography.

Volcanic outpourings form chains of volcanic mountains such as the Hawaiian Islands, where a plate moves over a hot spot. Some mountains such as the Cascade Range of the Pacific Northwest are made up almost entirely of volcanic rocks, and the mid-ocean ridges are also mountains. However, most mountains on land are composed of all rocks types and show clear evidence of deformation by compression.

oblique-slip fault A fault showing both dip-slip and strike-slip movement.

orogeny An episode of mountain building involving deformation, usually accompanied by igneous activity, and crustal thickening.

PLATE TECTONICS AND MOUNTAIN BUILDING

Geologists define the term **orogeny** as an episode of mountain building. Any theory that accounts for mountain building must adequately explain the characteristics of mountain ranges, such as their geometry and location; they tend to be long and narrow and at or near plate margins. Mountains also show intense deformation, especially compression-induced overturned and recumbent folds, as well as reverse and thrust faults. Furthermore, granitic plutons and regional metamorphism characterize the interiors or cores of mountain ranges. Another feature is sedimentary rocks that are now far above sea level but that were deposited in shallow and deep marine environments.

Deformation and associated activities at convergent plate boundaries are certainly important processes in mountain building. They account for a mountain system's location and geometry, as well as complex geologic structures, plutons, and metamorphism. Yet, the present-day topographic expression of mountains is also related to several surface processes, such as mass wasting (gravity-driven processes including landslides), glaciers, and running water. In other words, erosion also plays an important role in the evolution of mountains.

Orogenies at Oceanic–Oceanic Plate Boundaries

Deformation, igneous activity, and the origin of a volcanic island arc characterize orogenies that take place where oceanic lithosphere is subducted beneath oceanic lithosphere. Sediments derived from the island arc are deposited in an adjacent oceanic trench and then

Figure 10.15 Origins of Horsts and Grabens

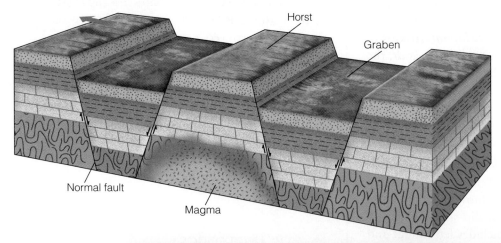

a. Block-faulting and the origin of horsts and grabens. Many of the mountain ranges in the Basin and Range Province of the western United States and northern Mexico formed in this manner.

b. The Egan Range in Nevada is one of the many ranges in that state that is bounded on one or both sides by normal faults.

deformed and scraped off against the landward side of the trench (Figure 10.16). These deformed sediments are part of a subduction complex, or an *accretionary wedge*, of intricately folded rocks cut by numerous thrust faults. In addition, orogenies in this setting are characterized by low-temperature, high-pressure metamorphism of the blueschist facies (see Figure 7.19).

Deformation caused largely by the emplacement of plutons also takes place in the island arc system, where many rocks show evidence of high-temperature,

low-pressure metamorphism. The overall effect of an island arc orogeny is the origin of two more-or-less parallel orogenic belts consisting of a landward volcanic island arc underlain by batholiths and a seaward belt of deformed trench rocks (Figure 10.16). The Japanese Islands are a good example.

In the area between an island arc and its nearby continent, the back-arc basin, volcanic rocks and sediments derived from the island arc and the adjacent continent are also deformed as the plates continue to

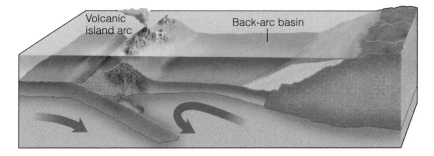

Volcanic island arc

Back-arc basin

Figure 10.16 Orogeny and the Origin of a Volcanic Island Arc at an Oceanic–Oceanic Plate Boundary

a. Subduction of an oceanic plate and the origin of a volcanic island arc and a back-arc basin.

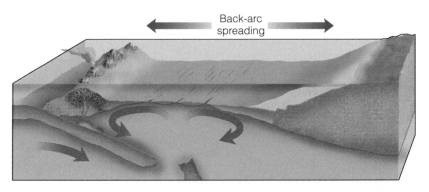

Back-arc spreading

b. Continued subduction and back-arc spreading.

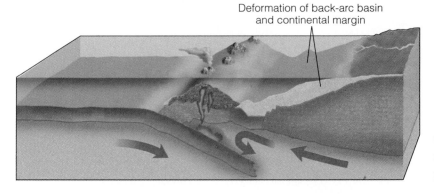

Deformation of back-arc basin and continental margin

c. Back-arc basin begins to close, resulting in deformation of back-arc basin and continental margin deposits.

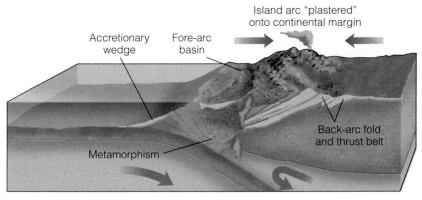

Island arc "plastered" onto continental margin

Accretionary wedge

Fore-arc basin

Back-arc fold and thrust belt

Metamorphism

d. Thrusting of back-arc sediments onto the adjacent continent and suturing of the island arc to the continent.

converge. The sediments are intensely folded and displaced toward the continent along low-angle thrust faults. Eventually, the entire island arc complex is fused to the edge of the continent, and the back-arc basin sediments are thrust onto the continent, forming a thick stack of thrust sheets (Figure 10.16).

Orogenies at Oceanic–Continental Plate Boundaries

The Andes of South America are the best example of continuing orogeny at an oceanic–continental plate boundary (see Figure 2.14b). Among the ranges of the

Figure 10.17 The Andes Mountains in South America

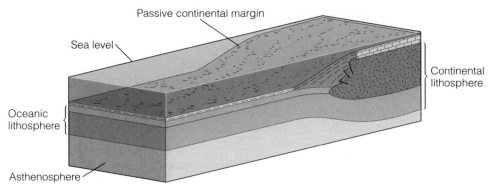

a. Before 200 million years ago, the western margin of South America was a passive continental margin.

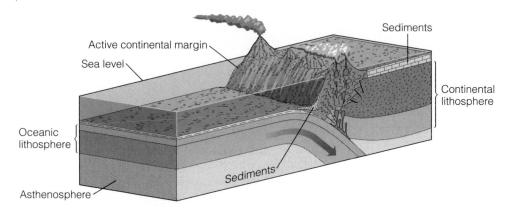

b. Orogeny began when this area became an active continental margin as the South American plate moved to the west and collided with oceanic lithosphere.

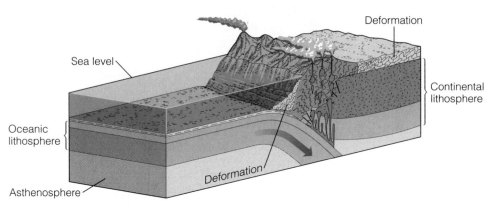

c. Continued deformation, plutonism, and volcanism.

Andes are the highest mountain peaks in the Americas and many active volcanoes. Furthermore, the west coast of South America is an extremely active segment of the circum-Pacific earthquake belt, and one of Earth's great oceanic trench systems, the Peru–Chile Trench, lies just off the coast.

Before 200 million years ago, the western margin of South America was a passive continental margin where sediments accumulated much as they do now along the East Coast of North America. However, when Pangaea split apart along what is now the Mid-Atlantic Ridge, the South American plate moved westward. As

a consequence, the oceanic lithosphere west of South America began subducting beneath the continent (Figure 10.17). Subduction resulted in partial melting of the descending plate, which produced the andesitic volcanic arc of composite volcanoes, and the West Coast became an active continental margin. Felsic magmas, mostly of granitic composition, were emplaced as large plutons beneath the arc (Figure 10.17).

As a result of the events just described, the Andes Mountains consist of a central core of granitic rocks capped by andesitic volcanoes. To the west of this central core along the coast are the deformed rocks of the accretionary wedge. And to the east of the central core are intensely folded sedimentary rocks that were thrust eastward onto the continent (Figure 10.17). Present-day subduction, volcanism, and seismicity along South America's west coast indicate that the Andes Mountains are still forming.

Orogenies at Continental–Continental Plate Boundaries

The best example of an orogeny along a continental–continental plate boundary is the Himalayas of Asia (see Figure 2.14c). The Himalayas began forming when India collided with Asia about 40 to 50 million years ago. Before that time, India was far south of Asia and separated from it by an ocean basin (Figure 10.18a). As the Indian plate moved northward, a subduction zone formed along the southern margin of Asia where oceanic lithosphere was consumed. Magma rose to form a volcanic arc, and large granite plutons were emplaced into what is now Tibet. At this stage, the activity along Asia's southern margin was similar to what is now occurring along the west coast of South America.

The ocean separating India from Asia continued to close, and India eventually collided with Asia (Figure 10.18a). As a result, two continental plates became welded, or sutured, together. Thus, the Himalayas are now within a continent rather than along a continental margin. The leading margin of India was thrust beneath Asia, causing crustal thickening, thrusting, and uplift. Sedimentary rocks that had been deposited in the sea south of Asia were thrust northward, and two major thrust faults carried rocks of Asian origin onto the Indian plate. Rocks deposited in the shallow seas along India's northern margin now form the higher parts of the Himalayas (Figure 10.18b). Since its collision with Asia, India has been thrust horizontally about 2,000 km beneath Asia and now moves north at several centimeters per year.

Figure 10.18 Orogeny at a Continental–Continental Plate Boundary and the Origin of the Himalayas of Asia

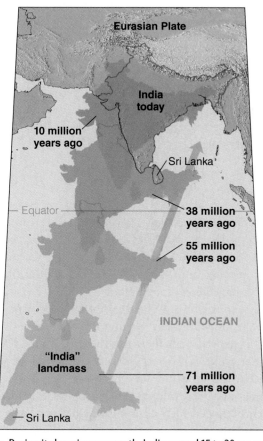

a. During its long journey north, India moved 15 to 20 cm per year; however, beginning 40 to 50 million years ago, its rate of movement decreased markedly as it collided with the Eurasian plate.

b. The Karakoram Range seen here from Karimaba, Pakistan, is within the Himalayan origin. The range lies on the border of Pakistan, China, and India.

x

Other mountain systems also formed as a result of collisions between two continental plates. The Urals in Russia and the Appalachians of North America formed by such collisions. In addition, the Arabian plate is now colliding with Asia along the Zagros Mountains of Iran.

TERRANES AND THE ORIGIN OF MOUNTAINS

In the preceding section, we discussed orogenies along convergent plate boundaries that result in adding material to a continent, a process termed **continental accretion**. Much of the material added to continental margins is eroded older continental crust, but some plutonic and volcanic rocks are new additions. During the 1970s and 1980s, however, geologists discovered that parts of many mountain systems are also made up of small, accreted lithospheric blocks that clearly originated elsewhere. These **terranes,*** are fragments of seamounts, island arcs, and small pieces of continents that were carried on oceanic plates that collided with continental plates, thus adding them to the continental margins.

LO5 Earth's Continental Crust

Continental crust stands higher than oceanic crust, but why should this be so? Also, why do mountains stand higher than surrounding areas? To answer these questions, we must examine continental crust in more detail. You already know that continental crust is granitic with an overall density of 2.7 g/cm^3, whereas oceanic crust is made up of basalt and gabbro and its density is 3.0 g/cm^3 (see Figure 8.18). In most places, continental crust is about 35 km thick,

continental accretion
Orogenics along convergent plate boundaries that result in adding material to a continent.

terrane Fragments of seamounts, island arcs, and small pieces of continents that were carried on oceanic plates that collided with continental plates.

---

*Some geologists prefer the terms *suspect terrane, exotic terrane,* or *displaced terrane.* Notice also the spelling of *terrane* as opposed to the more familiar *terrain,* the latter a geographic term indicating a particular area of land.

except beneath mountain systems, where it is much thicker. The oceanic crust, in contrast, varies from only 5 to 10 km thick. So, these differences, as well as variations in crustal thickness, account for why mountains stand high and why continents stand higher than ocean basins.

FLOATING CONTINENTS?

How is it possible for a solid (continental crust) to float in another solid (the mantle)? Floating brings to mind a ship at sea or a block of wood in water; however, continents do not behave in this manner. Or do they? Actually, they do float, in a manner of speaking, but a complete answer requires more discussion on the concept of gravity and on the principle of isostasy.

Isaac Newton formulated the law of universal gravitation in which the force of gravity (*F*) between two masses (m_1 and m_2) is directly proportional to the products of their masses and inversely proportional to the square of the distance between their centers of mass. This means that an attractive force exists between any two objects, and the magnitude of that force varies depending on the masses of the objects and the distance between their centers.

Gravitational attraction would be the same everywhere on the surface if Earth were perfectly spherical, homogeneous throughout, and not rotating. But because Earth varies in all of these aspects, the force of gravity varies from area to area.

PRINCIPLE OF ISOSTASY

Geologists realized long ago that mountains are not simply piles of materials on Earth's surface, and in 1865, George Airy proposed that, in addition to projecting high above sea level, mountains also project far below the surface and thus have a low-density root (Figure 10.19). In effect, he was saying that the thicker crust of mountains float on denser rock at depth, with their excess mass above sea level compensated for by low-density material at depth. Another explanation was proposed by J. H. Pratt, who thought that mountains were high because they were composed of rocks of lower density than those in adjacent regions.

Actually, both Airy and Pratt were correct, because there are places where density or thickness accounts for differences in the level of the crust. For example, Pratt's hyphothesis was confirmed because

Figure 10.19 Gravity Anomalies

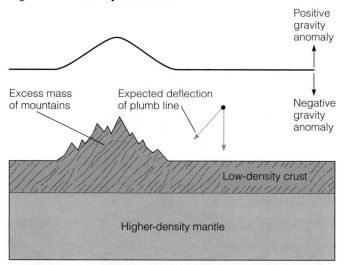

a. A plumb line (a cord with a suspended weight) is normally vertical, pointing to Earth's center of gravity. Near a mountain range, the plumb line should be deflected as shown if the mountains are simply thicker, low-density material resting on denser material, and a gravity survey across the mountains would indicate a positive gravity anomaly.

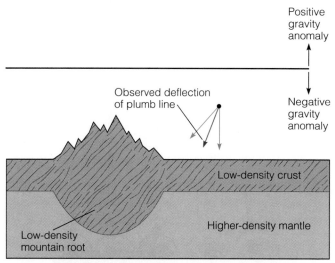

b. The actual deflection of the plumb line as measured by British surveyors during a survey in India was less than expected. It was explained by postulating that the Himalayas have a low-density root. A gravity survey, in this case, would show no anomaly because the mass of the mountains above the surface is compensated for at depth by low-density material displacing denser material.

(1) continental crust is thicker and less dense than oceanic crust and thus stands high, and (2) the mid-oceanic ridges stand high because the crust there is hot and less dense than cooler crust elsewhere. Airy, on the other hand, was correct in his claim that the crust, continental or oceanic, "floats" on the mantle, which has a density of 3.3g/cm³ in its upper part. However, we have not yet explained what we mean by one solid floating in another solid.

This phenomenon of Earth's crust floating in the denser mantle is now known as the **principle of isostasy**, which is easy to understand by an analogy to an iceberg. Ice is slightly less dense than water, so it floats. According to Archimedes' principle of buoyancy, an iceberg sinks in water until it displaces a volume of water whose weight is equal to that of the ice. When the iceberg has sunk to an equilibrium position, only about 10% of its volume is above water level. If some of the ice above water level should melt, the iceberg rises to maintain equilibrium with the same proportion of ice above and below the water.

Earth's crust is similar to the iceberg in that it sinks into the mantle to its equilibrium level. Where the crust is thickest, as beneath mountains, it sinks farther down into the mantle, and it also rises higher above the surface. And because continental crust is thicker and less dense than oceanic crust, it stands higher than the ocean basins. Remember, the mantle is hot, yet solid, and under tremendous pressure, so it behaves in a fluid-like manner.

Some of you might realize that crust floating on the mantle raises an apparent contradiction. In Chapter 8, we said that the mantle is a solid because it transmits S-waves, which do not move through a fluid. But according to the principle of isostasy, the mantle behaves as a fluid. When considered in terms of the brief time required for S-waves to pass through it, the mantle is indeed solid. But when subjected to stress over long periods, it yields by flowage; thus, at this time scale, it can be regarded as a viscous fluid.

ISOSTATIC REBOUND

What happens when a ship is loaded with cargo and then later unloaded? Of course, it first sinks lower in the water and then rises, but it always finds its equilibrium position. Earth's crust responds similarly to loading and unloading, but much more slowly. For example, if the crust is loaded, as when widespread glaciers accumulate, the crust sinks farther into the mantle to maintain equilibrium. The crust behaves

principle of isostasy The theoretical concept of Earth's crust "floating" on a dense underlying layer.

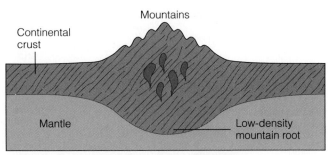

a. The crust and mantle before erosion and deposition.

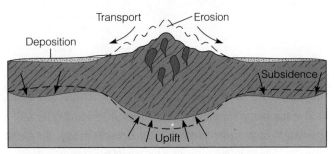

b. Erosion of the mountains and deposition in adjacent areas. Isostatic rebound begins.

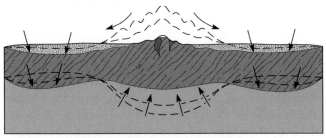

c. Continuing erosion, deposition, and isostatic rebound.

Figure 10.20 Isostatic Rebound
A diagrammatic representation showing the isostatic response of the crust to erosion (unloading) and widespread deposition (loading).

In isostatic rebound, Earth's crust returns to equilibrium.

similarly in areas where huge quantities of sediment accumulate.

If loading by glacial ice or sediment depresses Earth's crust farther into the mantle, it follows that when vast glaciers melt or where deep erosion takes place, the crust should rise back up to its equilibrium level—and in fact it does. This phenomenon, known as **isostatic rebound**, is taking place in Scandinavia, which was covered by a thick ice sheet until about 10,000 years ago; it is now rebounding at about 1 m per century. In fact, coastal cities in Scandinavia have rebounded rapidly enough that docks constructed several centuries ago are now far from shore. Isostatic rebound has also occurred in eastern Canada, where the crust has risen as much as 100 m in the last 6,000 years.

Figure 10.20 shows the response of Earth's continental crust to loading and unloading as

isostatic rebound
The phenomenon in which unloading of the crust causes it to rise until it attains equilibrium.

mountains form and evolve. Recall that during an orogeny, emplacement of plutons, metamorphism, and general thickening of the crust accompany deformation. However, as the mountains erode, isostatic rebound takes place and the mountains rise, whereas adjacent areas of sedimentation subside (Figure 10.20). If continued long enough, the mountains will disappear and then can be detected only by the plutons and metamorphic rocks that show their former existence.

CHAPTER 11
MASS WASTING

These slopes along the Grand Canyon of the Yellowstone River in Yellowstone National Park, Wyoming, have been oversteepened by river erosion. This oversteepening has resulted in slumping along the canyon walls, which contribute to the widening of the canyon.

Introduction

"Although water can play an important role, the relentless pull of gravity is the major force behind mass wasting."

Triggered by relentless torrential rains that began in December 1999, the floods and mudslides that devastated Venezuela were some of the worst ever to strike that country. Although an accurate death toll is impossible to determine, it is estimated that at least 19,000 people were killed, as many as 150,000 were left homeless, 35,000 to 40,000 homes were destroyed or buried by mudslides, and between $10 billion and $20 billion in damage was done before the rains and slides abated. It is easy to cite the numbers of dead and homeless, but the human side of the disaster was most vividly brought home by a mother who described standing helplessly by and watching her four small children buried alive in the family car as a raging mudslide carried it away.

This terrible tragedy illustrates how geology affects all of our lives. The underlying causes of the mudslides in Venezuela can be found anywhere in the world. In fact, *landslides* (a general term for mass movements of Earth materials) cause, on average, between 25 and 50 deaths and more than $2 billion in damage annually in the United States. By being able to recognize and understand how landslides occur and what the results may be, we can find ways to reduce the hazards and minimize damage in terms of both human life and property damage.

LEARNING OUTCOMES

After reading this unit, you should be able to do the following:

LO1 List the factors that influence mass wasting

LO2 Describe the types of mass wasting

LO3 Understand how to recognize and minimize the effects of mass wasting

---

Mass wasting (also called *mass movement*) is defined as the downslope movement of material under the direct influence of gravity. Most types of mass wasting are aided by weathering and usually involve surficial material. The material moves at rates ranging from almost imperceptible, as in the case of creep, to extremely fast, as in a rockfall or slide. Although water can play an important role, the relentless pull of gravity is the major force behind mass wasting.

LO1 Factors That Influence Mass Wasting

Mass wasting is an important geologic process that can occur at any time and almost any place. Although all major landslides have natural causes, many smaller ones are the result of human activity and could have been prevented or their damage minimized.

When the gravitational force acting on a slope exceeds its resisting force, slope failure (mass wasting) occurs. The resisting forces that help maintain slope stability include the slope material's strength and cohesion, the amount of internal friction between grains, and any external support of the slope (Figure 11.1a). These factors collectively define a slope's **shear strength**.

Opposing a slope's shear strength is the force of gravity. Gravity operates vertically but also has a component of force acting parallel to the slope, thereby causing instability (Figure 11.1a). The steeper a slope's angle, the greater the component of force acting parallel to the slope, and the greater the chance for mass wasting. The steepest angle that a slope can maintain without collapsing is its *angle of repose* (Figure 11.1b). At this angle, the shear strength of the slope's material exactly counterbalances the force of gravity. For

mass wasting The downslope movement of Earth materials under the influence of gravity.

shear strength The resisting forces that help maintain a slope's stability.

Figure 11.1 Slope Shear Strength

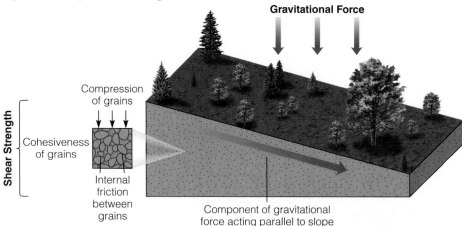

Gravitational Force

Shear Strength

Compression of grains

Cohesiveness of grains

Internal friction between grains

Component of gravitational force acting parallel to slope

a. A slope's shear strength depends on the slope material's strength and cohesion, the amount of internal friction between grains, and any external support of the slope. These factors promote slope stability. The force of gravity operates vertically, but also has a component of force acting parallel to the slope. When this force, which promotes instability, exceeds a slope's shear strength, slope failure occurs.

b. The angle of repose is a function of sheer strength. Dry sand usually has an angle of repose of about 30 degrees. With damp sand, shear strength is increased, and so much steeper angles of repose are possible.

unconsolidated material, the angle of repose normally ranges from 25 to 40 degrees. Slopes steeper than 40 degrees usually consist of unweathered solid rock.

All slopes are in a state of *dynamic equilibrium*, which means that they are constantly adjusting to new conditions. Although we tend to view mass wasting as a disruptive and usually destructive event, it is one of the ways that a slope adjusts to new conditions. Whenever a building or road is constructed on a hillside, the equilibrium of that slope is affected. The slope must then adjust, sometimes by mass wasting, to this new set of conditions.

Many factors can cause mass wasting: a change in slope angle, weakening of material by weathering, increased water content, changes in the vegetation cover, and overloading. Although most of these processes are interrelated, we will examine them separately for ease of discussion, but we will also show how they individually and collectively affect a slope's equilibrium.

SLOPE ANGLE

Slope angle is probably the major cause of mass wasting. Generally speaking, the steeper the slope, the less stable it is. Therefore, steep slopes are more likely to experience mass wasting than gentle ones.

Several processes can oversteepen a slope. One of the most common is undercutting by stream or wave action (Figure 11.2). This process removes the slope's base, increases the slope angle, and thereby increases the gravitational force acting parallel to the slope. Wave action, especially during storms, often results in mass movements along the shores of oceans or large lakes (Figure 11.3).

Excavations for road cuts and hillside building sites are another major cause of slope failure (Figure 11.4). Grading the slope too steeply or cutting into its side increases the stress in the rock or soil until it is no longer strong enough to remain at the steeper angle, and mass movement ensues.

Figure 11.2 Undercutting a Slope's Base by Stream Erosion

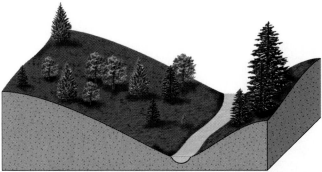

a. Undercutting by stream erosion removes a slope's base,

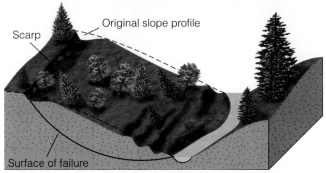

b. which increases the slope angle, and can lead to slope failure.

c. Undercutting by stream erosion caused slumping along this stream near Weidman, Michigan. Notice the scarp, which is the exposed surface of the underlying material following slumping.

WEATHERING AND CLIMATE

Mass wasting is more likely to occur in loose or poorly consolidated slope material than in bedrock. As soon as rock is exposed at Earth's surface, weathering begins

Mass wasting is more likely to occur in loose or poorly consolidated slope material than in bedrock.

to disintegrate and decompose it, reducing its shear strength and increasing its susceptibility to mass wasting. The deeper the weathering zone extends, the greater the likelihood of some type of mass movement.

Recall that some rocks are more susceptible to weathering than others and that climate plays an important role in the rate and type of weathering. In the tropics, where temperatures are high and considerable rain falls, the effects of weathering extend to depths of several tens of meters, and mass movements most commonly occur in the deep weathering zone. In arid and semiarid regions, the weathering zone is usually considerably shallower. Nevertheless, intense, localized cloudbursts can drop large quantities of water on an area in a short time. With little vegetation to absorb this water, runoff is rapid and frequently results in mudflows.

WATER CONTENT

The amount of water in rock or soil influences slope stability. Large quantities of water from melting snow or heavy rainfall greatly increase the likelihood of slope failure. The additional weight that water adds to a slope can be enough to cause mass movement. Furthermore, water percolating through a slope's material helps to decrease friction between grains, contributing to a loss of cohesion. For example, slopes composed of dry clay are usually quite stable, but when wetted, they quickly lose cohesiveness and internal friction and become an unstable slurry. This occurs because clay, which can hold large quantities of water, consists of platy particles that easily slide over each other when wet. For this reason, clay beds are frequently the slippery layer along which overlying rock units slide downslope.

Figure 11.3 Undercutting a Slope's Base by Wave Action
This sea cliff north of Bodega Bay, California, was undercut by waves during the winter of 1997–1998. As a result, part of the land slid into the ocean, damaging several houses.

COPYRIGHT AND PHOTOGRAPH BY DR. PARVINDER S. SETHI

Figure 11.4 Highway Excavation

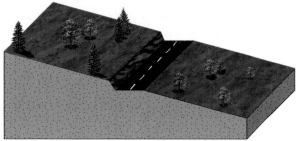

a. Highway excavations disturb the equilibrium of a slope by

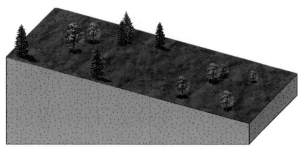

b. removing a portion of its support, as well as oversteepening it at the point of excavation, which can result in

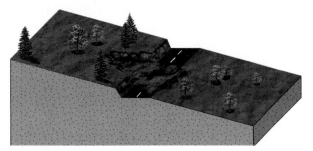

c. landslides along the highway.

COURTESY OF R. V. DIETRICH

d. Cutting into the hillside to construct this portion of the Pan-American Highway in Mexico resulted in a rockfall that completely blocked the road.

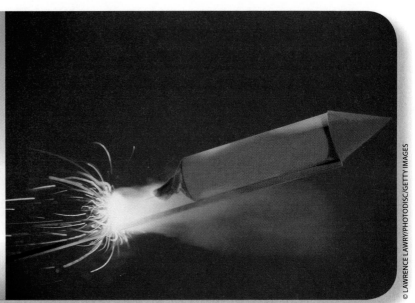

TRIGGERING MECHANISMS

Volcanic eruptions, explosions, and even loud claps of thunder may be enough to trigger a landslide if the slope is sufficiently unstable. Many *avalanches,* which are rapid movements of snow and ice down steep mountain slopes, are triggered by a loud gunshot or, in rare cases, even a person's shout.

VEGETATION

Vegetation affects slope stability in several ways. By absorbing the water from a rainstorm, vegetation decreases water saturation of a slope's material that would otherwise lead to a loss of shear strength. Vegetation's root system also helps stabilize a slope by binding soil particles together and holding the soil to bedrock.

The removal of vegetation by either natural or human activity is a major cause of many mass movements. Summer brush and forest fires in southern California frequently leave the hillsides bare of vegetation. Fall rainstorms saturate the ground, causing mudslides that do tremendous damage and cost millions of dollars to clean up.

OVERLOADING

Overloading is almost always the result of human activity and typically results from the dumping, filling, or piling up of material. Under natural conditions, a material's load is carried by its grain-to-grain contacts, with the friction between the grains maintaining a slope. The additional weight created by overloading increases the water pressure within the material, which in turn decreases its shear strength, thereby weakening the slope material. If enough material is added, the slope will eventually fail, sometimes with tragic consequences.

GEOLOGY AND SLOPE STABILITY

The relationship between the topography of an area and its geology is important in determining slope stability (Figure 11.5). If the rocks underlying a slope dip in the same direction as the slope, mass wasting is more likely to occur than if the rocks are horizontal or dip in the opposite direction. When the rocks dip in the same direction as the slope, water can percolate along the various bedding planes and decrease the cohesiveness and friction between adjacent rock units. This is particularly true when clay layers are present, because clay becomes slippery when wet.

Even if the rocks are horizontal or dip in a direction opposite to that of the slope, joints may dip in the same direction as the slope. Water migrating through them weathers the rock and expands these openings until the weight of the overlying rock causes it to fall.

TRIGGERING MECHANISMS

The factors discussed thus far all contribute to slope instability. Most, though not all, rapid mass movements are triggered by a force that temporarily disturbs slope equilibrium. The most common triggering mechanisms are strong vibrations from earthquakes and excessive amounts of water from a winter snow melt or a heavy rainstorm (Figure 11.6).

Figure 11.5 Geology, Slope Stability, and Mass Wasting
Rocks dipping in the same direction as a hill's slope are particularly susceptible to mass wasting.

1. Water percolates through
soil into clay-rich layers ▭
that become slippery, and may swell,
weakening the overlying rock ▭.

2. The clay-rich layer dips in the same direction
as the even more steeply dipping slope. Gravity
can therefore turn it into a skid surface, or
potential landslide plane.

3. Undercutting by the stream at the foot of the slope
exposes another watery, weak clay layer underlying
a heavy, strong limestone bed ▭. The heavy
limestone is now prone to slide across the clay,
carrying the rest of the overlying slope with it.

4. Layers on this side of the valley dip in an opposite
direction from the slope. Thus, gravity cannot easily act
to destabilize them, even if water percolation is
deep and undercutting occurs.

LO2 Types of Mass Wasting

Mass movements are generally classified on the basis of three major criteria (Table 11.1): (1) rate of movement (rapid or slow); (2) type of movement (primarily falling, sliding, or flowing); and (3) type of material involved (rock, soil, or debris). Even though many slope failures are combinations of different materials and movements, the resulting mass movements are typically classified according to their dominant behavior.

rapid mass movement
Any kind of mass wasting that involves a visible downslope displacement of material.

slow mass movement
Mass movement that advances at an imperceptible rate and is usually detectable only by the effects of its movement.

Rapid mass movements involve a visible movement of material. Such movements usually occur quite suddenly, and the material moves quickly downslope. Rapid mass movements are potentially dangerous and frequently result in loss of life and property damage. Most rapid mass movements occur on relatively steep slopes and can involve rock, soil, or debris.

Slow mass movements advance at an imperceptible rate and are usually detectable only by the effects of their movement, such as tilted trees and power poles or cracked foundations. Although rapid mass movements are more dramatic, slow mass movements are responsible for the downslope transport of a much greater volume of weathered material.

Figure 11.7 Rockfalls

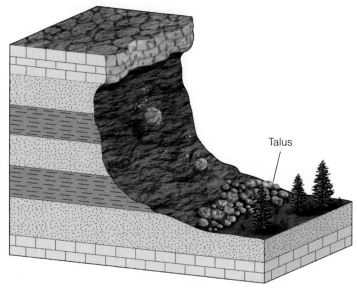

a. Rockfalls result from failure along cracks, fractures, or bedding planes in the bedrock and are common features in areas of steep cliffs.

Figure 11.6 Landslide Triggered by Heavy Rains, La Conchita California

La Conchita, California, is located at the base of a steep sloped terrace. Heavy rains and irrigation of an avocado orchard (visible at the top of the terrace) contributed to the landslide that destroyed nine homes in 1995.

b. A recent rockfall of granite in Yosemite National Park, California.

FALLS

Rockfalls are a common type of extremely rapid mass movement in which rocks of any size fall through the air (Figure 11.7). Rockfalls occur along steep canyons, cliffs, and road cuts and build up accumulations of loose rocks and rock fragments at their base called *talus*.

Rockfalls result from failure along joints or bedding planes in the bedrock and are commonly triggered by natural or human undercutting of slopes, or by earthquakes. Many rockfalls in cold climates are the result of frost wedging (see Figure 6.2). Chemical weathering caused by water percolating through the fissures in carbonate rocks (limestone, dolostone, and marble) is also responsible for many rockfalls.

Rockfalls range in size from small rocks falling from a cliff to massive falls involving millions of cubic meters of debris that destroy buildings, bury towns, and block highways (Figure 11.7b). Rockfalls are a particularly common hazard in mountainous areas where roads have been built by blasting and grading through steep hillsides of bedrock. Slopes that are particularly susceptible to rockfalls are sometimes covered with wire mesh in an effort to prevent dislodged rocks from falling onto the road below (Figure 11.8).

rockfall A type of extremely fast mass wasting in which rocks fall through the air.

Figure 11.8 Minimizing Damage from Rock Falls
Wire mesh has been used to cover this steep slope near Narvik in northern Norway. This is a common practice in mountainous areas to prevent rocks from falling on the road.

COPYRIGHT AND PHOTOGRAPH BY DR. PARVINDER S. SETHI

SLIDES

A **slide** involves movement of material along one or more surfaces of failure. The type of material may be soil, rock, or a combination of the two, and it may break apart during movement or remain intact. A slide's rate of movement can vary from extremely slow to very rapid (Table 11.1).

Two types of slides are generally recognized: (1) slumps or rotational slides, in which movement occurs along a curved surface; and (2) rock or block slides, which move along a more or less planar surface.

A **slump** involves the downward movement of material along a curved surface of rupture and is characterized by the backward rotation of the slump block (Figure 11.9). Slumps usually occur in unconsolidated or weakly

slide Mass wasting involving movement of material along one or more surfaces of failure.

slump Mass wasting that takes place along a curved surface of failure and results in the backward rotation of the slump mass.

TABLE 11.1

CLASSIFICATION OF MASS MOVEMENTS AND THEIR CHARACTERISTICS

TYPE OF MOVEMENT	SUBDIVISION	CHARACTERISTICS	RATE OF MOVEMENT
Falls	Rockfall	Rocks of any size fall through the air from steep cliffs, canyons, and road cuts	Extremely rapid
Slides	Slump	Movement occurs along a curved surface of rupture; most commonly involves unconsolidated or weakly consolidated material	Extremely slow to moderate
	Rock slide	Movement occurs along a generally planar surface	Rapid to very rapid
Flows	Mudflow	Consists of at least 50% silt- and clay-sized particles and up to 30% water	Very rapid
	Debris flow	Contains larger-sized particles and less water than mudflows	Rapid to very rapid
	Earthflow	Thick, viscous, tongue-shaped mass of wet regolith	Slow to moderate
	Quick clays	Composed of fine silt and clay particles saturated with water; when disturbed by a sudden shock, lose their cohesiveness and flow like a liquid	Rapid to very rapid
	Solifluction	Water-saturated surface sediment	Slow
	Creep	Downslope movement of soil and rock	Extremely slow
Complex movements		Combination of different movement types	Slow to extremely rapid

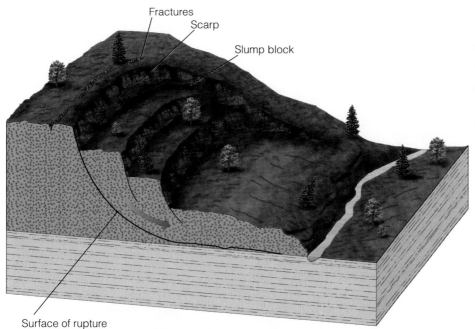

Fractures
Scarp
Slump block
Surface of rupture

COPYRIGHT AND PHOTOGRAPH BY DR. PARVINDER S. SETHI

Figure 11.9 Slumping
In a slump, material moves downward along the curved surface of a rupture, causing the slump block to rotate backward. Most slumps involve unconsolidated or weakly consolidated material and are typically caused by erosion along the slope's base.

Figure 11.10 Slumping in Point Fermin, California
Homes dangerously close to an oversteepened, eroding coastal bluff in Point Fermin, California. It is just a matter of time before these homes are destroyed by the effects of erosion or slumping of the unstable bluff.

consolidated material and range in size from small individual sets, such as occur along stream banks, to massive, multiple sets that affect large areas and cause considerable damage.

Slumps can be caused by a variety of factors, but the most common is erosion along the base of a slope, which removes support for the overlying material. This local steepening may be caused naturally by stream erosion along its banks (Figure 11.2c) or by

wave action at the base of a coastal cliff (Figure 11.10).

Slope oversteepening can also be caused by human activity, such as the construction of highways and housing developments. Slumps are particularly prevalent along highway cuts, where they are generally the most frequent type of slope failure observed.

Although many slumps are merely a nuisance, large-scale slumps in populated areas and along highways can cause extensive damage. Such is the case in coastal southern California, where slumping and sliding have been a constant problem, resulting in the destruction of many homes and the closing and relocation of numerous roads and highways (Figures 11.10).

A **rock slide** or *block slide* occurs when rocks move downslope along a more or less planar surface. Most rock slides take place because the local slopes and rock layers dip in the same direction (Figure 11.11), although they can also occur along fractures parallel to a slope. Rock slides are also common occurrences along the southern California coast. At Point Fermin, seaward-dipping rocks with interbedded slippery clay layers are undercut by waves, causing numerous slides.

rock slide Rapid mass wasting in which rocks move downslope along a more or less planar surface.

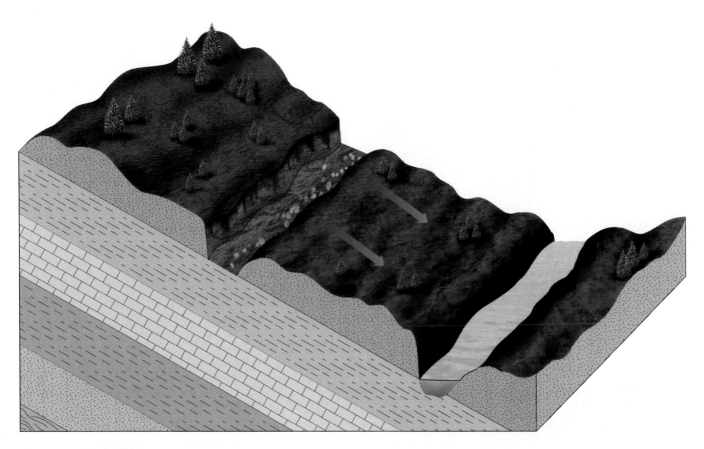

Figure 11.11 Rock Slides
Rock slides occur when material moves downslope along a generally planar surface. Most rock slides result when the underlying rocks dip in the same general angle as the slope of the land. Undercutting along the base of the slope and clay layers beneath porous rock or soil layers increase the chance of rock slides.

For example, the town of Laguna Beach was hit by rock slides and mudslides in 1978, 1998, and as recently as 2005, when numerous homes were destroyed or damaged and two people were killed (Figure 11.12).

Just as at Point Fermin, the rocks dip about 25 degrees in the same direction as the slope of the canyon walls and contain clay beds that "lubricate" the overlying rock layers, causing the rocks and the houses built on them to slide. Percolating water from heavy rains wets subsurface clayey siltstone, thus reducing its shear strength and helping to activate the slide. In addition, these slides are part of a larger ancient slide complex.

Not all rock slides are the result of rocks dipping in the same direction as a hill's slope. The rock slide at Frank, Alberta, Canada, on April 29, 1903, illustrates how nature and human activity can combine to create a situation with tragic results (Figure 11.13). It would appear at first glance that the coal-mining town of Frank, lying at the base of Turtle Mountain, was in no danger from a landslide (Figure 11.13). After all, many of the rocks dipped away from the mining valley. The joints in the massive limestone composing Turtle Mountain, however, dip steeply toward the valley and are essentially parallel with the slope of the mountain itself. Furthermore, Turtle Mountain is supported by weak siltstones, shales, and coal layers that underwent slow plastic deformation from the weight of the overlying massive limestone. Coal mining along the base of the valley also contributed to the stress on the rocks by removing some of the underlying support. All of these factors, as well as the frost action and chemical weathering that widened the joints, finally resulted in a massive rock slide. Approximately 40 million m³ of rock slid down Turtle Mountain along joint planes, killing 70 people and partially burying the town of Frank.

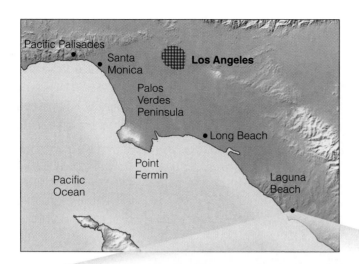

Figure 11.12 Rock Slide, Laguna Beach, California
A combination of interbedded clay layers that become slippery when wet, rocks dipping in the same direction as the slope of the sea cliffs, and undercutting of the sea cliffs by wave action activated a rock slide at Laguna Beach, California, that destroyed numerous homes and cars on October 2, 1978. This same area was hit by another rock slide in 2005.

FLOWS

Mass movements in which material flows as a viscous fluid or displays plastic movement are termed *flows*. Their rate of movement ranges from extremely slow to extremely rapid (Table 11.1). In many cases, mass movements begin as falls, slumps, or slides, and change into flows farther downslope.

Of the major mass movement types, **mudflows** are the most fluid and move most rapidly (at speeds up to 80 km per hour). They consist of at least 50% silt- and clay-sized material combined with a significant amount of water (up to 30%). Mudflows are common in arid and semiarid environments, where they are triggered by heavy rainstorms that quickly saturate the regolith,

> **mudflow** A flow consisting mostly of clay- and silt-sized particles and up to 30% water that moves downslope under the influence of gravity.

Figure 11.13 Rock Slide, Turtle Mountain, Canada

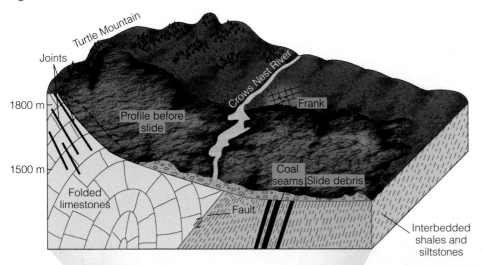

a. The tragic Turtle Mountain rock slide that killed 70 people and partially buried the town of Frank, Alberta, Canada, on April 29, 1903, was caused by a combination of factors. These included joints that dipped in the same direction as the slope of Turtle Mountain, a fault partway down the mountain, weak shale and siltstone beds underlying the base of the mountain, and mined-out coal seams.

b. Results of the 1903 rock slide at Frank.

turning it into a raging flow of mud that engulfs everything in its path. Mudflows can also occur in mountain regions (Figure 11.14) and in areas covered by volcanic ash, where they can be particularly destructive (see Chapter 5). Because mudflows are so fluid, they generally follow preexisting channels until the slope decreases or the channel widens, at which point they fan out.

Figure 11.14 Mudflow, Rocky Mountain National Park
Mudflows move swiftly downslope, engulfing everything in their path. Note how this mudflow in Rocky Mountain National Park has fanned out at the base of the hill. Also note the small lake adjacent to the mudflow that was formed after this mudflow created a dam across the stream.

Debris flows are composed of larger particles than mudflows and do not contain as much water. Consequently, they are usually more viscous than mudflows, typically do not move as rapidly, and rarely are confined to preexisting channels. Debris flows can be just as damaging, though, because they can transport large objects (Figure 11.15).

Earthflows move more slowly than either mudflows or debris flows. An earthflow slumps from the upper part of a hillside, leaving a scarp, and flows slowly downslope as a thick, viscous, tongue-shaped mass of wet regolith (Figure 11.16). Like mudflows and debris flows, earthflows can be of any size and are frequently destructive. They occur most commonly in humid climates on grassy, soil-covered slopes following heavy rains.

Some clays spontaneously liquefy and flow like water when they are disturbed. Such **quick clays** have caused serious damage and loss of lives in Sweden, Norway, eastern Canada, and Alaska. Quick clays are composed of fine silt and clay particles made by the grinding action of glaciers. Geologists think that these fine sediments were originally deposited in a marine environment, where their pore space was filled with saltwater. The ions in saltwater helped establish strong bonds between the clay particles, thus stabilizing and strengthening the clay. When the clays were subsequently uplifted above sea level, the saltwater was flushed out by fresh groundwater, reducing the effectiveness of the ionic bonds between the clay particles and thereby reducing the overall strength and cohesiveness of the clay. Consequently, when the clay is disturbed by a sudden shock or shaking, it essentially turns to a liquid and flows.

An excellent example of the damage that can be done by quick clays occurred in the Turnagain Heights area of Anchorage, Alaska, in 1964 (Figure 11.17). Underlying most of the Anchorage area is the Bootlegger Cove Clay, a massive clay unit of poor permeability. Because the Bootlegger Cove Clay

debris flow A type of mass wasting that involves a viscous mass of soil, rock fragments, and water that moves downslope; debris flows have larger particles than mudflows and contain less water.

earthflow A mass wasting process involving the downslope movement of water-saturated soil.

quick clay A clay deposit that spontaneously liquefies and flows like water when disturbed.

Figure 11.15 Debris Flow, Ophir Creek, Nevada
A debris flow and damaged house in lower Ophir Creek, western Nevada. Note the many large boulders that are part of the debris flow. Debris flows do not contain as much water as mudflows and typically are composed of larger particles.

Figure 11.16 Earthflow

a. Earthflows form tongue-shaped masses of wet regolith that move slowly downslope. They occur most commonly in humid climates on grassy, soil-covered slopes.

b. An earthflow near Baraga, Michigan.

Figure 11.17 Quick-Clay Slide, Anchorage, Alaska

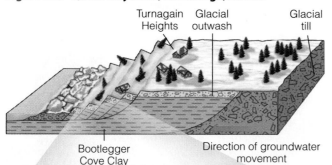

a. Ground shaking by the 1964 Alaska earthquake turned parts of the Bootlegger Cove Clay into a quick clay, causing numerous slides.

ALASKA EARTHQUAKE COLLECTION/USGS

b. Low-altitude photograph of the Turnagain Heights subdivision of Anchorage shows some of the numerous landslide fissures that developed, as well as the extensive damage to buildings in the area. The remains of the Four Seasons apartment building can be seen in the background.

forms a barrier that prevents groundwater from flowing through the adjacent glacial deposits to the sea, considerable hydraulic pressure builds up on the landward side of the clay. Some of this water has flushed out the saltwater in the clay and has saturated the lenses of sand and silt associated with the clay beds. When the magnitude-8.6 Good Friday earthquake struck on March 27, 1964, the shaking turned parts of the Bootlegger Cove Clay into a quick clay and precipitated a series of massive slides in the coastal bluffs that destroyed most of the homes in the Turnagain Heights subdivision (Figure 11.17b).

Solifluction is the slow downslope movement of water-saturated surface sediment. Solifluction can occur in any climate where the ground becomes saturated with water, but is most common in areas of

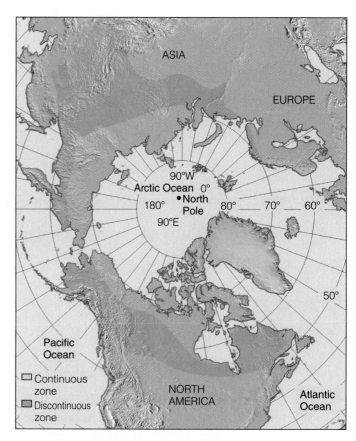

Figure 11.18 Permafrost and Solifluction
Distribution of permafrost areas in the Northern Hemisphere.

permafrost. **Permafrost**, ground that remains permanently frozen, covers nearly 20% of the world's land surface (Figure 11.18). During the warmer season, when the upper portion of the permafrost thaws, water and surface sediment form a soggy mass that flows by solifluction and produces a characteristic lobate topography.

As might be expected, many problems are associated with construction in a permafrost environment. For example, when an uninsulated building is constructed directly on permafrost, heat escapes through the floor, thaws the ground below, and turns it into a soggy, unstable mush. Because the ground is no longer solid, the building settles unevenly into the ground, and numerous structural problems result (Figure 11.19).

solifluction Mass wasting involving the slow downslope movement of water-saturated surface materials; especially at high elevations or high latitudes where the flow is underlain by frozen soil.

permafrost Ground that remains permanently frozen.

Figure 11.19 Permafrost Damage This house, south of Fairbanks, Alaska, has settled unevenly because the underlying permafrost in fine-grained silts and sands has thawed.

O.J. FERRAINS, JR./USGS

Creep, the slowest type of flow, is the most widespread and significant mass wasting process in terms of the total amount of material moved downslope and the monetary damage it does annually. Creep involves extremely slow downhill movement of soil or rock. Although it can occur anywhere and in any climate, it is most effective and significant as a geologic agent in humid regions.

Because the rate of movement is essentially imperceptible, we are frequently unaware of creep's existence until we notice its effects: tilted trees and power poles, broken streets and sidewalks, or cracked retaining walls or foundations (Figure 11.20). Creep usually involves the whole hillside and probably occurs, to some extent, on any weathered or soil-covered, sloping surface.

Creep is not only difficult to recognize, but also to control. Although engineers can sometimes slow or stabilize creep, many times the only course of action is to simply avoid the area if at all possible or, if the zone of creep is relatively thin, design structures that can be anchored into the bedrock.

creep A widespread type of mass wasting in which soil or rock moves slowly downslope.

complex movement A combination of different types of mass movements in which no single type is dominant; usually involves sliding and flowing.

COMPLEX MOVEMENTS

Recall that many mass movements are combinations of different movement types. When one type is dominant, the movement can be classified as one of those described thus far. If several types are more or less equally involved, however, it is called a **complex movement**.

The most common type of complex movement is the slide-flow, in which there is sliding at the head and then some type of flowage farther along its course. Most slide-flow landslides involve well-defined slumping at the head, followed by a debris flow or earthflow (Figure 11.21). Any combination of different mass movement types is a complex movement.

LO3 Recognizing and Minimizing the Effects of Mass Wasting

The most important factor in eliminating or minimizing the damaging effects of mass wasting is a thorough geologic investigation of the region in question. In this way, former landslides and areas susceptible to mass movements can be identified and perhaps avoided. By assessing the risks of possible mass wasting before con-

Figure 11.20 Creep

a. Some evidence of creep: (A) curved tree trunks; (B) displaced monuments; (C) tilted power poles; (D) displaced and tilted fences; (E) roadways moved out of alignment; (F) hummocky surface.

b. Trees, bent by creep, Wyoming.

c. Creep has bent these sandstone and shale beds of the Haymond Formation near Marathon, Texas.

d. Stone wall tilted due to creep in Champion, Michigan.

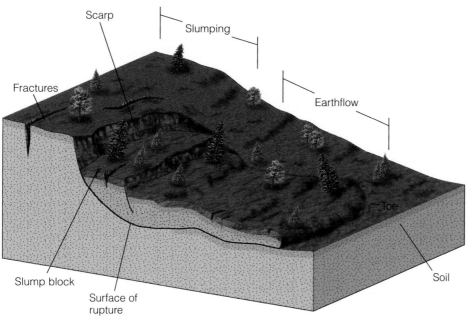

Figure 11.21 **Complex Movement**
A complex movement is one in which several types of mass wasting are involved. In this example, slumping occurs at the head, followed by an earthflow.

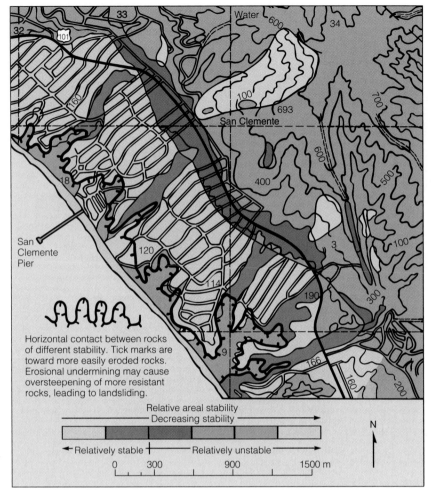

Horizontal contact between rocks
of different stability. Tick marks are
toward more easily eroded rocks.
Erosional undermining may cause
oversteepening of more resistant
rocks, leading to landsliding.

Relative areal stability
Decreasing stability

← Relatively stable ┼ Relatively unstable →

0   300   900   1500 m

N

Figure 11.22 Slope-Stability Map
This slope-stability map of part of San Clemente, California, shows areas delineated
according to relative stability. Such maps help planners and developers make decisions about
where to site roads, utility lines, buildings, and other structures.

struction begins, engineers can take steps to eliminate or minimize the effects of such events.

Identifying areas with a high potential for slope failure is important in any hazard assessment study; these studies include identifying former landslides, as well as sites of potential mass movement. Scarps, open fissures, displaced or tilted objects, a hummocky surface, and sudden changes in vegetation are some of the features that indicate former landslides or an area susceptible to slope failure. The effects of weathering, erosion, and vegetation may, however, obscure the evidence of previous mass wasting.

The information derived from a hazard assessment study can be used to produce *slope-stability maps* of the area (Figure 11.22). These maps allow planners and developers to make decisions about where to site roads, utility lines, and housing or industrial developments based on the relative stability or instability of a particular location. The maps also indicate the extent of an area's landslide problem and the type of mass movement that may occur.

Although most large mass movements usually cannot be prevented, geologists and engineers can use various methods to minimize the danger and damage resulting from them. Because water plays such an important role in many landslides, one of the most effective and inexpensive ways to reduce the potential for slope failure or to increase existing slope stability is surface and subsurface drainage of a hillside. Drainage serves two purposes. It reduces the weight of the material likely to slide, and increases the shear strength of the slope material by lowering pore pressure.

Surface waters can be drained and diverted by ditches, gutters, or culverts designed to direct water away from slopes. Drainpipes perforated along one surface and driven into a hillside can help remove subsurface water (Figure 11.23). Finally, planting vegetation on hillsides helps stabilize slopes by holding the soil together and reducing the amount of water in the soil.

Another way to help stabilize a hillside is to reduce its slope. Recall that overloading and oversteepening by grading are common causes of slope failure. Reducing the angle of a hillside decreases the potential for slope failure. Two methods are usually employed to reduce a slope's angle. In the *cut-and-fill* method, material is removed from the upper part of the slope and used as fill at the base, thus providing a flat surface for construction and reducing the slope (Figure 11.24). The second method, which is called *benching*, involves cutting a series of benches or steps into a hillside (Figure 11.25). This process reduces the overall average slope, and the benches serve as collecting sites for small landslides or rockfalls that might occur. Benching is most commonly used on steep hillsides in conjunction with a system of surface drains to divert runoff.

Figure 11.23 Using Drainpipes to Remove Subsurface Water

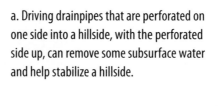

a. Driving drainpipes that are perforated on one side into a hillside, with the perforated side up, can remove some subsurface water and help stabilize a hillside.

Flow of groundwater

b. A drainpipe driven into the hillside at Point Fermin, California, helps to reduce the amount of subsurface water in these porous beds.

REED WICANDER

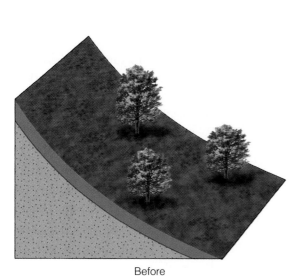

This material has been removed

Former slope

Material from upper slope added here

Before

After

Figure 11.24 Stabilizing a Hillside by the Cut-and-Fill Method
One common method used to help stabilize a hillside and reduce its slope is the cut-and-fill method. Material from the steeper upper part of the hillside is removed, thereby decreasing the slope angle, and is used to fill in the base. This provides some additional support at the base of the slope.

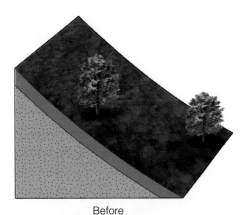

Before

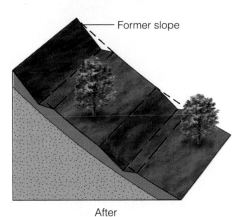

Former slope

After

Figure 11.25 **Stabilizing a Hillside by Benching**
Another common method used to stabilize a hillside and reduce its slope is benching. This process involves making several cuts along a hillside to reduce the overall slope. Furthermore, individual slope failures are now limited in size, and the material collects on the benches.

Figure 11.26 **Retaining Walls Help Reduce Landslides**

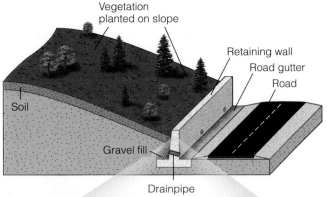

Vegetation planted on slope

Retaining wall
Road gutter
Road
Soil
Gravel fill
Drainpipe

a. Retaining walls anchored into bedrock, backfilled with gravel, and provided with drainpipes can support a slope's base and reduce landslides.

b. A steel retaining wall built to stabilize the slope and keep falling and sliding rocks off the highway.

In some situations, retaining walls are constructed to provide support for the base of the slope (Figure 11.26). The walls are usually anchored well into bedrock, backfilled with crushed rock, and provided with drain holes to prevent the buildup of water pressure in the hillside.

Recognition, prevention, and control of landslide-prone areas are expensive, but not nearly as expensive as the damage can be when such warning signs are ignored or not recognized. Unfortunately, there are numerous examples of landfill and dam collapses that serve as tragic reminders of the price paid in loss of lives and property damage when the warning signs of impending disaster are ignored.

CHAPTER 12
RUNNING WATER

The Firehole River in Yellowstone National Park in Wyoming. In this view, the river meanders through one of the park's hydrothermal areas.

Introduction

"The water in channels is, with few exceptions, the most important geologic agent in modifying the land surface."

The terrestrial planets share a similar early history of accretion, differentiation, and volcanism, but Earth is the only one with abundant surface water. The small size and high temperature of Mercury and the runaway greenhouse effect on Venus preclude the possibility of liquid water. Mars is too small and too cold for liquid water, although it does have some frozen water and trace amounts of water vapor in its atmosphere. However, it does have winding valleys and canyons that were probably eroded by running water during the planet's early history. In contrast, oceans and seas cover approximately 71% of Earth's surface.

The *hydrosphere* consists mostly of water in the oceans, but it also includes water vapor in the atmosphere, groundwater, water frozen in glaciers, and water on land in lakes, swamps, bogs, streams, and rivers. Our main concern in this chapter is with the small amount of running water confined to channels. It is

Figure 12.1 Aftermath of the Johnstown, Pennsylvania, Flood
On May 31, 1889, an 18-m-high wall of water destroyed Johnstown and killed at least 2,200 people.

NATIONAL PARK SERVICE

important to note that running water has a tremendous impact on most of the land surface.

You have probably experienced the power of running water if you have ever swam or canoed in a rapidly flowing stream or river, but to truly appreciate the energy of moving water, you need only read the vivid accounts of floods. For example, at 4:07 p.m. on May 31, 1889, residents of Johnstown, Pennsylvania, heard "a roar like thunder" and within 10 minutes the town was destroyed when an 18-m-high wall of water tore through the town at more than 60 km per hour, sweeping up houses, debris, and entire families (Figure 12.1). According to one account, "Thousands of people desperately tried to escape the wave. Those caught by the wave found themselves swept up in a torrent of oily, muddy water, surrounded by tons of grinding debris. . . . Many became hopelessly entangled in miles of barbed wire from the destroyed wire works."*

When the flood was over, at least 2,200 people were dead, some of them victims of a fire that broke out on floating debris on which they escaped the flood. The Johnstown flood, the most deadly river flood in U.S. history, resulted from heavy rainfall and the failure of a dam upstream from the town.

---

*National Park Service—U.S. Department of Interior, Shiretown Information Service Online.

LEARNING OUTCOMES

After reading this unit, you should be able to do the following:

LO1 Identify sources of water on Earth

LO2 Describe the role of running water

LO3 Explain how running water erodes and transports sediment

LO4 Describe deposition by running water

LO5 Question whether floods can be predicted and controlled

LO6 Understand drainage systems

LO7 Recognize the significance of base level

LO8 Understand the evolution of valleys

Every year, floods cause extensive property damage and fatalities, and yet we derive many benefits from running water, even from some floods. Running water—that is, water confined to channels—is one source of freshwater for agriculture, industry, domestic use, and recreation. About 8% of all electricity used in North America is generated by falling water at hydroelectric generating plants. Large waterways throughout the world are avenues of commerce, and when Europeans explored the interior of North America, they followed the St. Lawrence, Mississippi, Missouri, and Ohio rivers.

All water derived from the oceans eventually makes it back to the oceans and can thus begin the hydrologic cycle again.

does the moisture for rain and snow come from in the first place? You might immediately suspect that the oceans are the ultimate source of precipitation. In fact, water is continuously recycled from the oceans, through the atmosphere, to the continents, and back to the oceans. This **hydrologic cycle**, as it is called (Figure 12.2), is powered by solar radiation and is possible because water changes easily from liquid to gas (water vapor) About 85% of all water entering the atmosphere evaporates from the oceans. The remaining 15% comes from water on land, but this water originally came from the oceans as well.

LO1 Water on Earth

Most of Earth's 1.33 billion km³ of water (97.2%) is in the oceans, and nearly all of the rest is frozen in glaciers on land (2.15%). That leaves only 0.65% in the atmosphere, groundwater, lakes, swamps, bogs, and a tiny important amount in stream and river channels. Nevertheless, the water in channels is, with few exceptions, the most important geologic agent in modifying the land surface.

Much of our discussion of running water is descriptive, but always be aware that streams and rivers are dynamic systems that must continuously respond to change. For example, paving in urban areas increases surface runoff to waterways, and other human activities such as building dams and impounding reservoirs also alter the dynamics of stream and river systems. Natural changes, too, affect the complex interacting parts of these systems.

Regardless of its source, water vapor rises into the atmosphere where the complex processes of cloud formation and condensation take place. About 80% of all precipitation falls directly back into the oceans, so the hydrologic cycle is a three-step process of evaporation, condensation, and precipitation. For the 20% of precipitation that falls on land, the hydrologic cycle is more complex, involving evaporation, condensation, movement of water vapor from the oceans to land, precipitation, and runoff. Some precipitation evaporates as it falls and reenters the cycle, but about 36,000 km³ of the precipitation falls on land and returns to the oceans by **runoff**, the surface flow in streams and rivers.

Some precipitation is temporarily stored in lakes and swamps, snowfields and glaciers, or seeps below the surface where it enters the groundwater system. Water might remain in some of these reservoirs for thousands of years, but eventually glaciers melt, lakes and groundwater feed streams and rivers, and this water returns to the oceans. Even the water used by plants evaporates, a process known as *evapotranspiration*, and returns to the atmosphere.

hydrologic cycle The continuous recycling of water from the oceans, through the atmosphere, to the continents, and back to the oceans, or from the oceans, through the atmosphere, and back to the oceans.

runoff The surface flow in streams and rivers.

infiltration capacity The maximum rate at which soil or sediment absorbs water.

THE HYDROLOGIC CYCLE

The connection between precipitation and clouds is obvious, but where

FLUID FLOW

Solids are rigid substances that retain their shapes unless deformed by a force, but fluids—that is, liquids and gases—have no strength, so they flow in response to any force, no matter how slight. Liquid water flows downslope in response to gravity.

Runoff during a rainstorm depends on **infiltration capacity**, the maximum rate at which surface ma-

Figure 12.2 Stages of the Hydrologic Cycle
Water is recycled from the oceans to land and back to the oceans.

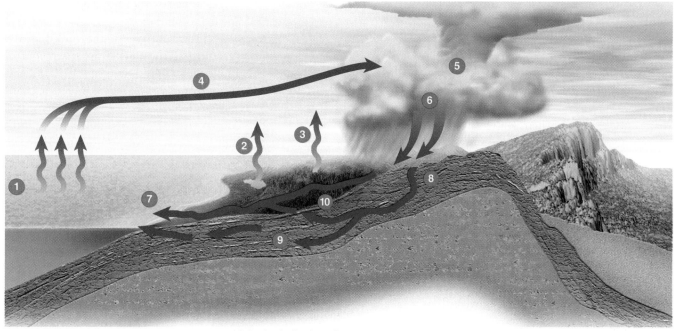

Stages of the hydrologic cycle

1. Evaporation from the sea
2. Evaporation from streams, rivers, lakes, and soil
3. Evapotranspiration from plants
4. Movement by wind of moisture-laden air masses
5. Condensation into clouds as air rises

6. Precipitation
7. Runoff of rainfall, snow, and ice-melt back to the sea via rivers
8. Infiltration of some surface waters
9. Groundwater flow back toward the sea
10. Emergence of some groundwater as springs that feed streams and rivers

terials absorb water. Several factors control infiltration capacity, including intensity and duration of rainfall. If rain is absorbed as fast as it falls, no surface run-off takes place. Loosely packed dry soil absorbs water faster than tightly packed wet soil, and thus more rain must fall on loose dry soil before runoff begins. Regardless of the initial condition of surface materials, once they are saturated, excess water collects on the surface and, if on a slope, it moves downhill.

LO2 Running Water

The term *running water* applies to any surface water that moves from higher to lower areas in response to gravity. We have already noted that running water is very effective in modifying Earth's land surface by erosion and that it is the primary geologic process responsible for sediment transport and deposition in many areas. Indeed, it is responsible for everything from the

tiniest rills in farmer's fields to scenic wonders such as the Grand Canyon in Arizona, as well as vast deposits such as the Mississippi River delta.

SHEET FLOW AND CHANNEL FLOW

Even on steep slopes, flow is initially slow and hence causes little or no erosion. As water moves downslope, though, it accelerates and may move by *sheet flow*, a more or less continuous film of water flowing over the surface. Sheet flow is not confined to depressions, and it accounts for *sheet erosion*, a particular problem on some agricultural lands.

In *channel flow*, surface runoff is confined to troughlike depressions that vary in size from tiny rills with a trickling stream of water to huge river channels. We describe flow in channels with terms such as *rill*, *brook*, *creek*, *stream*, and *river*, most of which are distinguished by size and volume. Here we use the terms *stream* and *river* more or less interchangeably, although the latter usually refers to a larger body of running water.

Streams and rivers receive water from sheet flow and rain that falls directly into their channels, but far more important is the water supplied by soil moisture and groundwater. In areas where groundwater is plentiful, streams and rivers maintain a fairly stable flow year-round, because their water supply is continuous. In contrast, the amount of water in streams and rivers of arid and semiarid regions fluctuates widely, because they depend more on infrequent rainstorms and surface runoff for their water.

© BANANASTOCK/JUPITERIMAGES

GRADIENT, VELOCITY, AND DISCHARGE

Water in any channel flows downhill over a slope known as its **gradient**. Suppose a river has its headwaters (source) 1,000 m above sea level and it flows 500 km to the sea, so it drops vertically 1,000 m over a horizontal distance of 500 km. Its gradient is found by dividing the vertical drop by the horizontal distance, which in this example is 1,000 m/500 km = 2 m/km on average (Figure 12.3a).

The **velocity** of running water is a measure of the downstream distance water travels in a given time. It is usually expressed in meters per second (m/sec) or feet per second (ft/sec), and it varies across a channel's width as well as along its length. Water moves more slowly and with greater turbulence near a channel's bed and banks, because friction is greater there than it is some distance from these boundaries (Figure 12.3b). Channel shape and roughness also influence flow velocity. Broad, shallow channels and narrow, deep channels have proportionately more water in contact with their perimeters than do channels with semicircular cross sections (Figure 12.3c). As one would expect, rough channels, such as those strewn with boulders, offer more frictional resistance to flow than do channels with a bed and banks composed of sand or mud.

Intuitively, you might think that the gradient is the most important control on velocity—the steeper the gradient, the greater the velocity. In fact, a channel's average velocity actually increases downstream even though its gradient decreases! Keep in mind that we are talking about average velocity for a long segment of a channel, not velocity at a single point. Two factors account for this downstream increase in velocity. First, the upstream reaches of channels tend to be boulder-strewn, broad, and shallow, so frictional resistance to flow is high, whereas downstream segments of the same channels are more semicircular and have banks composed of finer materials. And second, the number of smaller tributaries joining a larger channel increases downstream. Thus, the total volume of water (discharge) increases, and increasing discharge results in greater velocity.

Specifically, **discharge** is the volume of water that passes a particular point in a given period of time. Discharge is found from the dimensions of a water-filled channel—that is, its cross-sectional area (A) and flow velocity (V). Discharge (Q) is then calculated with the formula $Q = VA$ and is expressed in cubic meters per second (m³/sec) or cubic feet per second (ft³/sec).

LO3 How Does Running Water Erode and Transport Sediment?

Streams and rivers possess two kinds of energy: potential and kinetic. *Potential energy* is the energy of position, such as the energy of water at high elevation. During stream flow, potential energy is converted to *kinetic energy*, the energy of motion. Much of this kinetic energy is used up in fluid turbulence, but some is available for erosion and transport. The materials transported by a stream include a dissolved load, and a load of solid particles (mud, sand, and gravel).

Because the **dissolved load** of a stream is invisible, it is commonly overlooked, but it is an important part of the total sediment load. Some of it is acquired from a stream's bed and banks, where soluble rocks such as limestone are present, but much of it is carried into waterways by sheet flow and by groundwater.

gradient The slope over which a stream or river flows expressed in m/km or ft/mi.

velocity A measure of distance traveled per unit of time, as in the flow velocity in a stream or river.

discharge The volume of water in a stream or river moving past a specific point in a given interval of time; expressed in cubic meters per second (m³/sec) or cubic feet per second (ft³/sec).

dissolved load The part of a stream's load consisting of ions in solution.

Figure 12.3 Gradient and Flow Velocity

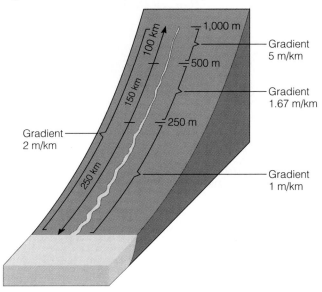

a. The average gradient of this stream is 2 m/km; however, gradient can be calculated for any segment of a stream, as shown in this example. Notice that the gradient is steepest in the headwaters area and decreases downstream.

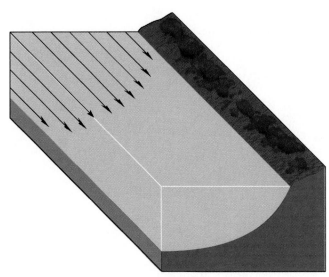

b. The maximum flow velocity is near the center and top of a straight channel, where the least friction takes place. The arrows are proportional to velocity.

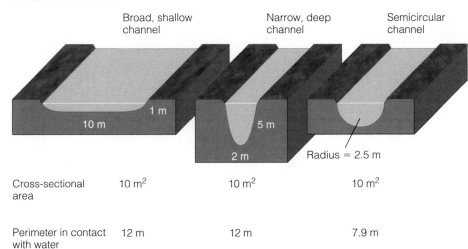

	Broad, shallow channel	Narrow, deep channel	Semicircular channel
Cross-sectional area	10 m²	10 m²	10 m²
Perimeter in contact with water	12 m	12 m	7.9 m

c. These three differently shaped channels have the same cross-sectional area; however, the semicircular one has less water in contact with its perimeter and thus less frictional resistance to flow.

A stream's solid load is made up of particles ranging from clay sized (> 1/256 mm) to huge boulders, much of it supplied by mass wasting, but some is eroded directly from a stream's bed and banks. The direct impact of running water, **hydraulic action**, is sufficient to set particles in motion.

Running water carrying sand and gravel erodes by **abrasion**, as exposed rock is worn and scraped by the impact of these particles (Figure 12.4a). Circular to oval depressions called *potholes* in streambeds are one manifestation of abrasion (Figure 12.4b). They form where swirling currents with sand and gravel eroded the rock.

Once materials are eroded, they are transported from their source and eventually deposited. The dissolved load is transported in the water itself, but the load of solid particles moves as *suspended load* or *bed load*. The **suspended load** consists of the smallest particles of silt and clay, which are kept suspended above the channel's bed by fluid turbulence (Figure 12.5).

The **bed load** of larger particles, mostly sand and gravel, cannot

hydraulic action The removal of loose particles by the power of moving water.

abrasion The process whereby rock is worn smooth by the impact of sediment transported by running water, glaciers, waves, or wind.

suspended load The smallest particles (silt and clay) carried by running water, which are kept suspended by fluid turbulence.

bed load That part of a stream's sediment load, mostly sand and gravel, transported along its bed.

Figure 12.4 Abrasion by Running Water Carrying Sand and Gravel

b. These potholes in the bed of the Chippewa River in Ontario, Canada, are about 1 m across. Two potholes have merged to form a larger composite pothole.

a. The rocks just above water level have been smoothed and polished by abrasion.

Figure 12.5 Sediment Transport

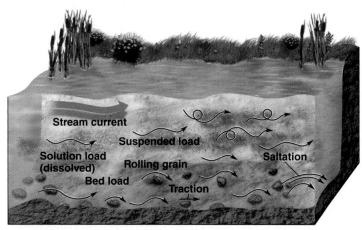

a. Sediment transport as bed load, suspended load, and dissolved load. Flow velocity is highest near the surface, but gravel- and sand-sized particles are too large to be lifted far from the streambed so they make up the bed load, whereas silt and clay are in the suspended load.

b. The Snake River in Grand Teton National Park, Wyoming. The suspended load gives the water its murky appearance.

be kept suspended by fluid turbulence, so that it is transported along the bed. However, some of the sand may be temporarily suspended by currents that swirl across the streambed and lift grains into the water. The grains move forward with the water, but also settle and finally come to rest and then again move by the same process of intermittent bouncing and skipping, a phenomenon known as *saltation* (Figure 12.5). Particles too large to be even temporarily suspended are transported by traction; that is, they simply roll or slide along a channel's bed.

LO4 Deposition by Running Water

Rivers and streams constantly erode, transport, and deposit sediment, but they do most of their geologic work when they flood. Consequently, their deposits, collectively called **alluvium**, do not represent the day-to-day activities of running water, but rather the periodic sedimentation that takes place during floods. Recall from Chapter 6 that sediments accumulate in continental, transitional, and marine *depositional environments* (see Figure 6.12). Deposits of rivers and streams are found mostly in the first two of these set-

tings; however, much of the detrital sediment found on continental margins is derived from the land and transported to the oceans by running water.

The Deposits of Braided and Meandering Channels

A **braided stream** has an intricate network of dividing and rejoining channels separated from one another by sand and gravel bars (Figure 12.6). Braided channels develop when the sediment supply exceeds the transport capacity of running water, resulting in the deposition of sand and gravel bars. Braided streams have broad, shallow channels and are characterized as bed-load transport streams, because they transport and deposit mostly sand and gravel.

Meandering streams have a single sinuous channel with broadly looping curves known as *meanders* (Figure 12.7). Channels of meandering streams are semicircular in cross section along straight reaches, but markedly asymmetric at meanders, where they vary from shallow to deep across the meander. The deeper side of the channel is known as the *cut bank*, because greater velocity and fluid turbulence erode it. In contrast, flow velocity is at a minimum on the opposite bank, which slopes gently into the channel. As a result of this unequal distribution of flow

alluvium A collective term for all detrital sediment transported and deposited by running water.

braided stream A stream with multiple dividing and rejoining channels.

meandering stream A stream that has a single, sinuous channel with broadly looping curves.

Figure 12.6 Braided Streams and Their Deposits

a. A braided stream in Denali National Park, Alaska.

b. This small braided stream is near Grindelwald, Switzerland. The deposits of both streams are mostly sand and gravel.

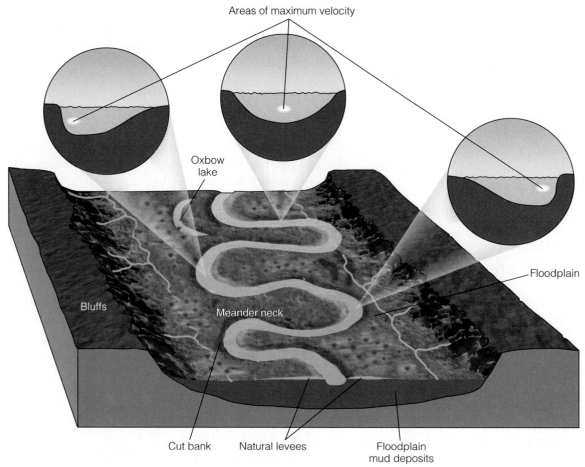

Figure 12.7 Diagrammatic View of a Meandering Stream
Notice the areas of maximum flow velocity.

velocity across meanders, the cut bank erodes, and a **point bar** is deposited on the gently sloping inner bank (Figure 12.8).

> **point bar** The sediment body deposited on the gently sloping side of a meander loop.
>
> **oxbow lake** A cutoff meander filled with water.
>
> **floodplain** A low-lying, flat area adjacent to a channel that is partly or completely water-covered when a stream or river overflows its banks.
>
> **natural levee** A ridge of sandy alluvium deposited along the margins of a channel during floods.

Meanders commonly become so sinuous that the thin neck of land between adjacent ones is cut off during a flood. Many of the floors of valleys with meandering channels are marked by crescent-shaped **oxbow lakes**, which are simply cutoff meanders (Figures 12.7 and 12.9). Oxbow lakes may persist for a long time, but they eventually fill with organic matter and fine-grained sediments carried by floods.

FLOODPLAIN DEPOSITS

Channels periodically receive more water than they can accommodate, so they overflow their banks and spread across adjacent low-lying, relatively flat **floodplains** (Figure 12.7 and 12.10a). Floodplain sediments might be sand and gravel, but more commonly, fine-grained sediments, mostly mud, are deposited. During a flood, a stream overtops its banks and water pours onto the floodplain, but as it does so, its velocity and depth rapidly decrease. As a result, ridges of sandy alluvium known as **natural levees** are deposited along the channel margins, and mud is carried beyond the natural levees into the floodplain, where it settles from suspension (Figure 12.10 b,c).

DELTAS

Where a river or stream flows into a lake or the ocean, its flow velocity rapidly diminishes, and any sediment in transport is deposited. Under some circumstances,

JAMES S. MONROE

Point bars

Out bank

Figure 12.8 Point Bars
Point bars form on the gently sloping side of
meanders, where flow velocity is lowest. Two
point bars of sand in Otter Creek in Yellowstone
National Park in Wyoming. Note that the point
bars are inclined into the deeper part of the
channel.

Figure 12.9 Oxbow Lakes
Meandering streams become so sinuous
that meanders get cut off.

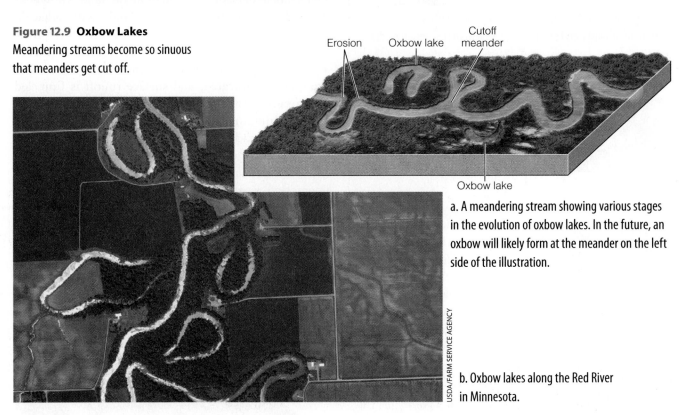

Erosion   Oxbow lake   Cutoff
meander

Oxbow lake

USDA/FARM SERVICE AGENCY

a. A meandering stream showing various stages
in the evolution of oxbow lakes. In the future, an
oxbow will likely form at the meander on the left
side of the illustration.

b. Oxbow lakes along the Red River
in Minnesota.

this deposition creates a delta, an alluvial deposit that
causes the shoreline to build outward into the lake or
sea, a process called *progradation*. The simplest prograding deltas have a characteristic vertical sequence
of *bottomset beds* overlain successively by *foreset beds*
and *topset beds* (Figure 12.11). This vertical sequence
develops when a river or stream enters another body of
water where the finest sediment (silt and clay) is carried some distance out into the lake or sea, where it
settles to form bottomset beds. Nearer shore, foreset
beds are deposited as gently inclined layers, and topset
beds, consisting of the coarsest sediments, are deposited in a network of *distributary channels* traversing
the top of the delta (Figure 12.11).

Figure 12.10 Floodplain Deposits

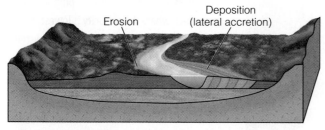

Erosion · Deposition (lateral accretion)

a. Floodplain deposits formed by lateral accretion of point bars are made up mostly of sand.

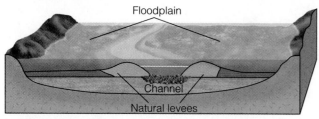

Floodplain · Channel · Natural levees

b. The origin of vertical accretion deposits. During floods, streams deposit natural levees, and silt and mud settle from suspension on the floodplain.

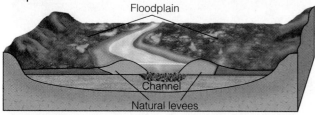

Floodplain · Channel · Natural levees

c. After flooding.

alluvial fan A cone-shaped accumulation of mostly sand and gravel deposited where a stream flows from a mountain valley onto an adjacent lowland.

Small deltas in lakes may have the three-part sequence described previously, but deltas deposited along seacoasts are much larger, far more complex, and considerably more

important as potential areas of natural resources. Depending on the relative importance of stream (or river), wave, and tide processes, geologists identify three main types of marine deltas (Figure 12.12). *Stream-dominated deltas* have long fingerlike sand bodies, each deposited in a distributary channel that progrades far seaward. *Wave-dominated deltas* also have distributary channels, but the seaward margin of the delta consists of islands reworked by waves, and the entire margin of the delta progrades. *Tide-dominated deltas* are continuously modified into tidal sand bodies that parallel the direction of tidal flow.

ALLUVIAL FANS

Fan-shaped deposits of alluvium on land known as **alluvial fans** form best on lowlands with adjacent highlands in arid and semiarid regions where little vegetation exists to stabilize surface materials (Figure 12.13). During periodic rainstorms, surface materials are quickly saturated, and surface runoff is funneled into a mountain canyon leading to adjacent lowlands. In the mountain canyon, the runoff is confined so that it cannot spread laterally, but when it discharges onto the lowlands, it quickly spreads out, its velocity diminishes, and deposition ensues. Repeated episodes of sedimentation result in the accumulation of a fan-shaped body of alluvium.

Deposition by running water in the manner just described is responsible for many alluvial fans. In some cases, though, the water flowing through a canyon picks up so much sediment that it becomes a viscous debris flow. Consequently, some alluvial fans consist mostly of debris-flow deposits that show little or no layering.

Figure 12.11 Prograding Delta
Internal structure of the simplest type of prograding delta. Small deltas in lakes may have this structure, but marine deltas are much more complex.

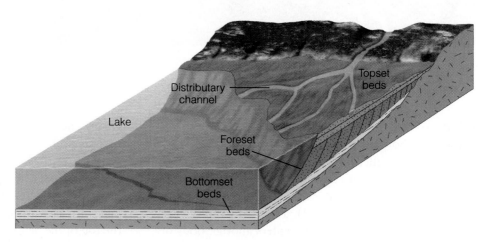

Distributary channel · Lake · Topset beds · Foreset beds · Bottomset beds

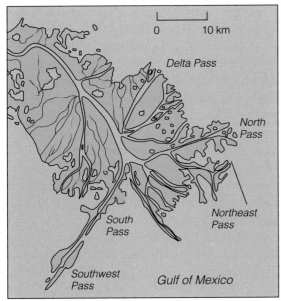

a. The Mississippi River delta on the U.S. Gulf Coast is stream dominated.

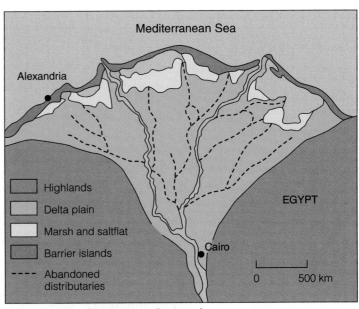

b. The Nile delta of Egypt is wave dominated.

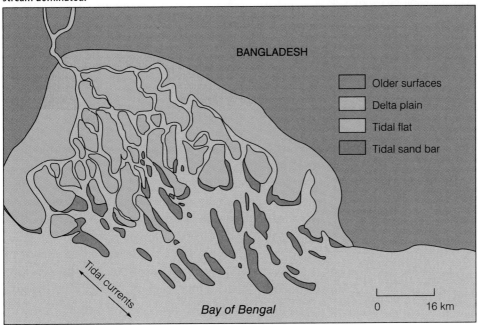

c. The Ganges–Brahmaputra delta of Bangledesh is tide dominated.

Figure 12.12 Marine Deltas

COPYRIGHT AND PHOTOGRAPH BY DR. PARVINDER S. SETHI

Figure 12.13 Alluvial Fans and Their Deposits
These alluvial fans at the base of the Panamint Range in Death Valley, California, were deposited where streams discharged from mountain canyons onto adjacent lowlands.

LO5 Can Floods Be Predicted and Controlled?

When a river or stream receives more water than its channel can handle, it floods, occupying part or all of its floodplain. People have tried to control floods for thousands of years. Common practices are to construct dams that impound reservoirs and to build levees along stream banks (Figure 12.14a, b). Levees raise the banks of a stream, thereby restricting flow during floods. Unfortunately, deposition within the channel raises the streambed, making the levees useless unless they, too, are raised. Levees along the banks of the Huang He in China caused the streambed to rise more than 20 m above its surrounding floodplain in 4,000 years. When the Huang He breached its levees in 1887, more than 1 million people were killed. Sacramento, California, lying at the junction of two rivers, is among the most flood-prone cities in the United States. Some of the levees that protect the city are 150 years old and in poor condition; the cost of repairing them has risen to as much as $250,000 per 100 m.

A floodway is a channel constructed to divert part of the excess water in a stream around populated areas or areas of economic importance (Figure 12.14c). Some communities build *floodwalls* to protect them from floods. Floodwalls have gates that permit access to the waterway but can be closed when the water rises (Figure 12.14d). Reforestation of cleared land also reduces the potential for flooding, because vegetated soil helps prevent runoff by absorbing more water.

When flood-control projects are well planned and constructed, they are functional. What many people fail to realize is that these projects are designed to contain floods of a given size; should larger floods occur, rivers spill onto floodplains anyway. Furthermore, dams occasionally collapse, and reservoirs eventually fill with sediment unless they are dredged.

As for predicting floods, the best that can be done is to monitor streams, evaluate their past behavior, and anticipate floods of a given size in a specified period. Most people have heard of 10-year floods, 20-years floods, and so on, but how are such determinations made? The U.S. Geologic Survey, as well as state agencies, record and analyze stream behavior through time and anticipate floods of a specified size. So a 20-year flood, for example, is the period during which a flood of a given magnitude can be expected. It does not mean that the river in question will have a flood of that size every 20 years, only that over a long period of time, it will average 20 years. Or we can say that the chances of a 10-year flood taking place in any one year are 1 in 10 (1/10). In fact, it is possible that two 10-year floods could take place in successive years, but then not occur again for several decades.

LO6 Drainage Systems

Thousands of waterways, which are parts of larger drainage systems, flow directly or indirectly into the oceans. The only exceptions are some rivers and streams that flow into desert basins surrounded by higher areas. But even these are parts of larger systems consisting of a main channel with all its tributaries—that is, streams that contribute water to another stream. The Mississippi River and its tributaries, such as the Ohio, Missouri, Arkansas, and Red rivers and thousands of smaller ones, carry runoff from an area known as a **drainage basin**. A topographically high area called a

drainage basin The surface area drained by a stream or river and its tributaries.

In August 2007, a tragic flood took place in Southeast Asia, particularly in India and Bangladesh. More than 2,100 people perished, and millions were displaced by swollen rivers, overflowing from monsoon rains. As displaced people fled to high ground, so did the wildlife, including poisonous snakes that claimed dozens of lives.

© POLKA DOT IMAGES/JUPITERIMAGES

Figure 12.14 Flood Control
Dams and reservoirs, levees, floodways, and floodwalls are some of the structures used to control floods.

a. Oroville Dam in California, at 235 m high, is the highest dam in the United States. It helps control floods, provides water for irrigation, and produces electricity at its power plant.

b. This levee, an artificial embankment along a waterway, helps protect nearby areas from floods. A university campus lies out of view just to the right of the levee.

c. This floodway carries excess water from a river (not visible) around a small community.

d. This floodwall on the bank of the Danube River at Mohács, Hungary, helps protect the city from floods.

divide separates a drainage basin from adjoining ones (Figure 12.15). The continental divide along the crest of the Rocky Mountains in North America, for instance, separates drainage in opposite directions; drainage to the west goes to the Pacific, whereas drainage to the east eventually reaches the Gulf of Mexico.

The arrangements of channels within an area are types of **drainage patterns**. The most common is *dendritic drainage*, which consists of a network of channels resembling tree branching (Figure 12.16a). It develops on gently sloping surfaces composed of materials that respond more or less homogeneously to erosion, such as areas underlain by nearly horizontal sedimentary rocks.

In dendritic drainage, tributaries join larger channels at various angles, but *rectangular drainage* is characterized by right-angle bends and tributaries joining larger channels at right angles (Figure 12.16b). Such regularity in channels is strongly controlled by geologic structures, particularly regional joint systems that intersect at right angles.

Trellis drainage, consisting of a network of nearly parallel main streams with tributaries joining them at right

divide A topographically high area that separates adjacent drainage basins.

drainage pattern The regional arrangement of channels in a drainage system.

Figure 12.15 Drainage Basins
A detailed view of the Wabash River's drainage basin, a tributary of the Ohio River. All tributary streams within the drainage basin, such as the Vermilion River, have their own smaller drainage basins. Divides are shown by red lines.

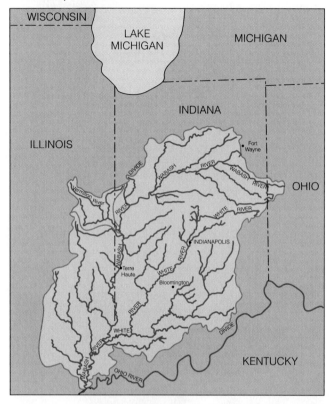

angles, is common in some parts of the eastern United States. In Virginia and Pennsylvania, erosion of folded sedimentary rocks developed a landscape of alternating ridges on resistant rocks and valleys underlain by easily eroded rocks. Main waterways follow the valleys, and short tributaries flowing from the nearby ridges join the main channels at nearly right angles (Figure 12.16c).

In *radial drainage*, streams flow outward in all directions from a central high point, such as a large volcano (Figure 12.16d). Many of the volcanoes in the Cascade Range of western North America have radial drainage patterns.

In the types of drainage mentioned so far, some kind of pattern is easily recognized. *Deranged drainage*, in contrast, is characterized by irregularity, with streams flowing into and out of swamps and lakes, streams with only a few short tributaries, and vast swampy areas between channels (Figure 12.16e). This kind of drainage developed recently and has not yet formed a fully organized drainage system. In parts of Minnesota, Wisconsin, and Michigan, where glaciers obliterated the previous drainage, only 10,000 years have elapsed since the glaciers melted. As a result, drainage systems have not yet fully developed, and large areas remain undrained.

Figure 12.16 Drainage Patterns

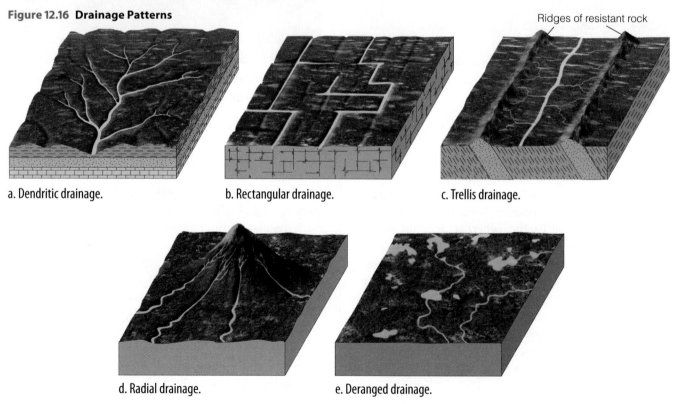

a. Dendritic drainage.

b. Rectangular drainage.

c. Trellis drainage.

d. Radial drainage.

e. Deranged drainage.

LO7 The Significance of Base Level

Base level is the lowest limit to which a stream or river can erode. With the exception of streams that flow into closed depressions in deserts, all others are restricted ultimately to sea level. That is, they can erode no lower than sea level, because they must have some gradient to maintain flow. So *ultimate base level* is sea level, which is simply the lowest level of erosion for any waterway that flows into the sea (Figure 12.17). Ultimate base level applies to an entire stream or river system, but channels may also have a *local* or *temporary base level*. For example, a local base level may be a lake or another stream, or where a stream or river flows across particularly resistant rocks and a waterfall develops (Figure 12.17).

Ultimate base level is sea level, but suppose that sea level dropped or rose with respect to the land, or suppose that the land rose or subsided? In these cases, base level would change and bring about changes in stream and river systems. During the Pleistocene Epoch (Ice Age), sea level was about 130 m lower than it is now, and streams adjusted by eroding deeper valleys and extending well out onto the continental shelves.

Rising sea level at the end of the Ice Age accounted for a rising base level, decreased stream gradients, and deposition within channels.

Geologists and engineers are well aware that building a dam to impound a reservoir creates a local base level. A stream entering a reservoir deposits sediment, so unless they are dredged, reservoirs eventually fill with sediment.

Draining a lake may seem like a small change and well worth the time and expense to expose dry land for agriculture or commercial development. But draining a lake eliminates a local base level, and a stream that originally flowed into the lake responds by rapidly eroding a deeper valley as it adjusts to a new base level.

WHAT IS A GRADED STREAM?

The *longitudinal profile* of any waterway shows the elevations of a channel along its length as viewed in cross section (Figure 12.18). For some rivers and streams, the longitudinal profile is smooth, but others show irregularities such as lakes and

> **base level** The level below which a stream or river cannot erode; sea level is ultimate base level.

Figure 12.17 Base Level

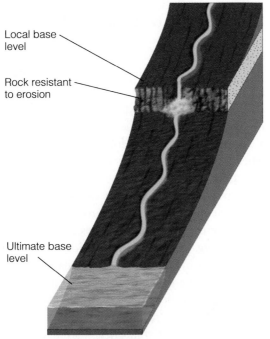

a. Sea level is ultimate base level, but a resistant rock layer over which a waterfall plunges forms a local base level.

- Local base level
- Rock resistant to erosion
- Ultimate base level

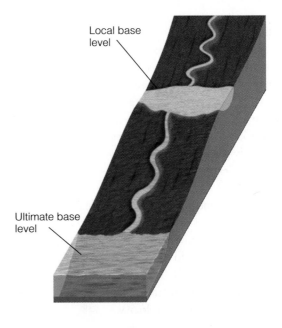

b. Local base level where a stream flows into a lake.

- Local base level
- Ultimate base level

Figure 12.18 Longitudinal Profiles of Streams

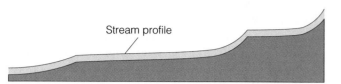

a. An ungraded stream has irregularities in its longitudinal profile.

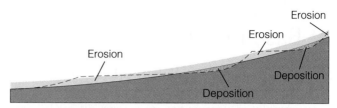

b. Erosion and deposition along the course of a stream eliminate irregularities and cause it to develop the smooth, concave profile typical of a graded stream.

graded stream A stream that has an equilibrium profile in which a delicate balance exists among gradient, discharge, flow velocity, channel characteristics, and sediment load so that neither significant deposition nor erosion takes place within its channel.

valley A linear depression bounded by higher areas such as ridges or mountains.

waterfalls, all of which are local base levels. Over time, these irregularities tend to be eliminated, because deposition takes place where the gradient is insufficient to maintain sediment transport, and erosion decreases the gradient where it is steep. So, given enough time, rivers and streams develop a smooth, concave longitudinal profile of equilibrium, meaning that all parts of the system dynamically adjust to one another.

A **graded stream** is one with an equilibrium profile in which a delicate balance exists among gradient, discharge, flow velocity, channel shape, and sediment load so that neither significant erosion nor deposition takes place within its channel (Figure 12.18b). Such a delicate balance is rarely attained, so the concept of a graded stream is an ideal. Nevertheless, the graded condition is closely approached in many streams, although only temporarily and not necessarily along their entire lengths.

Even though the concept of a graded stream is an ideal, we can anticipate the response of a graded stream to changes that alter its equilibrium. For instance, a change in base level would cause a stream to adjust as previously discussed. Increased rainfall in a stream's drainage basin would result in greater discharge and flow velocity. In short, the stream would now possess greater energy—energy that must be dissipated within the stream system by, for example, a change from a semicircular to a broad, shallow channel that would dis-

sipate more energy by friction. On the other hand, the stream may respond by eroding a deeper valley, effectively reducing its gradient until it is once again graded.

LO8 The Evolution of Valleys

Valleys are low areas on land bounded by higher land, and most of them have a river or stream running their length, with tributaries draining the nearby high areas. With few exceptions, valleys form and evolve in response to erosion by running water, although other processes, especially mass wasting, contribute. The shapes and sizes of valleys vary from small, steep-sided *gullies* to those that are broad with gently sloping valley walls (Figure 12.19). Steep-walled, deep valleys of vast size are *canyons*, and particularly narrow and deep ones are *gorges*.

A valley might start to erode where runoff has sufficient energy to dislodge surface materials and excavate a small rill. Once formed, a rill collects more runoff and becomes deeper and wider and continues to do so until a full-fledged valley develops. *Downcutting* takes place when a river or stream has more energy than it needs to transport sediment, so some of its excess energy is used to deepen its valley. In most cases, the valley walls are simultaneously undercut, a process called *lateral erosion*, creating unstable slopes that may fail by *mass wasting*. Furthermore, erosion by sheetwash and erosion by tributary streams carry materials from the valley walls into the main stream in the valley.

Valleys not only become deeper and wider, but they also become longer by *headward erosion*, a phenomenon involving erosion by entering runoff at the upstream end of a valley (Figure 12.20a).

Figure 12.19 Gullies and Valleys

b. This valley has steep walls that descend to a narrow valley bottom.

a. Gullies are small valleys, but they are narrow and deep. This gully in New Mexico measures about 15 m across.

Figure 12.20 Two Stages in the Evolution of a Valley

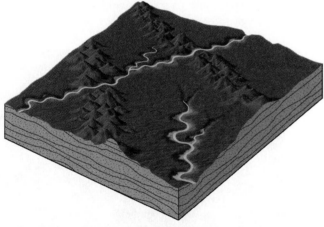

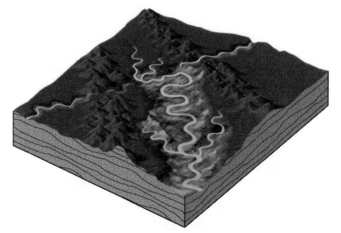

a. The stream widens its valley by lateral erosion and mass wasting while simultaneously extending its valley by headward erosion.

b. As the larger stream continues to erode headward, stream piracy takes place when it captures some of the drainage of the smaller stream. Notice also that the valley is wider in (b) than it was in (a).

Continued headward erosion may result in *stream piracy*, the breaching of a drainage divide and diversion of part of the drainage of another stream (Figure 12.20b).

STREAM TERRACES

Adjacent to many channels are erosional remnants of floodplains that formed when the streams were flowing at a higher level. These **stream terraces** consist of a fairly flat upper surface and a steep slope descending to the level of the lower, present-day floodplain (Figure 12.21). Some streams have several steplike surfaces above their present-day floodplains, indicating that terraces formed several times.

> **stream terrace** An erosional remnant of a floodplain that formed when a stream was flowing at a higher level.

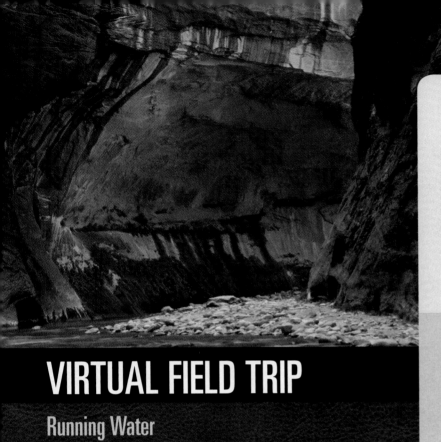

VIRTUAL FIELD TRIP

Running Water

GOALS OF THE TRIP

There are many geologic features to see in Zion National Park, but our main goals on this trip are to study:

1. Erosion.
2. Sediment transport.
3. Deposition by running water.
4. The origin and evolution of the park's canyons.

FOLLOW-UP QUESTIONS

1. Why do meandering streams form an alternating series of cut banks and point bars along their course?
2. Give a brief history of the origin and development of the canyons in Zion National Park.
3. What processes, other than downcutting by the Virgin River, contribute to the ongoing evolution of the canyons in Zion National Park?

Ready to Go!

Zion National Park in southwestern Utah was made a national park in November, 1919. Majestic cliffs and towers made up of colorful rocks and deep canyons dominate the landscape. Indeed, the rocks seem to change color throughout the day. Several geologic processes, such as mass wasting and mechanical and chemical weathering, have contributed to the magnificent scenery in the park, but erosion and deposition by running water is our focus in this virtual field trip. The Virgin River is the main waterway in the park, and it, along with its tributaries, is largely responsible for erosion of the mostly sandstone terrain.

The rocks exposed in the park's cliffs constitute the geologic record for this area and preserve evidence of ancient shallow seas, deposition in rivers, lakes, and ponds, as well as coastal sand dunes. The Navajo Sandstone consists of a lower reddish-brown, iron-rich sandstone and an overlying light-colored, iron-poor sandstone. All of the park's rock formations are Mesozoic in age and represent about 150 million years of geologic time. About 13 million years ago, uplift of the entire region steepened the gradient of the Virgin River which began downcutting, a process that would eventually yield the cliffs and deep canyons now present in Zion National Park.

Let's start our trip trough Zion National Park where we can make many observations relevant to our discussions of running water in this chapter.

What to See When You Go

This image shows both erosion and deposition by a meandering stream, that is, a stream with a single, sinuous channel (see Figure 12.7 in this chapter). Notice the steep bank on the right side of the image, which is called the cut bank because, as discussed in this chapter's section "The Deposits of Braided and Meandering Channels," the water velocity is highest here and erosion takes place. On the opposite bank, however, velocity is least and deposition occurs, forming a point bar deposit mostly of sand that slopes gently toward the cut bank. Point bars are the most distinctive features of ancient meandering stream deposits. It is important to note that further downstream the maximum velocity switches from right to left, thus forming a succession alternating point bars and cut banks (see Figure 12.8 in this chapter).

n the photo showing
a point bar (on the
previous page) notice
that the water in the
river is turbid (murky) because it
s carrying mud (silt and clay) and probably sand along
ts bed. The photo above shows a deposit of gravel and
sand. Some of the gravel is of boulder size (greater than
25.6 cm). From observations here we can conclude that
the stream was much higher when this deposit formed,
and it must have been flowing very rapidly to transport
such large particles. Also the gravel particles are well
rounded, meaning that their sharp corners and edges
were worn smooth by abrasion during transport. And
finally, the deposit is poorly sorted because it contains
a wide range of particle sizes.

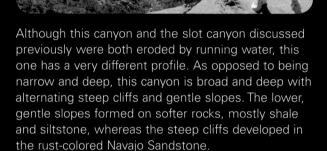

Although this canyon and the slot canyon discussed
previously were both eroded by running water, this
one has a very different profile. As opposed to being
narrow and deep, this canyon is broad and deep with
alternating steep cliffs and gentle slopes. The lower,
gentle slopes formed on softer rocks, mostly shale
and siltstone, whereas the steep cliffs developed in
the rust-colored Navajo Sandstone.

Great view of a slot canyon, that is, a canyon that is
much deeper than it is wide and usually with vertical
or near vertical sides. Most slot canyons form where
running water erodes into sandstone (as here) or
limestone, although they may form in other rocks,
too. Past floods are indicated by the scour marks and
polished rock surface several meters above water level.
Be very careful if you hike in a slot canyon, because a
summer thunderstorm can cause the water level to rise
several meters in just a few minutes.

As we learned in this chapter in the section "The
Evolutions of Valleys," downcutting by running water
accounts for canyon deepening, but canyons also
widen by a combination of processes including erosion
by tributaries, weathering, and mass wasting. This
blind arch, as it is called, formed when the softer rock
below was eroded and part of the overlying sandstone
collapsed. In any event, the debris from this event
eventually makes it to the river only to be carried away
and deposited elsewhere.

Figure 12.21 Origin of Stream Terraces

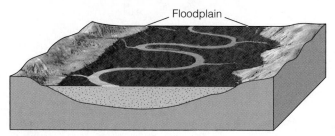

a. A stream has a broad floodplain.

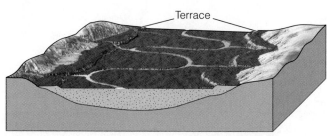

b. The stream erodes downward and establishes a new floodplain at a lower level. Remnants of its old, higher floodplain are stream terraces.

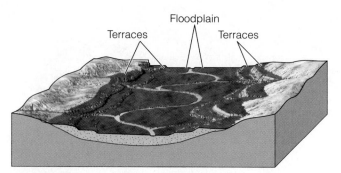

c. Another level of stream terraces forms as the stream erodes downward again.

d. Stream terraces along the Madison River in Montana.

Although stream terraces result from erosion, they are preceded by an episode of floodplain formation and sediment deposition. Subsequent erosion causes the stream to cut downward until it is once again graded (Figure 12.21). Then it begins to erode laterally and establishes a new floodplain at a lower level. Several such episodes account for the multiple terrace levels adjacent to some channels.

Renewed erosion and the formation of stream terraces are usually attributed to a change in base level. Either uplift of the land that a stream flows over or lowering of sea level yields a steeper gradient and increased flow velocity, thus initiating an episode of downcutting. When the stream reaches a level at which it is once again graded, downcutting ceases.

incised meander A deep, meandering canyon cut into bedrock by a stream or river.

INCISED MEANDERS

Some streams are restricted to deep, meandering canyons cut into bedrock, where they form features called **incised meanders** (Figure 12.22). Streams restricted by rock walls usually cannot erode laterally; thus, they lack a floodplain and occupy the entire width of the canyon floor.

It is not difficult to understand how a stream can cut downward into rock, but how a stream forms a meandering pattern in bedrock is another matter. Because lateral erosion is inhibited once downcutting begins, one must infer that the meandering course was established when the stream flowed across an area covered by alluvium. For example, suppose that a stream near base level has established a meandering pattern. If the land that the stream flows over is uplifted, then erosion begins and the meanders become incised into the underlying bedrock.

Figure 12.22 Incised Meanders
The Colorado River at Dead Horse State Park in Utah is incised to a depth of 600 m.

JAMES S. MONROE

SUPERPOSED STREAMS

Water flows downhill in response to gravity, so the direction of flow in streams and rivers is determined by topography. Yet a number of waterways seem, at first glance, to have defied this fundamental control. For instance, several rivers in the eastern United States flow in valleys that cut directly through ridges that lie in their paths. These are examples of **superposed streams**, all of which once flowed on a surface at a higher level, but as they eroded downward, they eroded into resistant rocks and cut narrow canyons, or what geologists call *water gaps* (Figure 12.23).

A water gap has a stream flowing through it, but if the stream is diverted elsewhere, perhaps by stream piracy, the abandoned gap is then called a *wind gap*. The Cumberland Gap in Kentucky is a good example; it was the avenue through which settlers migrated from Virginia to Kentucky from 1790 until well into the 1800s. Furthermore, several water gaps and wind gaps played important strategic roles during the Civil War (1861–1865).

> **superposed stream** A stream that once flowed on a higher surface and eroded downward into resistant rocks while maintaining its course.

Figure 12.23 Origin of a Superposed Stream

a. As a stream erodes down and removes the surface layers of rock, it is lowered onto ridges that form when resistant rocks in the underlying structure are exposed.

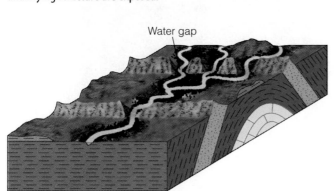

b. The narrow valleys through the ridges are water gaps.

MARIL BRYANT MILLER

c. View of a water gap cut by the Jefferson River in Montana.

CHAPTER 13
GROUNDWATER

A variety of cave deposits such as stalactites, stalagmites, and curtains can be seen in caves such as Shenandoah Caverns in Virginia.

Introduction

"As the world's population and industrial development expand, the demand for water, particularly groundwater, will increase."

Within the limestone region of western Kentucky lies the largest cave system in the world. In 1941, approximately 51,000 acres were set aside and designated as Mammoth Cave National Park. In 1981, it became a World Heritage Site. From ground level, the topography of the area is unimposing, with gently rolling hills. Beneath the surface, however, are more than 540 km of interconnected passageways whose spectacular geologic features have been enjoyed by millions of cave explorers and tourists.

During the War of 1812, approximately 180 metric tons of saltpeter, used in the manufacture of gunpowder, was mined from Mammoth Cave. At the end of the war, the saltpeter market collapsed, and Mammoth Cave was developed as a tourist attraction, easily overshadowing the other caves in the area. During the next 150 years, the discovery of new passageways and links to other caverns helped establish Mammoth Cave as the world's premier cave and the standard against which all others were measured.

The colorful cave deposits are the primary reason that millions of tourists have visited Mammoth Cave over the years. Hanging down from the ceiling and growing up from the floor are spectacular icicle-like structures, as well as columns and curtains in a variety of colors. Moreover, intricate passageways connect rooms of various sizes. The cave is also home to more than 200 species of insects and other animals, including about 45 blind species.

In addition to the beautiful caves, caverns, and cave deposits produced by groundwater movement, groundwater is also an important natural resource. Although groundwater constitutes only 0.6% of the world's water, it is, nonetheless, a significant source of freshwater for agriculture, industry, and domestic users. More than 65% of the groundwater used in the United States each year goes for irrigation, with industrial use second, followed by domestic needs. In fact, of the largest 100 cities in the United States, 34 depend solely on local groundwater supplies. These demands have severely depleted the groundwater supply in many areas and have led to such problems as ground subsidence and saltwater contamination. In other areas, pollution from landfills, toxic waste, and agriculture has rendered the groundwater supply unsafe.

As the world's population and industrial development expand, the demand for water, particularly groundwater, will increase. Not only must new groundwater sources be located, but once found, these sources must be protected from pollution and managed properly to ensure that users do not withdraw more water than can be replenished.

LEARNING OUTCOMES

After reading this unit, you should be able to do the following:

LO1 Describe groundwater and its role in the hydrologic cycle

LO2 Define porosity and permeability

LO3 Describe the water table

LO4 Understand patterns of groundwater movement

LO5 Identify springs, water wells, and artesian systems

LO6 Discuss erosion and deposition caused by groundwater

LO7 Describe the effects of modifications to the groundwater system

LO8 Describe hydrothermal activity

It is important that people become aware of what a valuable resource groundwater is, so they can ensure that future generations have a clean and adequate supply of water from this source.

LO1 Groundwater and the Hydrologic Cycle

Groundwater—water that fills open spaces in rocks, sediment, and soil beneath Earth's surface—is one reservoir in the hydrologic cycle, accounting for approximately 22% (8.4 million km³) of the world's supply of freshwater. Like all other water in the hydrologic cycle, the ultimate source of groundwater is the oceans; however, its more immediate source is the precipitation that infiltrates the ground and seeps down through the voids in soil, sediment, and rocks. Groundwater may also come from water infiltrating from streams, lakes, swamps, artificial recharge ponds, and water-treatment systems.

Regardless of its source, groundwater moving through the tiny openings between soil and sediment particles and the spaces in rocks filters out many impurities, such as disease-causing microorganisms and many pollutants. However, not all soils and rocks are good filters, and sometimes so much undesirable material may be present that it contaminates the groundwater. Groundwater movement and its recovery from wells depend on two critical aspects of the materials that it moves through: *porosity* and *permeability*.

LO2 Porosity and Permeability

Porosity and permeability are important physical properties of Earth materials and are largely responsible for the amount, availability, and movement of groundwater. Water soaks into the ground because soil, sediment, and rock have open spaces or pores. **Porosity** is the percentage of a material's total volume that is pore space. Porosity most often consists of the spaces between particles in soil, sediment, and sedimentary rocks, but other types of porosity include cracks, fractures, faults, and vesicles in volcanic rocks (Figure 13.1).

Porosity varies among different rock types and is dependent on the size, shape, and arrangement of the material composing the rock (Table 13.1). Most igneous and metamorphic rocks, as well as many limestones and dolostones, have very low

groundwater Underground water stored in the pore spaces of soil, sediment, and rock.

porosity The percentage of a material's total volume that is pore space.

Table 13.1

POROSITY VALUES FOR DIFFERENT MATERIALS

MATERIAL	PERCENTAGE POROSITY
Unconsolidated sediment	
Soil	55
Gravel	20–40
Sand	25–50
Silt	35–50
Clay	50–70
Rocks	
Sandstone	5–30
Shale	0–10
Solution activity in limestone, dolostone	10–30
Fractured basalt	5–40
Fractured granite	10

Source: U.S. Geological Survey, Water Supply Paper 2220 (1983) and others.

© DON MASON/BRAND X PICTURES/JUPITERIMAGES

Groundwater provides 80% of the water used for rural livestock and domestic use, as well as providing 40% of public water supplies.

porosity, because they consist of tightly interlocking crystals. Their porosity can be increased, however, if they have been fractured or weathered by groundwater. This is particularly true for massive limestone and dolostone, whose fractures can be enlarged by acidic groundwater.

By contrast, detrital sedimentary rocks composed of well-sorted and well-rounded grains can have high porosity, because any two grains touch at only a single point, leaving relatively large open spaces between the grains (Figure 13.1a). Poorly sorted sedimentary rocks, on the other hand, typically have low porosity, because smaller grains fill in the spaces between the larger grains, further reducing porosity (Figure 13.1b). In addition, the amount of cement between grains can decrease porosity.

Porosity determines the amount of groundwater that Earth materials can hold, but it does not guarantee that the water can be easily extracted. So, in addition to being porous, Earth materials must have the capacity to transmit fluids, a property known as **permeability**. Thus, both porosity and permeability play important roles in groundwater movement and recovery.

Permeability is dependent not only on porosity, but also on the size of the pores or fractures and their interconnections. For example, deposits of silt or clay are typically more porous than sand or gravel, but they have low permeability, because the pores between the particles are very small, and molecular attraction between the particles and water is great, thereby preventing movement of the water. In contrast, the pore spaces between grains in sandstone and conglomerate are much larger, and molecular attraction on the water is therefore low. Chemical and biochemical sedimentary rocks, such as limestone and dolostone, and many igneous and metamorphic rocks that are highly fractured can also be very permeable provided that the fractures are interconnected.

A permeable layer transporting groundwater is an *aquifer*, from the Latin *aqua*, "water." The most effective aquifers are deposits of well-sorted and well-rounded sand and gravel. Limestones in which fractures and bedding planes have been enlarged by solution are also good aquifers. Shales and many igneous and metamorphic rocks make poor aquifers because they are typically impermeable, unless fractured. Rocks such as these and any other materials that prevent the movement of groundwater are *aquicludes*.

LO3 The Water Table

Some of the precipitation on land evaporates, and some enters streams and returns to the oceans by surface runoff; the remainder seeps into the ground. As this water moves down from the surface, a small amount adheres to the material it moves through and halts its downward progress. With the exception of this *suspended water*, however, the rest seeps further downward and collects until it fills all of the available pore spaces. Thus, two zones are defined by whether

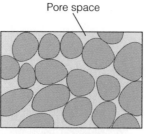

a. A well-sorted sedimentary rock has high porosity, whereas

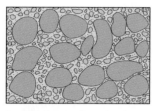

b. a poorly sorted one has lower porosity.

c. In soluble rocks such as limestone, porosity can be increased by solution, whereas

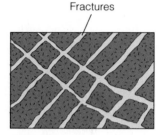

d. crystalline metamorphic and igneous rocks can be rendered porous by fracturing.

Figure 13.1 Porosity
A rock's porosity depends on the size, shape, and arrangement of the material composing the rock.

their pore spaces contain mostly air, the **zone of aeration**, or mostly water, the underlying **zone of saturation**. The surface that separates these two zones is the **water table** (Figure 13.2).

The base of the zone of saturation varies from place to place, but usually extends to a depth where an impermeable layer is encountered or to a depth where confining pressure closes all open space. Extending irregularly upward a few centimeters to several meters from the zone of saturation is the *capillary fringe*. Water moves upward in this region because of surface tension, much as water moves upward through a paper towel.

In general, the configuration of the water table is a subdued replica of the overlying land surface; that is, it rises beneath hills and has its lowest elevations beneath valleys. Several factors contribute to the surface

permeability A material's capacity to transmit fluids.

zone of aeration The zone above the water table that contains both air and water within the pore spaces of soil, sediment, or rock.

zone of saturation The area below the water table in which all pore spaces are filled with water.

water table The surface that separates the zone of aeration from the underlying zone of saturation.

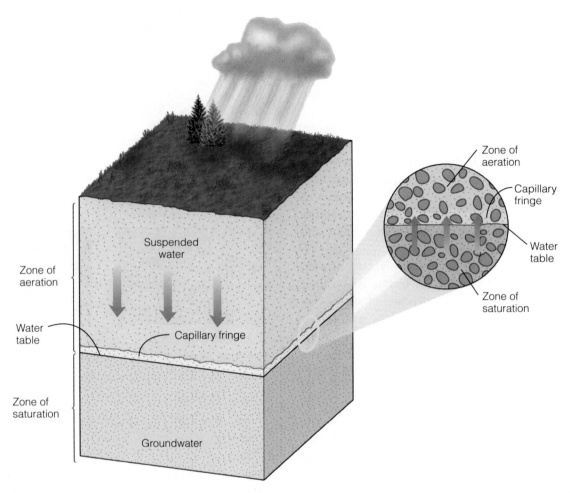

Figure 13.2 Water Table

The zone of aeration contains both air and water within its pore spaces, whereas all pore spaces in the zone of saturation are filled with groundwater. The water table is the surface separating the zones of aeration and saturation. Within the capillary fringe, water rises by surface tension from the zone of saturation into the zone of aeration.

configuration of a region's water table, including regional differences in the amount of rainfall, permeability, and rate of groundwater movement.

LO4 Groundwater Movement

Gravity provides the energy for the downward movement of groundwater. Water entering the ground moves through the zone of aeration to the zone of saturation (Figure 13.3). When water reaches the water table, it continues to move through the zone of saturation from areas where the water table is high toward areas where it is lower, such as streams, lakes, or swamps. Only some of the water follows the direct route along the slope of the water table. Most of it takes longer

curving paths down and then enters a stream, lake, or swamp from below, because it moves from areas of high pressure toward areas of lower pressure within the saturated zone.

Groundwater velocity varies greatly and depends on many factors. Velocities range from 250 m per day in some extremely permeable material to less than a few centimeters per year in nearly impermeable material. In most ordinary aquifers, the average velocity of groundwater is a few centimeters per day.

LO5 Springs, Water Wells, and Artesian Systems

You can think of the water in the zone of saturation much like a reservoir whose surface rises or falls

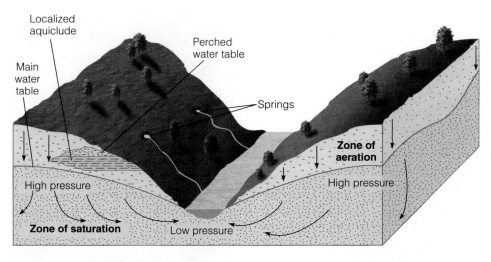

Figure 13.3 **Groundwater Movement**
Groundwater moves down through the zone of aeration to the zone of saturation. Then some of it moves along the slope of the water table, and the rest moves through the zone of saturation from areas of high pressure toward areas of low pressure. Some water might collect over a local aquiclude, such as a shale layer, thus forming a perched water table.

depending on additions as opposed to natural and artificial withdrawals. *Recharge*—that is, additions to the zone of saturation—comes from rainfall or melting snow, or water might be added artificially at wastewater-treatment plants or recharge ponds constructed for just this purpose. But if groundwater is discharged naturally or withdrawn at wells without sufficient recharge, the water table drops—just as a savings account diminishes if withdrawals exceed deposits. Withdrawals from the groundwater system take place where groundwater flows laterally into streams, lakes, or swamps, where it discharges at the surface as *springs*, and where it is withdrawn from the system at water wells.

SPRINGS

Places where groundwater flows or seeps out of the ground as **springs** have always fascinated people. The

water flows out of the ground for no seemingly apparent reason and from no readily identifiable source. So it is not surprising that springs have long been regarded with superstition and revered for their supposed medicinal value and healing powers. Nevertheless, there is nothing mystical or mysterious about springs.

Although springs can occur under a wide variety of geologic conditions, they all form in basically the same way (Figure 13.4a). When percolating water reaches the water table or an impermeable layer, it flows laterally, and if this flow intersects the surface, the water discharges as a spring (Figure 13.4b).

Springs can also develop wherever a *perched water table* (Figure 13.3)—a local aquiclude present within a larger aquifer, such as a lens of shale within sandstone—intersects the surface. As water migrates through the zone of aeration, it is stopped by the local aquiclude, and a localized zone of saturation "perched" above the main water table forms. Water moving laterally along the perched water table may intersect the surface to produce a spring.

WATER WELLS

Water wells are openings made by digging or drilling down into the zone of saturation. Once the zone of saturation has been penetrated,

spring A place where groundwater flows or seeps out of the ground.

water well A well made by digging or drilling into the zone of saturation.

In many parts of the United States and Canada, one can see windmills from times past that used wind power to pump water. Most of these are no longer in use, having been replaced by more efficient electric pumps.

Figure 13.4 Springs

Springs form wherever laterally moving groundwater intersects Earth's surface.

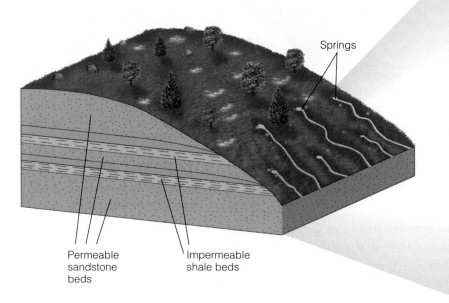

Springs

Permeable
sandstone
beds

Impermeable
shale beds

JOHN KARACHEWSKI

a. Most commonly, springs form when percolating water reaches an impermeable layer and migrates laterally until it seeps out at the surface.

b. Thunder River Spring in the Grand Canyon, Arizona, issues from rocks along a wall of the Grand Canyon. Water percolating downward through permeable rocks is forced to move laterally when it encounters an impermeable zone, and thus gushes out along this cliff. Notice the vegetation parallel to and below the springs, indicating that enough water flows from springs along the cliff wall to support the vegetation.

water percolates into the well, filling it to the level of the water table. A few wells are free flowing; however, for most, the water must be brought to the surface by pumping.

When groundwater is pumped from a well, the water table in the area around the well is lowered, forming a **cone of depression** (Figure 13.5). A cone of depression forms because the rate of water withdrawal from the well exceeds the rate of water inflow to the well, thus lowering the water table around the well. A cone of depression's gradient—that is, whether it is steep or gentle—depends to a great extent on the permeability of the aquifer being pumped. A highly permeable aquifer produces a gentle gradient in the cone of depression, whereas a low-permeability aquifer results in a steep cone of depression, because water cannot easily flow to the well to replace the water being withdrawn.

The formation of a cone of depression does not normally pose a problem for the average domestic well, provided that the well is drilled deep enough into the zone of saturation. However, the tremendous amounts of water used by industry and for irrigation may create a large cone of depression that

cone of depression A cone-shaped depression around a well where water is pumped from an aquifer faster than it can be replaced.

lowers the water table sufficiently to cause shallow wells in the immediate area to go dry (Figure 13.5). Lowering of the regional water table, because more groundwater is being withdrawn than is being replenished, is becoming a serious problem in many areas, particularly in the southwestern United States, where rapid growth has placed tremendous demands on the groundwater system. As mentioned earlier, some of the largest cities in the United States depend entirely on groundwater for their municipal needs. Furthermore, some of the largest agricultural states are withdrawing groundwater from regional aquifers that are not being sufficiently replenished.

ARTESIAN SYSTEMS

The word *artesian* comes from the French town and province of Artois (called Artesium during Roman times) near Calais, where the first European artesian well was drilled in 1126 and is still flowing today. The

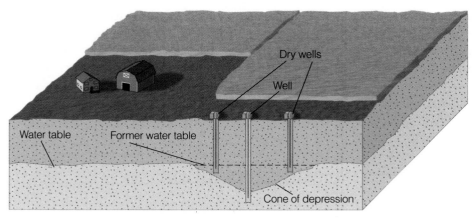

Figure 13.5 Cone of Depression
A cone of depression forms whenever water is withdrawn from a well. If water is withdrawn faster than it can be replenished, the cone of depression will grow in depth and circumference, lowering the water table in the area and causing nearby shallow wells to go dry.

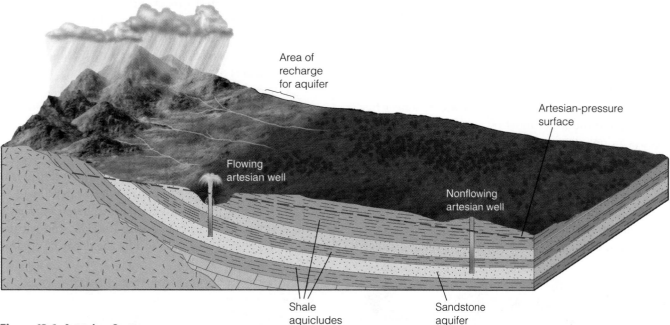

Figure 13.6 Artesian System
An artesian system (1) must have an aquifer confined above and below by aquicludes; (2) the aquifer must be exposed at the surface; and (3) the rock units are typically tilted so as to build up hydrostatic pressure within the aquifer. The elevation of the water table in the recharge area, which is indicated by a sloping dashed line (the artesian-pressure surface), defines the highest level to which well water can rise. A wellhead below the elevation of the artesian-pressure surface will be free-flowing, because the water will rise toward the artesian-pressure surface, which is at a higher elevation than the wellhead. Conversely, a well will be nonflowing if the elevation of the wellhead is at or above that of the artesian-pressure surface.

term **artesian system** can be applied to any system in which groundwater is confined and builds up high hydrostatic (fluid) pressure (Figure 13.6). Water in such a system is able to rise above the level of the aquifer if a well is drilled through the confining layer, thereby reducing the pressure and forcing the water upward. An artesian system can develop (1) when an aquifer is confined above and below by aquicludes; (2) the rock sequence is usually tilted to build up hydrostatic pressure; and (3) the aquifer is exposed at the surface, thus enabling it to be recharged.

The elevation of the water table in the recharge area and the distance of the well from the recharge area determine the height to which artesian water rises in a well. The surface defined by the water table in the recharge area, called the *artesian-pressure surface*, is indicated by the sloping dashed line in Figure 13.6. Friction slightly reduces the pressure of the aquifer water and consequently the level to which artesian water rises, which is why the pressure surface slopes.

An artesian well will flow freely at the ground surface only if the

> **artesian system** A confined groundwater system with high hydrostatic pressure that causes water to rise above the level of the aquifer.

wellhead is at an elevation below the artesian-pressure surface. In this situation, the water flows out of the well, because it rises toward the artesian-pressure surface, which is at a higher elevation than the wellhead. In a nonflowing artesian well, the wellhead is above the artesian-pressure surface, and the water will rise in the well only as high as the artesian-pressure surface.

One of the best-known artesian systems in the United States underlies South Dakota and extends southward to central Texas. The majority of the artesian water from this system is used for irrigation. The aquifer of this artesian system, the Dakota Sandstone, is recharged where it is exposed along the margins of the Black Hills of South Dakota. The hydrostatic pressure in this system was originally great enough to produce free-flowing wells and to operate waterwheels. However, because of the extensive use of this groundwater for irrigation over the years, the hydrostatic pressure in many of the wells is so low that they are no longer free-flowing, and the water must be pumped.

As a final comment on artesian systems, we should mention that it is not unusual for advertisers to tout the quality of artesian water as somehow being superior to other groundwater. Some artesian water might in fact be of excellent quality, but its quality is not dependent on the fact that water rises above the surface of an aquifer. Rather, its quality is a function of dissolved minerals and any introduced substances, so artesian water really is no different from any other groundwater. The myth of its superiority probably arises from the fact that people have always been fascinated by water that flows freely from the ground.

LO6 Groundwater Erosion and Deposition

When rainwater begins to seep into the ground, it immediately starts to react with the minerals it contacts, weathering them chemically. In an area underlain by soluble rock, groundwater is the principal agent of erosion and is responsible for the formation of many major features of the landscape.

Limestone, a common sedimentary rock composed primarily of the mineral calcite ($CaCO_3$), underlies large areas of Earth's surface (Figure 13.7). Although limestone is practically insoluble in pure water, it readily dissolves if a small amount of acid is present. Carbonic acid (H_2CO_3) is a weak acid that forms when carbon dioxide combines with water ($H_2O + CO_2 \rightarrow H_2CO_3$). Because the atmosphere contains a small amount of carbon dioxide (0.03%) and carbon dioxide is also produced in soil by the decay of organic

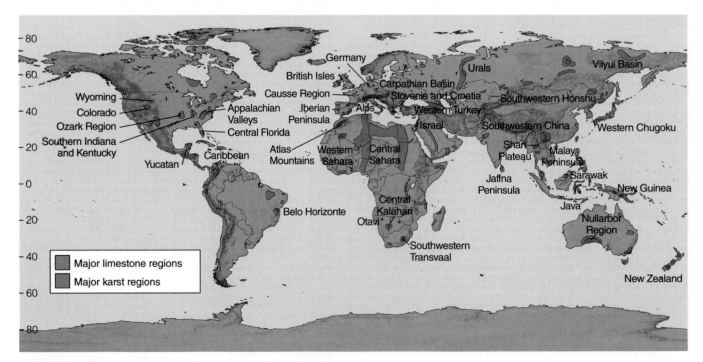

Figure 13.7 Major Limestone and Karst Areas of the World
Distribution of the major limestone and karst areas of the world. Karst topography develops largely by groundwater erosion in areas underlain by soluble rocks.

FRANK KUJAWA/UNIVERSITY OF CENTRAL FLORIDA

Figure 13.8 Sinkholes

This sinkhole formed on May 8 and 9, 1981, in Winter Park, Florida. It formed in previously dissolved limestone following a drop in the water table. The 100-m-wide, 35-m-deep sinkhole destroyed a house, numerous cars, and a municipal swimming pool.

matter, most groundwater is slightly acidic. When groundwater percolates through the various openings in limestone, the slightly acidic water readily reacts with the calcite to dissolve the rock by forming soluble calcium bicarbonate, which is carried away in solution (see Chapter 6).

SINKHOLES AND KARST TOPOGRAPHY

In regions underlain by soluble rock, the ground surface may be pitted with numerous depressions that vary in size and shape. These depressions, called **sinkholes** or merely *sinks*, mark areas with underlying soluble rock (Figure 13.8). Most sinkholes form in one of two ways. The first is when soluble rock below the soil is dissolved by seeping water, and openings in the rock are enlarged and filled in by the overlying soil. As the groundwater continues to dissolve the rock, the soil is eventually removed, leaving shallow depressions with gently sloping sides. When adjacent sinkholes merge, they form a network of larger, irregular, closed depressions called *solution valleys*.

Sinkholes also form when a cave's roof collapses, usually producing a steep-sided crater. Sinkholes formed in this way are a serious hazard, particularly in populated areas.

Karst topography, or simply *karst*, develops largely by groundwater erosion in many areas underlain by soluble rocks (Figure 13.9). The name *karst* is derived from the plateau region of the border area of Slovenia, Croatia, and northeastern Italy, where this type of topography is well developed. In the United States, regions of karst topography include large areas of southwestern Illinois, southern Indiana, Kentucky, Tennessee, southern Missouri, Alabama, and central and northern Florida (Figure 13.7).

Karst topography is characterized by numerous caves, springs, sinkholes, solution valleys, and disappearing streams (Figure 13.9). *Disappearing streams* are so named because they typically flow only a short distance at the surface and then disappear into a sinkhole. The water continues flowing underground through fractures or caves until it surfaces again at a spring or other stream.

Karst topography varies from the spectacular high-relief landscapes of China (Figure 13.10) to the subdued and pockmarked landforms of Kentucky. Common to all karst topography, though, is the presence of thick-bedded, readily soluble rock at the surface or just below the soil, and enough water for solution activity to occur. Karst topography is therefore typically restricted to humid and temperate climates.

CAVES AND CAVE DEPOSITS

Caves are perhaps the most spectacular examples of the combined effects of weathering and erosion by groundwater. As groundwater percolates through carbonate rocks, it dissolves and enlarges fractures and openings to form a complex interconnecting system of crevices, caves, caverns, and underground streams. A **cave** is usually defined as a

sinkhole A depression in the ground that forms by the solution of the underlying carbonate rocks or by the collapse of a cave roof.

karst topography Landscape consisting of numerous caves, sinkholes, and solution valleys formed by groundwater solution of rocks such as limestone and dolostone.

cave A natural subsurface opening generally connected to the surface and large enough for a person to enter.

naturally formed subsurface opening that is generally connected to the surface and is large enough for a person to enter. A *cavern* is a very large cave or a system of interconnected caves.

More than 17,000 caves are known in the United States. Some of the more famous ones are Mammoth Cave, Kentucky; Carlsbad Caverns, New Mexico; Lewis and Clark Caverns, Montana; Lehman Cave, Nevada; and Meramec Caverns, Missouri, which Jesse James and his outlaw band often used as a hideout. Canada also has many famous caves, including the 536-m-deep Arctomys Cave in Mount Robson Provincial Park, British Columbia, the deepest known cave in North America.

Caves and caverns form as a result of the dissolution of carbonate rocks by weakly acidic groundwater (Figure 13.11). Groundwater percolating through the zone of aeration slowly dissolves the carbonate rock and enlarges its fractures and bedding planes. On reaching the water table, the groundwater migrates toward the region's surface streams. As the groundwater moves through the zone of saturation, it continues to dissolve the rock and gradually forms a system of horizontal passageways through which the dissolved rock is carried to the streams. As the surface streams erode deeper valleys, the water table drops in response to the lower elevation of the streams. The water that flowed through the system of horizontal passageways now percolates to the lower water table, where a new system of passageways begins to form. The abandoned channelways form an interconnecting system of caves and caverns. Caves eventually become unstable and collapse, littering the floor with fallen debris.

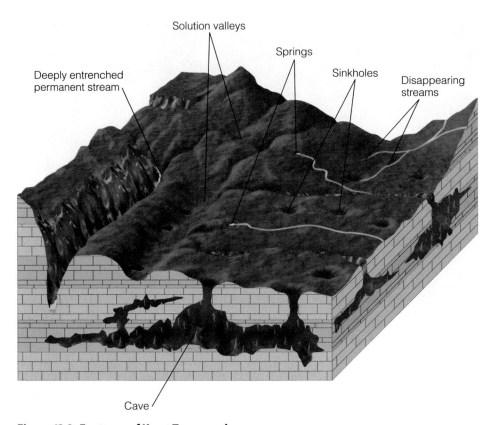

Figure 13.9 Features of Karst Topography
Erosion of soluble rock by groundwater produces karst topography. Features commonly found include solution valleys, springs, sinkholes, and disappearing streams.

Figure 13.10 Karst Landscape southeast of Kunming, China
The Stone Forest, 125 km southeast of Kunming, China, is a high-relief karst landscape formed by the dissolution of carbonate rocks.

Figure 13.11 Cave Formation

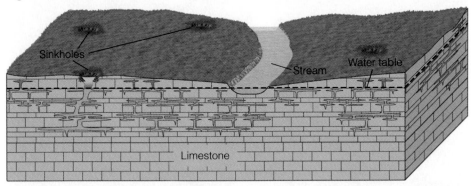

a. As groundwater percolates through the zone of aeration and flows through the zone of saturation, it dissolves the carbonate rocks and gradually forms a system of passageways.

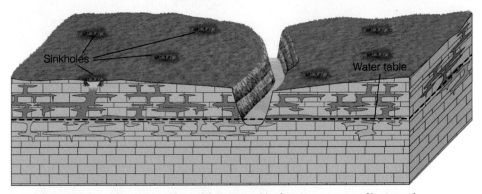

b. Groundwater moves along the surface of the water table, forming a system of horizontal passageways through which dissolved rock is carried to the surface streams, thus enlarging the passageways.

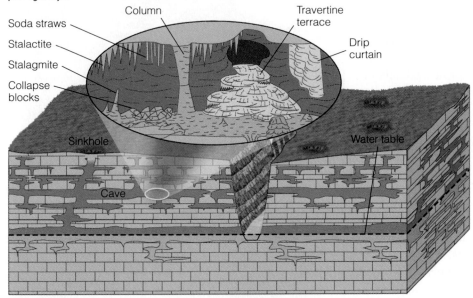

c. As the surface streams erode deeper valleys, the water table drops, and the abandoned channelways form an interconnecting system of caves and caverns.

When most people think of caves, they think of the seemingly endless variety of colorful and bizarre-shaped deposits found in them. Although a great many different types of cave deposits exist, most form in essentially the same manner and are collectively known as *dripstone*. As water seeps into a cave, some of the dissolved carbon dioxide in the water escapes, and a small amount of calcite is precipitated. In this manner, the various dripstone deposits are formed (Figure 13.11c).

Stalactites are icicle-shaped structures hanging from cave ceilings that form as a result of precipitation from dripping water (Figure 13.12). With each drop of water, a thin layer of calcite is deposited over the previous layer, forming a cone-shaped projection that grows down from the ceiling. The water that drips from a cave's ceiling also precipitates a small amount of calcite when it hits the floor. As additional calcite is deposited, an upward-growing projection called a *stalagmite* forms (Figure 13.12). If a stalactite and stalagmite meet, they form a *column*. Groundwater seeping from a crack in a cave's ceiling may form a vertical sheet of rock called a *drip curtain*, and water flowing across a cave's floor may produce *travertine terraces* (Figure 13.11c).

Figure 13.12 Cave Deposits

Stalactites are the icicle-shaped structures hanging from a cave's ceiling, whereas the upward-pointing structures on the floor are stalagmites. Columns result when stalactites and stalagmites meet. All three structures are visible in Shenandoah Caverns, Virginia.

LO7 Modifications of the Groundwater System and Its Effects

Groundwater is a valuable natural resource that is rapidly being exploited with seemingly little regard to the effects of overuse and misuse. Currently, approximately 20% of all water used in the United States is groundwater. This percentage is rapidly increasing, and unless this resource is used more wisely, sufficient amounts of clean groundwater will not be available in the future. Modifications of the groundwater system may have many consequences, including (1) lowering of the water table, causing wells to dry up; (2) saltwater incursion; (3) subsidence; and (4) contamination.

LOWERING THE WATER TABLE

Withdrawing groundwater at a significantly greater rate than it is replaced by either natural or artificial recharge can have serious effects. For example, the High Plains aquifer is one of the most important aquifers in the United States. It underlies more than 450,000 km², including most of Nebraska, large parts of Colorado and Kansas, portions of South Dakota, Wyoming, and New Mexico, as well as the panhandle regions of Oklahoma and Texas, and accounts for approximately 30% of the groundwater used for irrigation in the United States (Figure 13.13).

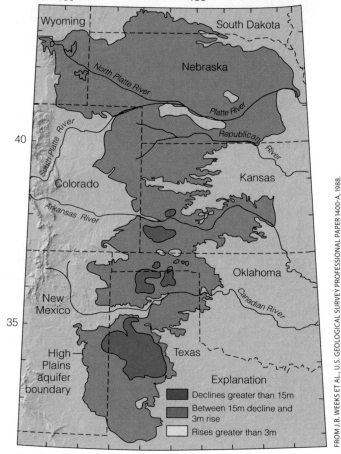

Explanation

- Declines greater than 15m
- Between 15m decline and 3m rise
- Rises greater than 3m

0   50   100 Kilometers

Figure 13.13 High Plains Aquifer

The geographic extent of the High Plains aquifer and changes in water level from predevelopment through 1993. Irrigation from the High Plains aquifer is largely responsible for the region's agricultural productivity.

Although the High Plains aquifer has contributed to the high agricultural productivity of the region, it cannot continue to provide the quantities of water that it has in the past. In some parts of the High Plains, from 2 to 100 times more water is being pumped annually than is being recharged, causing a substantial drop in the water table in many areas (Figure 13.13).

Most users of the aquifer realize that they cannot continue to withdraw the quantities of groundwater that they have in the past, and thus are turning to greater conservation, monitoring of the aquifer, and using new technologies to try to better balance withdrawal with recharge rates.

Water supply problems certainly exist in many areas, but on the positive side, water use in the United States actually declined during the five years following 1980 and has remained nearly constant since then, even though the population has increased. This downturn in demand resulted largely from improved techniques in irrigation, more efficient industrial water use, and a general public awareness of water problems coupled with conservation practices.

SALTWATER INCURSION

The excessive pumping of groundwater in coastal areas has resulted in *saltwater incursion*, which has become a major problem in many rapidly growing coastal communities. Along coastlines where permeable rocks or sediments are in contact with the ocean, the fresh groundwater, being less dense than seawater, forms a lens-shaped body above the underlying saltwater (Figure 13.14a). The weight of the freshwater exerts pressure on the underlying saltwater. As long as rates of recharge equal rates of withdrawal, the contact between the fresh groundwater and the seawater remains the same. If excessive pumping occurs, however, a deep cone of depression forms in the fresh groundwater (Figure 13.14b). Because some of the pressure from the overlying freshwater has been removed, saltwater forms a *cone of ascension* as it rises to fill the pore space that formerly contained freshwater. When this occurs, wells become contaminated with saltwater and remain contaminated until

Figure 13.14 Saltwater Incursion

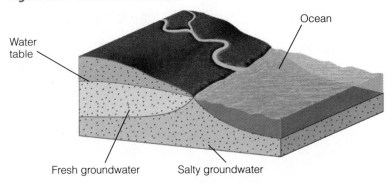

a. Because freshwater is not as dense as saltwater, it forms a lens-shaped body above the underlying saltwater.

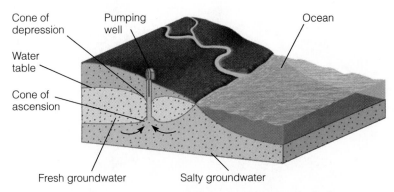

b. If excessive pumping occurs, a cone of depression develops in the fresh groundwater, and a cone of ascension forms in the underlying salty groundwater, which may result in saltwater contamination of the well.

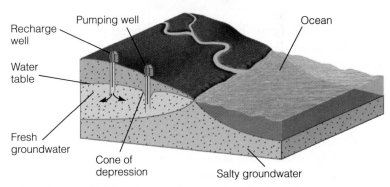

c. Pumping water back into the groundwater system through recharge wells can help lower the interface between the fresh groundwater and the salty groundwater and reduce saltwater incursion.

recharge by freshwater restores the former level of the fresh-groundwater water table.

To counteract the effects of saltwater incursion, recharge wells are often drilled to pump water back into

A spectacular example of continuing subsidence is taking place in Mexico City, which is built on a former lake bed. As groundwater is removed for the increasing needs of the city's 17.8 million people, the water table has been lowered up to 10 m. As a result, the fine-grained lake deposits are compacting, and Mexico City is slowly and unevenly subsiding. Its opera house has settled more than 3 m, and half of the first floor is now below ground level. Other parts of the city have subsided as much as 7.5 m, creating similar problems for other structures. The fact that 72% of the city's water comes from the aquifer beneath the metropolitan area ensures that problems of subsidence will continue.

the groundwater system (Figure 13.14c). Recharge ponds that allow large quantities of fresh surface water to infiltrate the groundwater supply may also be constructed.

SUBSIDENCE

As excessive amounts of groundwater are withdrawn from poorly consolidated sediments and sedimentary rocks, the water pressure between grains is reduced, and the weight of the overlying materials causes the grains to pack more closely together, resulting in *subsidence* of the ground.

The San Joaquin Valley of California is a major agricultural region that relies largely on groundwater for irrigation. Between 1925 and 1977, groundwater withdrawals in parts of the valley caused subsidence of nearly 9 m (Figure 13.15). Other areas in the United States that have experienced subsidence due to groundwater withdrawal are New Orleans, Louisiana, and Houston, Texas, both of which have subsided more than 2 m, and Las Vegas, Nevada, where 8.5 m of subsidence has taken place.

Looking elsewhere in the world, the tilt of the Leaning Tower of Pisa in Italy is partly a result of groundwater withdrawal. The tower started tilting soon after construction began in 1173 because of differential compaction of the foundation. During the 1960s, the city of Pisa withdrew ever-larger amounts of groundwater, causing the ground to subside further; as a result, the tilt of the tower increased until it was in danger of falling over. Strict control of groundwater withdrawal, stabilization of the foundation, and recent renovations have reduced the amount of tilting to about 1 mm per year, thus ensuring that the tower should stand for several more centuries.

The extraction of oil can also cause subsidence. Long Beach, California, has subsided 9 m as a result of many decades of oil production. More than $100 million in damages was done to the pumping, transportation, and harbor facilities in this area because of subsidence and encroachment of the sea (Figure 13.16). Once water was pumped back into the oil reservoir, thus stabilizing it, subsidence virtually stopped.

GROUNDWATER CONTAMINATION

A major problem facing our society is the safe disposal of the numerous pollutant byproducts of an industrialized economy. The most common sources of groundwater contamination are sewage, landfills, toxic waste disposal sites, and agriculture. Once pollutants get into the groundwater system, they spread wherever groundwater travels, which can make their containment difficult, and because groundwater moves so slowly, it takes a long time to cleanse a groundwater reservoir once it has become contaminated.

In many areas, septic tanks are the most common way of disposing of sewage. A septic tank slowly releases sewage into the ground, where it is decomposed by oxidation and microorganisms and filtered by the sediment as it percolates through the zone of aeration. In most situations, by the time the water from the sewage reaches the zone of saturation, it has been cleansed of any impurities and is safe to use (Figure 13.17a). If the water table is close to the surface or if the rocks are very permeable, however, water entering the zone of saturation may still be contaminated and unfit to use.

Landfills are also potential sources of groundwater contamination (Figure 13.17b). Not only does liquid waste seep into the ground, but rainwater also carries

Figure 13.15 Subsidence in the San Joaquin Valley, California

The dates on this power pole dramatically illustrate the amount of subsidence in the San Joaquin Valley, California. Because of groundwater withdrawals and subsequent sediment compaction, the ground subsided nearly 9 m between 1925 and 1977. For a time, surface water use reduced subsidence, but during the drought of 1987 to 1992, it started again as more groundwater was withdrawn.

Figure 13.16 Oil Field Subsidence, Long Beach, California

The withdrawal of petroleum from the Long Beach, California, oil field resulted in up to 9 m of ground subsidence in some areas because of sediment compaction. In this photograph, note that the ground has settled around the well stems (the white "posts"), leaving the wellheads above the ground. The levee on the left edge of the photograph was built to keep seawater in the adjacent marina from flooding the oil field. It was not until water was pumped back into the reservoir to replace the extracted petroleum that ground subsidence finally ceased.

Toxic Waste: Dump now, pay the piper later!

dissolved chemicals and other pollutants down into the groundwater reservoir. Unless the landfill is carefully designed and lined with an impermeable layer such as clay, many toxic compounds such as paints, solvents, cleansers, pesticides, and battery acid will find their way into the groundwater system.

Toxic waste sites where dangerous chemicals are either buried or pumped underground are an increasing source of groundwater contamination. The United States alone must dispose of several thousand metric tons of hazardous chemical waste per year. Unfortunately, much of this waste has been, and still is, being improperly dumped and is contaminating the surface water, soil, and groundwater.

Figure 13.17 Groundwater Contamination

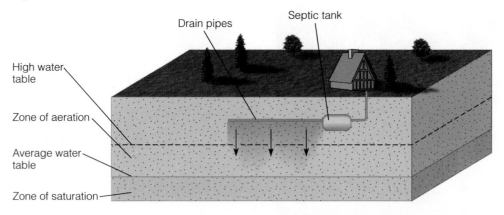

a. A septic system slowly releases sewage into the zone of aeration. Oxidation, bacterial degradation, and filtering usually remove impurities before they reach the water table. However, if the rocks are very permeable or the water table is too close to the septic system, contamination of the groundwater can result.

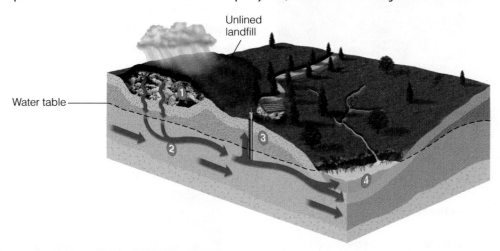

b. Unless there is an impermeable barrier between a landfill and the water table, pollutants can be carried into the zone of saturation and contaminate the groundwater supply: (1) Infiltrating water leaches contaminates from the landfill; (2) the polluted water enters the water table and moves away from the landfill; (3) wells may tap the polluted water and thus contaminate drinking water supplies; and (4) the polluted water may emerge into streams and other water bodies downslope from the landfill.

LO8 Hydrothermal Activity

hydrothermal A term referring to hot water, as in hot springs and geysers.

hot spring A spring in which the water temperature is warmer than the temperature of the human body (37°C).

Hydrothermal is a term referring to hot water. Some geologists restrict the meaning to include only water heated by magma, but here we use it to refer to any hot subsurface water and the surface activity that results from its discharge. One manifestation of hydrothermal activity in areas of active or recently active volcanism is the discharge of gases, such as steam, at vents known as *fumeroles* (Figure 13.18). Of more immediate concern here is the groundwater that rises to the surface as *hot springs* or *geysers*. It may be heated by its proximity to magma or by Earth's geothermal gradient because it circulates deeply.

HOT SPRINGS

A **hot spring** (also called a *thermal spring* or *warm spring*) is any spring in which the water temperature is higher than 37°C, the temperature of the human

Figure 13.18 Fumeroles
Gases emitted from vents (fumeroles) at Yellowstone National Park, Wyoming.

Figure 13.19 Hot Springs

a. One of the more colorful hot springs in Yellowstone National Park, the Morning Glory hot spring is fringed with multicolored mats of heat-loving, cyanobacteria and algal mats. Each color represents a certain temperature range that allows for specific bacterial species to thrive in this extreme environment.

b. The water in this hot spring at Bumpass Hell in Lassen Volcanic National Park, California, is boiling.

body (Figure 13.19a). Some hot springs are much hotter, with temperatures up to the boiling point in many instances (Figure 13.19b). Of the approximately 1,100 known hot springs in the United States, more than 1,000 are in the far West, with the others in the Black Hills of South Dakota, Georgia, the Ouachita region of Arkansas, and the Appalachian region.

Hot springs are also common in other parts of the world. One of the most famous is in Bath, England, where shortly after the Roman conquest of Britain in A.D. 43, numerous bathhouses and a temple were built around the hot springs (Figure 13.20).

The heat for most hot springs comes from magma or cooling igneous rocks. The geologically recent igneous activity in the western United States accounts for the large number of hot springs in that region. The water in some hot springs, however, circulates deep into Earth, where it is warmed by the normal

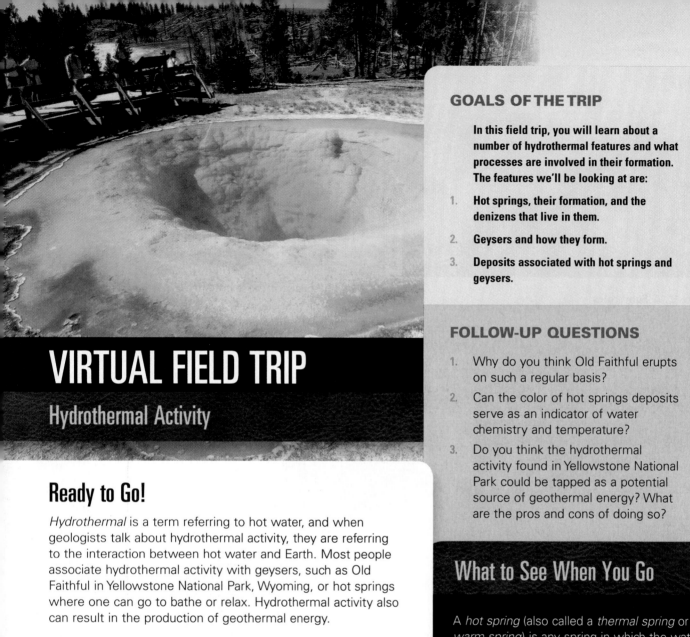

VIRTUAL FIELD TRIP

Hydrothermal Activity

Ready to Go!

Hydrothermal is a term referring to hot water, and when geologists talk about hydrothermal activity, they are referring to the interaction between hot water and Earth. Most people associate hydrothermal activity with geysers, such as Old Faithful in Yellowstone National Park, Wyoming, or hot springs where one can go to bathe or relax. Hydrothermal activity also can result in the production of geothermal energy.

There are three areas in the world known for their hydrothermal activity. These are Iceland, New Zealand, and Yellowstone National Park, Wyoming. In this virtual field trip, we'll be visiting what is probably the best known area to view hydrothermal activity and its results. So, get ready to visit the crown-jewel of the U.S. National Park System, Yellowstone National Park.

What to See When You Go

A *hot spring* (also called a *thermal spring* or *warm spring*) is any spring in which the water temperature is higher than the temperature of the human body. The heat for most hot springs comes from magma or cooling igneous rock. The large number of hot springs in the western United States is due to geologically recent igneous activity. The underground system of fractures and opening that are associated with hot springs are not as constrictive, or usually as deep, as in a geyser, hence the water can bubble up and spill out onto the surface. Nevertheless, there is still steam and volcanic gases associated with hot springs as seen in this photo.

This photograph shows the vent of the Pump Geyser. Hot, acidic water containing silica and dissolved gases flows down the slope from the vent, providing an environment for *extremophiles* (organisms that live in extreme environments), in this case *thermophiles* species that thrive in extremely hot water, and are typically bacteria or cyanobacteria).

The beautiful red colors are rust-colored iron-oxide deposits resulting from the metabolism of iron by some of these bacteria. The yellow-colored deposits are sulfur, which is formed by sulfur-loving species of bacteria that reduce the hydrogen sulfide gas being emitted at the vent. The brownish bacterial mats contain bacteria that live in the cooler waters (below 140°F).

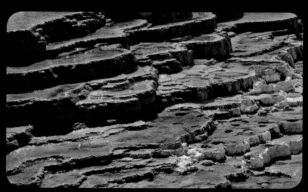

Hot springs that intermittently eject hot water and steam with tremendous force are known as *geysers*. Recall from your reading of this chapter, that geysers are the surface expression of an extensive underground system of deep interconnected fractures within hot rocks. Groundwater in these fractures is heated above the boiling point of water. Because of the pressure at depth, this water does not boil, but if there is a slight drop in pressure, such as from escaping gas, the hot water will instantly change to steam, which will then push the water upwards to the ground and into the air, producing a geyser.

Old Faithful Geyser, shown here in eruption, is probably the most famous geyser in the world. Eruptions occur approximately every 90 minutes.

Both hot springs and geysers typically contain large quantities of dissolved minerals. When the highly mineralized waters of hot springs and geysers reach the surface, some of the dissolved mineral matter is precipitated, forming various types of deposits. The amount and type of precipitated minerals depends on the solubility and composition of the groundwater flowing through the underground fractures.

The terraces you see here are composed of travertine ($CaCO_3$) as opposed to sinter (SiO_2). These deposits are the result of groundwater flowing through limestones and dissolving the calcium carbonate. As the water reaches the surface and begins to flow over it, calcium carbonate is precipitated as small crystals and thin sheets, which, over time, form the massive terraces seen here.

When hot water containing dissolved silica is erupted from a geyser, it cools and deposits silica (commonly called *sinter*) around the vent. Sinter comes in a variety of shapes and sizes. One of the largest deposits in the world is found at The Castle Geyser, so named for its resemblance to a castle ruin. The Castle Geyser is thought to be thousands of years old and the sinter it is currently depositing is being laid down over even older and thicker deposits of sinter.

Figure 13.20 Bath, England
One of the many bathhouses in Bath, England, that were built around hot springs shortly after the Roman conquest in A.D. 43.

Figure 13.21 Old Faithful Geyser
Old Faithful Geyser in Yellowstone National Park, Wyoming, is one of the world's most famous geysers, erupting faithfully every 30 to 90 minutes and spewing water 32 to 56 m high.

geyser A hot spring that periodically ejects hot water and steam.

increase in temperature, the geothermal gradient. For example, the spring water of Warm Springs, Georgia, is heated in this manner. This hot spring was a health and bathing resort long before the Civil War (1861–1865); later, with the establishment of the Georgia Warm Springs Foundation, it was used to help treat polio victims.

GEYSERS

Hot springs that intermittently eject hot water and steam with tremendous force are known as **geysers**.

The word comes from the Icelandic *geysir*, "to gush" or "to rush forth." One of the most famous geysers in the world is Old Faithful in Yellowstone National Park, Wyoming (Figure 13.21). With a thunderous roar, it erupts a column of hot water and steam every 30 to 90 minutes. Other well-known geyser areas are found in Iceland and New Zealand.

Geysers are the surface expression of an extensive underground system of interconnected fractures within hot igneous rocks (Figure 13.22). Groundwater percolating down into the network of fractures is heated as it comes into contact with the hot rocks. Because the water near the bottom of the fracture system is under higher pressure than the water near the top, it must be heated to a higher temperature before

Figure 13.22 Anatomy of a Geyser

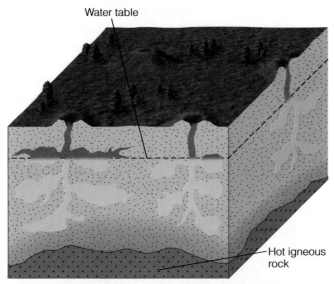

Water table

Hot igneous rock

a. The eruption of a geyser starts when groundwater percolates down into a network of interconnected openings and is heated by the hot igneous rocks. The water near the bottom of the fracture system is under higher pressure than the water near the top and consequently must be heated to a higher temperature before it will boil.

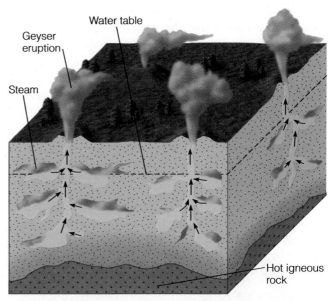

Water table

Geyser eruption

Steam

Hot igneous rock

b. Any rise in the temperature of the water above its boiling point or a drop in pressure will cause the water to change to steam, which quickly pushes the water above it up and out of the ground, producing a geyser eruption.

it will boil. Thus, when the deeper water is heated to near the boiling point, a slight rise in temperature or a drop in pressure, such as from escaping gas, will instantly change it to steam. The expanding steam quickly pushes the water above it out of the ground and into the air, producing a geyser eruption. After the eruption, relatively cool groundwater starts to seep back into the fracture system, where it heats to near its boiling temperature and the eruption cycle begins again. Such a process explains how geysers erupt with some regularity.

Hot spring and geyser water typically contains large quantities of dissolved minerals, because most minerals dissolve more rapidly in warm water than in cold water. Because of this high mineral content, some believe that the waters of many hot springs have medicinal properties. Numerous spas and bathhouses have been built at hot springs throughout the world to take advantage of these supposed healing properties.

When the highly mineralized water of hot springs or geysers cools at the surface, some of the material in solution is precipitated, forming various types of deposits. The amount and type of precipitated minerals depend on the solubility and composition of the

material that the groundwater flows through. If the groundwater contains dissolved calcium carbonate ($CaCO_3$), then *travertine* or *calcareous tufa* (both of which are varieties of limestone) are precipitated. Spectacular examples of hot spring travertine deposits are found at Pamukkale in Turkey and at Mammoth Hot Springs in Yellowstone National Park (Figure 13.23). Groundwater containing dissolved silica will, upon reaching the surface, precipitate a soft, white, hydrated mineral called *siliceous sinter* or *geyserite*, which can accumulate around a geyser's opening.

geothermal energy
Energy that comes from steam and hot water trapped within Earth's crust.

Geothermal Energy

Geothermal energy is any energy produced from Earth's internal heat. In fact, the term *geothermal* comes from *geo*, "Earth," and *thermal*, "heat." Several forms of internal heat are known, such as hot dry rocks and magma, but so far, only hot water and steam are used.

As oil reserves decline, geothermal energy is becoming an attractive alternative. Approximately 1% to 2% of the world's current energy needs could be met by geothermal energy. In those areas where it is plentiful, geothermal energy can supply most, if not all, of the energy needs, sometimes at a fraction of the cost of other types of energy. Some of the countries currently using geothermal energy in one form or another are Iceland, the United States, Mexico, Italy, New Zealand, Japan, the Philippines, and Indonesia.

In the United States, the first commercial geothermal electricity-generating plant was built in 1960 at The Geysers, about 120 km north of San Francisco, California. Here, wells were drilled into the numerous near-vertical fractures underlying the region. As pressure on the rising groundwater decreases, the water changes to steam, which is piped directly to electricity-generating turbines and generators.

Figure 13.23 Hot-Spring Deposits in Yellowstone National Park, Wyoming
Minerva Terrace, formed when calcium-carbonate-rich hot-spring water cooled, precipitating travertine.

GET FLASHCARDS

THEY DID

Our extensive research shows that almost all students use flashcards to reinforce their understanding of core concepts. To help you prepare flashcards that fit your specific study needs, **GEOL** provides three downloadable varieties:

(1) term only
(2) definition only
(3) term and definition

Pick the style that works best for you!

Visit **4ltrpress.cengage.com** for flashcards and many other additional resources!

CHAPTER 14
GLACIERS AND GLACIATION

Face of Matanuska Glacier, Alaska. Glaciers, which are moving bodies of ice on land, effectively modified Earth's surface by erosion and deposition of sediment.

Introduction

"What causes an Ice Age?"

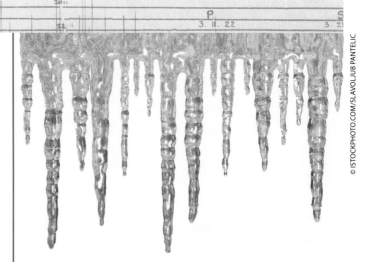

e know from the geologic record that an Ice Age took place between 1.8 million and 10,000 years ago, and since that time, Earth has experienced several entirely natural climatic changes. During the Holocene Maximum about 6,000 years ago, the average temperatures were slightly higher than they are now, and some of today's arid regions were much more humid. The Sahara Desert of North Africa had sufficient precipitation to support lush vegetation, swamps, and lakes. Indeed, Egypt's only arable land today is along the Nile River, but until only a few thousands of years ago, much of North Africa was covered by grasslands.

After the Holocene Maximum was a time of cooler temperatures, but from about A.D. 1000 to 1300, Europe experienced the Medieval Warm Period, during which wine grapes grew 480 km farther north than they do now. Then a cooling trend beginning in about A.D. 1300 led to the **Little Ice Age**, which lasted from 1500 to the middle or late 1800s. The Little Ice Age was a time of expansion of glaciers; cooler, wetter summers; colder winters; and the persistence of sea ice for long periods in Greenland, Iceland, and the

Canadian Arctic islands. And because of the cooler, wetter summers, growing seasons were shorter, which accounts for several widespread famines.

During the coldest part of the Little Ice Age (1680–1730), the growing season in England was five weeks shorter than during the 1900s, and in 1695, Iceland was surrounded by sea ice for most of the year. The canals in Holland froze over in some winters. In 1608, the first Frost Fair was held in London, England, on the Thames River, which began to freeze over nearly every winter. In the late 1700s, New York Harbor froze over, and 1816 is known as the "year without a summer," when unusually cold temperatures persisted into June and July in New England and northern Europe. (The eruption of Tambora in 1815 contributed to the cold spring and summer of 1816.)

Many of you have probably heard of the Ice Age and have some idea of what a glacier is, but it is doubtful that you know much about the dynamics of glaciers, how they form, and what may cause ice ages. In any case, *glaciers* are moving bodies of ice on land that are particularly effective at erosion, sediment transport, and deposition. They deeply scour the surfaces they move over, producing many easily recognizable landforms, and they deposit huge amounts of sediment, much of it important sources of sand and gravel. Glaciers today cover about 10% of Earth's land surface, but during the Ice Age they were much more widespread.

LEARNING OUTCOMES

After reading this unit, you should be able to do the following:

LO1 Identify the different kinds of glaciers

LO2 Recognize that glaciers are moving bodies of ice on land

LO3 Understand the glacial budget of accumulation and wastage

LO4 Identify features resulting from erosion and transport by glaciers

LO5 Identify the types of glacial deposits

LO6 Identify the kinds of landforms composed of stratified drift

LO7 Explain what causes ice ages

Little Ice Age An interval from about 1500 to the mid- to late-1800s during which glaciers expanded to their greatest historic extent.

Figure 14.1 Valley Glaciers
A valley glacier in Alaska. Notice the tributaries that unite to form a larger glacier.

LO1 The Kinds of Glaciers

Geologists define a **glacier** as a moving body of ice on land that flows downslope or outward from an area of accumulation. Our definition of a glacier excludes frozen sea-water, as in the North Polar region, and sea ice that forms yearly adjacent to Greenland and Iceland. Drifting icebergs are not glaciers either, although they may have come from glaciers that flowed into lakes or the sea. The critical points in the defini-

glacier A mass of ice on land that moves by plastic flow and basal slip.

valley glacier A glacier confined to a mountain valley or an interconnected system of mountain valleys.

continental glacier A glacier that covers a vast area (at least 50,000 km²) and is not confined by topography; also called an *ice sheet*.

tion are *moving* and *on land*. Accordingly, permanent snowfields in high mountains, though on land, are not glaciers because they do not move. All glaciers share several characteristics, but they differ enough in size and location for geologists to define two specific types, valley glaciers and continental glaciers, and several subvarieties.

VALLEY GLACIERS

Valley glaciers are confined to mountain valleys where they flow from higher to lower elevations (Figure 14.1), whereas continental glaciers cover vast areas, they are not confined by the underlying topography, and they flow outward in all directions from areas of snow and ice accumulation. We use the term **valley glacier**, but some geologists prefer the synonyms *alpine glacier* and *mountain glacier*. Valley glaciers commonly have tributaries, just as streams do, thereby forming a network of glaciers in an interconnected system of mountain valleys.

Valley glaciers are common in the mountains of all continents except Australia. In fact, Australia is the only continent that has no glaciers. A valley glacier's shape is controlled by the shape of the valley it occupies, so it tends to be a long, narrow tongue of moving ice. Valley glaciers that flow into the ocean are called *tidewater glaciers;* they differ from other valley glaciers only in that their terminus is in the sea rather than on land

Valley glaciers are small compared with the much more extensive continental glaciers, but even so, they may be as large as several kilometers across, 200 km long, and hundreds of meters thick. Erosion and deposition by valley glaciers were responsible for much of the spectacular scenery in several U.S. and Canadian national parks.

CONTINENTAL GLACIERS

Continental glaciers, also known as *ice sheets*, are vast, covering at least 50,000 km², and they are unconfined by topography; that is, their shape and movement are not controlled by the underlying landscape. Valley glaciers are long, narrow tongues of ice that conform to the shape of the valley they occupy, and the existing slope determines their direction of flow. In contrast, continental glaciers flow outward in all directions from a central area or areas of accumulation in response to variations in ice thickness.

Figure 14.2 Continental Glaciers and Ice Caps

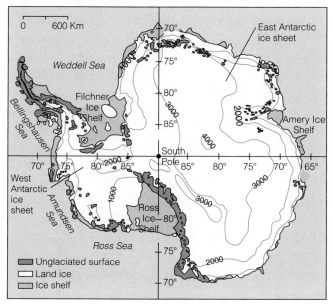

a. The West and East Antarctic ice sheets merge to form a nearly continuous ice cover that averages 2,160 m thick. The blue lines are lines of equal thickness.

b. View of part of the Antarctic ice sheet. Notice the nunatak, which is a peak extending above the glacial ice.

In Earth's two areas of continental glaciation, Greenland and Antarctica, the ice is more than 3,000 m thick and covers all but the highest mountains (Figure 14.2). The continental glacier in Greenland covers about 1,800,000 km², and in Antarctica the East and West Antarctic Glaciers merge to form a continuous ice sheet covering more than 12,650,000 km². The glaciers in Antarctica flow into the sea, where the buoyant effect of water causes the ice to float in vast *ice shelves*; the Ross Ice Shelf alone covers more than 547,000 km² (Figure 14.2).

An **ice cap**, a dome-shaped mass of glacial ice, is similar to but smaller than a continental glacier,

covering less than 50,000 km². Some ice caps form when valley glaciers grow and overtop the divides and passes between adjacent valleys and coalesce to form a continuous ice cover. They also form on fairly flat terrain in Iceland and some of the islands in the Canadian Arctic.

LO2 Glaciers: Moving Bodies of Ice on Land

We use the term **glaciation** to indicate all glacial activity, including the origin, expansion, and retreat of glaciers, as well as their impact on Earth's surface. Presently, glaciers cover nearly 15 million km², or about 10% of Earth's land surface.

At first glance, glaciers appear static. Even briefly visiting a glacier may not dispel this impression because, although glaciers move, they usually do so slowly. Nevertheless, they do move, and just like other geologic agents such as running water, glaciers are dynamic systems that continuously adjust to changes. For example, a glacier may flow slower or more rapidly depending on decreased or increased amounts of snow or the absence or presence of water at its base.

GLACIERS: PART OF THE HYDROLOGIC CYCLE

Glaciers make up one reservoir in the hydrologic cycle, where water is stored for long periods, but even this water eventually returns to its original source, the oceans. Glaciers at high latitudes, as in Alaska, northern Canada, and Scandinavia, flow directly into the oceans where they melt, or icebergs break off (a process known as *calving*) and drift out to sea, where they eventually melt. At

ice cap A dome-shaped mass of glacial ice that covers less than 50,000 km².

glaciation Refers to all aspects of glaciers, including their origin, expansion, and retreat, and their impact on Earth's surface.

low latitudes or areas remote from the oceans, glaciers flow to lower elevations, where they melt and the liquid water enters the groundwater system (another reservoir in the hydrologic cycle), or it returns to the seas by surface runoff.

In addition to melting, glaciers lose water by *sublimation*, when ice changes to water vapor without an intermediate liquid phase. The water vapor so derived enters the atmosphere, where it may condense and fall as rain or snow, but in the long run, this water also returns to the oceans.

How Do Glaciers Originate and Move?

Ice is a crystalline solid with characteristic physical properties and a specific chemical composition, and thus is a mineral. Accordingly, glacial ice is a type of metamorphic rock, but one that is easily deformed. Glaciers form in any area where more snow falls than melts during the warmer seasons and a net accumulation takes place. Freshly fallen snow has about 80% air-filled pore space and 20% solids, but it compacts as it accumulates, partially thaws, and refreezes, converting to a granular type of snow known as **firn**. As more snow accumulates, the firn is buried and further compacted and recrystallized until it is transformed into **glacial ice**, consisting of about 90% solids and 10% air (Figure 14.3).

Now you know how glacial ice forms, but we still have not addressed how glaciers move. At this time, it is useful to recall some terms from Chapter 10. Remember that *stress* is force per unit area, and *strain* is a change in the shape or volume of solids. When accumulating snow and ice reach a critical thickness of about 40 m, the stress on the ice at depth is great enough to induce **plastic flow**, a type of permanent deformation involving no fracturing. Glaciers move mostly by plastic flow, but they

firn Granular snow formed by partial melting and refreezing of snow; transitional material between snow and glacial ice.

glacial ice Water in the solid state within a glacier; forms as snow partially melts and refreezes and compacts so that it is transformed first to firn and then to glacial ice.

plastic flow The flow that takes place in response to pressure and causes deformation with no fracturing.

basal slip Movement involving a glacier sliding over its underlying surface.

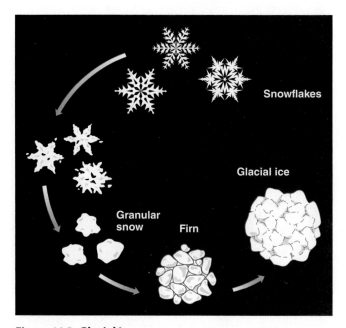

Figure 14.3 Glacial Ice
The conversion of freshly fallen snow to firn and then to glacial ice.

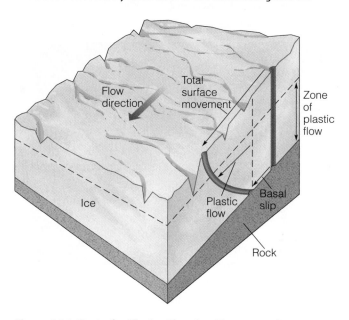

Figure 14.4 Part of a Glacier Showing Movement by a Combination of Plastic Flow and Basal Slip
Plastic flow involves internal deformation within the ice, whereas basal slip is sliding over the underlying surface. If a glacier is solidly frozen to its bed, it moves only by plastic flow. Notice that the top of the glacier moves farther in a given time than the bottom does.

may also slide over their underlying surface by **basal slip** (Figure 14.4). Liquid water facilitates basal slip because it reduces friction between a glacier and the surface over which it moves.

The total movement of a glacier in a given time is a consequence of plastic flow and basal slip, although

Figure 14.5 Crevasses

Crevasses are common in the upper parts of glaciers when the ice is subjected to tension.

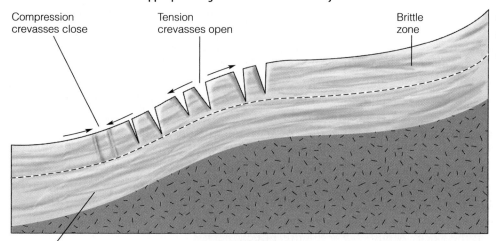

Compression crevasses close

Tension crevasses open

Brittle zone

Zone of plastic flow

a. Crevasses open where the brittle part of a glacier is stretched as it moves over a steeper slope in its valley.

b. These crevasses are on the Tidewater Glacier in Prince William Sound, Alaska.

the former occurs continuously, whereas the latter varies depending on the season, latitude, and elevation. Indeed, if a glacier is solidly frozen to the surface below, as in the case of many polar environments, it moves only by plastic flow. Furthermore, basal slip is far more important in valley glaciers as they flow from higher to lower elevations, whereas continental glaciers need no slope for flow.

Although glaciers move by plastic flow, the upper 40 m or so of ice behaves like a brittle solid and fractures if subjected to stress. Large crevasses commonly develop in glaciers, where they flow over an increase in slope of the underlying surface or where they flow around a corner (Figure 14.5). In either case, the ice is stretched (subjected to tension) and crevasses open, which extend down to the zone of plastic flow. In some cases, a glacier descends over such a steep precipice that crevasses break up the ice into a jumble of blocks and spires, and an icefall develops.

DISTRIBUTION OF GLACIERS

As you might suspect, the amount of snowfall and temperature are important factors in determining where glaciers form. Parts of northern Canada are cold enough to support glaciers but receive too little snowfall, whereas some mountain areas in California receive huge amounts of snow but are too warm for glaciers. Of course, temperature varies with elevation and latitude, so we would expect to find glaciers in high mountains and at high latitudes, if these areas receive enough snow.

Many small glaciers are present in the Sierra Nevada of California, but only at elevations exceeding 3,900 m. In fact, the high mountains in California, Oregon, and Washington all have glaciers because they receive so much snow. Mount Baker in Washington had almost 29 m of snow during the winter of 1998–1999, and average accumulations of 10 m or more are common in many parts of these mountains.

Glaciers are also found in the mountains along the Pacific Coast of Canada, which also receive considerable snowfall, and of course, they are farther north. Some of the higher peaks in the Rocky Mountains in both the United States and Canada also support glaciers. At even higher latitudes, as in Alaska, northern Canada, and Scandinavia, glaciers exist at sea level.

LO3 The Glacial Budget: Accumulation and Wastage

We describe a glacier's behavior in terms of a **glacial budget**, which is essentially a balance sheet of accumulation and wastage. For instance, the upper part of a valley glacier is a **zone of accumulation**, where additions exceed losses and the surface is perennially snow covered. In contrast, the lower part of the same glacier is a **zone of wastage**, where losses from melting, sublimation, and calving of icebergs exceed the rate of accumulation (Figure 14.6).

At the end of winter, a glacier's surface is covered with the accumulated seasonal snowfall. During the spring and summer, the snow begins to melt, first at lower elevations and then progressively higher up the glacier. The elevation to which snow recedes during a wastage season is the *firn limit* (Figure 14.6). You can easily identify the zones of accumulation and wastage by noting the location of the firn limit.

The firn limit on a glacier may change yearly, but if it does not change or shows only minor fluctuations, the glacier has a balanced budget. That is, additions in the zone of accumulation are exactly balanced by losses in the zone of wastage, and the distal end, or terminus, of the glacier remains stationary (Figure 14.6a). If the firn limit moves up the glacier, indicating a negative budget, the glacier's terminus retreats (Figure 14.6b). If the firn limit moves down the glacier, however, the glacier has a positive budget, additions exceed losses, and its terminus advances (Figure 14.6c).

Even though a glacier may have a negative budget and a retreating terminus, the glacial ice continues to move toward the terminus by plastic flow and basal slip. If a negative budget persists long enough, though, the glacier continues to recede, and it thins until it is no longer thick enough to maintain flow. It then ceases moving and becomes a *stagnant glacier;* if wastage continues, the glacier eventually disappears.

We used a valley glacier as an example, but the same budget considerations control the flow of ice caps and continental glaciers as well. The entire Antarctic ice sheet is in the zone of accumulation, but it flows into the ocean, where wastage occurs.

How Fast Do Glaciers Move?

Valley glaciers usually move more rapidly than continental glaciers, but the rates for both vary from centimeters to tens of meters per day. Valley glaciers moving down steep slopes flow more rapidly than glaciers of comparable size on gentle slopes, assuming that all other variables are the same. The main glacier in a valley glacier system contains a greater volume of ice and thus has a greater discharge and flow velocity than its tributaries. Temperature exerts a seasonal control on valley glaciers because, although plastic flow remains rather constant year-round, basal slip is more important during warmer months when meltwater is abundant.

Flow rates also vary within the ice itself. For example, flow velocity increases downslope in the zone of accumulation until the firn limit is reached; from that point, the velocity becomes progressively lower toward the glacier's terminus. Valley glaciers are similar to streams, in that the valley walls and floor cause frictional resistance to flow, so the ice in contact with the

glacial budget The balance between expansion and contraction of a glacier in response to accumulation versus wastage.

zone of accumulation The part of a glacier where additions exceed losses and the glacier's surface is perennially covered with snow. Also refers to horizon B in soil where soluble material leached from horizon A accumulates as irregular masses.

zone of wastage The part of a glacier where losses from melting, sublimation, and calving of icebergs exceed the rate of accumulation.

Figure 14.6 Response of a Hypothetical Glacier to Changes in Its Budget

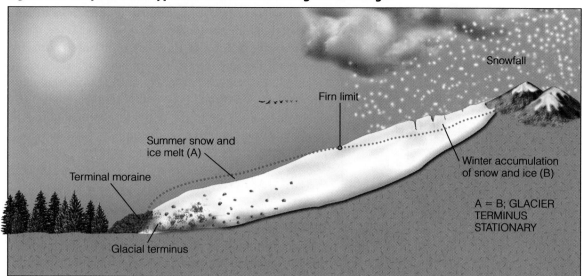

a. Winter accumulation (B) and summer snow and ice melt (A) are equal. That is, additions and losses are equal, so the glacier's terminus remains stationary. The terminal moraine is deposited at the terminus of a glacier.

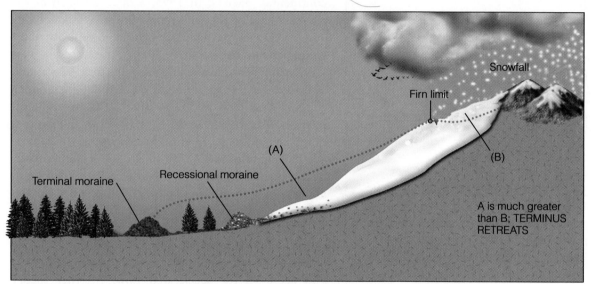

b. Summer snow and ice melt (A) are much greater than winter accumulation (B), and the glacier's terminus retreats, although the glacier continues to move by plastic flow and basal slip. The recessional moraine is deposited at the glacier's new terminus.

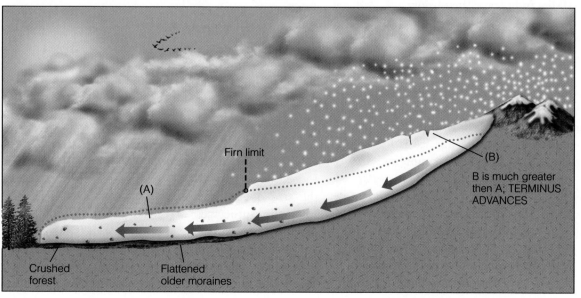

c. Winter accumulation (B) is much greater than summer snow and ice melt (A), so the glacier's terminus advances. As it does so, it overrides and modifies its previously deposited moraines.

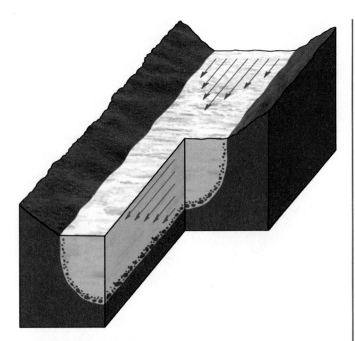

Figure 14.7 Flow Velocity in a Valley Glacier
Flow velocity in a valley glacier varies both horizontally and vertically. Velocity is greatest at the top center of the glacier, because friction with the walls and floor of the trough slows the flow adjacent to these boundaries. The lengths of the arrows in the figure are proportional to velocity.

walls and floor moves more slowly than the ice some distance away (Figure 14.7).

Notice in Figure 14.7 that flow velocity in the interior of a glacier increases upward until the top few tens of meters of ice are reached, but little or no additional increase occurs after that point. This upper ice layer constitutes the rigid part of the glacier that is moving as a result of basal slip and plastic flow below.

Continental glaciers ordinarily flow at a rate of centimeters to meters per day. One reason continental glaciers move comparatively slowly is that they exist at higher latitudes and are frozen to the underlying surface most of the time, which limits the amount of basal slip. Nevertheless, some parts of continental glaciers manage to achieve extremely high flow rates. Near the margins of the Greenland ice sheet, the ice is forced between mountains in what are called *outlet glaciers*. In some of these outlets, flow velocities exceed 100 m per day.

In parts of the continental glacier covering West Antarctica, scien-

glacial surge A time of greatly accelerated flow in a glacier. Commonly results in displacement of the glacier's terminus by several kilometers.

tists have identified ice streams in which flow rates are considerably higher than in adjacent glacial ice. Drilling has revealed a 5-m-thick layer of water-saturated sediment beneath these ice streams, which acts to facilitate movement of the ice above. Some geologists think that geothermal heat from subglacial volcanism melts the underside of the ice, thus accounting for the layer of water-saturated sediment.

GLACIAL SURGES

A **glacial surge** is a short-lived episode of accelerated flow in a glacier, during which its surface breaks into a maze of crevasses and its terminus advances noticeably. Glacial surges are best documented in valley glaciers, although they also take place in ice caps and in continental glaciers. In 1995, a huge ice shelf at the northern end of the Antarctic Peninsula broke apart, and several ice streams from the Antarctic ice sheet surged toward the ocean.

During a surge, a glacier may advance several tens of meters per day for weeks or months and then return to its normal flow rate. The fastest glacial surge ever recorded was in 1953 in the Kutiah Glacier in Pakistan; the glacier advanced 12 km in three months, for an average daily rate of about 130 m. In 1986, the terminus of the Hubbard Glacier in Alaska began advancing at about 10 m per day, and in 1993, Alaska's Bering Glacier advanced more than 1.5 km in just three weeks.

One theory for glacial surges holds that thickening in the zone of accumulation with concurrent thinning in the zone of wastage increases the glacier's slope and accounts for accelerated flow. Another theory holds that pressure on soft sediment beneath a glacier squeezes fluids through the sediment, thereby allowing the overlying glacier to slide more effectively.

LO4 Erosion and Transport by Glaciers

As moving solids, glaciers erode, transport, and eventually deposit huge quantities of sediment and soil. Indeed, they have the capacity to transport boulders the size of a house, as well as clay-sized particles. Important processes of erosion include bulldozing, plucking, and abrasion.

Although *bulldozing* is not a formal geologic term, it is fairly self-explanatory; glaciers shove or push unconsolidated materials in their paths. *Plucking*, also

Figure 14.8 Glacial Striations and Polish
Abrasion produced glacial polish and striations, the straight scratches, on basalt at Devils Postpile National Monument in California.

called quarrying, results when glacial ice freezes in the cracks and crevices of a bedrock projection and eventually pulls it loose.

Bedrock over which sediment-laden glacial ice moves is effectively eroded by **abrasion** and develops a **glacial polish**, a smooth surface that glistens in reflected light (Figure 14.8). Abrasion also yields **glacial striations**, consisting of rather straight scratches rarely more than a few millimeters deep on rock surfaces. Abrasion thoroughly pulverizes rocks, yielding an aggregate of clay- and silt-sized particles that have the consistency of flour—hence, the name *rock flour.*

Continental glaciers derive sediment from mountains projecting through them, and windblown dust settles on their surfaces, but most of their sediment comes from the surface they move over. As a result, most sediment is transported in the lower part of the ice sheet. In contrast, valley glaciers carry sediment in all parts of the ice, but it is concentrated at the base and along the margins (Figure 14.9). Some of the marginal sediment

Figure 14.9 Sediment Transport by Valley Glaciers
Debris on the surface of the Mendenhall Glacier in Alaska. The largest boulder is about 2 m across. Notice the icefall in the background. The person left of center provides scale.

is derived by abrasion and plucking, but much of it is supplied by mass wasting, as when soil, sediment, or rock falls or slides onto the glacier's surface.

EROSION BY VALLEY GLACIERS

When mountains are eroded by valley glaciers, they take on a unique appearance of angular ridges and peaks in the midst of broad, smooth

abrasion The process whereby rock is worn smooth by the impact of sediment transported by running water, glaciers, waves, or wind.

glacial polish A smooth, glistening rock surface formed by the movement of sediment-laden ice over bedrock.

glacial striation A straight scratch rarely more than a few millimeters deep on a rock caused by the movement of sediment-laden glacial ice.

valleys with near-vertical walls. The erosional landforms produced by valley glaciers are easily recognized and enable us to appreciate the tremendous erosive power of moving ice.

U-Shaped Glacial Troughs

A **U-shaped glacial trough** is one of the most distinctive features of valley glaciation. Mountain valleys eroded by running water are typically V-shaped in cross section; that is, they have valley walls that descend to a narrow valley bottom (Figure 14.10). In contrast, valleys scoured by glaciers are deepened, widened, and straightened so that they have very steep or vertical walls, but broad, rather flat valley floors; thus they exhibit a U-shaped profile (Figure 14.10b, 14.11).

During the Pleistocene, when glaciers were more extensive, sea level was as much as 130 m lower than at present, so glaciers flowing into the sea eroded their valleys below present sea level. When the glaciers melted at the end of the Pleistocene, sea level rose and the ocean filled the lower ends of the glacial troughs, so that now they are long, steep-walled embayments called **fiords**.

Fiords are restricted to high latitudes where glaciers exist at low elevations, such as Alaska, western Canada, Scandinavia, Greenland, southern New Zealand, and southern Chile. Lower sea level during the Pleistocene was not entirely responsible for the formation of all fiords. Unlike running water, glaciers can erode a considerable distance below sea level. In fact, a glacier 500 m thick can stay in contact with the seafloor and effectively erode it to a depth of about 450 m before the buoyant effects of water cause the glacial ice to float!

U-shaped glacial trough
A valley with steep or vertical walls and a broad, rather flat floor formed by the movement of a glacier through a stream valley.

fiord An arm of the sea extending into a glacial trough eroded below sea level.

hanging valley A tributary glacial valley whose floor is at a higher level than that of the main glacial valley.

Hanging Valleys

Some of the world's highest and most spectacular waterfalls are found in recently glaciated areas. Nevada Falls in Yosemite National Park, California, plunge from a **hanging valley**, which is a tributary valley whose floor is at a higher level than that of the main valley. Where the two valleys meet, the mouth

Figure 14.10 Erosional Landforms Produced by Valley Glaciers

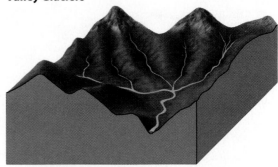

a. A mountain area before glaciation.

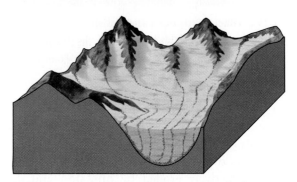

b. The same area during the maximum extent of valley glaciers.

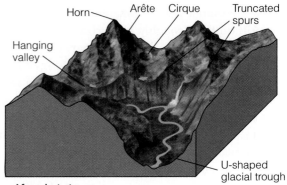

c. After glaciation.

of the hanging valley is perched far above the main valley's floor (Figure 14.10c, 14.12a). Accordingly, streams flowing through hanging valleys plunge over vertical or steep precipices.

Although not all hanging valleys form by glacial erosion, many do. As Figure 14.10 shows, the large glacier in the main valley vigorously erodes, whereas the smaller glaciers in tributary valleys are less capable of erosion. When the glaciers disappear, the smaller tributary valleys remain as hanging valleys.

Figure 14.11 Landforms Produced by Valley Glacier Erosion
U-shaped glacial troughs, cirques, and arêtes are visible in this view of the Chugach Mountains in Alaska.

USGS

Figure 14.12 Landforms Produced by Valley Glacier Erosion

SUE MONROE

SWISS NATIONAL TOURIST OFFICE

b. The Matterhorn in Switzerland is a well-known horn.

a. Nevada Falls plunges 181 m from a hanging valley in Yosemite National Park in California. The valley in the foreground is a huge U-shaped glacial trough.

Figure 14.13 An Ice-Scoured Plain in Northwest Territories of Canada
This low-relief surface is an ice-scoured plain in the Northwest Territories of Canada. Numerous lakes, little or no soil, and extensive bedrock exposures are typical of these areas eroded by continental glaciers.

ALLAN KELLELHELM/MARY PAT ZITER, JLM VISUALS

Cirques, Arêtes, and Horns

Perhaps the most spectacular erosional landforms in areas of valley glaciation are at the upper ends of glacial troughs and along the divides that separate adjacent glacial troughs. Valley glaciers form and move out from steep-walled, bowl-shaped depressions called **cirques** at the upper end of their troughs (Figure 14.10c). Cirques are typically steep-walled on three sides, but one side is open and leads into the glacial trough. Some cirques slope continuously into the glacial trough, but many have a lip or threshold at their lower end.

The details of cirque origin are not fully understood, but they probably form by erosion of a preexisting depression on a mountainside. As snow and ice accumulate in the depression, frost wedging and plucking, combined with glacial erosion, enlarge and transform the head of a steep mountain valley into a typical amphitheater-shaped cirque. Small lakes of meltwater, called *tarns*, often form on the floors of cirques behind such thresholds.

Arêtes—narrow, serrated ridges—form in two ways. In many cases, cirques form on opposite sides of a ridge, and headward erosion reduces the ridge until only a thin partition of rock remains (Figure 14.10c). The same effect occurs when erosion in two parallel glacial troughs reduces the intervening ridge to a thin spine of rock.

The most majestic of all mountain peaks are **horns**, steep-walled, pyramidal peaks formed by headward erosion of cirques. For a horn to form, a mountain peak must have at least three cirques on its flanks, all of which erode headward (Figure 14.10c, 14.12b)

cirque A steep-walled, bowl-shaped depression on a mountainside at the upper end of a glacial valley.

arête A narrow, serrated ridge between two glacial valleys or adjacent cirques.

horn A steep-walled, pyramid-shaped peak formed by the headward erosion of at least three cirques.

glacial drift A collective term for all sediment deposited directly by glacial ice (till) and by meltwater streams (outwash).

CONTINENTAL GLACIERS AND EROSIONAL LANDFORMS

Areas eroded by continental glaciers tend to be smooth and rounded, because these glaciers bevel and abrade high areas that project into the ice. Rather than yielding the sharp, angular landforms typical of valley glaciation, they produce a landscape of subdued topography interrupted by rounded hills, because they bury landscapes entirely during their development.

In a large part of Canada, particularly the vast Canadian shield region, continental glaciers have stripped off the soil and unconsolidated surface sediment, revealing extensive exposures of striated and polished bedrock. These areas have deranged drainage, numerous lakes and swamps, low relief, extensive bedrock exposures, and little or no soil. They are referred to as *ice-scoured plains* (Figure 14.13). Similar though smaller bedrock exposures are also widespread in the northern United States from Maine through Minnesota.

LO5 Deposits of Glaciers

GLACIAL DRIFT

Both valley and continental glaciers deposit sediment as **glacial drift**, a general term for all deposits resulting

Figure 14.14 Glacial Erratics

a. A glacial erratic in the making. This boulder on the surface of the Mendenhall Glacier in Alaska will eventually be deposited far from its source.

JAMES S. MONROE

COPYRIGHT AND PHOTOGRAPH BY DR. PARVINDER S. SETHI

b. This glacial drift from the Matanuska Glacier, near Palmer, Alaska, is till, because it is unsorted and shows no stratification.

from glacial activity. A vast sheet of Pleistocene glacial drift is present in the northern tier of the United States and adjacent parts of Canada. The appearance of these deposits may not be as inspiring as some landforms resulting from glacial erosion, but they are important as reservoirs of groundwater, and in many areas, they are exploited for their sand and gravel.

One conspicuous aspect of glacial drift is rock fragments of various sizes that were obviously not derived from the underlying bedrock. These **glacial erratics** were derived from some distant source and transported to their present location (Figure 14.14). Some erratics are gigantic.

As noted, *glacial drift* is a general term, and geologists define two types of drift: till and stratified drift. **Till** consists of sediments deposited directly by glacial ice. They are not sorted by particle size or density, and they show no stratification (Figure 14.14b). The till of both valley and continental glaciers is similar, but that of continental glaciers is much more extensive and usually has been transported much farther.

glacial erratic A rock fragment carried some distance from its source by a glacier and usually deposited on bedrock of a different composition.

till All sediment deposited directly by glacial ice.

Figure 14.15 End Moraine

End moraine

JAMES S. MONROE

JAMES S. MONROE

a. An end moraine deposited by a valley glacier. This particular end moraine is also a terminal moraine, because it is the one most distant from the glacier's source.

b. Closeup of an end moraine. Notice that the deposit is not sorted by particle size, and it shows no layering or stratification.

As opposed to till, **stratified drift** is layered—that is, stratified—and it invariably exhibits some degree of sorting by particle size. As a matter of fact, most stratified drift is actually layers of sand and gravel or mixtures thereof that accumulated in braided stream channels that issue from melting glaciers.

stratified drift Glacial deposits that show both stratification and sorting.

end moraine A pile or ridge of rubble deposited at the terminus of a glacier.

ground moraine The layer of sediment released from melting ice as a glacier's terminus retreats.

recessional moraine An end moraine that forms when a glacier's terminus retreats, then stabilizes, and a ridge or mound of till is deposited.

LANDFORMS COMPOSED OF TILL

Landforms composed of till include several types of *moraines* and elongated hills known as *drumlins*.

End Moraines

If a glacier has a balanced budget, its terminus may become stabilized in one position for some period of time, perhaps a few years or even decades. When an ice front is stationary, flow within the glacier continues, and any sediment transported within or upon the ice is dumped as a pile of rubble at the glacier's terminus (Figure 14.15). These deposits are **end moraines**, which continue to grow as long as the ice front remains stationary. End moraines of valley glaciers are crescent-shaped ridges of till spanning the valley occupied by the glacier. Those of continental glaciers similarly parallel the ice front but are much more extensive.

Following a period of stabilization, a glacier may advance or retreat, depending on changes in its budget. If it advances, the ice front overrides and modifies its former moraine. If it has a negative budget, though, the ice front retreats toward the zone of accumulation. As the ice front recedes, till is deposited as it is liberated from the melting ice and forms a layer of **ground moraine**. Ground moraine has an irregular, rolling topography, whereas end moraine consists of long, ridgelike accumulations of sediment.

After a glacier has retreated for some time, its terminus may once again stabilize, and it deposits another end moraine. Because the ice front has receded, such moraines are called **recessional moraines**. The outermost end moraines, marking the

Figure 14.16 Lateral and Medial Moraines
The types of medial and lateral moraines shown here are defined by their position. These moraines are on the Bernard Glacier in the St. Elias Mountains in Alaska.

Image credit (vertical text): JOURNAL OF ENGINEERING MECHANICS, VIRGINIA POLYTECHNIC INSTITUTE AND STATE UNIVERSITY

greatest extent of the glaciers, go by the special name **terminal moriane**.

Lateral and Medial Moraines

Valley glaciers transport considerable sediment along their margins, much of it abraded and plucked from the valley walls, but a significant amount falls or slides onto the glacier's surface by mass wasting processes. In any case, this sediment is transported and deposited as long ridges of till called **lateral moraines** along the margin of the glacier (Figure 14.16).

Where two lateral moraines merge, as when a tributary glacier flows into a larger glacier, a **medial moraine** forms (Figure 14.16). A large glacier will often have several dark stripes of sediment on its surface, each of which is a medial moraine. One can determine how many tributaries a valley glacier has by the number of its medial moraines.

Drumlins

In many areas where continental glaciers deposited till, the till has been reshaped into elongated hills known as **drumlins** (Figure 14.17). Some drumlins are 50 m high and 1 km long, but most are much smaller. From the side, a drumlin looks like an inverted spoon, with the steep end on the side from which the glacial ice advanced and the gently sloping end pointing in the direction of ice movement. Drumlins are rarely found as single, isolated hills; instead, they occur in *drumlin fields* that contain hundreds or thousands of drumlins.

According to one hypothesis, drumlins form when till beneath a glacier is reshaped into streamlined hills as the ice moves over it by plastic flow. Another hypothesis holds that huge floods of glacial meltwater modify till into drumlins.

LANDFORMS COMPOSED OF STRATIFIED DRIFT

Stratified drift exhibits sorting and layering, both indications that it was deposited by running water. In fact, it is deposited by streams discharging from valley and continental glaciers, but as you would expect, it is more extensive in areas of continental glaciation.

Outwash Plains and Valley Trains

Sediment-laden meltwater discharges from glaciers most of the time, except perhaps during the coldest months.

---

terminal moraine An end moraine consisting of a ridge or mound of rubble marking the farthest extent of a glacier.

lateral moraine Ridge of sediment deposited along the margin of a valley glacier.

medial moraine A moraine carried on the central surface of a glacier; formed where two lateral moraines merge.

drumlin An elongate hill of till formed by the movement of a continental glacier or by floods.

outwash plain The sediment deposited by meltwater discharging from a continental glacier's terminus.

valley train A long, narrow deposit of stratified drift confined within a glacial valley.

Figure 14.17 Stages in the Development of Major Features Associated with Past Continental Glaciation

a-c. Moraines, eskers, and drumlins form during glaciation, though eskers and drumlins originate under the ice cover. Kames and kettles develop at the end of glaciation.

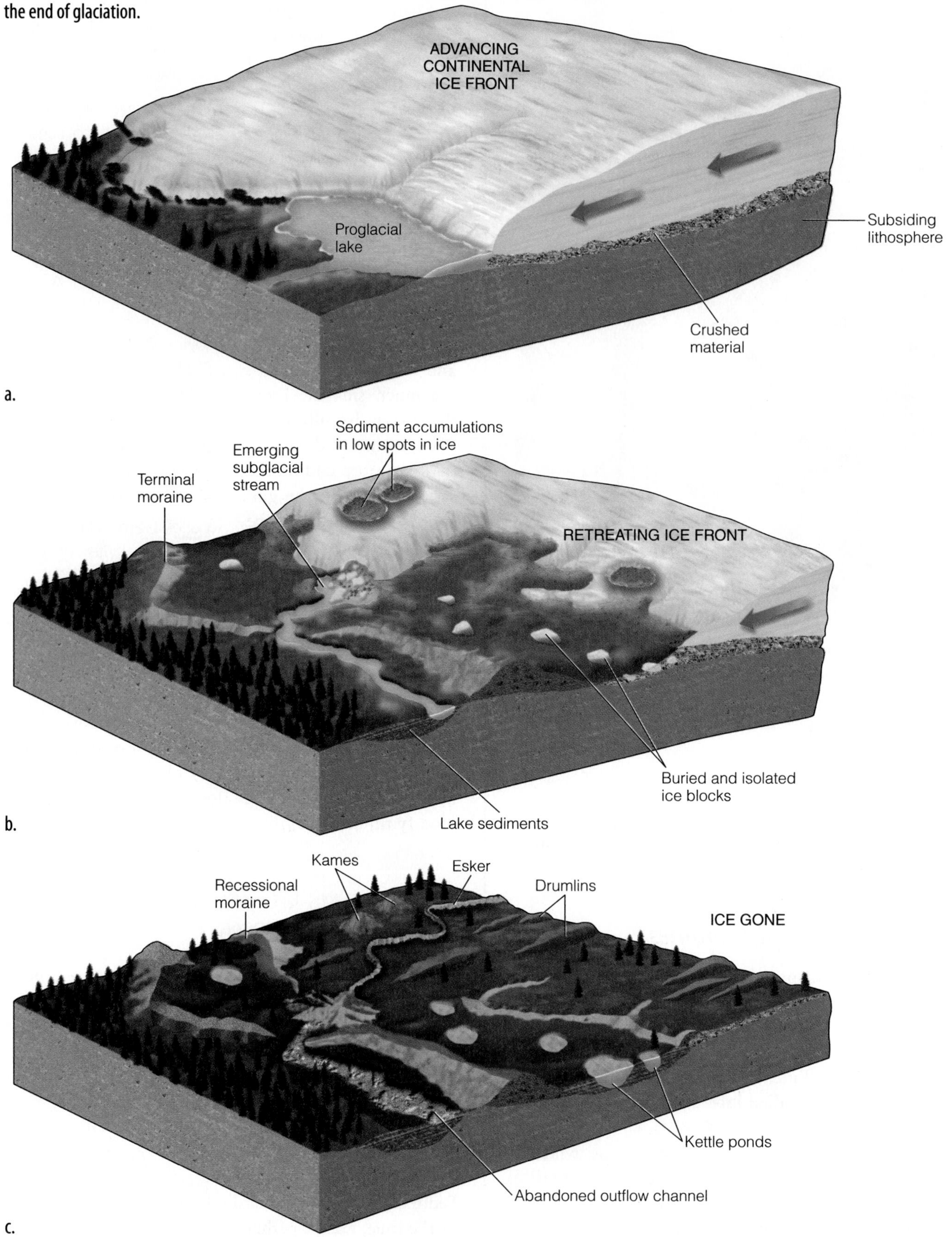

ADVANCING CONTINENTAL ICE FRONT

Proglacial lake

Subsiding lithosphere

Crushed material

a.

Sediment accumulations in low spots in ice

Emerging subglacial stream

Terminal moraine

RETREATING ICE FRONT

Buried and isolated ice blocks

Lake sediments

b.

Kames

Esker

Drumlins

Recessional moraine

ICE GONE

Kettle ponds

Abandoned outflow channel

c.

This meltwater forms a series of braided streams that radiate out from the front of continental glaciers over a wide region. So much sediment is supplied to these streams that much of it is deposited within their channels as sand and gravel bars, thereby forming an **outwash plain**. Valley glaciers also discharge large amounts of meltwater and have braided streams extending from them. However, these streams are confined to the lower parts of glacial troughs, and their long, narrow deposits of stratified drift are known as **valley trains** (Figure 14.18a).

Outwash plains, valley trains, and some moraines commonly contain numerous circular to oval depressions, many of which contain small lakes. These depressions, or *kettles*, form when a retreating glacier leaves a

Figure 14.18 Valley Train and a Kettle

a. A kettle in a moraine in Alaska.

b. A valley train in Alaska made up of stratified drift.

block of ice that is subsequently partly or wholly buried (Figures 14.17 and 14.18b). When the ice block eventually melts, it leaves a depression, and if the depression extends below the water table, it becomes the site of a small lake. Some outwash plains have so many kettles that they are called *pitted outwash plains*.

Kames and Eskers

Kames are conical hills of stratified drift up to 50 m high (Figure 14.17). Many form when a stream deposits sediment in a depression on a glacier's surface; as the ice melts, the deposit is lowered to the land surface. Kames also form in cavities within or beneath stagnant ice.

Long sinuous ridges of stratified drift, many of which meander and have tributaries, are **eskers** (Figure 14.17). Most eskers have sharp crests and sides that slope at about 30 degrees. Some are as high as 100 m and can be traced for more than 500 km. The sorting and stratification of the sediments in eskers clearly indicate deposition by running water. The features of ancient eskers and observations of present-day glaciers show that they form in tunnels beneath stagnant ice.

DEPOSITS IN GLACIAL LAKES

Some lakes in areas of glaciation formed as a result of glaciers scouring out depressions; others occur where a stream's drainage was blocked; and others are the result of water accumulating behind moraines or in kettles. Regardless of how they formed, glacial lakes, like all lakes, are areas of deposition. Sediment may be carried into them and deposited as small deltas, but of special interest are the fine-grained deposits. Mud deposits in glacial lakes are commonly finely laminated (having layers less than 1 cm thick) and consist of alternating light and dark layers known as *varves* (Figure 14.19), which represents an annual episode of deposition. The light layer formed during the spring and summer and consists of silt and clay; the dark layer formed during the winter when the smallest particles of clay and organic matter settled from suspension as the lake froze over.

Another distinctive feature of glacial lakes with varves is *dropstones* (Figure 14.19). These are pieces of gravel, some of boulder size, in otherwise very fine-grained

kame Conical hill of stratified drift originally deposited in a depression on a glacier's surface.

esker A long, sinuous ridge of stratified drift deposited by running water in a tunnel beneath stagnant ice.

Figure 14.19 Varves and a Dropstone in Glacial Deposits
These varves have a dropstone that was probably liberated from floating ice.

deposits. The presence of varves indicates that currents and turbulence in these lakes were minimal; otherwise, clay and organic matter would not have settled from suspension. How then can we account for dropstones in a low-energy environment? Most of them were probably carried into the lakes by icebergs that eventually melted and released sediment contained in the ice.

LO6 What Causes Ice Ages?

We discussed the conditions necessary for a glacier to form earlier in this chapter: More snow falls than melts during the warm season, thus accounting for a net accumulation of snow and ice over the years. But this really does not address the broader question of what causes ice ages. Actually, we need to address not only what causes ice ages, but also why there have been so few episodes of widespread glaciation in all of Earth history.

For more than a century, scientists have attempted to develop a comprehensive theory explaining all aspects of ice ages, but they have not yet been completely successful. One reason for their lack of success is that the climatic changes responsible for glaciation, the cyclic occurrence of glacial–interglacial episodes, and short-term events such as the Little Ice Age operate on vastly different time scales.

Milankovitch theory
An explanation for the cyclic variations in climate and the onset of ice ages as a result of irregularities in Earth's rotation and orbit.

Only a few periods of glaciation are recognized in the geologic record, each separated from the others by long intervals of mild climate. Such long-term climatic changes probably result from slow geographic changes related to plate tectonic activity. Moving plates carry continents to high latitudes where glaciers exist, provided they receive enough precipitation as snow. Plate collisions, the subsequent uplift of vast areas far above sea level, and the changing atmospheric and oceanic circulation patterns caused by the changing shapes and positions of plates also contribute to long-term climate change.

THE MILANKOVITCH THEORY

During the 1920s, the Serbian astronomer Milutin Milankovitch proposed that minor irregularities in Earth's rotation and orbit are sufficient to alter the amount of solar radiation received at any given latitude and hence bring about climate changes. Now called the **Milankovitch theory**, it was initially ignored but has received renewed interest since the 1970s and is now widely accepted.

Milankovitch attributed the onset of the Pleistocene Ice Age to variations in three aspects of Earth's orbit. The first is *orbital eccentricity*, which is the degree to which Earth's orbit around the sun changes over time (Figure 14.20a). When the orbit is nearly circular, both the Northern and Southern Hemispheres have similar contrasts between the seasons. However, if the orbit is more elliptical, hot summers and cold winters will occur in one hemisphere, whereas warm summers and cool winters will take place in the other hemisphere. Calculations indicate a roughly 100,000-year cycle between times of maximum eccentricity, which corresponds closely to the 20 warm–cold climatic cycles that took place during the Pleistocene.

Milankovitch also pointed out that the angle between Earth's axis and a line perpendicular to the plane of Earth's orbit shifts about 1.5 degrees from its current value of 23.5 degrees during a 41,000-year cycle (Figure 14.20b). Although changes in *axial tilt* have little effect on equatorial latitudes, they strongly affect the amount of solar radiation received at high latitudes and the duration of the dark period at and near Earth's poles. Coupled with the third aspect of Earth's orbit, precession of the equinoxes (Figure 14.20c), high latitudes might receive as much as 15% less solar radiation, certainly enough to affect glacial growth and melting.

Precession of the equinoxes, the last aspect of Earth's orbit that Milankovitch cited, refers to a change in the time of the equinoxes (Figure 14.20c). At present, the equinoxes take place on, or about, March 21 and September 21, when the Sun is directly over the equator. But as Earth rotates on its axis, it also wobbles as its axial tilt varies 1.5 degrees from its current value, thus changing the time of the equinoxes. Taken alone, the time of the equinoxes has little climatic effect, but changes in Earth's axial tilt also change the times of *aphelion* and *perihelion*, which are, respectively, when Earth is farthest from and closest to the Sun during its orbit (Figure 14.20c). Earth is now at perihelion, closest to the Sun, during Northern Hemisphere winters, but in about 11,000 years, perihelion will be in July. Accordingly, Earth will be at aphelion, farthest from the Sun, in January and have colder winters.

Continuous variations in Earth's orbit and axial tilt cause the amount of solar heat received at any latitude to vary slightly through time. The total heat received by the planet changes little, but according to Milankovitch, and now many scientists agree, these changes cause complex climatic variations and provided the triggering mechanism for the glacial–interglacial episodes of the Pleistocene.

SHORT-TERM CLIMATIC EVENTS

Climatic events with durations of several centuries, such as the Little Ice Age, are too short to be accounted for by plate tectonics or Milankovitch cycles. Several hypotheses have been proposed, including variations in solar energy and volcanism.

Variations in solar energy could result from changes within the Sun itself or from anything that would reduce the amount of energy Earth receives from the Sun. The latter could result from the solar system passing through clouds of interstellar dust and gas or from substances in the atmosphere reflecting solar radiation back into space. Records kept during the past century indicate that during this time the amount of solar radiation has varied only slightly. Although variations in solar energy may influence short-term climatic events, such a correlation has not been demonstrated.

During large volcanic eruptions, tremendous amounts of ash and gases are spewed into the atmosphere, where they reflect incoming solar radiation and thus reduce atmospheric temperatures. Small droplets of sulfur gases remain in the atmosphere for years and can have a significant effect on climate. Several large-scale volcanic events have occurred, such as the 1815 eruption of Tambora, and are known to have had climatic effects. However, no relationship between periods of volcanic activity and periods of glaciation has yet been established.

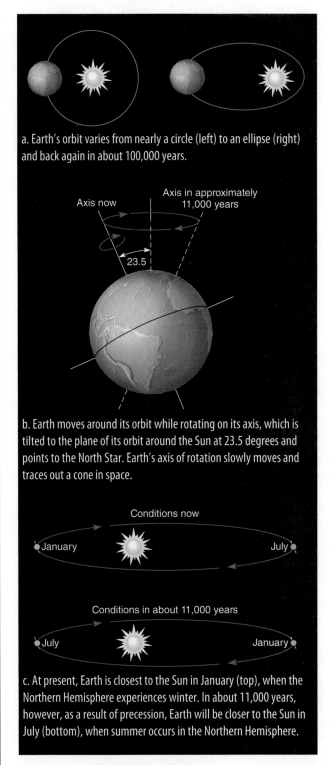

a. Earth's orbit varies from nearly a circle (left) to an ellipse (right) and back again in about 100,000 years.

b. Earth moves around its orbit while rotating on its axis, which is tilted to the plane of its orbit around the Sun at 23.5 degrees and points to the North Star. Earth's axis of rotation slowly moves and traces out a cone in space.

c. At present, Earth is closest to the Sun in January (top), when the Northern Hemisphere experiences winter. In about 11,000 years, however, as a result of precession, Earth will be closer to the Sun in July (bottom), when summer occurs in the Northern Hemisphere.

Figure 14.20 Milankovitch Theory
According to the Milankovitch Theory, minor irregularities in Earth's rotation and orbit may affect climatic changes.

CHAPTER 15
THE WORK OF WIND AND DESERTS

The Mesquite Flat sand dunes in Death Valley, California, are a mix of dominantly transverse-type dunes with some crescent-type dunes and star-type dunes.

Introduction

"Because of the relentless advance of deserts, hundreds of thousands of people have died of starvation or been forced to migrate."

During the past several decades, deserts have been advancing across millions of acres of productive land, destroying rangeland, croplands, and even villages. Such expansion, estimated at 70,000 km² per year, has exacted a terrible toll in human suffering. Because of the relentless advance of deserts, hundreds of thousands of people have died of starvation or been forced to migrate as "environmental refugees" from their homelands to camps, where the majority are severely malnourished. This expansion of deserts into formerly productive lands is called **desertification** and is a major problem in many countries.

Most regions undergoing desertification lie along the margins of existing deserts, where a delicately balanced ecosystem serves as a buffer between the desert on one side and a more humid environment on the other. These regions have limited potential to adjust to increasing environmental pressures from natural causes as well as human activity. Ordinarily, desert regions expand and contract gradually in response to natural processes such as climatic change, but much of the recent desertification has been greatly accelerated by human activities.

In many areas, the natural vegetation has been cleared as crop cultivation has expanded into increasingly drier desert fringes to support growing populations. Because grasses are the dominant natural vegetation in most fringe areas, raising livestock is a common economic activity. However, increasing numbers of livestock in many areas have greatly exceeded the land's ability to support them. Consequently, the vegetation cover that protects the soil has diminished, causing the soil to crumble and be stripped away by wind and water, which results in increased desertification.

One particularly hard-hit area of desertification is the Sahel of Africa (a belt 300 to 1,100 km wide, lying south of the Sahara). Because drought is common in the Sahel, the region can support only a limited population of livestock and human. Unfortunately, expanding human and animal populations and more intensive agriculture have increased the demands on the lands. Plagued with periodic droughts, this region has suffered tremendously as crops have failed and livestock has overgrazed the natural vegetation, resulting in thousands of deaths, displaced people, and the encroachment of the Sahara.

There are many important reasons to study deserts and the processes that are responsible for their formation. First, deserts cover large regions of Earth's surface. More than 40% of Australia is desert, and the Sahara occupies a vast part of northern Africa. Although deserts are generally sparsely populated, some desert regions are experiencing an influx of people, such as Las Vegas, Nevada, the high desert area of southern California, and various locations in Arizona. Many of these places already have problems with population growth and the strains it places on the environment, particularly the need for greater amounts of groundwater (see Chapter 13).

Furthermore, with the current debate about global warming, it is important to understand

desertification The expansion of deserts into formerly productive lands.

LEARNING OUTCOMES

After reading this unit, you should be able to do the following:

LO1 Discuss the role wind plays in transporting sediment

LO2 Explain the two processes of wind erosion

LO3 Identify the types of wind deposits

LO4 Describe air-pressure belts and global wind patterns

LO5 Describe the distribution of deserts

LO6 Identify the various characteristics of deserts

LO7 Identify the different types of desert landforms

how desert processes operate and how global climate changes affect the various Earth systems and subsystems. Learning about the underlying causes of climate change by examining ancient desert regions may provide insight into the possible duration and severity of future climatic changes. This can have important ramifications in decisions about whether burying nuclear waste in a desert, such as Yucca Mountain, Nevada, is as safe as some claim and is in our best interests as a society.

LO1 Sediment Transport by Wind

Wind is a turbulent fluid and therefore transports sediment in much the same way as running water. Although wind typically flows at a greater velocity than water, it has a lower density and thus can carry only clay- and silt-size particles as *suspended load*. Sand and larger particles are moved along the ground as *bed load*.

BED LOAD

Sediments that are too large or heavy to be carried in suspension by water or wind are moved as bed load either by *saltation* or by rolling and sliding. Saltation on land occurs when wind starts sand grains rolling and lifts and carries some grains short distances before they fall back to the surface. As the descending sand grains hit the surface, they strike other grains, causing them to bounce along (Figure 15.1). Wind-tunnel experiments show that once sand grains begin moving, they continue to move, even if the wind drops below the speed necessary to start them moving! This happens because once saltation begins, it sets off a chain reaction of collisions between sand grains that keeps the grains in constant motion.

Saltating sand usually moves near the surface, and even when winds are strong, grains are rarely lifted higher than about a meter. If the winds are very strong,

More than 6,000 years ago, the Sahara was a fertile savannah supporting a diverse fauna and flora, including humans. Then the climate changed, and the area became a desert. How did this happen? Will this region change back again in the future? These are some of the questions geoscientists hope to answer by studying deserts.

these wind-whipped grains can cause extensive abrasion. A car's paint can be removed by sandblasting in a short time, and its windshield will become completely frosted and translucent from pitting.

SUSPENDED LOAD

Silt- and clay-sized particles constitute most of a wind's suspended load. Even though these particles are much smaller and lighter than sand-sized particles, wind usually starts the latter moving first. The reason for this phenomenon is that a very thin layer of motionless air lies next to the ground, where the small silt and clay particles remain undisturbed. The larger sand grains, however, stick up into the turbulent air zone, where they can be moved. Unless the stationary air

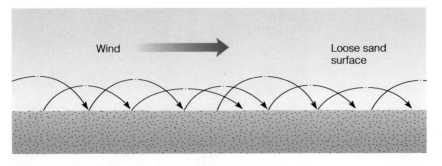

Figure 15.1 Saltation

Most sand is moved near the ground surface by saltation. Sand grains are picked up by the wind and carried a short distance before falling back to the ground, where they usually hit other grains, causing them to bounce and move in the direction of the wind.

layer is disrupted, the silt and clay particles remain on the ground, providing a smooth surface.

This phenomenon can be observed on a dirt road on a windy day. Unless a vehicle travels over the road, little dust is raised even though it is windy. When a vehicle moves over the road, it breaks the calm boundary layer of air and disturbs the smooth layer of dust, which is picked up by the wind and forms a dust cloud in the vehicle's wake.

In a similar manner, when a sediment layer is disturbed, silt- and clay-sized particles are easily picked up and carried in suspension by the wind, creating clouds of dust or even dust storms. Once these fine particles are lifted into the atmosphere, they may be carried thousands of kilometers from their source.

LO2 Wind Erosion

Although wind action produces many distinctive erosional features and is an extremely efficient sorting agent, running water is responsible for most erosional landforms in arid regions, even though stream channels are typically dry (Figure 15.2). Wind erodes material in two ways: *abrasion* and *deflation*.

ABRASION

Abrasion involves the impact of saltating sand grains on an object and is analogous to sandblasting. The effects of abrasion are usually minor, because sand, the most common agent of abrasion, is rarely carried more than a meter above the surface. Rather than creating major erosional features, wind abrasion typically modifies existing features by etching, pitting, smoothing, or polishing. Nonetheless, wind abrasion can produce many strange-looking and bizarre-shaped features.

Ventifacts are a common product of wind

> **abrasion** The process whereby rock is worn smooth by the impact of sediment transported by running water, glaciers, waves, or wind.
>
> **ventifact** A stone with a surface polished, pitted, grooved, or faceted by wind abrasion.

Figure 15.2 Erosion

Natural forces of wind and water have eroded the rocks in Monument Valley, Arizona, to create "mitten" buttes such as this one.

Figure 15.3 Ventifacts

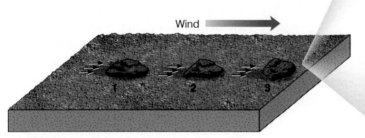

a. A ventifact forms when windborne particles (1) abrade the surface of a rock, (2) forming a flat surface. If the rock is moved, (3) additional flat surfaces are formed.

b. Numerous ventifacts are visible in this photo, which also shows desert pavement in Death Valley, California. Desert pavement prevents further erosion and transport of a desert's surface materials by forming a protective layer of close-fitting, larger rocks.

abrasion; these are stones whose surfaces have been polished, pitted, grooved, or faceted by the wind (Figure 15.3). If the wind blows from different directions, or if the stone is moved, the ventifact will have multiple facets. Ventifacts are most common in deserts, yet they can form wherever stones are exposed to saltating sand grains, as on beaches in humid regions and some outwash plains in New England.

> **deflation** The removal of loose surface sediment by wind.

DEFLATION

Another important mechanism of wind erosion is **deflation**, which is the removal of loose surface sediment by wind. Among the characteristic features of deflation in many arid and semiarid regions are *deflation hollows* or *blowouts* (Figure 15.4). These shallow depressions of variable dimensions result from differential erosion of surface materials. Ranging in size from several kilometers in diameter and tens of meters deep to small depressions only a few meters wide and less than a meter deep, deflation hollows are common in the southern Great Plains region of the United States.

In many dry regions, the removal of sand-sized

Figure 15.4 Deflation Hollow

A deflation hollow, the low area, between two sand dunes in Death Valley, California. Deflation hollows result when loose surface sediment is differentially removed by wind.

Figure 15.5 Desert Pavement

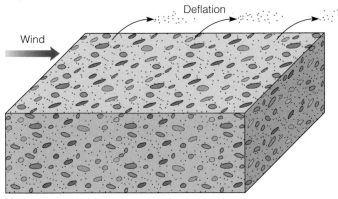

a. Fine-grained material is removed by wind,

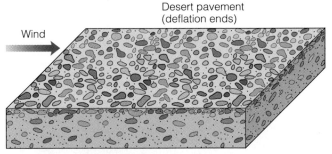

b. leaving a concentration of larger particles that form desert pavement.

and smaller particles by wind leaves a surface of pebbles, cobbles, and boulders. As the wind removes the fine-grained material from the surface, the effects of gravity and occasional heavy rain, and even the swelling of clay grains, rearrange the remaining coarse

particles into a mosaic of close-fitting rocks called **desert pavement** (Figure 15.3b and Figure 15.5). Once desert pavement forms, it protects the underlying material from further deflation.

LO3 Wind Deposits

Although wind is of minor importance as an erosional agent, it is responsible for impressive deposits, which are primarily of two types. The first, *dunes*, occur in several distinctive types, all of which consist of sand-sized particles that are usually deposited near their source. The second is *loess*, which consists of layers of windblown silt and clay deposited over large areas downwind and commonly far from their source.

THE FORMATION AND MIGRATION OF DUNES

The most characteristic features in sand-covered regions are **dunes**, which are mounds or ridges of wind-deposited sand (Figure 15.6). Dunes form when wind flows over and around an obstruction, resulting in the deposition of

> **desert pavement** A surface mosaic of close-fitting pebbles, cobbles, and boulders found in many dry regions; results from wind erosion of sand and smaller particles.
>
> **dune** A mound or ridge of wind-deposited sand.

Figure 15.6 Sand Dunes

Large sand dunes in Death Valley, California. The prevailing wind direction is from left to right, as indicated by the sand dunes in which the gentle windward side is on the left and the steeper leeward slope is on the right.

Figure 15.7 Dune Migration

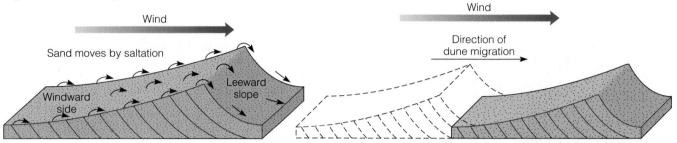

a. Profile of a sand dune.

b. Dunes migrate when sand moves up the windward side and slides down the leeward slope. Such movement of the sand grains produces a series of cross-beds that slope in the direction of wind movement.

sand grains, which accumulate and build up a deposit of sand. As they grow, these sand deposits become self-generating in that they form ever-larger wind barriers that further reduce the wind's velocity, resulting in more sand deposition and growth of the dune.

Most dunes have an asymmetrical profile, with a gentle windward slope and a steeper downwind or leeward slope that is inclined in the direction of the prevailing wind (Figure 15.7a). Sand grains move up the gentle windward slope by saltation and accumulate on the leeward side, forming an angle of 30 to 34 degrees from the horizontal, which is the angle of repose of dry sand. When this angle is exceeded by accumulating sand, the slope collapses, and the sand slides down the leeward slope, coming to rest at its base. As sand moves from a dune's windward side and periodically slides down its leeward slope, the dune slowly migrates in the direction of the prevailing wind (Figure 15.7b). When preserved in the geologic record, dunes help geologists determine the prevailing direction of ancient winds (Figure 15.8)

Figure 15.8 Cross-Bedding
Ancient cross-bedding in sandstone beds in Zion National Park, Utah, helps geologists determine the prevailing direction of the wind that formed these ancient sand dunes.

DUNE TYPES

Geologists recognize four major dune types (barchan, longitudinal, transverse, and parabolic), although intermediate forms also exist. The size, shape, and arrangement of dunes result from the interaction of such factors as sand supply, the direction and velocity of the prevailing wind, and the amount of vegetation. Although dunes are usually found in deserts, they can also develop wherever sand is abundant, such as along the upper parts of many beaches.

Barchan dunes are crescent-shaped dunes whose tips point downwind (Figure 15.9). They form in areas that have a generally flat, dry surface with little vegetation, a limited supply of sand, and a nearly constant wind direction. Most barchans are small, with

barchan dune A crescent-shaped sand dune with its tips pointing downwind.

Figure 15.9 Barchan Dunes

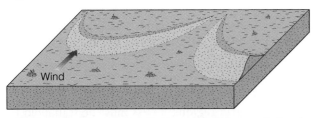

a. Barchan dunes form in areas that have a limited amount of sand, a nearly constant wind direction, and a generally flat, dry surface with little vegetation. The tips of barchan dunes point downward.

b. A ground-level view of several barchan dunes.

JOHN KARACHEWSKI

Figure 15.10 Longitudinal Dunes

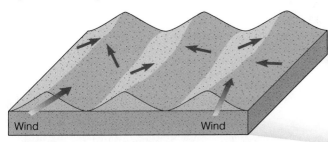

a. Longitudinal dunes form long, parallel ridges of sand aligned roughly parallel to the prevailing wind direction. They typically form where sand supplies are limited.

the largest reaching about 30 m high. Barchans are the most mobile of the major dune types, moving at rates that can exceed 10 m per year.

Longitudinal dunes (also called *seif dunes*) are long, parallel ridges of sand aligned generally parallel to the direction of the prevailing winds; they form where the sand supply is somewhat limited (Figure 15.10). Longitudinal dunes result when winds converge from slightly different directions to produce the prevailing wind. They range in height from about 3 m to more than 100 m, and some stretch for more than 100 km. Longitudinal dunes are especially well developed in central Australia, where they cover nearly one-fourth of the continent.

Transverse dunes form long ridges perpendicular to the prevailing wind direction in areas that have abundant sand and little or no vegetation (Figure 15.11). When viewed from the air, transverse dunes have a wavelike appearance and are therefore sometimes called *sand seas*. The crests of transverse dunes can be as high as 200 m, and the dunes may be as wide as 3 km.

Parabolic dunes are most common in coastal areas with abundant sand, strong onshore winds, and a partial cover of vegetation (Figure 15.12). Although parabolic dunes

longitudinal dune A long ridge of sand generally parallel to the direction of the prevailing wind.

transverse dune A ridge of sand with its long axis perpendicular to the wind direction.

parabolic dune A crescent-shaped dune with its tips pointing upwind.

© 1994 CNES. PROVIDED BY SPOT IMAGE CORPORATION

b. Longitudinal dunes, 15 m high, in the Gibson Desert, west central Australia. The bright blue areas between the dunes are shallow pools of rainwater, and the darkest patches are areas where the Aborigines have set fires to encourage the growth of spring grasses.

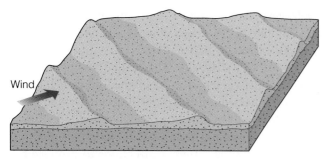

Figure 15.11 Transverse Dunes
Transverse dunes form long ridges of sand that are perpendicular to the prevailing wind direction in areas of little or no vegetation and abundant sand.

have a crescent shape like barchan dunes, their tips point upwind. Parabolic dunes form when the vegetation cover is broken and deflation produces a deflation hollow or blowout. As the wind transports the sand out of the depression, it builds up on the convex downwind dune crest. The central part of the dune is excavated by the wind, while vegetation holds the ends and sides fairly well in place.

loess Wind-blown deposits of silt and clay.

LOESS

Wind-blown silt and clay deposits composed of angular quartz grains, feldspar, micas, and calcite are known as **loess**. The distribution of loess shows that it is derived from three main sources: deserts, Pleistocene glacial outwash deposits, and the floodplains of rivers in semiarid regions. Loess must be stabilized by moisture and vegetation in order to accumulate. Consequently, loess is not found in deserts, even though deserts provide much of its material. Because of its unconsolidated nature, loess is easily eroded, and as a result, eroded loess areas are characterized by steep cliffs and rapid lateral and headward stream erosion.

At present, loess deposits cover approximately 10% of Earth's land surface and 30% of the United States. The most extensive and thickest loess deposits are found in northeast China, where accumulations greater than 30 m thick are common. Loess-derived soils are some of the world's most fertile. It is therefore not surprising that the world's major grain-producing regions correspond to large loess deposits, such as the North European Plain, Ukraine, and the Great Plains of North America.

Figure 15.12 Parabolic Dunes

a. Parabolic dunes typically form in coastal areas that have a partial cover of vegetation, a strong onshore wind, and abundant sand.

b. A parabolic dune developed along the Lake Michigan shoreline west of St. Ignace, Michigan.

REED WICANDER

LO4 Air-Pressure Belts and Global Wind Patterns

To understand the work of wind and the distribution of deserts, we need to consider the global pattern of air-pressure belts and winds, which are responsible for Earth's atmospheric circulation patterns. Air pressure is the density of air exerted on its surroundings (i.e., its weight). When air is heated, it expands and rises, reducing its mass for a given volume and causing a decrease in air pressure. Conversely, when air is cooled, it contracts and air pressure increases. Therefore, those areas of Earth's surface that receive the most solar radiation, such as the equatorial regions, have low air pressure, whereas the colder areas, such as the polar regions, have high air pressure.

Air flows from high-pressure zones to low-pressure zones. If Earth did not rotate, winds would move in a straight line from one zone to another. Because Earth rotates, however, winds are deflected to the right of their direction of motion (clockwise) in the Northern Hemisphere and to the left of their direction of motion (counterclockwise) in the Southern Hemisphere. This deflection of air between latitudinal zones resulting from Earth's rotation is known as the **Coriolis effect**. The combination of latitudinal pressure differences and the Coriolis effect produces a worldwide pattern of east-west–oriented wind belts (Figure 15.13).

Earth's equatorial zone receives the most solar energy, which heats the surface air and causes it to rise. As the air rises, it cools and releases moisture that falls as rain in the equatorial region (Figure 15.13).

> **Coriolis effect** The apparent deflection of a moving object from its anticipated course because of Earth's rotation. Winds and oceanic currents are deflected clockwise in the Northern Hemisphere and counterclockwise in the Southern Hemisphere.

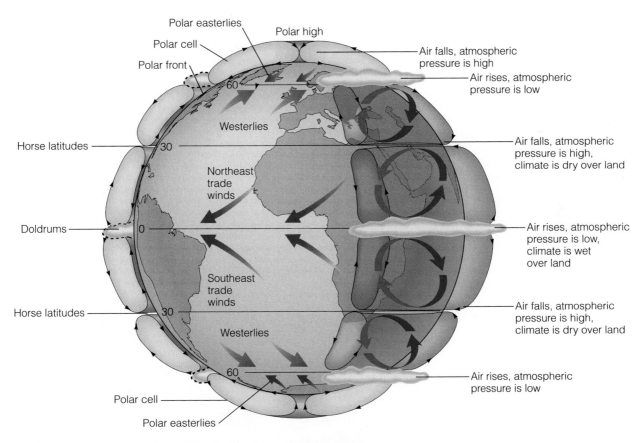

Figure 15.13 The General Circulation Pattern of Earth's Atmosphere
Air flows from high-pressure zones to low-pressure zones, and the resulting winds are deflected to the right of their direction of movement (clockwise) in the Northern Hemisphere and to the left of their direction of movement (counterclockwise) in the Southern Hemisphere. This deflection of air between latitudinal zones resulting from Earth's rotation is known as the Coriolis effect.

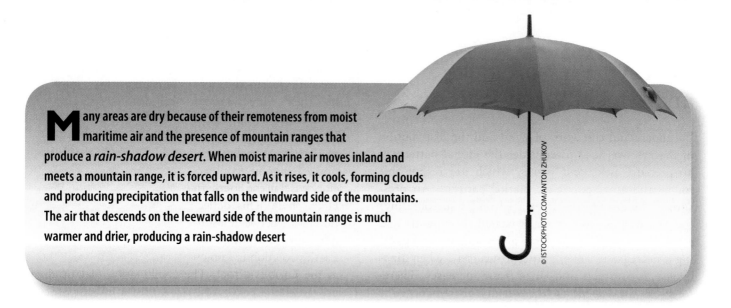

Many areas are dry because of their remoteness from moist maritime air and the presence of mountain ranges that produce a *rain-shadow desert*. When moist marine air moves inland and meets a mountain range, it is forced upward. As it rises, it cools, forming clouds and producing precipitation that falls on the windward side of the mountains. The air that descends on the leeward side of the mountain range is much warmer and drier, producing a rain-shadow desert

The rising air is now much drier as it moves northward and southward toward each pole. By the time it reaches 20 to 30 degrees north and south latitudes, the air has become cooler and denser and begins to descend. Compression of the atmosphere warms the descending air mass and produces a warm, dry, high-pressure area, the perfect conditions for the formation of the low-latitude deserts of the Northern and Southern Hemispheres (Figure 15.14).

LO5 The Distribution of Deserts

Dry climates occur in the low and middle latitudes, where the potential loss of water by evaporation may exceed the yearly precipitation (Figure 15.14). Dry climates cover 30% of Earth's land surface and are subdivided into semiarid and arid regions. *Semiarid regions* receive more precipitation than arid regions,

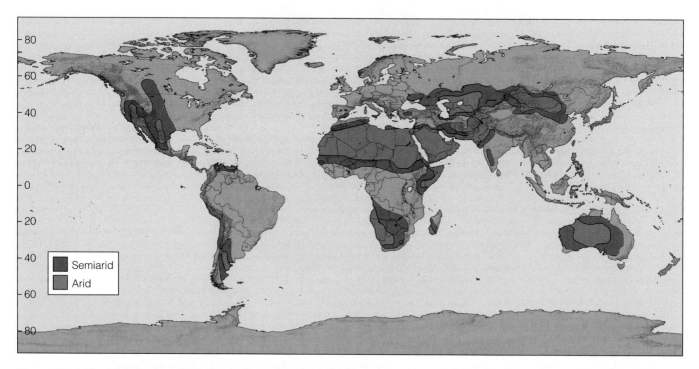

Figure 15.14 The Distribution of Earth's Arid and Semiarid Regions
Semiarid regions receive more precipitation than arid regions, yet they are still moderately dry. Arid regions, generally described as deserts, are dry and receive less than 25 cm of rain per year. The majority of the world's deserts are located in the dry climates of the low and middle latitudes.

yet are moderately dry. Their soils are usually well developed and fertile and support a natural grass cover. *Arid regions*, generally described as **deserts**, are dry; they receive less than 25 cm of rain per year, have high evaporation rates, typically have poorly developed soils, and are mostly or completely devoid of vegetation.

The majority of the world's deserts are in the dry climates of the low and middle latitudes (Figure 15.14). The remaining dry climates of the world are found in the middle and high latitudes, mostly within continental interiors in the Northern Hemisphere (Figure 15.14).

LO6 Characteristics of Deserts

To people who live in humid regions, deserts may seem stark and inhospitable. Instead of a landscape of rolling hills and gentle slopes with an almost continuous cover of vegetation, deserts are dry, have little vegetation, and consist of nearly continuous rock exposures, desert pavement, or sand dunes. Yet despite the great contrast between deserts and more humid areas, the same geologic processes are at work, only operating under different climatic conditions.

Figure 15.15 Desert Vegetation
Desert vegetation is typically sparse, widely spaced, and characterized by slow growth rates. The vegetation shown here is in Death Valley, California.

COPYRIGHT AND PHOTOGRAPH BY DR. PARVINDER S. SETHI

TEMPERATURE, PRECIPITATION, AND VEGETATION

The heat and dryness of deserts are well known. Many of the deserts of the low latitudes have average summer temperatures that range between 32°C and 38°C. It is not uncommon for some low-elevation inland deserts to record daytime highs of 46°C to 50°C for weeks at a time. During the winter months, when the Sun's angle is lower and there are fewer daylight hours, daytime temperatures average between 10°C and 18°C.

Although deserts are defined as regions that receive, on average, less than 25 cm of rain per year, the amount of rain that falls each year is unpredictable and unreliable. It is not uncommon for an area to receive more than an entire year's average rainfall in one cloudburst and then to receive little rain for several years. Thus, yearly rainfall averages can be misleading.

Deserts display a wide variety of vegetation (Figure 15.15). Although the driest deserts, or those with large areas of shifting sand, are almost devoid of vegetation, most deserts support at least a sparse plant cover. Desert plants are widely spaced, typically small, and grow slowly. Their stems and leaves are usually hard and waxy to minimize water loss by evaporation and to protect the plant from sand erosion. Most plants have a widespread shallow root system to absorb the dew that forms each morning in all but the driest deserts and to help anchor the plant in what little soil there may be. In extreme cases, many plants lie dormant during particularly dry years and spring to life after the first rain shower with a beautiful profusion of flowers.

WEATHERING AND SOILS

Mechanical weathering is dominant in desert regions. Daily temperature fluctuations and frost wedging are the primary forms

VIRTUAL FIELD TRIP

Desert Environments

GOALS OF THE TRIP

Although there are lots of things to see and learn about deserts and how they form, we will limit ourselves in this virtual field trip to three main desert processes and features. These are:

1. The effects of wind erosion.
2. The features of wind deposition.
3. A desert landform.

FOLLOW-UP QUESTIONS

1. What type of dunes would probably form if there were far less sand at Mesquite Flat Dunes than there presently is?
2. What type of sedimentary structure forms as a result of dune formation? [Hint: see Figure 15.8 in this chapter and the Sedimentary Rocks Virtual Field Trip]. What can those sedimentary structures tell you about the environment at that location in the geologic past?
3. Why is it so difficult to travel across Death Valley? Imagine what the emigrants must have felt like after crossing the country, only to be confronted with salt pans, sand dunes, and unrelenting heat.

Ready to Go!

What better place to learn about desert processes and features than Death Valley National Park, California. Aptly named, Death Valley National Park is the hottest, driest, and lowest of the nation's national parks. With summer daytime temperatures often above 120°F, scant rainfall (average annual precipitation is less than 2 inches), and signs reminding visitors that sea level is 282 feet above them, Death Valley, for most people, is a very inhospitable place. But for geologists, it is a wonderful location where the geological processes and products of an arid environment can be studied up close and personal. In this virtual field trip, you too will get to experience what it is like to walk on a salt pan and see the results of wind erosion and deposition. However, you will get to do this from the comfort of your living room or wherever you are connected to the internet.

The oldest rocks in Death Valley consist of 1.8 billion year old gneisses exposed in the Black Mountain range. During the Paleozoic Era, the area in what is now Death Valley was below a warm, shallow sea, where the sandstones, limestones, and dolostones exposed in the Panamint and Funeral Mountains today, were deposited. Volcanic activity dominated the region during the Paleogene and Neogene periods, and the results of this activity can be seen in the beautiful colored strata in the Artist's Palette area of Artist Drive. Death Valley came into its own about three million years ago when tensional forces associated with the formation of the Basin and Range Province (see Chapter 18) produced a series of valleys and mountain ranges. Death Valley is one such valley that formed between the Panamint Range on the west and the Black Mountains on the east.

It is now time to see all of the beauty and geology that Death Valley National Park, California offers. So, put on a hat, lather up with the sunscreen so you don't get sunburned, and bring plenty of water while we explore and learn about desert processes and features.

Interestingly, most people associate desert features with wind action. However, although wind produces many distinctive erosional features, running water is responsible for most erosional landforms in arid regions. Nonetheless, wind abrasion typically modifies existing features by etching, pitting, smoothing, and polishing. In this photo, we can see two beautifully formed *ventifacts*, which are stones whose surfaces have been polished, pitted, grooved, or faceted by the wind. In addition, these ventifacts are sitting on what geologists call *desert pavement*, which results when wind removes the fine-grained material from the surface, leaving behind a close-fitting mosaic of coarse particles that forms a pavement that protects the underlying material from further erosion (See Figures 15.3 and 15.5 of this chapter).

Playa lakes form in low areas following rainstorms. When a playa lake evaporates, the dry lake bed is called a *playa* or *salt pan*. As seen in this panorama of Badwater Basin, salt pans are appropriately named. Here, salt deposits and salt ridges cover the floor of this area. The photo shows a close up of a typical salt polygon, in which salt ridges define the polygon and mudcracks are visible within the polygon itself. Formed from the evaporation of briny water, some of these deposits were at one time commercially viable, such as the borates which were hauled out of Death Valley by 20 mule team wagons during the late 1800s.

Just as wind can erode and produce some unusual-looking structures, it also deposits loose material as it moves over a landscape. The most characteristic features in sand-covered regions are dunes, which are mounds or ridges of wind-deposited sand. Dunes form when wind flows over and around an obstruction, resulting in the deposition of sand grains. Most dunes have an asymmetrical profile, with a gentle windward slope and a steeper downwind slope that is inclined in the direction of the wind. Geologists recognize four major dune types, which form from the interaction of such factors as sand supply, the direction, and velocity of the prevailing wind, and the amount of vegetation.

At this location, numerous *transverse dunes* are visible. Transverse dunes form long undulating ridges perpendicular to the prevailing wind direction in areas that have abundant sand and little or no vegetation. In the white oval on the Virtual Field Trip screen, click on the windward part of the sand dune. Can you tell what the prevailing direction of wind is in this photo?

of mechanical weathering (see Chapter 6). The breakdown of rocks by roots and from salt crystal growth is of minor importance. Some chemical weathering does occur, but its rate is greatly reduced by aridity and the scarcity of organic acids produced by the sparse vegetation. Most chemical weathering takes place during the winter months when there is more precipitation, particularly in the mid-latitude deserts.

Desert soils, if developed, are usually thin and patchy, because the limited rainfall and the resultant scarcity of vegetation reduce the efficiency of chemical weathering and hence soil formation. Furthermore, the sparseness of the vegetative cover enhances wind and water erosion of what little soil actually forms.

MASS WASTING, STREAMS, AND GROUNDWATER

When traveling through a desert, most people are impressed by such wind-formed features as moving sand, sand dunes, and sand and dust storms. They may also notice the dry washes and dry streambeds. Because of the lack of running water, most people would conclude that wind is the most important erosional agent in deserts. They would be wrong! Running water, even though it occurs infrequently, causes most of the erosion in deserts. The dry conditions and sparse vegetation characteristic of deserts enhance water erosion.

Most of a desert's average annual rainfall of 25 cm or less comes in brief, heavy, localized cloudbursts. During these times, considerable erosion takes place, because the ground cannot absorb all of the rainwater. With so little vegetation to hinder the flow of water, runoff is rapid, especially on moderately to steeply sloping surfaces, resulting in flash floods and sheet flows. Dry stream channels quickly fill with raging torrents of muddy water and mudflows, which carve out steep-sided gullies and overflow their banks. During these times, a tremendous amount of sediment is rapidly transported and deposited far downstream.

Most desert streams are poorly integrated and flow only intermittently. Many of them never reach the sea, because the water table is usually far deeper than the channels of most streams, so they cannot draw upon groundwater to replace water lost to evaporation and absorption into the ground. This type of drainage in which a stream's load is deposited

playa A dry lakebed found in deserts.

within the desert is called *internal drainage* and is common in most arid regions.

Although most deserts have internal drainage, some deserts have permanent through-flowing streams, such as the Nile and Niger rivers in Africa, the Rio Grande and Colorado River in the southwestern United States, and the Indus River in Asia. These streams can flow through desert regions, because their headwaters are well outside the desert, and water is plentiful enough to offset losses resulting from evaporation and infiltration.

WIND

Although running water does most of the erosional work in deserts, wind can also be an effective geologic agent capable of producing a variety of distinctive erosional (Figure 15.2) and depositional features (Figures 15.9 through 15.12). Wind is effective in transporting and depositing unconsolidated sand-, silt-, and dust-sized particles. Contrary to popular belief, most deserts are not sand-covered wastelands, but rather vast areas of rock exposures and desert pavement. Sand-covered regions, or sandy deserts, constitute less than 25% of the world's deserts. The sand in these areas has accumulated primarily by the action of wind.

LO7 Desert Landforms

Because of differences in temperature, precipitation, and wind, as well as the underlying rocks and recent tectonic events, landforms in arid regions vary considerably. Running water, although infrequent in deserts, is responsible for producing and modifying many distinctive landforms found there.

After an infrequent and particularly intense rainstorm, excess water not absorbed by the ground may accumulate in low areas and form *playa lakes*. These lakes are temporary, lasting from a few hours to several months. Most of them are shallow and have rapidly shifting boundaries as water flows in or leaves by evaporation and seepage into the ground. The water is often very saline.

When a playa lake evaporates, the dry lake bed is called a **playa** or *salt pan* and is characterized by mud cracks and precipitated salt crystals (Figure 15.16). Salts in some playas are thick enough to be mined commercially. For example, borates have been mined in Death Valley, California, for more than 100 years.

Figure 15.16 Playas and Playa Lakes

a. A playa lake formed after a rainstorm near Badwater, Death Valley National Park, California. Playa lakes are ephemeral features, lasting from a few hours to several months.

b. Salt deposits and salt ridges cover the floor of this playa in the Mojave Desert, California. Salt crystals and mud cracks are characteristic features of playas.

Another common feature of deserts, particularly in the Basin and Range Province of the western United States, are alluvial fans. **Alluvial fans** form when sediment-laden streams flowing out from the generally straight, steep mountain fronts deposit their load on the relatively flat desert floor. Once beyond the mountain front where no valley walls confine streams, the sediment spreads out laterally, forming a gently sloping and poorly sorted fan-shaped sedimentary deposit (Figure 15.17). Alluvial fans are similar in origin and shape to deltas (see Chapter 12) but are formed entirely on land.

Most mountains in desert regions, including those of the Basin and Range Province, rise abruptly from gently sloping surfaces called **pediments**. Pediments are erosional bedrock surfaces of low relief that slope gently away from mountain bases. Most pediments are covered by a thin layer of debris, or alluvial fans.

Other easily recognized erosional remnants common to arid and semiarid regions are mesas and buttes (Figure 15.18). A **mesa** is a broad, flat-topped erosional remnant bounded

alluvial fan A cone-shaped accumulation of mostly sand and gravel deposited where a stream flows from a mountain valley onto an adjacent lowland.

pediment An erosion surface of low relief gently sloping away from the base of a mountain range.

mesa A broad, flat-topped erosional remnant bounded on all sides by steep slopes.

Figure 15.17 Alluvial Fan
A ground view of an alluvial fan, Death Valley, California. Alluvial fans form when sediment-laden streams flowing out from a mountain deposit their load on the desert floor, forming a gently sloping, fan-shaped, sedimentary deposit.

Figure 15.18 Buttes
Right Mitten Butte and Merrick Butte in Monument Valley Navajo Tribal Park on the border of Arizona and Utah.

on all sides by steep slopes. Continued weathering and stream erosion form isolated pillar-like structures known as **buttes**. Buttes and mesas consist of relatively easily weathered sedimentary rocks capped by nearly horizontal, resistant rocks such as sandstone, limestone, or basalt. They form when the resistant rock layer is breached, which allows rapid erosion of the less resistant underlying sediment.

butte An isolated, steep-sided, pinnacle-like hill formed when resistant cap rock is breached allowing erosion of less resistant underlying rocks.

CHAPTER 16
SHORELINES AND SHORELINE PROCESSES

This part of the Pacific Coast near Big Sur, California., is rocky and ragged.

COPYRIGHT AND PHOTOGRAPH BY DR. PARVINDER S. SETHI

Introduction

"Many centers of commerce and much of Earth's population are concentrated in a narrow band at or near shorelines."

No doubt you know that a **shoreline** is the area of land in contact with the sea or a lake; however, we can expand this definition by noting that ocean shorelines include the land between low tide and the highest level on land affected by storm waves. Our main concern in this chapter is ocean shorelines, or seashores, but waves and nearshore currents are also effective in large lakes. Waves and nearshore currents are certainly more vigorous along seashores, and even the largest lakes have insignificant tides. Lake Superior has a tidal rise and fall of only about 2.5 cm, whereas tidal fluctuations on seashores may be several meters.

You already know that the hydrosphere consists of all water on Earth, most of which is in the oceans. In this enormous body of water, wave energy is transferred through the water to shorelines, where it has a tremendous impact (Figure 16.1). Accordingly, understanding the geologic processes operating on shorelines is important to many people. Indeed, many centers of

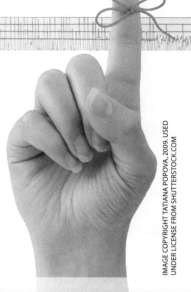

IMAGE COPYRIGHT TATIANA POPOVA, 2009. USED UNDER LICENSE FROM SHUTTERSTOCK.COM

Remember, a *shoreline* differs from a *coast*, because coast is a more inclusive term that includes the shoreline as well as an area of indefinite width both seaward and landward of the shoreline. For instance, in addition to the shoreline area, a coast includes nearshore sandbars and islands, and areas on land such as wind-blown sand dunes, marshes, and cliffs.

LEARNING OUTCOMES

After reading this unit, you should be able to do the following:

LO1 Describe the behavior of tides, waves, and nearshore currents

LO2 Understand the characteristics of shoreline erosion

LO3 Explain the different types of shoreline deposits

LO4 Define the nearshore sediment budget

LO5 Identify the different types of coasts and their charactistics

LO6 Discuss the effects of storm waves and coastal flooding

LO7 Describe how coastal areas are managed as sea levels rises

commerce and much of Earth's population are concentrated in a narrow band at or near shorelines. In addition, coastal communities depend heavily on tourists visiting their seashores.

Geologists, oceanographers, coastal engineers, and marine biologists, among others, are interested in the dynamics of shorelines. So are elected officials and city planners of coastal communities, because they must be familiar with shoreline processes in order to develop policies and zoning regulations that serve the public while protecting the fragile shoreline environment. Understanding shorelines is especially important now because sea level is rising in many areas, so buildings that were far inland are now in peril or have already been destroyed. Furthermore, hurricanes expend much of their energy on shorelines, resulting in extensive coastal flooding, numerous fatalities, and widespread property damage.

shoreline The area between mean low tide and the highest level on land affected by storm waves.

Figure 16.1 Impact of Waves on Shoreline
Wave energy is transferred through the water to shorelines, as it is here along the Pacific Coast, near Jenner, California.

The study of shorelines provides another excellent example of systems interactions—in this case, between part of the hydrosphere and the solid Earth. The atmosphere is also involved in transferring energy from wind to water, thereby causing waves, which in turn generate nearshore currents. And, of course, the gravitational attraction of the Moon and Sun on ocean waters is responsible for the rhythmic rise and fall of tides.

LO1 Tides, Waves, and Nearshore Currents

In contrast to running water, wind, and glaciers, which operate over vast areas, shoreline processes are restricted to a narrow zone at any particular time. However, shorelines might migrate landward or seaward depending on changing sea level or uplift or subsidence of coastal regions.

tide The regular fluctuation of the sea's surface in response to the gravitational attraction of the Moon and Sun.

In the marine realm, several biological, chemical, and physical processes are operating continuously. Organisms change the local chemistry of seawater and contribute their skeletons to nearshore sediments, and temperature and salinity changes and internal waves occur in the oceans. However, the processes most important for modifying shorelines are purely physical ones, especially waves, tides, and nearshore currents.

TIDES

The surface of the oceans rises and falls twice daily in response to the gravitational attraction of the Moon and Sun. These regular fluctuations in the ocean's surface, or **tides**, result in most seashores having two daily high tides and two low tides as sea level rises and falls from a few centimeters to more than 15 m (Figure 16.2). A complete tidal cycle includes a *flood tide* that progressively covers more and more of a nearshore area until high tide is reached, followed by an *ebb tide*, during which the nearshore area is once again exposed (Figure 16.2).

Both the Moon and the Sun have sufficient gravitational attraction to exert tide-generating forces strong

a. Low tide.

Figure 16.2 Low and High Tides
Low tide (a) and high tide (b) in Turnagain Arm, part of Cook Inlet in Alaska. The tidal range here is about 10 m. Turnagain Arm is a huge fiord now being filled with sediment carried in by rivers. Notice the mudflats in (a).

b. High tide.

enough to deform the solid body of Earth, but they have a much greater influence on the oceans. The Sun is 27 million times more massive than the Moon, but it is 390 times as far from Earth, and its tide-generating force is only 46% as strong as that of the Moon. Accordingly, the tides are dominated by the Moon, but the Sun plays an important role as well.

If we consider only the Moon acting on a spherical, water-covered Earth, its tide-generating forces produce two bulges on the ocean surface (Figure 16.3). One bulge points toward the Moon, because it is on the side of Earth where the Moon's gravitational attraction is greatest. The other bulge is on the opposite side of Earth; it points away from the Moon, because of centrifugal force due to Earth's rotation, and the Moon's gravitational attraction is less. These two bulges always point toward and away from the Moon (Figure 16.3a), so as Earth rotates and the Moon's position changes, an observer at a particular shoreline location experiences the rhythmic rise and fall of tides twice daily, but the heights of two successive high tides may vary depending on the Moon's inclination with respect to the equator.

The Moon revolves around Earth every 28 days, so its position with respect to any latitude changes slightly each day. That is, as the Moon moves in its orbit and Earth rotates on its axis, it takes the Moon 50 minutes longer each day to return to the same position it was in

the previous day. Thus, an observer would experience a high tide at 1:00 P.M. on one day, for example, and at 1:50 P.M. on the following day.

Tides are also complicated by the combined effects of the Moon and the Sun. When the two are aligned every two weeks, their forces added together generate *spring tides*, which are about 20% higher than average tides (Figure 16.3b). When the Moon and Sun are at right angles to each another, also at two-week intervals, the Sun's tide-generating force cancels some of the Moon's, and *neap tides*, which are about 20% lower than average, occur (Figure 16.3c).

Tidal ranges are also affected by shoreline configuration. Broad, gently sloping continental shelves as in the Gulf of Mexico have low tidal ranges, whereas steep, irregular shorelines experience much greater rise and fall of tides. Tidal ranges are greatest in some narrow, funnel-shaped bays and inlets. The Bay of Fundy in Nova Scotia has a tidal range of 16.5 m, and ranges greater than 10 m occur in several other areas.

Tides have an important impact on shorelines, because the area of wave attack constantly shifts onshore and offshore as the tides rise and fall. Tidal currents themselves, however, have little modifying effect on shorelines, except in narrow passages where tidal current velocity is great enough to erode and transport sediment.

a. Tidal bulges if only the Moon caused them.

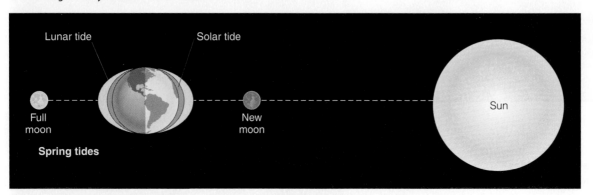

b. When the Moon is new or full, the solar and lunar tides reinforce one another, causing spring tides, the highest high tides and lowest low tides.

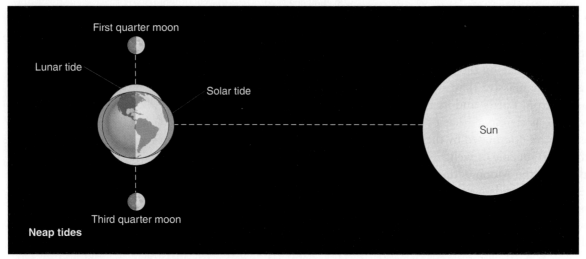

c. During the Moon's first and third quarters, the Moon, Sun, and Earth form right angles, causing neap tides, the lowest high tides and highest low tides.

Figure 16.3 Tidal Bulges
The gravitational attraction of the Moon and Sun causes tides. The sizes of the tidal bulges are greatly exaggerated.

wave An undulation on the surface of a body of water, resulting in the water surface rising and falling.

crest The highest part of a wave.

trough The lowest point between wave crests.

WAVES

You can see **waves**, or oscillations of a water surface, on all bodies of water, but they are best developed in the oceans, where they have their greatest impact on sea-shores. In fact, waves are directly or indirectly responsible for most erosion, sediment transport, and deposition in coastal areas. Wave terminology is illustrated with a typical series of waves in Figure 16.4a. A **crest**, as you would expect, is the highest part of a wave, whereas the low area between crests is a **trough**. The distance from crest to crest (or trough to trough) is the *wavelength*, and the vertical distance from trough

Figure 16.4 Waves and Wave Terminology

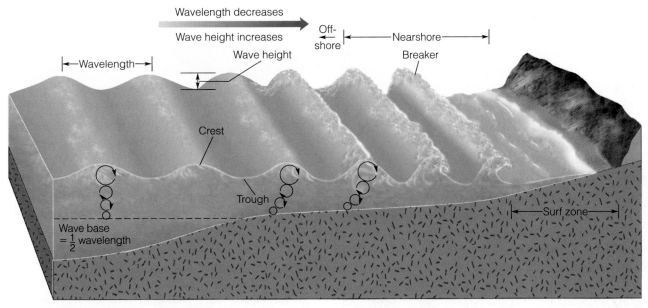

a. Waves and the terminology applied to them.

TOM GARRISON

b. When swells in the deep water move toward the shore, the orbital motion of water within them is disrupted as they enter water shallower than wave base. Wavelength decreases while wave height increases, causing the waves to oversteepen and plunge forward as breakers.

SUE MONROE

c. For scale, notice the surfers waiting for this wave that is beginning to break.

to crest is *wave height*. You can calculate the speed at which a wave advances, called *celerity* (C), by the formula

$$C = L/T$$

where *L* is wavelength and *T* is wave period—that is, the time it takes for two successive wave crests, or troughs, to pass a given point.

The speed of wave advance (C) is actually a measure of the velocity of the wave form rather than the speed of the molecules of water in a wave. When waves move across a water surface, the water moves in circular orbits but shows little or no net forward movement (Figure 16.4a). Only the wave form moves forward, and as it does it transfers energy in the direction of wave movement.

The diameters of the orbits that water follows in waves diminish rapidly with depth, and at a depth of about one-half wavelength (*L/2*), called **wave base**, they are essentially zero. Thus, at a depth exceeding

wave base The depth corresponding to about one-half wavelength, below which water in unaffected by surface waves.

Figure 16.5 Wave Base and Breakers

a. The waves in this lake have a wavelength of about 2 m, so we can infer that wave base is 1 m deep. Where the waves encounter wave base, they stir up the bottom sediment, thus accounting for the nearshore turbid water.

JAMES S. MONROE

IAN MASTERSON/NOAA

b. A plunging breaker on the north shore of Oahu, Hawaii.

wave base, the water and seafloor or lake floor are unaffected by surface waves (Figure 16.4a).

Wave Generation

Most geologic work on shorelines is accomplished by wind-generated waves, especially storm waves. When wind blows over water—that is, one fluid (air) moves over another fluid (water)—friction between the two transfers energy to the water, causing the water surface to oscillate.

In areas where waves are generated, such as beneath a storm center at sea, sharp-crested, irregular waves called *seas* develop. Seas have various heights and lengths, and one wave cannot be easily distinguished from another. But as seas move

fetch The distance the wind blows over a continuous water surface.

out from their area of generation, they are sorted into broad *swells* with rounded, long crests, and all are about the same siz (Figure 16.4a).

The harder and longer the wind blows, the larger are the waves. Wind velocity and duration, however, are not the only factors that control wave size. High-velocity wind blowing over a small pond will never generate large waves regardless of how long it blows. In fact, waves on ponds and most lakes appear only while the wind is blowing. In contrast, the surface of the ocean is always in motion, and waves with heights of 34 m have been recorded during storms in the open sea.

The reason for the disparity between wave sizes on ponds and lakes and on the oceans is the **fetch**, which is the distance the wind blows over a continuous water surface. Fetch is limited by the available water surface, so on ponds and lakes it corresponds to their length or width, depending on wind direction. To produce waves of greater length and height, more energy must be transferred from wind to water; hence large waves form beneath large storms at sea.

Shallow-Water Waves and Breakers

Swells moving out from an area of wave generation lose little energy as they travel long distances across the ocean. In these deepwater swells, the water surface oscillates and water moves in circular orbits, but little net displacement of water takes place in the direction of wave travel (Figure 16.4a). Of course, wind blows some water from wave crests, thus forming white-caps with foamy white crests, and surface currents transport water great distances, but deepwater waves accomplish little actual water movement. When these

Under some circumstances, two wave crests merge to form *rogue waves* that are three or four times higher than the average. These waves can rise unexpectedly out of an otherwise comparatively calm sea and threaten even the largest ships. During the last two decades, more than 200 supertankers and container ships have been lost at sea, many of them apparently hit by these huge waves. As recently as April 16, 2005, the Norwegian cruise ship *Dawn* was damaged by a 21-m-high rogue wave that flooded more than 60 cabins and injured four passengers.

IMAGE COPYRIGHT ELENA ELISSEEVA, 2010. USED UNDER LICENSE FROM SHUTTERSTOCK.COM

waves enter progressively shallower water, however, the wave shape changes, and water is displaced in the direction of wave advance.

Broad, undulating deepwater waves are transformed into sharp-crested waves as they enter shallow water. This transformation begins at a water depth corresponding to wave base—that is, one-half wavelength (Figure 16.4a, 16.5a). At this point, the waves "feel" the seafloor, and the orbital motion of water within the waves is disrupted. As waves continue moving shoreward, the speed of wave advance and wavelength decrease, but wave height increases. Thus, as they enter shallow water, waves become oversteepened as the wave crest advances faster than the wave form, and eventually the crest plunges forward as a **breaker** (Figure 16.4b). Breaking waves might be several times higher than their deepwater counterparts, and when they break, they expend their kinetic energy on the shoreline.

The waves described above are the classic *plunging breakers* (Figure 16.5b) that crash onto shorelines with steep offshore slopes, such as those on the north shore of Oahu in the Hawaiian Islands. In contrast, shorelines where the offshore slope is more gentle usually have *spilling breakers*, where the waves build up slowly and the wave's crest spills down the wave front.

NEARSHORE CURRENTS

The area extending seaward from the upper limit of the shoreline to just beyond the area of breaking waves is conveniently designated as the *nearshore zone*. Within the nearshore zone are the breaker zone and a surf zone, where water from breaking waves rushes forward and then flows seaward as backwash. The nearshore zone's width varies depending on the length of approaching waves, because long waves break at a greater depth, and thus farther offshore, than do short waves. Incoming waves are responsible for two types of currents in the nearshore zone: *longshore currents* and *rip currents*.

Wave Refraction and Longshore Currents

Deepwater waves have long, continuous crests, but rarely are their crests parallel with the shoreline (Figure 16.6). In other words, they seldom approach a shoreline head-on, but rather at some angle. Thus, one part of a wave enters shallow water, where it encounters wave base and begins breaking before other parts of the same wave. As a wave begins to break, its velocity diminishes, but the part of the wave still in deep water races ahead until it too encounters wave base. The net effect of this oblique approach is that waves bend so that they more nearly parallel the shoreline, a phenomenon known as **wave refraction** (Figure 16.6).

breaker A wave that steepens as it enters shallow water until its crest plunges forward.

wave refraction The bending of waves so that they move nearly parallel to the shoreline.

Figure 16.6 Wave Refraction

Wave refraction (wave crests are indicated by dashed lines). These waves are refracted as they enter shallow water and more nearly parallel the shoreline. The waves generate a longshore current that flows in the direction of wave approach, from upper left to lower right (*arrow*) in this example.

Even though waves are refracted, they still usually strike the shoreline at some angle, causing the water between the breaker zone and the beach to flow parallel to the shoreline. These **longshore currents**, as they are called, are long and narrow and flow in the same general direction as the approaching waves. These currents are particularly important in transporting and depositing sediment in the nearshore zone.

longshore current A current resulting from wave refraction found between the breaker zone and a beach that flows parallel to the shoreline.

rip current A narrow surface current that flows out to sea through the breaker zone.

wave-cut platform A beveled surface that slopes gently seaward; formed by the erosion and retreat of a sea cliff.

Rip Currents

Waves carry water into the nearshore zone, so there must be a mechanism for mass transfer of water back out to sea. One way in which water moves seaward from the nearshore zone is in **rip currents**, narrow surface currents that flow out to sea through the breaker zone (Figure 16.7). Surfers commonly take advantage of rip

currents for an easy ride out beyond the breaker zone, but these currents pose a danger to inexperienced swimmers. Some rip currents flow at several kilometers per hour, so if a swimmer is caught in one, it is useless to try to swim directly back to shore. Instead, because rip currents are narrow and usually nearly perpendicular to the shore, one can swim parallel to the shoreline for a short distance and then turn shoreward with little difficulty.

Rip currents are circulating cells fed by longshore currents that increase in velocity from midway between each rip current (Figure 16.7). When waves approach a shoreline, the amount of water builds up until the excess moves out to sea through the breaker zone.

Rip currents commonly develop where wave heights are lower than in adjacent areas, and differences in wave height are controlled by variations in water depth. For instance, if waves move over a depression, the height of the waves over the depression tends to be less than in adjacent areas, forming the ideal environment for rip currents.

LO2 Shoreline Erosion

Beaches, the most familiar coastal landforms, are absent, poorly developed, or restricted to protected areas on seacoasts where erosion rather than deposition predominates. Erosion creates steep or vertical slopes known as *sea cliffs*. During storms, these cliffs are pounded by waves (hydraulic action), worn by the impact of sand and gravel (abrasion) (Figure 16.8), and more or less continuously eroded by dissolution involving the chemical breakdown of rocks by the solvent action of seawater.

WAVE-CUT PLATFORMS

Wave intensity and the resistance of shoreline materials to erosion determine the rate at which a sea cliff

Figure 16.7 Rip Currents

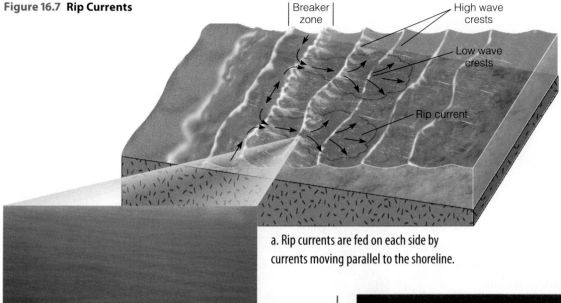

Breaker zone

High wave crests

Low wave crests

Rip current

a. Rip currents are fed on each side by currents moving parallel to the shoreline.

PHOTO BY STEVE ELGAR, PLANE PILOTED BY KIMBALL MILLIKAN

b. Suspended sediment, indicated by discolored water, is being carried seaward in these rip currents.

retreats landward. A sea cliff of glacial drift on Cape Cod, Massachusetts, erodes as much as 30 m per century. By comparison, sea cliffs of dense igneous or metamorphic rocks erode and retreat much more slowly.

Sea cliffs erode mostly as a result of hydraulic action and abrasion at their bases. As a sea cliff is undercut by erosion, the upper part is left unsupported and susceptible to mass wasting processes. Thus, sea cliffs retreat little by little, and as they do, they leave a beveled surface called a **wave-cut platform** that slopes gently seaward (Figure 16.9). Broad wave-cut platforms exist in many areas, but invariably the water over them is shallow, because the abrasive planing action of waves is effective to a depth of only about 10 m. The sediment eroded from sea cliffs is transported seaward until it reaches deeper water at the edge of the wave-cut platform. There it is deposited and forms a *wave-built platform*, which is a seaward extension

JAMES S. MONROE

Figure 16.8 Wave Erosion by Abrasion and Hydraulic Action

These waves pounding the California coast carry sand and gravel and effectively erode by abrasion, that is, the impact of solid particles against shoreline materials. In many areas, sea cliffs undercut by abrasion fail by mass wasting processes (see Figure 11.3).

Figure 16.9 Origin of a Wave-Cut Platform

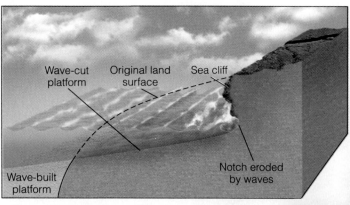

Wave-cut platform
Original land surface
Sea cliff
Wave-built platform
Notch eroded by waves

a. Wave erosion causes a sea cliff to migrate landward, leaving a gently sloping surface, called a wave-cut platform. A wave-built platform originates by deposition at the seaward margin of the wave-cut platform.

JAMES S. MONROE

b. This gently sloping surface along the coast of Oregon is a marine terrace. Notice the sea stacks rising above the terrace.

of the wave-cut platform (Figure 16.9a). Wave-cut platforms now above sea level are known as **marine terraces**.

SEA CAVES, ARCHES, AND STACKS

Sea cliffs do not retreat uniformly, because some shoreline materials are more resistant to erosion than others. **Headlands** are seaward-projecting parts of the shoreline that are eroded on both sides by wave refraction (Figure 16.10). *Sea caves* form on opposite sides of a headland, and if these caves join, they form a *sea arch* (Figure 16.10a). Continued erosion causes the span of an arch to collapse, creating isolated *sea stacks* on wave-cut platforms (Figure 16.10).

In the long run, shoreline processes tend to straighten an initially irregular shoreline. Wave refraction causes more wave energy to be expended on headlands and less on embayments. Thus, headlands erode, and some of the sediment yielded by erosion is deposited in the embayments.

LO3 Deposition Along Shorelines

We mentioned that longshore currents are effective at transporting sediment, and indeed they are. In fact, we can think of the area from the breaker zone to the upper limit of the surf zone as a "river" that flows along the shoreline. Unlike rivers on land, though, this shoreline river's direction of flow changes if waves approach from a different direction. Nevertheless, the analogy is apt, and just like rivers on land, a longshore current's capacity for transport varies with flow velocity and water depth.

Wave refraction and the resulting longshore currents are the primary agents of sediment transport and

marine terrace A wave-cut platform now above sea level.

headland Part of a shoreline commonly bounded by cliffs that extends out into the sea or a lake.

Figure 16.10 Erosion of a Headland

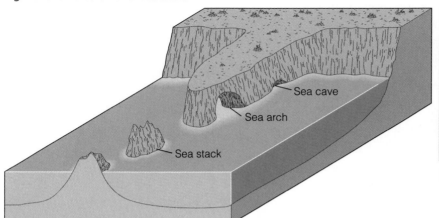

b. Sea stacks at Ruby Beach in Washington State.

a. Erosion of a headland and the origin of a sea cave, a sea arch, and sea stacks.

deposition on shorelines, but tides also play a role, because, as they rise and fall, the position of wave attack shifts onshore and offshore. Rip currents play no role in shoreline deposition, but they do transport fine-grained sediment (silt and clay) offshore through the breaker zone.

BEACHES

Beaches are the most familiar coastal landforms, attracting millions of visitors each year and providing the economic base for many communities. Depending on shoreline materials and wave intensity, beaches may be discontinuous, existing as only *pocket beaches* in protected areas such as embayments, or they may be continuous for long distances.

By definition, a **beach** is a deposit of unconsolidated sediment extending landward from low tide to a change in topography, such as a line of sand dunes, a sea cliff, or the point where permanent vegetation begins. Typically, a beach has several component parts (Figure 16.11), including a **backshore** that is usually dry, being covered by water only during storms or exceptionally high tides. The backshore consists of one or more **berms**, platforms composed of sediment deposited by waves; the berms are nearly horizontal or slope gently landward. The sloping area below a berm exposed to wave swash is the **beach face**. The beach face is part of the **foreshore**, an area covered by water during high tide but exposed during low tide.

Some of the sediment on beaches is derived from weathering and wave erosion of the shoreline, but most of it is transported to the coast by streams and redistributed along the shoreline by longshore currents. As

we noted, waves usually strike beaches at some angle, causing the sand grains to move up the beach face at a similar angle; as the sand grains are carried seaward in the backwash, however, they move perpendicular to the long axis of the beach. Thus, individual sand grains move in a zigzag pattern in the direction of longshore currents. This movement is not restricted to the beach; it extends seaward to the outer edge of the breaker zone.

In an attempt to widen a beach or prevent erosion, shoreline residents often build *groins*, structures that project seaward at right angles from the shoreline. Groins interrupt the flow of longshore currents, causing sand deposition on their upcurrent sides and widening of the beach at that location. However, erosion inevitably occurs on the downcurrent side of a groin.

Quartz is the most common mineral in most beach sands, but there are some notable exceptions. For example, the black sand beaches of Hawaii are composed of

beach Any deposit of sediment extending landward from low tide to a change in topography or where permanent vegetation begins.

backshore That part of a beach that is usually dry, being water-covered only by storm waves or especially high tides.

berm A platform of sediment with a steeply sloping seaward face deposited by waves; some beaches have no berm, whereas others may have several.

beach face The sloping area of a beach that is exposed to wave swash.

foreshore That part of a beach covered by water at high tide but exposed during low tide.

Figure 16.11 Beaches

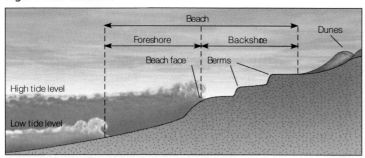

a. Diagram of a beach showing its component parts..

MICHAEL SLEAR

b. The Grand Strand of South Carolina, shown here at Myrtle Beach, is 100 km of nearly continuous beach.

c. A pocket beach near Seal Rock, Oregon.

JAMES S. MONROE

basalt rock fragments or small grains of volcanic glass, and some Florida beaches are composed of the fragmented shells of marine organisms. In short, beaches are composed of whatever material is available.

SEASONAL CHANGES IN BEACHES

The loose grains on beaches are constantly moved by waves, but the overall configuration of a beach remains unchanged as long as equilibrium conditions persist. We can think of the beach profile consisting of a berm or berms and a beach face, as a profile of equilibrium; that is, all parts of the beach are adjusted to the prevailing conditions of wave intensity, nearshore currents, and materials composing the beach (Figure 16.11).

Tides and longshore currents affect the configuration of beaches to some degree, but storm waves are by far the most important agent modifying their equilibrium profile. In many areas, beach profiles change with the seasons, so we recognize *summer beaches* and *winter beaches*, each of which is adjusted to the conditions prevailing at those times. Summer beaches are sand-covered and have a wide berm, a gently sloping beach face, and a smooth offshore profile. Winter beaches, in contrast, tend to be coarser grained and steeper; they have a small berm or none at all, and their offshore profiles reveal sandbars paralleling the shoreline (Figure 16.11a).

Seasonal changes in beach profiles are related to changing wave intensity (Figure 16.12). During the winter, energetic storm waves erode the sand from

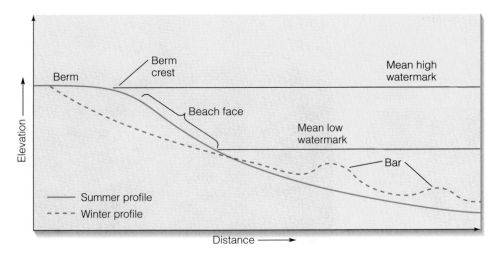

Figure 16.12 Seasonal Changes in Beach Profiles

beaches and transport it offshore, where it is stored in sandbars. The same sand that was eroded from a beach during the winter returns the next summer when it is driven onshore by more gentle swells. The volume of sand in the system remains more or less constant; it simply moves farther offshore or onshore depending on wave energy.

The terms *winter beach* and *summer beach*, though widely used, are somewhat misleading. A winter beach profile can develop at any time if there is a large storm, and likewise a summer beach profile can develop during a prolonged winter calm period.

Spits, Baymouth Bars, and Tombolos

Beaches are the most familiar depositional features of coasts, but spits, baymouth bars, and tombolos are common, too. In fact, these features are simply continuations of a beach. A **spit**, for instance, is a fingerlike projection of a beach into a body of water such as a bay, and a **baymouth bar** is a spit that has grown until it completely closes off a bay from the open sea (Figure 16.13). Both are composed of sand, more rarely gravel, that was transported and deposited by longshore currents where they weakened as they entered the deeper water of a bay's opening. Some spits are modified by waves so that their free ends are curved; they go by the name *hook* or *recurved spit* (Figure 16.13a).

A rarer type of spit is a **tombolo** that extends out from the shoreline to an island (Figure 16.13d). A tombolo forms on the shoreward side of an island as wave refraction around the island creates converging

currents that turn seaward and deposit a sandbar. So, in contrast to spits and baymouth bars, which are usually nearly parallel with the shoreline, the long axes of tombolos are nearly at right angles to the shoreline.

Spits and baymouth bars constitute a continuing problem where bays must be kept open for pleasure boating, commercial shipping, or both. Obviously, a bay closed off by a sandbar is of little use for either endeavor, so a bay must be regularly dredged or protected from deposition by longshore currents.

Barrier Islands

Long, narrow islands of sand lying a short distance offshore from the mainland are **barrier islands** (Figure 16.14). Everyone agrees that barrier islands form on gently sloping continental shelves where abundant sand is available and where both wave energy and the tidal range are low. For these reasons, many barrier islands are located along the United States' Atlantic and Gulf Coasts. But even though it is well known where barrier islands form, the details of their origin are still unresolved. According to one model, they formed as spits that became detached from

spit A fingerlike projection of a beach into a body of water such as a bay.

baymouth bar A spit that has grown until it closes off a bay from the open sea or lake.

tombolo A type of spit that extends out from the shoreline and connects the mainland with an island.

barrier island A long, narrow island of sand parallel to a shoreline but separated from the mainland by a lagoon.

Figure 16.13 Spits, Baymouth Bars, and Tombolos

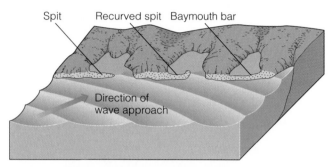

a. Spits form where longshore currents deposit sand in deeper water, as at the entrance to a bay. A baymouth bar is simply a spit that has grown until it extends across the mouth of a bay.

b. A spit at the mouth of the Russian River near Jenner, California.

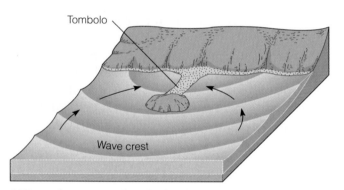

c. Rodeo Beach north of San Francisco, California, is a baymouth bar.

d. Wave refraction around an island and the origin of a tombolo.

land, whereas another model holds that they formed as beach ridges that subsequently subsided.

Most barrier islands are migrating landward as a result of erosion on their seaward sides and deposition on their landward sides. This is a natural part of barrier island evolution, and it takes place rather slowly. However, it takes place fast enough to cause many problems for island residents and communities.

nearshore sediment budget The balance between additions and losses of sediment in the nearshore zone.

LO4 The Nearshore Sediment Budget

We can think of the gains and losses of sediment in the nearshore zone in terms of a **nearshore sediment budget** (Figure 16.15). If a nearshore system has a balanced budget, sediment is supplied as fast as it is removed, and the volume of sediment remains more or less constant, although sand may shift offshore and onshore with the changing seasons (Figure 16.12). A positive budget means that gains exceed losses,

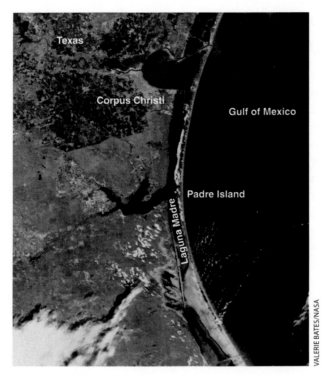

Figure 16.14 Barrier Islands
View from space of the barrier islands along the Gulf Coast of Texas. Notice that a lagoon up to 20 km wide separates the long, narrow barrier islands from the mainland.

Texas
Corpus Christi
Gulf of Mexico
Padre Island
Laguna Madre

VALERIE BATES/NASA

INPUT V^+ Sediment added from erosion by longshore transport onto beach
OUTPUTS V^- Sediment carried down-coast from the beach by longshore transport
          W^- Sediment blown inland by wind
          O^- Sediment cascading down the submarine slope
STABLE BEACH: $(V^+) + (V^- + W^- + O^-) = 0$

River
Inland dunes
V^+
W^-
Beach
V^-
O^-
Submarine landslide

Figure 16.15 The Nearshore Sediment Budget
The long-term sediment budget can be assessed by considering inputs versus outputs. If inputs and outputs are equal, the system is in a steady state, or equilibrium. If outputs exceed inputs, however, the beach has a negative budget and erosion occurs. Accretion takes place when the beach has a positive budget with inputs exceeding outputs.

whereas a negative budget means that losses exceed gains. If a negative budget prevails long enough, a nearshore system is depleted, and beaches may disappear.

Although there are a few exceptions, most sediment on typical beaches is transported to the shoreline by streams and then redistributed along the shoreline by longshore currents. Thus, longshore currents also play a role in the nearshore sediment budget, because they continuously move sediment into and away from beach systems.

The primary ways that a nearshore system loses sediment are offshore transport, wind, and deposition in submarine canyons. Offshore transport mostly involves fine-grained sediment carried seaward, where it eventually settles in deeper water. Wind is an important process, because it removes sand from beaches and blows it inland, where it piles up as sand dunes.

If the heads of submarine canyons are nearshore, huge quantities of sand are funneled into them and deposited in deeper water. La Jolla and Scripps submarine canyons off the coast of southern California funnel off an estimated 2 million m^3 of sand each year. In most areas, however, submarine canyons are too far offshore to interrupt the flow of sand in the nearshore zone.

It should be apparent from this discussion that if a nearshore system is in equilibrium, its incoming supply of sediment exactly offsets its losses. Such a delicate balance tends to continue unless the system is somehow disrupted. One change that affects this balance is the construction of dams across the streams that supply sand. Once dams have been built, all sediment from the upper reaches of the drainage systems is trapped in reservoirs and thus cannot reach the shoreline.

LO5 Types of Coasts

Coasts are difficult to classify because of variations in the factors that control their development and variations in their composition and configuration. Rather than attempt to categorize all coasts, we shall simply note that two types of coasts have already been discussed: those dominated by deposition and those dominated by erosion.

DEPOSITIONAL AND EROSIONAL COASTS

Depositional coasts, such as the U.S. Gulf Coast, are characterized by an abundance of detrital sediment and such depositional landforms as wide sandy beaches, deltas, and barrier islands. In contrast, erosional coasts are steep and irregular and typically lack well-developed beaches, except in protected areas. They are further characterized by sea cliffs, wave-cut platforms, and sea stacks. Many of the coasts along the West Coast of North America fall into this category.

The following section examines coasts in terms of their changing relationship to sea level. But note that although some coasts, such as those in southern California, are described as emergent (uplifted), these same coasts may be erosional as well. In other words, coasts commonly possess features that allow them to be classified in more than one way.

SUBMERGENT AND EMERGENT COASTS

If sea level rises with respect to the land or the land subsides, coastal regions are flooded and said to be **submergent coasts** or *drowned coasts* (Figure 16.16). Much of the East Coast of North America from Maine southward through South Carolina was flooded during the rise in sea level following the Pleistocene Epoch, so it is extremely irregular. Recall that during the expansion of glaciers during the Pleistocene, sea level was as much as 130 m lower than at present, and that streams eroded their valleys more deeply and extended across continental shelves. When sea level rose, the lower ends of these valleys were drowned, forming *estuaries* such as Delaware and Chesapeake bays (Figure 16.16). Estuaries are simply the seaward ends of river valleys where seawater and freshwater mix.

Emergent coasts are found where the land has risen with respect to sea level (Figure 16.17). Emergence takes place when water is withdrawn from the oceans, as occurred during the Pleistocene expansion of glaciers. Presently, coasts are emerging as a result of isostasy or tectonism. In northeastern Canada and the Scandinavian countries, for instance, the coasts are irregular, because isostatic rebound is elevating formerly glaciated terrain from beneath the sea.

Figure 16.16 Submergent Coasts

Submergent coasts tend to be extremely irregular, with estuaries such as the Chesapeake and Delaware bays. They formed when the East Coast of the United States was flooded as sea level rose following the Pleistocene Epoch.

Coasts that form in response to tectonism, on the other hand, tend to be straighter, because the seafloor topography being exposed by uplift is smooth. The west coasts of North and South America are rising as a consequence of plate tectonics. Distinctive features of these coasts are marine terraces (Figures 16.9 and 16.17), which are wave-cut platforms now elevated above sea level. Uplift in these areas appears to be episodic rather than continuous, as indicated by the multiple levels of terraces in some places.

LO6 Storm Waves and Coastal Flooding

What causes most fatalities during hurricanes? Many people think it is strong wind, but actually, coastal flooding caused by storm waves and heavy rainfall are more dangerous. Coastal flooding during hurricanes results when large waves are driven onshore, and by as much as 60 cm of rain in as little as 24 hours. In addition, as a hurricane moves over the ocean, low atmospheric pressure causes the ocean surface to bulge upward as much as 0.5 m. When the eye of the storm reaches shoreline, the bulge, and wind-driven waves,

Figure 16.17 An Emergent Coast in California

a. Emergent coasts tend to be steep and straighter than submergent coasts. Notice the several sea stacks and the sea arch. Also, a marine terrace is visible in the distance.

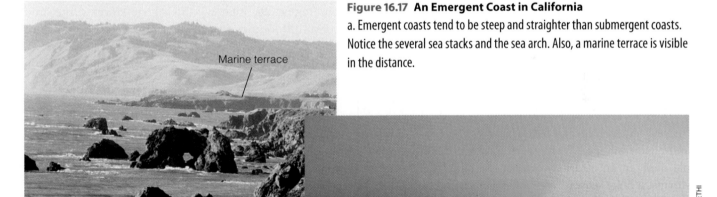

b. An emergent beach near Bandon Beach, Oregon. Notice the number of sea stacks.

Figure 16.18 Hurricane Katrina, 2005

NOAA

a. On August 29, 2005, Hurricane Katrina made landfall along the Gulf Coast. In this image, the eye of the storm is passing just to the east of New Orleans, Louisiana.

SUE MONROE

b. This is a repaired section of floodwall in New Orleans. This barrier is along one of the canals leading into the city from Lake Pontchartrain. Failure of the floodwalls during and following Hurricane Katrina caused flooding of about 80% of New Orleans.

storm surge The surge of water onto a shoreline as a result of a bulge in the ocean's surface beneath the eye of a hurricane and wind-driven waves.

piles up in a **storm surge** that may rise several meters above normal high tide and inundate areas far inland.

In 1900, hurricane-driven waves surged inland, eventually covering the entire island that Galveston, Texas, was built on. Buildings and other structures near the seashore were battered to pieces and "great beams and railway ties were lifted by the [waves] and driven like battering rams into dwellings and business houses."* Between 6,000 and 8,000 people died. In an effort to protect the city from future storms, a huge seawall was constructed, and the entire city was elevated to the level of the top of the seawall.

Although Galveston, Texas, has been largely protected from more recent storm surges, the same is not true for several other areas. In 1989, Charleston, South Carolina, and nearby areas were flooded by a storm surge generated by Hurricane Hugo, which caused 21 deaths and more than $7 billion in property damages.

*L. W. Bates, Jr., "Galveston—A City Built upon Sand," *Scientific American*, 95 (1906), p. 64.

The examples just cited were certainly disasters of great magnitude, but the U.S. Gulf Coast was hardest hit in 2005, first in August by Hurricane Katrina (Figure 16.18a), and then again in September by Hurricane Rita. When Hurricane Katrina roared ashore on August 29, high winds, a huge storm surge, and coastal flooding destroyed nearly everything in an area of 230,000 km². Areas in Gulfport and Biloxi, Mississippi, were leveled, but most of the public's attention has focused on the damage in New Orleans, Louisiana.

Even when New Orleans was founded in 1718, engineers warned that building a community on swampy, subsiding land between Lake Pontchartrain to the north and the Mississippi River to the south was risky; now most of the city lies below sea level. In any case, New Orleans was nearly surrounded by levees (earthen embankments) and floodwalls (structures of concrete and steel) to protect it from the lake and river.

When Hurricane Katrina came ashore, the levees initially held, but on the next day, some of the floodwalls were breached, and about 80% of the city was flooded. Because New Orleans is mostly below sea level, the floodwaters could not drain out naturally. In fact, the city has 22 pumping stations to remove water that builds up from normal rainstorms, but as the city flooded, the pumps were overwhelmed, and when the electricity failed, they were useless. All in all, Hurricane Katrina was the most expensive natural disaster in U.S. history; the property damages exceeded $100 billion, and more than 1,800 people died. Then, in September, Hurricane Rita hit the Gulf Coast, mostly in Texas, and caused about 120 more fatalities and nearly $12 billion in damages.

Scientists, engineers, and some politicians had warned of a Katrina-type disaster in New Orleans for many years. They were aware that during a large hurricane the levees or floodwalls would likely fail (Figure 16.18b). In fact, the political leaders in Louisiana had pleaded for years for funds to strengthen the levees, but for one reason or another, the funds were never allocated. "What do we do now?" is a question that will be debated for many years. Another factor to consider is that coastal flooding is exacerbated by rising sea level that results from global warming.

LO7 How Are Coastal Areas Managed as Sea Level Rises?

During the last century, sea level rose about 12 cm worldwide, and all indications are that it will continue to rise. The absolute rate of sea-level rise in a shoreline region depends on two factors. The first is the volume of water in the ocean basins, which is increasing as a result of glacial ice melting and the thermal expansion of near-surface seawater.

The second factor is the rate of uplift or subsidence of a coastal area. In some areas, uplift is occurring fast enough that sea level is actually falling with respect to the land. In other areas, sea level is rising while the coastal region is simultaneously subsiding, resulting in a net change in sea level of as much as 30 cm per century. Perhaps such a "slow" rate of sea-level change seems insignificant; after all, it amounts to only a few millimeters per year. But in gently sloping coastal areas, as in the eastern United States from New Jersey southward, even a slight rise in sea level will eventually have widespread effects.

Many of the nearly 300 barrier islands along the East and Gulf Coasts of the United States are migrating landward as sea level rises (Figure 16.19). Landward migration of barrier islands would pose few

Figure 16.19 Barrier Island Migration

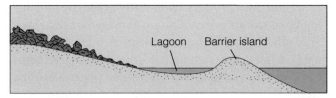

a. A barrier island.

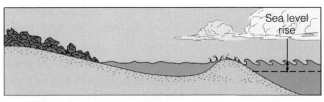

b. A barrier island migrates landward as sea level rises and storm waves carry sand from its seaward side into its lagoon.

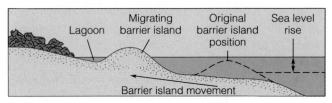

c. Over time, the entire island shifts toward the land.

problems if it were not for the numerous communities, resorts, and vacation homes located on them. Moreover, barrier islands are not the only threatened areas. For example, Louisiana's coastal wetlands—an important wildlife habitat and seafood-producing area—are currently being lost at a rate of about 90 km^2 per year. Much of this loss results from sediment compaction, but rising sea level exacerbates the problem.

Rising sea level also directly threatens many beaches upon which communities depend for revenue. The beach at Miami Beach, Florida, for instance, was disappearing at an alarming rate until the Army Corps of Engineers began replacing the eroded beach sand. The problem is even more serious in other countries. A rise in sea level of only 2 m would inundate large areas of the U.S. East and Gulf Coasts, but it would cover 20% of the entire country of Bangladesh. Other problems associated with rising sea level include increased coastal flooding during storms and saltwater incursions that may threaten groundwater supplies (see Chapter 13).

Armoring shorelines with *seawalls* (embankments of reinforced concrete or stone) and using *riprap* (piles of stones, Figure 16.20) help protect beachfront structures, but both projects are initially expensive and during large storms are commonly damaged or destroyed. Seawalls do afford some protection and are seen in many coastal areas along the oceans and large lakes, but some states, including North and South Carolina, Rhode Island, Oregon, and Maine, no longer allow their construction.

Because we can do nothing to prevent sea level from rising, engineers, scientists, planners, and political leaders must examine what they can do to prevent or minimize the effects of shoreline erosion. At present, only a few viable options exist. One is to put strict controls on coastal development. North Carolina, for example, permits large structures to be built no closer to the shoreline than 60 times the annual erosion rate. Although a growing awareness of shoreline processes has resulted in similar legislation elsewhere, some states have virtually no restrictions on coastal development.

Regulating coastal development is commendable, but it has no impact on existing structures and coastal communities. A general retreat from the shoreline may be possible, but expensive, for individual dwellings and small communities, but it is impractical for large population centers. Such communities as Atlantic City, New Jersey; Miami Beach, Florida; and Galveston, Texas, have adopted one of two strategies to combat coastal erosion. One is to build protective barriers such as seawalls. Seawalls, such as the one at Galveston, Texas, are effective, but they are tremendously expensive to construct and maintain. More than $50

Figure 16.20 Riprap and Sand Berm
Riprap made up of large pieces of basalt was piled on this beach to protect a luxury hotel just to the left of this image.

million was spent in just five years to replenish the beach and build a seawall at Ocean City, Maryland. Furthermore, barriers retard erosion only in the area directly behind them; Galveston Island west of its seawall has been eroded back about 45 m.

Another option, adopted by both Atlantic City, New Jersey, and Miami Beach, Florida, is to pump sand onto the beaches to replace that lost to erosion. This, too, is expensive as the sand must be replenished periodically, because erosion is a continuous process. In many areas, groins are constructed to preserve beaches, but unless additional sand is artificially supplied to the beaches, longshore currents invariably erode sand from the downcurrent sides of the groins.

LISTEN UP!

SHE DID

GEOL was designed for students just like you—busy people who want choices, flexibility, and multiple learning options.

GEOL delivers concise, focused information in a fresh and contemporary format. And, **GEOL** gives you a variety of online learning materials designed with you in mind.

At **4ltrpress.cengage.com,** you'll find electronic resources such as **videos,** an **eBook,** and **interactive quizzes** for each chapter.

These resources will help supplement your understanding of core concepts in a format that fits your busy lifestyle. Visit **4ltrpress.cengage.com** to learn more about the multiple resources available to help you succeed!

CHAPTER 17
GEOLOGIC TIME: CONCEPTS AND PRINCIPLES

The Grand Canyon, Arizona. Major John Wesley Powell led two expeditions down the Colorado River and through the canyon in 1869 and 1871. He was struck by the seemingly limitless time represented by the rocks exposed in the canyon walls and by the recognition that these rock layers, like the pages in a book, contain the geologic history of this region.

COPYRIGHT AND PHOTOGRAPH BY DR. PARVINDER S. SETHI

Introduction

"Vast periods of time set geology apart from most of the other sciences."

In 1869, Major John Wesley Powell, a Civil War veteran who lost his right arm in the battle of Shiloh, led a group of hardy explorers down the uncharted Colorado River through the Grand Canyon. With no maps or other information, Powell and his group ran the many rapids of the Colorado River in fragile wooden boats, hastily recording what they saw. Powell wrote in his diary that "all about me are interesting geologic records. The book is open and I read as I run."

Probably no one has contributed as much to the understanding of the Grand Canyon as Major Powell. In recognition of his contributions, the Powell Memorial was erected on the South Rim of the Grand Canyon in 1969 to commemorate the 100th anniversary of this history-making first expedition.

Most tourists today, like Powell and his fellow explorers in 1869, are astonished by the seemingly limitless time represented by the rocks exposed in the walls of the Grand Canyon. For most visitors, viewing a 1.5-km-deep cut into Earth's crust is the only encounter they'll ever have with the magnitude of geologic time. When standing on the rim and looking down into the Grand Canyon, we are really looking far back in time, more than 1 billion years—all the way back to the early history of our planet.

Vast periods of time set geology apart from most of the other sciences, and an appreciation of the immensity of geologic time is fundamental to understanding the physical and biological history of our planet.

LO1 How Is Geologic Time Measured?

Geologists use two different frames of reference when discussing geologic time. **Relative dating** is placing geologic events in a sequential order as determined from their position in the geologic record. Relative dating will not tell us how long ago a particular event took place, only that one event preceded another.

> **relative dating** The process of determining the age of an event as compared to other events; involves placing geologic events in their correct chronological order, but does not involve the consideration of when the events occurred in number of years ago.

LEARNING OUTCOMES

After reading this unit, you should be able to do the following:

LO1 Explain how geologic time is measured

LO2 Review early concepts of geologic time and the age of Earth

LO3 Understand James Hutton's contribution to the recognition of geologic time

LO4 Describe relative dating methods

LO5 Explain the process of correlating rock units

LO6 Explain absolute dating methods

LO7 Describe the development of the geologic time scale

LO8 Describe the methods used to determine geologic time and climate change

© ISTOCKPHOTO.COM/PIXHOOK

The principles used to determine relative dating were discovered hundreds of years ago, and since then, they have been used to construct the *relative geologic time scale* (Figure 17.1). These principles are still widely used by geologists today.

Absolute dating provides specific dates for rock units or events expressed in years before the present. Radiometric dating is the most common method of obtaining absolute ages. Dates are calculated from the natural decay rates of various radioactive elements present in trace amounts in some rocks. It was not until the discovery of radioactivity near the end of the 19th century that absolute ages could be accurately applied to the relative geologic time scale.

Today, the geologic time scale is really a dual scale: a relative scale based on rock sequences with radiometric dates expressed as years before the present (Figure 17.1).

LO2 Early Concepts of Geologic Time and the Age of Earth

The concept of geologic time and its measurement have changed throughout human history. Some early Christian scholars and clerics tried to establish the date of creation by analyzing historical records and the genealogies found in Scripture. Based on their analyses, they generally believed that Earth and all its features were no more than about 6,000 years old. The idea of a very young Earth provided the basis for most Western chronologies of Earth history prior to the 18th century.

During the 18th and 19th centuries, several attempts were made to determine Earth's age on the basis of scientific evidence rather than revelation. The French zoologist Georges Louis de Buffon (1707–1788)

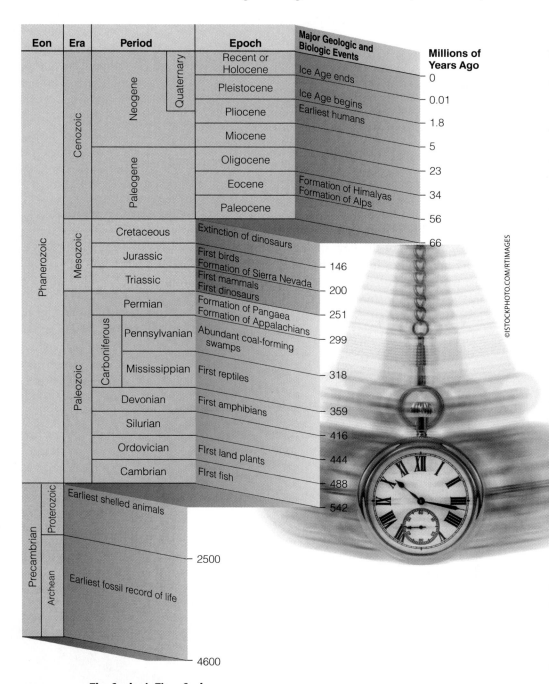

Figure 17.1 The Geologic Time Scale
Some of the major geologic and biologic events are indicated for the various eras, periods, and epochs.

DATES ARE FROM F. GRADSTEIN, J. OGG, AND A. SMITH, *A GEOLOGIC TIME SCALE* 2004 (CAMBRIDGE, UK: CAMBRIDGE UNIVERSITY PRESS, 2005), FIGURE 1.2.

absolute dating Uses various radioactive decay dating techniques to assign ages in years before the present to rocks.

© ISTOCKPHOTO.COM/GREG EPPERSON

assumed that Earth gradually cooled to its present condition from a molten beginning. To simulate this history, he melted iron balls of various diameters and allowed them to cool to the surrounding temperature. By extrapolating their cooling rate to a ball the size of Earth, he determined that Earth was at least 75,000 years old. Although this age was much older than that derived from Scripture, it was vastly younger than we now know our planet to be.

Other scholars were equally ingenious in attempting to calculate Earth's age. For example, if deposition rates could be determined for various sediments, geologists reasoned that they could calculate how long it would take to deposit any rock layer. They could then extrapolate how old Earth was from the total thickness of sedimentary rock in its crust. Rates of deposition vary, however, even for the same type of rock. Furthermore, it is impossible to estimate how much of a rock has been removed by erosion, or how much a rock sequence has been reduced by compaction. As a result of these variables, estimates of Earth's age ranged from younger than 1 million years to older than 2 billion years.

Besides trying to determine Earth's age, the naturalists of the 18th and 19th centuries were formulating some of the fundamental geologic principles that are used in deciphering Earth history. From the evidence preserved in the geologic record, it was clear to them that Earth is very old and that geologic processes have operated over long periods of time.

LO3 James Hutton and the Recognition of Geologic Time

Many consider the Scottish geologist James Hutton (1726–1797) to be the father of modern geology. His detailed studies and observations of rock exposures and present-day geologic processes were instrumental in establishing the principle of uniformitarianism (see Chapter 1), the concept that the same processes seen today have operated over vast amounts of time. Because Hutton relied on known processes to account for Earth history, he concluded that Earth must be very old and wrote that "we find no vestige of a beginning, and no prospect of an end."

In 1830, Charles Lyell published a landmark book, *Principles of Geology*, in which he championed Hutton's concept of uniformitarianism. Instead of relying on catastrophic events to explain various Earth features, Lyell recognized that imperceptible changes brought about by present-day processes could, over long periods of time, have tremendous cumulative effects. Through his writings, Lyell firmly established uniformitarianism as the guiding principle of geology.

LO4 Relative Dating Methods

Recall that relative dating places events in sequential order, but does not tell us how long ago an event took place.

FUNDAMENTAL PRINCIPLES OF RELATIVE DATING

The 17th century was an important time in the development of geology as a science because of the widely circulated writings of the Danish anatomist Nicolas Steno (1638–1686). Steno observed that when streams flood, they spread out across their floodplains and deposit layers of sediment that bury organisms dwelling on the floodplain. Subsequent floods produce new layers of sediments that are deposited or superposed over previous deposits. When lithified, these layers of sediment become sedimentary rock.

Thus, in an undisturbed succession of sedimentary rock layers, the oldest layer is at the bottom and the youngest layer is at the top. This **principle of superposition** is the basis for relative-age determinations of strata and their contained fossils (Figure 17.2a).

Steno also observed that, because sedimentary particles settle from water under the influence of gravity, sediment is deposited in essentially horizontal layers, thus illustrating the **principle of original horizontality** (Figure 17.2a). Therefore, a sequence of sedimentary rock layers that is steeply inclined from the horizontal must have been tilted after deposition and lithification (Figure 17.2b).

Steno's third principle, the **principle of lateral continuity**, states that sediment extends laterally in all directions until it thins and pinches out or terminates against the edge of the depositional basin (Figure 17.2a).

James Hutton is credited with formulating the **principle of crosscutting relationships**. Based on his detailed studies and observations of rock exposures in Scotland, Hutton recognized that an igneous intrusion or fault must be younger than the rocks it intrudes or displaces (Figure 17.3).

Another way to determine relative ages is by using the **principle of inclusions**. This principle holds that inclusions, or fragments of one rock contained within a layer

principle of superposition A principle holding that in a vertical sequence of undeformed sedimentary rocks, the relative ages of the rocks can be determined by their position in the sequence—oldest at the bottom followed by successively younger layers.

principle of original horizontality According to this principle, sediments are deposited in horizontal or nearly horizontal layers.

principle of lateral continuity A principle holding that rock layers extend outward in all directions until they terminate.

principle of cross-cutting relationships A principle holding that an igneous intrusion or fault must be younger than the rocks it intrudes or cuts across.

principle of inclusions A principle holding that inclusions or fragments in a rock unit are older than the rock unit itself; for example, granite inclusions in sandstone are older than the sandstone.

©ISTOCKPHOTO.COM/GREG EPPERSON

Figure 17.2 The Principles of Original Horizontality, Superposition, and Lateral Continuity

COPYRIGHT AND PHOTOGRAPH BY DR. PARVINDER S. SETHI

a. The sedimentary rocks of Bryce Canyon National Park, Utah, were originally deposited horizontally in a variety of continental environments (principle of original horizontality). The oldest rocks are at the bottom of this highly dissected landscape, and the youngest rocks are at the top, forming the rims (principle of superposition). The exposed rock layers extend laterally in all directions for some distance (principle of lateral continuity).

REED WICANDER

b. These shales and limestones of the Postolonnec Formation, at Postolonnec Beach, Crozon Peninsula, France, were originally deposited horizontally, but have been significantly tilted since their formation.

REED WICANDER

Figure 17.3 The Principle of Cross-Cutting Relationships
A small fault (arrows show direction of movement) cuts across, and thus displaces, tilted sedimentary beds along the Templin Highway in Castaic, California. The fault is therefore younger than the youngest beds that are displaced.

of another, are older than the rock layer itself. The batholith shown in Figure 17.4a contains sandstone inclusions, and the sandstone unit shows the effects of baking. Accordingly, we conclude that the sandstone is older than the batholith. In Figure 17.4b, however, the sandstone contains granite rock pieces, indicating that the batholith was the source rock for the inclusions and is therefore older than the sandstone.

Fossils have been known for centuries, yet their utility in relative dating and geologic mapping was not fully appreciated until the early 19th century. William Smith (1769–1839), an English civil engineer involved in surveying and building canals in southern England, independently recognized the principle of superposition by

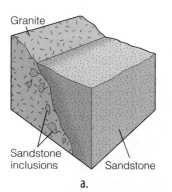
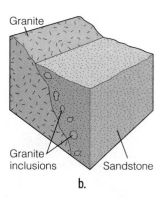

a.   b.

Figure 17.4 The Principle of Inclusions
a. The batholith is younger than the sandstone because the sandstone has been baked at its contact with the granite and the granite contains sandstone inclusions. b. Granite inclusions in the sandstone indicate that the batholith was the source of the sandstone and therefore is older. c. Outcrop in northern Wisconsin showing basalt inclusions (dark gray) in granite (white). Accordingly, the basalt inclusions are older than the granite.

c.

reasoning that the fossils at the bottom of a sequence of strata are older than those at the top of the sequence.

This recognition served as the basis for the **principle of fossil succession** or the *principle of faunal and floral succession*, as it is sometimes called (Figure 17.5).

UNCONFORMITIES

Our discussion so far has been concerned with conformable strata—sequences in which no depositional breaks of any consequence occur. A bedding plane between strata may represent a depositional break of anywhere from minutes to tens of years, but it is inconsequential in the context of geologic time. However, in some sequences of strata, surfaces known as **unconformities** may be present,

principle of fossil succession A principle holding that fossils, and especially groups or assemblages of fossils, succeed one another through time in a regular and predictable order.

unconformity A break in the geologic record represented by an erosional surface separating younger strata from older rocks.

disconformity An unconformity above and below which the rock layers are parallel.

angular unconformity An unconformity below which older rocks dip at a different angle (usually steeper) than overlying strata.

representing times of nondeposition, erosion, or both. Unconformities encompass long periods of geologic time, perhaps millions or tens of millions of years. Accordingly, the geologic record is incomplete wherever an unconformity is present, just as a book with missing pages is incomplete, and the interval of geologic time not represented by strata is called a *hiatus* (Figure 17.6).

Geologists recognize three types of unconformities. A **disconformity** is a surface of erosion or nondeposition separating younger from older rocks, both of which are parallel with one another (Figure 17.7). Unless the erosional surface separating the older from the younger parallel beds is well defined or distinct, the disconformity frequently resembles an ordinary bedding plane. Hence, many disconformities are difficult to recognize and must be identified on the basis of fossil assemblages.

An **angular unconformity** is an erosional surface on tilted or folded strata over which younger strata were deposited (Figure 17.8). The strata below the unconformable surface generally dip more steeply than those above, producing an angular relationship.

The most famous angular unconformity in the world is probably the one at Siccar Point, Scotland, where James Hutton realized that severe upheavals had tilted the lower rocks and formed mountains that were then worn away and covered by younger, flat-lying rocks. The

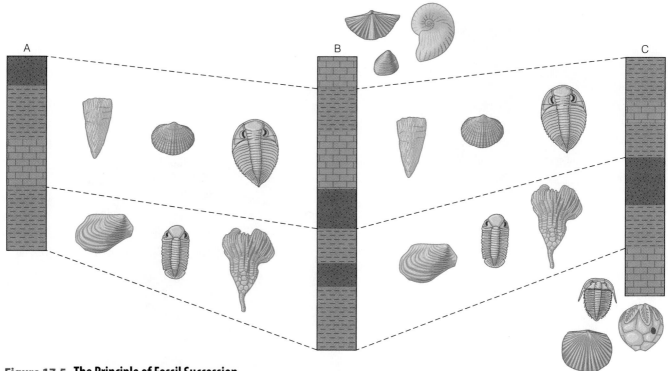

Figure 17.5 The Principle of Fossil Succession

This generalized diagram shows how geologists use the principle of fossil succession to identify strata of the same age in different areas. The rocks in the three sections encompassed by the dashed lines contain similar fossils and are therefore the same age. Note that the youngest rocks in this region are in section B, whereas the oldest rocks are in section C.

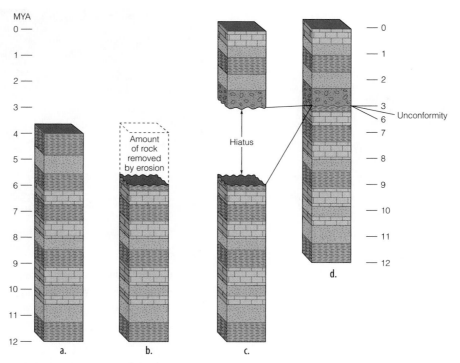

Figure 17.6 The Development of a Hiatus and an Unconformity

(a) Deposition began 12 million years ago (mya) and continued more or less uninterrupted until 4 mya. (b) Between 3 and 4 mya, an episode of erosion occurred. During that time, some of the strata deposited earlier was eroded. (c) A hiatus of 3 million years thus exists between the older strata and the strata that formed during a renewed episode of deposition that began 3 mya. (d) The actual stratigraphic record as seen in an outcrop today. The unconformity is the surface separating the strata and represents a major break in our record of geologic time.

Figure 17.7 Formation of a Disconformity

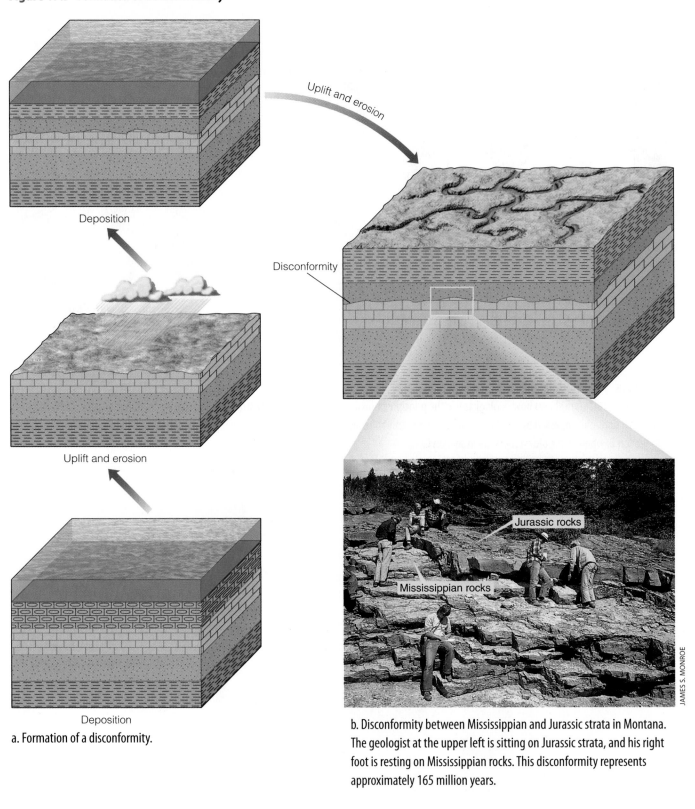

Uplift and erosion

Deposition

Uplift and erosion

Deposition

Disconformity

a. Formation of a disconformity.

JAMES S. MONROE

Jurassic rocks

Mississippian rocks

b. Disconformity between Mississippian and Jurassic strata in Montana. The geologist at the upper left is sitting on Jurassic strata, and his right foot is resting on Mississippian rocks. This disconformity represents approximately 165 million years.

Figure 17.8 Formation of an Angular Unconformity

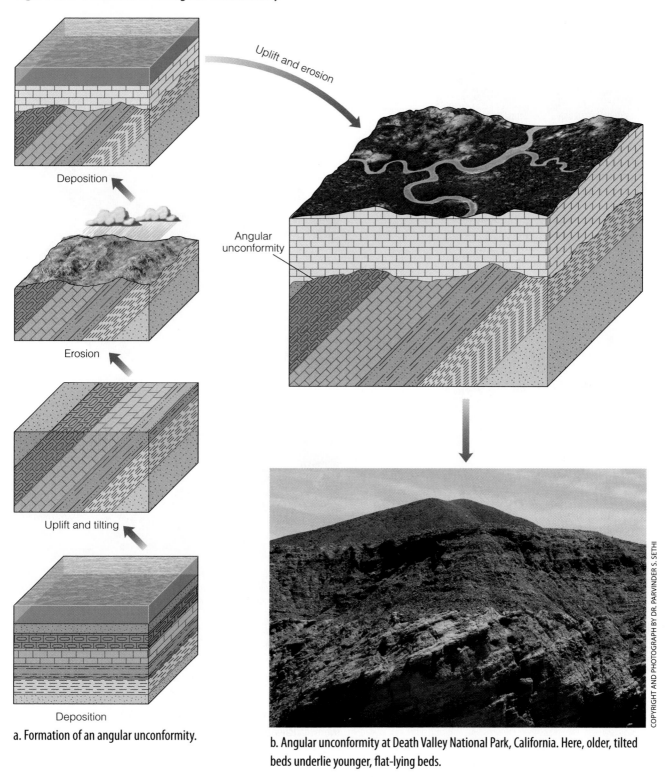

Deposition

Uplift and erosion

Erosion

Angular
unconformity

Uplift and tilting

Deposition

a. Formation of an angular unconformity.

b. Angular unconformity at Death Valley National Park, California. Here, older, tilted beds underlie younger, flat-lying beds.

COPYRIGHT AND PHOTOGRAPH BY DR. PARVINDER S. SETHI

Figure 17.9 Formation of a Nonconformity

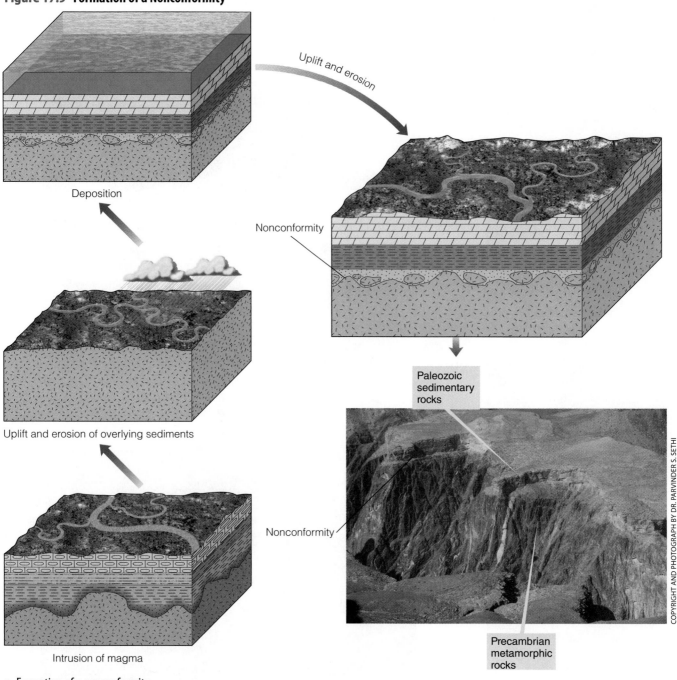

Deposition

Uplift and erosion

Nonconformity

Uplift and erosion of overlying sediments

Intrusion of magma

Paleozoic
sedimentary
rocks

Nonconformity

Precambrian
metamorphic
rocks

COPYRIGHT AND PHOTOGRAPH BY DR. PARVINDER S. SETHI

a. Formation of a nonconformity.

b. Nonconformity between Precambrian metamorphic rocks and overlying Paleozoic sedimentary rocks in the Grand Canyon, Arizona.

nonconformity An unconformity in which stratified sedimentary rocks overlie an erosion surface cut into igneous or metamorphic rocks.

erosional surface between the older tilted rocks and the younger flat-lying strata meant that a significant gap existed in the geologic record. Although Hutton did not use the term *unconformity*, he

was the first to understand and explain the significance of such discontinuities in the geologic record.

A **nonconformity** is the third type of unconformity. Here, an erosional surface cut into metamorphic or igneous rocks is covered by sedimentary rocks (Figure 17.9). This type of unconformity closely resembles an intrusive igneous contact with sedimentary rocks. The principle of inclusions (Figure 17.4) is helpful in determining whether the relationship between the underlying igneous rocks and the overlying sedimentary rocks is the result of an intrusion or erosion. In the case of an intrusion, the igneous rocks are younger, whereas in the case of erosion, the sedimentary rocks are younger. Being able to distinguish between a nonconformity and an intrusive contact is important because they represent different sequences of events.

LO5 Correlating Rock Units

To decipher Earth history, geologists must demonstrate the time equivalency of rock units in different areas. This process is known as **correlation**.

If surface exposures are adequate, units may simply be traced laterally (principle of lateral continuity), even if occasional gaps exist (Figure 17.10). Other criteria used to correlate units are similarity of rock type, position in a sequence, and key beds. Key beds are units, such as coal beds or volcanic ash layers, that are sufficiently distinctive to allow identification of the same unit in different areas (Figure 17.10).

Generally, no single location in a region has a geologic record of all events that occurred during its history; therefore, geologists must correlate from one area to another to determine the complete geologic history of the region. An excellent example is the history of the Colorado Plateau (Figure 17.11). This region provides a record of events occurring over approximately 2 billion years. Because of the forces of erosion, the entire record is not preserved at any single location. Within the walls of the Grand Canyon are rocks of the Precambrian and Paleozoic eras, whereas Paleozoic and Mesozoic Era rocks are found in Zion National Park, and Mesozoic and Cenozoic Era rocks are exposed in Bryce Canyon (Figure 17.11). By correlating the uppermost rocks at one location with the lowermost equivalent rocks of another area, geologists can decipher the history of the entire region.

Although geologists match up rocks on the basis of similar rock type and superposition, correlation of this type can be done only in a limited area where beds can be traced from one site to another. To correlate rock units over a large area or to correlate age-equivalent units of different composition, fossils and the principle of fossil succession must be used.

Fossils are useful as relative time indicators because they are the remains of organisms that lived for a certain length of time during the geologic past. Fossils that are easily identified, are geographically widespread, and existed for a rather short interval of geologic time are particularly useful. Such

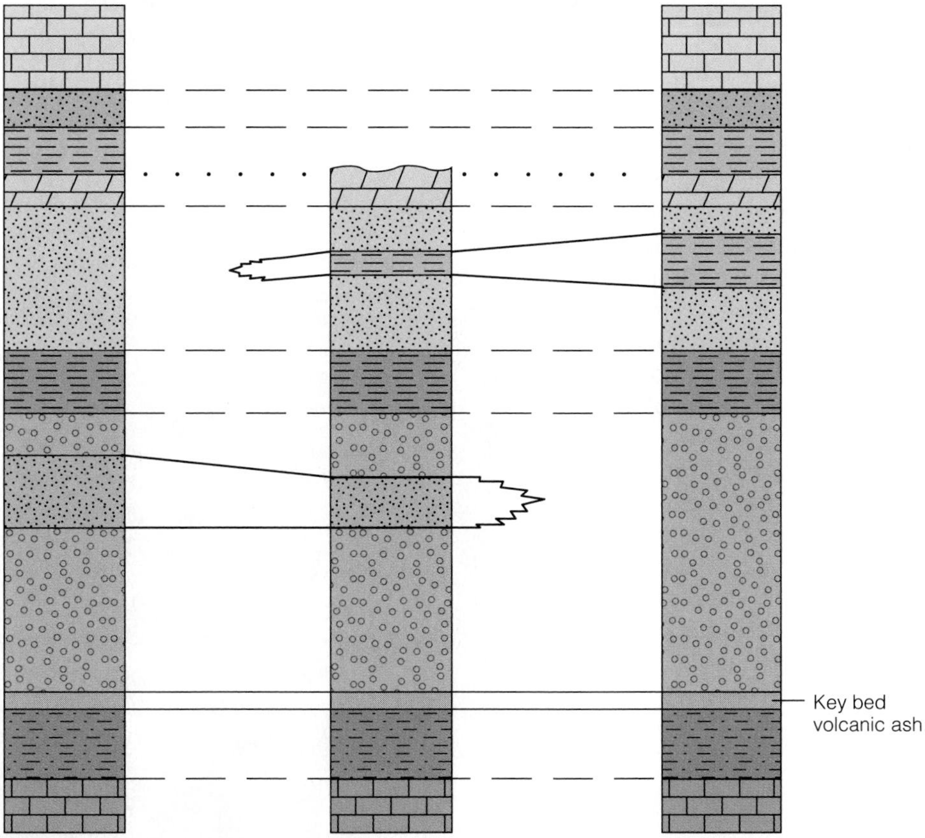

Figure 17.10 Correlating Rock Units
In areas of adequate exposure, rock units can be traced laterally, even if occasional gaps exist, and correlated on the basis of similarity in rock type and position in a sequence. Rocks can also be correlated by a key bed—in this case, volcanic ash.

Key bed
volcanic ash

correlation Demonstration of the physical continuity of rock units or biostratigraphic units, or demonstration of time equivalence as in time-stratigraphic correlation.

guide fossil Any easily identified fossil with an extensive geographic distribution and short geologic range useful for determining the relative ages of rocks in different areas.

VIRTUAL FIELD TRIP

Geologic Time

Ready to Go!

During this virtual field trip, you will be visiting Grand Canyon National Park, Arizona, one of the seven wonders of the world. Established as a national park in 1919, the Grand Canyon is visited by more than five million people every year! It is here that most people are first exposed to the idea of "deep time," or geologic time—that is, the concept of seemingly limitless amounts of time. When standing on the South Rim and looking down into the 1.5 km-deep canyon cut by the Colorado River, one can't help but be impressed with the magnitude of time represented by the rocks exposed in the canyon walls. It is here in the Grand Canyon that we will see examples of the various relative dating principles used to place geologic events in their correct sequential order, as well as examples of the three types of unconformities, all of which are discussed in this chapter. In addition to the Grand Canyon, we will also be visiting Capitol Reef National Park, Utah, where you will be able to see a variety of sedimentary features and examples of the various relative dating principles you have already read about.

It is now time to see actual examples of the concepts and geologic features discussed in your textbook. And, there is no better place to start this journey then the Grand Canyon! See you there.

GOALS OF THE TRIP

In this field trip you will come face to face with such important concepts as:

1. The various principles of relative dating.
2. Unconformities.
3. Correlation of rock units.

FOLLOW-UP QUESTIONS

1. Using the principles of relative dating and the types of unconformities you have learned, reconstruct the geologic history of the rock units in the Grand Canyon as seen and numbered in the "Reconstructing Geologic Events" slide.
2. According to Figure 17.11, the rocks forming the rim of the Grand Canyon, Arizona include the Moenkopi Formation. The oldest rock unit in Capitol Reef National Park is the Moenkopi Formation. Discuss how you can use the various principles of relative dating to reconstruct a more complete geologic history of the Colorado Plateau from these two locations.
3. Why is the Grand Canyon such an ideal location to understand what is meant by "deep time" or geologic time?

What to See When You Go

Long before radioactivity was discovered and geologists could assign actual dates to rock units and geologic events, earth scientists used a variety of relative dating methods (articulated as principles) to place rock sequences and geologic events in sequential order. In this spectacular sunset photo of the Grand Canyon, taken from Hopi Point, one can easily see the three interrelated principles of relative dating – the principle of original horizontality, superposition, and lateral continuity. Here, the rocks of the Grand Canyon extend in all directions (*principle of lateral continuity*), were deposited horizontally (*principle of original horizontality*), with the oldest rocks at the bottom of the canyon, and the youngest rocks forming the canyon rim (*principle of superposition*).

Recall from your reading in this chapter that *unconformities* are breaks in the geologic record represented by an erosional surface or surface of nondeposition separating younger from older rocks. There are three types of unconformities, all of which can be seen in the Grand Canyon. In the photo above of the Great Unconformity, sedimentary rocks of Cambrian age overlie metamorphic rocks of Precambrian age. This surface represents more than a billion years of geologic time for which we don't have a record at this location. It is an example of a *nonconformity*.

An excellent example of an *angular unconformity* can be observed at Lipan Point, a few miles from Zuni Point. Here, older rocks can be seen dipping at a different angle than the younger overlying rocks.

This photo shows the rocks of the Grand Canyon and illustrates a *disconformity*, that is, an unconformity between parallel rock layers, in this case the Muav Limestone (Cambrian age, Layer 3) and the overlying Redwall Limestone (Mississippian age, Layer 4). Note that the bedding plane between the Muav and Redwall Limestone looks just like the other bedding planes. We know this is a disconformity because the rocks have been dated by fossils (principle of fossil succession) and indicate that rocks representing the Ordovician through Devonian periods are missing here.

In order to decipher Earth history, geologists had to demonstrate time equivalency of rock units in differe areas, a process known as *correlation*. This view of Capitol Reef National Park was photographed from the Scenic Drive. You can clearly see the red-colored sedimentary rocks at this location and you should note that the reddish-brown layers near the bottom of the sequence are slightly tilted. This is due to the rocks being uplifted by tectonic forces. When surface exposures, such as here, are adequate, rocks can easily be traced laterally on the basis of similarity of rock types and position in a sequence.

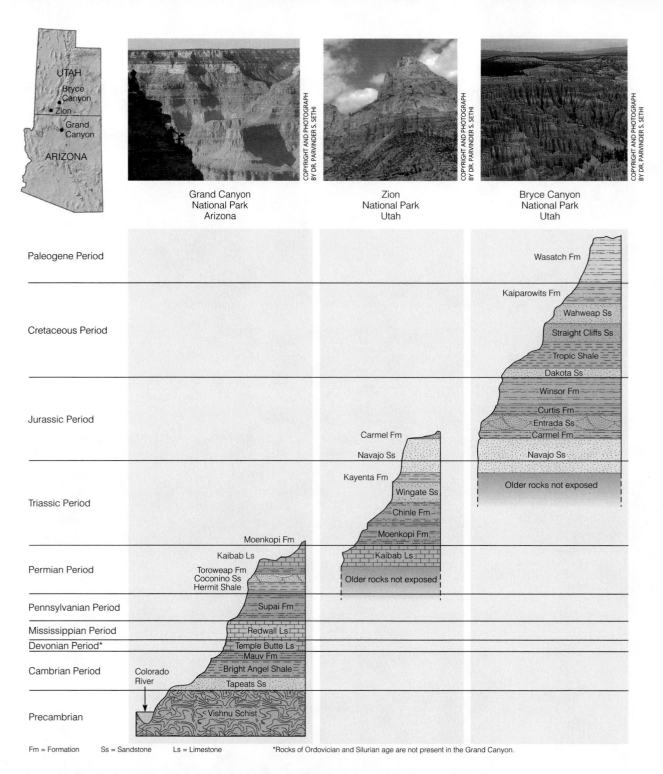

Grand Canyon National Park — Arizona
Zion National Park — Utah
Bryce Canyon National Park — Utah

Paleogene Period

Cretaceous Period

Jurassic Period

Triassic Period

Permian Period

Pennsylvanian Period

Mississippian Period

Devonian Period*

Cambrian Period

Precambrian

Wasatch Fm
Kaiparowits Fm
Wahweap Ss
Straight Cliffs Ss
Tropic Shale
Dakota Ss
Winsor Fm
Curtis Fm
Entrada Ss
Carmel Fm
Navajo Ss
Older rocks not exposed

Carmel Fm
Navajo Ss
Kayenta Fm
Wingate Ss
Chinle Fm
Moenkopi Fm
Kaibab Ls
Older rocks not exposed

Moenkopi Fm
Kaibab Ls
Toroweap Fm
Coconino Ss
Hermit Shale
Supai Fm
Redwall Ls
Temple Butte Ls
Mauv Fm
Bright Angel Shale
Tapeats Ss
Colorado River
Vishnu Schist

Fm = Formation   Ss = Sandstone   Ls = Limestone   *Rocks of Ordovician and Silurian age are not present in the Grand Canyon.

Figure 17.11 Correlation of Rock Units Within the Colorado Plateau

At each location, only a portion of the geologic record of the Colorado Plateau is exposed. By correlating the youngest rocks at one exposure with the oldest rocks at another exposure, geologists can determine the entire history of the region. For example, the rocks forming the rim of the Grand Canyon, Arizona, are the Kaibab Limestone and Moenkopi Formation—the youngest rocks exposed in the Grand Canyon. The Kaibab Limestone and Moenkopi Formation are the oldest rocks exposed in Zion National Park, Utah, and the youngest rocks are the Navajo Sandstone and Carmel Formation. The Navajo Sandstone and Carmel Formation are the oldest rocks exposed in Bryce Canyon National Park, Utah. By correlating the Kaibab Limestone and Moenkopi Formation between the Grand Canyon and Zion National Park, geologists have extended the geologic history from the Precambrian to the Jurassic. And by correlating the Navajo Sandstone and Carmel Formation between Zion and Bryce Canyon National Parks, geologists can extend the geologic history through the Paleogene Period. Thus, by correlating the rock exposures between these areas and applying the principle of superposition, geologists can reconstruct the geologic history of the region.

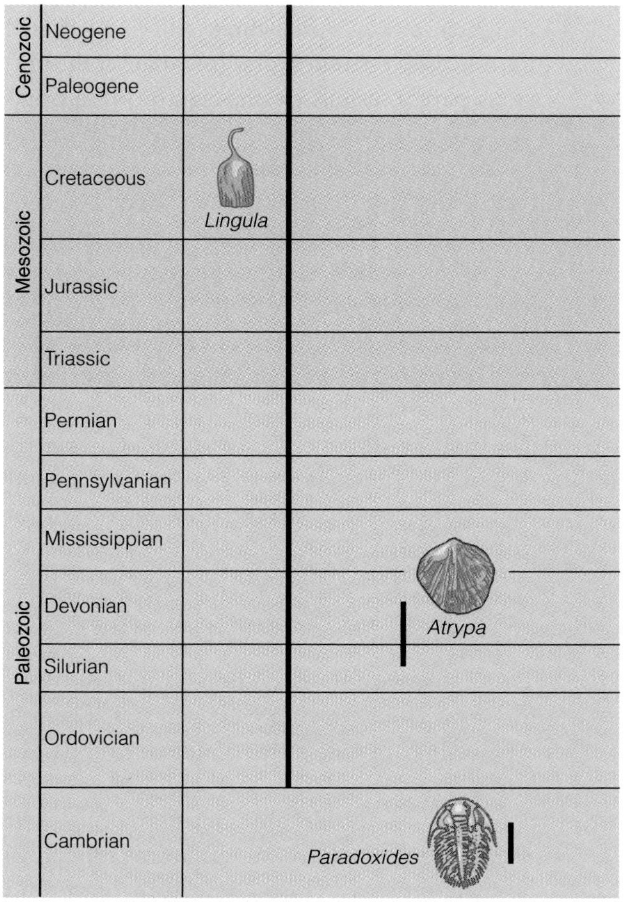

	Neogene		
Cenozoic	Paleogene		
Mesozoic	Cretaceous	*Lingula*	
	Jurassic		
	Triassic		
Paleozoic	Permian		
	Pennsylvanian		
	Mississippian		
	Devonian	*Atrypa*	
	Silurian		
	Ordovician		
	Cambrian	*Paradoxides*	

Figure 17.12 Guide Fossils
Comparison of the geologic ranges (heavy vertical lines) of three marine invertebrate animals. *Lingula* is of little use in correlation because it has such a long range. But *Atrypa* and *Paradoxides* are good guide fossils because both are widespread, easily identified, and have short geologic ranges. Thus, both can be used to correlate rock units that are widely separated and to establish the relative age of a rock that contains them.

fossils are **guide fossils** or *index fossils* (Figure 17.12). The trilobite *Paradoxides* and the brachiopod *Atrypa* meet these criteria and are therefore good guide fossils. In contrast, the brachiopod *Lingula* is easily identified and widespread, but its long geologic range of Ordovician to Recent makes it of little use in correlation.

LO6 Absolute Dating Methods

Although most of the isotopes of the 92 naturally occurring elements are stable, some are radioactive and spontaneously decay to other more stable isotopes of elements, releasing energy in the process. The discovery in 1903 by Pierre and Marie Curie that radioactive decay produces heat meant that geologists finally had a mechanism for explaining Earth's internal heat that did not rely on residual cooling from a molten origin. Furthermore, geologists now had a powerful tool to date geologic events accurately and to verify the long time periods postulated by Hutton and Lyell.

RADIOACTIVE DECAY AND HALF-LIVES

Radioactive decay is the process whereby an unstable atomic nucleus is spontaneously transformed into an atomic nucleus of a different element. Three types of radioactive decay are recognized, all of which result in a change of atomic structure (Figure 17.13).

In *alpha decay*, 2 protons and 2 neutrons are emitted from the nucleus, resulting in the loss of 2 atomic numbers and 4 atomic mass numbers. In *beta decay*, a fast-moving electron is emitted from a neutron in the nucleus, changing that neutron to a proton and consequently increasing the atomic number by 1, with no resultant atomic mass number change. *Electron capture* is when a proton captures an electron from an electron shell and thereby converts to a neutron, resulting in the loss of 1 atomic number, but not changing the atomic mass number.

Some elements undergo only one decay step in the conversion from an unstable form to a stable form. For example, rubidium 87 decays to strontium 87 by a single beta emission, and potassium 40 decays to argon 40 by a single electron capture. Other radioactive elements undergo several decay steps. Uranium 235 decays to lead 207 by 7 alpha and 6 beta steps, whereas uranium 238 decays to lead 206 by 8 alpha and 6 beta steps (Figure 17.14).

When discussing decay rates, it is convenient to refer to them in terms of half-lives. The **half-life** of a radioactive element is the time it takes for half of the atoms of the original unstable *parent element* to decay to atoms of a new, more stable *daughter element*. The half-life of a given radioactive

radioactive decay The spontaneous change of an atom to an atom of a different element by emission of a particle from its nucleus (alpha and beta decay) or by electron capture.

half-life The time necessary for half of the original number of radioactive atoms of an element to decay to a stable daughter product; for example, the half-life for potassium 40 is 1.3 billion years.

Figure 17.13 **Three Types of Radioactive Decay**

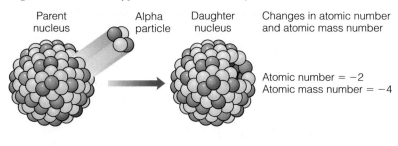

a. Alpha decay, in which an unstable parent nucleus emits 2 protons and 2 neutrons.

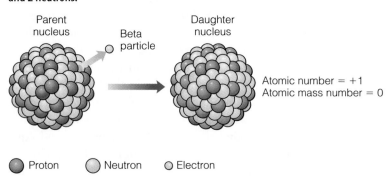

b. Beta decay, in which an electron is emitted from the nucleus.

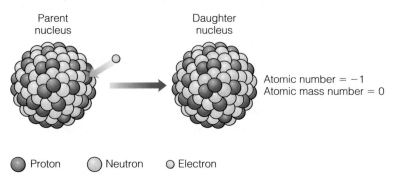

c. Electron capture, in which a proton captures an electron and is thereby converted to a neutron.

element is constant and can be precisely measured. Half-lives of various radioactive elements range from less than a billionth of a second to 49 billion years.

Radioactive decay occurs at a geometric rate rather than a linear rate (Figure 17.15). Therefore, a graph of the decay rate produces a curve rather than a straight line (Figure 17.15b). For example, an element with 1,000,000 parent atoms will have 500,000 parent atoms and 500,000 daughter atoms after one half-life. After two half-lives, it will have 250,000 parent atoms (one-half of the previous parent atoms, which is equivalent to one-fourth of the original parent atoms) and 750,000

daughter atoms. After three half-lives, it will have 125,000 parent atoms (one-half of the previous parent atoms, or one-eighth of the original parent atoms) and 875,000 daughter atoms, and so on, until the number of parent atoms remaining is so few that they cannot be accurately measured by present-day instruments.

By measuring the parent–daughter ratio and knowing the half-life of the parent (which has been determined in the laboratory), geologists can calculate the age of a sample that contains the radioactive element. The parent–daughter ratio is usually determined by a *mass spectrometer*, an instrument that measures the proportions of atoms of different masses.

SOURCES OF UNCERTAINTY

The most accurate radiometric dates are obtained from igneous rocks. As magma cools and begins to crystallize, radioactive parent atoms are separated from previously formed daughter atoms. Because they are the right size, some radioactive parent atoms are incorporated into the crystal structure of certain minerals. The stable daughter atoms, though, are a different size from the radioactive parent atoms and consequently cannot fit into the crystal structure of the same mineral as the parent atoms. Therefore, a mineral crystallizing in cooling magma will contain radioactive parent atoms but no stable daughter atoms (Figure 17.16). Thus, the time that is being measured is the time of crystallization of the mineral that contains the radioactive atoms and *not* the time of formation of the radioactive atoms.

To obtain accurate radiometric dates, geologists must be sure that they are dealing with a *closed system*, meaning that neither parent nor daughter atoms have been added or removed from the system since crystallization, and that the ratio between them results only from radioactive decay. Otherwise, an inaccurate date will result. If daughter atoms have leaked out of the mineral being analyzed, the calculated age will be too young; if parent atoms have been removed, the calculated age will be too great.

Leakage may take place if the rock is heated or subjected to intense pressure as can sometimes occur during metamorphism. If this happens, some of the parent or daughter atoms may be driven from the

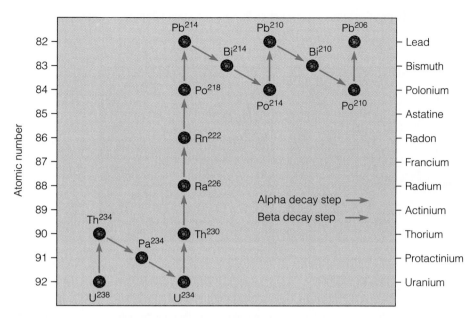

Figure 17.14 **Radioactive Decay Series for Uranium 238 to Lead 206**   Radioactive uranium 238 decays to its stable daughter product, lead 206, by 8 alpha and 6 beta decay steps. A number of different isotopes are produced as intermediate steps in the decay series.

SOURCE: BASED ON DATA FROM S. M. RICHARDSON AND H. Y. MCSWEEN, JR., *GEOCHEMISTRY—PATHWAYS AND PROCESSES*, PRENTICE-HALL.

Figure 17.15 **Uniform, Linear Change Compared to Geometric Radioactive Decay**

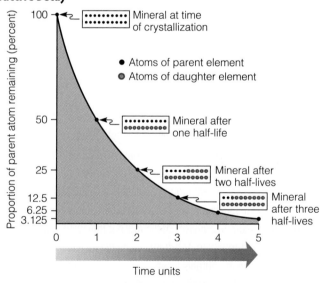

a. Uniform, linear change is characteristic of many familiar processes. In this example, water is being added to a glass at a constant rate.

b. A geometric radioactive decay curve, in which each time unit represents one half-life, and each half-life is the time it takes for half of the parent element to decay to the daughter element.

mineral being analyzed, resulting in an inaccurate age determination. If the daughter product was completely removed, then one would be measuring the time since metamorphism (a useful measurement itself) and not the time since crystallization of the mineral (Figure 17.17).

Because heat and pressure affect the parent–daughter ratio, metamorphic rocks are difficult to date accurately. Remember that although the resulting parent–daughter ratio of the sample being analyzed may have been affected by heat, the decay rate of the parent element remains constant, regardless of any physical or chemical changes.

Figure 17.16 Crystallization of Magma Containing Radioactive Parent and Stable Daughter Atoms

Magma

Mineral crystallizing from magma

Igneous rock

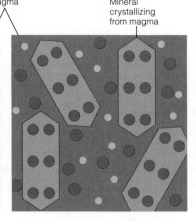

● Radioactive parent atoms
● Stable daughter atoms

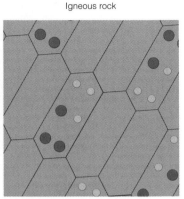

a. Magma contains both radioactive parent atoms and stable daughter atoms. The radioactive parent atoms are larger than the stable daughter atoms.

b. As magma cools and begins to crystallize, some of the radioactive atoms are incorporated into certain minerals because they are the right size and can fit into the crystal structure. In this example, only the larger radioactive parent atoms fit into the crystal structure. Therefore, at the time of crystallization, minerals in which the radioactive parent atoms can fit into the crystal structure will contain 100% radioactive parent atoms and 0% stable daughter atoms.

c. After one half-life, 50% of the radioactive parent atoms will have decayed to stable daughter atoms, such that those minerals that had radioactive parent atoms in their crystal structure will now have 50% radioactive parent atoms and 50% stable daughter atoms.

Figure 17.17 Effects of Metamorphism on Radiometric Dating
The effect of metamorphism in driving out daughter atoms from a mineral that crystallized 700 million years ago (mya). The mineral is shown immediately after crystallization (a), then at 400 million years (b), when some of the parent atoms had decayed to daughter atoms. Metamorphism at 350 mya (c), drives the daughter atoms out of the mineral into the surrounding rock. If the rock has remained a closed chemical system throughout its history, dating the mineral today (d), yields the time of metamorphism, whereas dating the whole rock provides the time of its crystallization, 700 mya.

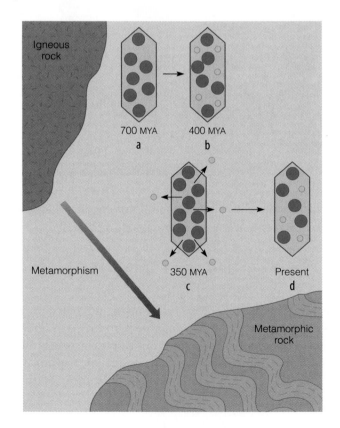

LONG-LIVED RADIOACTIVE ISOTOPE PAIRS

Table 17.1 shows the five common, long-lived parent–daughter isotope pairs used in radiometric dating. Long-lived pairs have half-lives of millions or billions of years. All of these were present when Earth formed and are still present in measurable quantities. Other shorter-lived radioactive isotope pairs have decayed to the point that only small quantities near the limit of detection remain.

The most commonly used isotope pairs are the uranium–lead and thorium–lead series, which are used principally to date ancient igneous intrusives, lunar samples, and some meteorites. The rubidium–strontium pair is also used for very old samples and has been effective in dating the oldest rocks on Earth, as well as meteorites.

The potassium–argon method is typically used for dating fine-grained volcanic rocks from which individual crystals cannot be separated; hence, the whole rock is analyzed. Because argon is a gas, great care must be taken to ensure that the sample has not been subjected to heat, which would allow argon to escape; such a sample would yield an age that is too young. Other long-lived radioactive isotope pairs exist, but they are rather rare and are used only in special situations.

CARBON-14 DATING METHOD

Carbon is an important element in nature and is one of the basic elements found in all forms of life. It has three isotopes; two of these, carbon 12 and 13, are stable, whereas carbon 14 is radioactive (see Figure 3.3). Carbon 14 has a half-life of 5,730 years plus or minus 30 years. The **carbon-14 dating technique** is based on the ratio of carbon 14 to carbon 12 and is used to date formerly living material.

The short half-life of carbon 14 makes this dating technique practical only for specimens younger than about 70,000 years. Consequently, the carbon-14 dating method is especially useful in archaeology and has greatly helped unravel the events of the latter portion of the Pleistocene Epoch.

Carbon 14 is constantly formed in the upper atmosphere when cosmic rays, which are high-energy particles (mostly protons), strike the atoms of upper-atmospheric gases, splitting their nuclei into protons and neutrons. When a neutron strikes the nucleus of a nitrogen atom (atomic number 7, atomic mass number 14), it may be absorbed into the nucleus and a proton emitted. Thus, the atomic number of the atom decreases by 1, whereas the atomic mass number stays the same. Because the atomic number has changed, a new element, carbon 14 (atomic number 6, atomic mass number 14), is formed. The newly formed carbon 14 is rapidly assimilated into the

carbon-14 dating technique Absolute dating technique relying on the ratio of C^{14} to C^{12} in an organic substance; useful back to about 70,000 years ago.

TABLE 17.1

FIVE OF THE PRINCIPAL LONG-LIVED RADIOACTIVE ISOTOPE PAIRS USED IN RADIOMETRIC DATING

ISOTOPES		Half-Life of Parent (years)	Effective Dating Range (years)	Minerals and Rocks That Can Be Dated	
Parent	Daughter				
Uranium 238	Lead 206	4.5 billion	10 million to 4.6 billion	Zircon Uraninite	
Uranium 235	Lead 207	704 million			
Thorium 232	Lead 208	14 billion			
Rubidium 87	Strontium 87	48.8 billion	10 million to 4.6 billion	Muscovite Biotite Potassium feldspar Whole metamorphic or igneous rock	
Potassium 40	Argon 40	1.3 billion	100,000 to 4.6 billion	Glauconite Muscovite Biotite	Hornblende Whole volcanic rock

carbon cycle and, along with carbon 12 and 13, is absorbed in a nearly constant ratio by all living organisms (Figure 17.18). When an organism dies, however, carbon 14 is not replenished, and the ratio of carbon 14 to carbon 12 decreases as carbon 14 decays back to nitrogen by a single beta decay step (Figure 17.18).

Currently, the ratio of carbon 14 to carbon 12 is remarkably constant both in the atmosphere and in living organisms. There is good evidence, however, that the production of carbon 14, and thus the ratio of carbon 14 to carbon 12, has varied somewhat during the past several thousand years. This was determined by comparing ages established by carbon-14 dating of wood samples with ages established by counting annual tree rings in the same samples. As a result, carbon-14 ages have been corrected to reflect such variations in the past.

LO7 Development of the Geologic Time Scale

The geologic time scale is a hierarchical scale in which the 4.6-billion-year history of Earth is divided into time units of varying duration (Figure 17.1). It did not result from the work of any one individual, but rather evolved, primarily during the 19th century, through the efforts of many people.

By applying relative-dating methods to rock outcrops, geologists in England and western Europe defined the major geologic time units without the benefit of radiometric dating techniques. Using the principles of superposition and fossil succession, they correlated various rock exposures and pieced together a composite geologic section. This composite section is, in effect, a relative time scale because the rocks are arranged in their correct sequential order.

By the beginning of the 20th century, geologists had developed a relative geologic time scale, but did not yet have any absolute dates for the various time–unit boundaries. Following the discovery of radioactivity near the end of the 19th century, radiometric dates were added to the relative geologic time scale (Figure 17.1).

Because sedimentary rocks, with rare exceptions, cannot be radiometrically dated, geologists have had to rely on interbedded volcanic rocks and igneous intrusions to apply absolute dates to the boundaries of the various subdivisions of the geologic time scale (Figure 17.19).

An ashfall or lava flow provides an excellent marker bed that is a time-equivalent surface, supplying

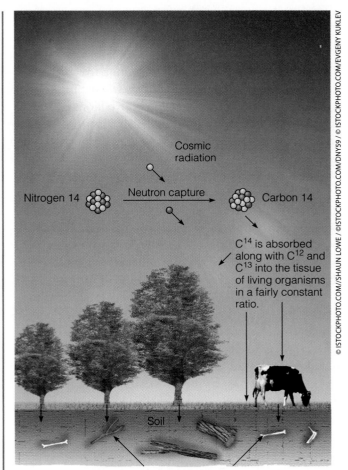

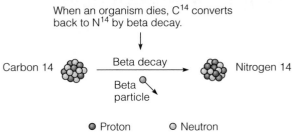

Figure 17.18 Carbon-14 Dating Method
The carbon cycle showing the formation of carbon 14 in the upper atmosphere, its dispersal and incorporation into the tissues of all living organisms, and its decay back to nitrogen 14 by beta decay.

a minimum age for the sedimentary rocks below and a maximum age for the rocks above. Ashfalls are particularly useful because they may fall over both marine and nonmarine sedimentary environments and can provide a connection between these different environments.

Thousands of absolute ages are now known for sedimentary rocks of known relative ages, and these absolute dates have been added to the relative time scale. In this way, geologists have been able to determine both the absolute ages of the various geologic periods

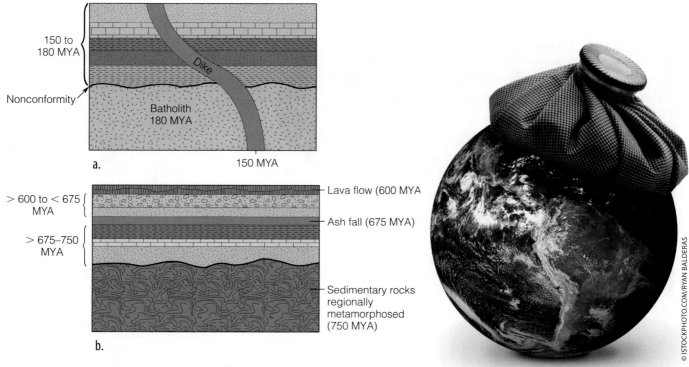

Figure 17.19 **Determining Absolute Dates for Sedimentary Rocks**
The absolute ages of sedimentary rocks can be determined by dating associated igneous rocks. In a. and b., sedimentary rocks are bracketed by rock bodies for which absolute ages have been determined.

and their durations (Figure 17.1). In fact, the dates for the era, period, and epoch boundaries of the geologic time scale are still being refined as more accurate dating methods are developed and new exposures dated.

LO8 Geologic Time and Climate Change

Given the debate concerning global warming and its possible implications, it is extremely important to be able to reconstruct past climatic regimes as accurately as possible. To model how Earth's climate system has responded to changes in the past and to use that information for simulations of future climate scenarios, geologists must have a geologic calendar that is as precise and accurate as possible. The ability to accurately determine when past climate changes occurred helps geologists correlate these changes with regional and global geologic events to see if there are any possible connections.

One interesting method that is becoming more common in reconstructing past climates is to analyze stalagmites from caves. Recall that stalagmites are icicle-shaped structures rising from a cave floor and formed of calcium carbonate precipitated from evaporating water (see Chapter 13). A stalagmite therefore records a layered history because each newly precipitated layer of calcium carbonate is younger than the previously precipitated layer (Figure 17.20).

Thus, a stalagmite's layers are oldest in the center at its base and progressively younger as they move outward (principle of superposition). Using techniques based on ratios of uranium 234 to thorium 230, and reliable back to approximately 500,000 years, geologists can achieve very precise radiometric dates on individual layers of a stalagmite.

A study of stalagmites from Crevice Cave in Missouri revealed a history of climatic and vegetation change in the midcontinent region of the United States during the interval between 75,000 and 25,000 years ago. Dates obtained from the Crevice Cave stalagmites were correlated with major changes in vegetation and average temperature fluctuations obtained from carbon-13 and oxygen-18 isotope profiles to reconstruct a detailed picture of climate changes during this time period.

Thus, precise dating techniques in stalagmite studies provide an accurate chronology that allows geologists to model climate systems of the past.

Figure 17.20 Stalagmites and Climate Change

Stalactites

Stalagmite

1 Newly forming layers of calcite in a stalagmite contain U^{234} (substituting for calcium).

2 The inside of a stalagmite is layered like an onion, showing its incremental growth.

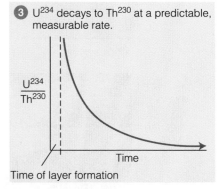

3 U^{234} decays to Th^{230} at a predictable, measurable rate.

$\dfrac{U^{234}}{Th^{230}}$

Time

Time of layer formation

4 The age of each layer can be determined by measuring its ratio of U^{234}/Th^{230}. Layer 5 (to the left) is older than layer 2. It has a lower U^{234}/Th^{230} ratio.

a. Stalagmites are icicle-shaped structures rising from the floor of a cave and are formed by the precipitation of calcium carbonate from evaporating water. A stalagmite is thus layered, with the oldest layer in the center and the youngest layers on the outside. Uranium 234 frequently substitutes for the calcium ion in the calcium carbonate of the stalagmite. Uranium 234 decays to thorium 230 at a predictable and measurable rate. Therefore, the age of each layer of the stalagmite can be dated by measuring its ratio of uranium 234 to thorium 230.

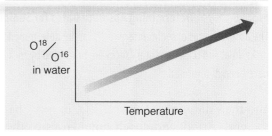

$\dfrac{O^{18}}{O^{16}}$ in water

Temperature

b. There are two isotopes of oxygen, a light one, oxygen 16, and a heavy one, oxygen 18. Because the oxygen 16 isotope is lighter than the oxygen 18 isotope, it vaporizes more readily than the oxygen 18 isotope when water evaporates. Therefore, as the climate becomes warmer, evaporation increases, and the O^{18}/O^{16} ratio becomes higher in the remaining water. Water in the form of rain or snow percolates into the ground and becomes trapped in the pores between the calcite forming the stalagmites.

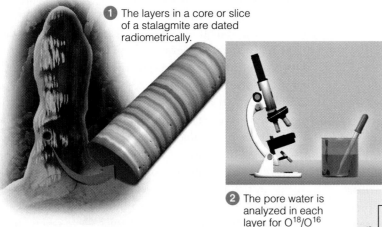

1 The layers in a core or slice of a stalagmite are dated radiometrically.

2 The pore water is analyzed in each layer for O^{18}/O^{16} and species of plants (from the pollen).

3 A record of climatic change is put together for the area of the caves.

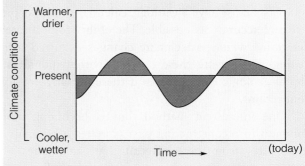

Warmer, drier

Present

Cooler, wetter

Climate conditions

Time ⟶   (today)

c. The layers of a stalagmite can be dated by measuring the U^{234}/Th^{230} ratio, and the O^{18}/O^{16} ratio determined for the pore water trapped in each layer. Thus, a detailed record of climatic change for the area can be determined by correlating the climate of the area as determined by the O^{18}/O^{16} ratio to the time period determined by the U^{234}/Th^{230} ratio.

SPEAK UP!

THEY DID

GEOL was built on a simple principle: to create a new textbook that reflects the way today's faculty teach and the way you learn.

Through conversations, focus groups, surveys, and interviews, we collected data that led to the creation of GEOL that you are using today. But it doesn't stop there – in order to make GEOL an even better learning experience, we'd like you to SPEAK UP and tell us how GEOL works for you.

What did you like about it? What would you change? Are there additional ideas you have that would help us build a better product for next semester's students?

At **4ltrpress.cengage.com** you'll find all of the resources you need to succeed—**Virtual Field Trips in Geology, the GEOL eBook, videos, flashcards, interactive quizzes** and more!

Visit 4ltrpress.cengage.com.

CHAPTER 18
EARTH HISTORY

JAMES S. MONROE

These hoodoos are oddly shaped pillars and spires of the Jurassic-age Entrada Sandstone in Goblin Valley State Park in Utah. The Entrada Sandstone consists of sandstone, siltstone, and shale, which in this area was deposited on ancient tidal flats. The hoodoos resulted from differential weathering and erosion along fractures in the rock. Another factor that may have contributed to their origin is differences in the amount of chemical cement within the formation, thus making some of the rock more resistant to weathering and erosion.

Introduction

> *"Imagine a barren, lifeless, waterless, hot planet with a poisonous atmosphere . . . what Earth was like shortly after it formed."*

Imagine a barren, lifeless, waterless, hot planet with a poisonous atmosphere. Cosmic radiation is intense, meteorites and comets crash to the ground, and volcanoes erupt almost continuously. Storms form in the turbulent atmosphere, lightning flashes much of the time, but no rain falls, because all water is in the form of vapor due to the high temperature, day and night. And because the atmosphere has no free oxygen, nothing burns, yet pools and streams of molten rock radiate a continuous red glow. This may seem like a description from a science fiction novel, but it is probably a reasonably accurate account of what Earth was like shortly after it formed.

This chapter has the ambitious task of sketching the history of Earth from its original state to its current and very different conditions. In Chapter 1 we introduced the concept of a *system* as a combination of related parts that operate in an organized fashion, and we gave examples of interactions in discussing volcanism, plate tectonics, running water, and glaciation. After Earth formed some 4.6 billion years ago, its systems became operative, although not all at the same time. For instance, Earth did not differentiate into a core and mantle until millions of years after it initially formed, and we know of no crust older than 4.0 billion years. However, once Earth did differentiate into layers, internal heat drove plate movements and the crust began evolving—as it continues to do.

By 3.5 billion years ago, Earth's systems were fully operative and interacting much as they do today. In short, Earth became a dynamic and changing planet early in its history. Evidence of many of these changes is preserved in the geologic record.

LEARNING OUTCOMES

After reading this unit, you should be able to do the following:

LO1 Describe Earth's Precambrian history

LO2 Describe the Paleozoic geography of Earth

LO3 Describe the Paleozoic evolution of North America

LO4 Examine the history of Paleozoic mobile belts

LO5 Explain the role of microplates in Paleozoic geology

LO6 Describe the stages in the breakup of Pangaea

LO7 Explain the Mesozoic history of North America

LO8 Explain Cenozoic Earth history

LO1 Precambrian Earth History

Precambrian is a widely used term that refers to both time and rocks. As a time term, it includes all geologic time from Earth's origin 4.6 billion years ago to the beginning of the Phanerozoic Eon 542 million years ago. The term also refers to all rocks lying below rocks of the Cambrian system. Unfortunately, no rocks are known for the first 600 million years of geologic time, so our geologic record begins about 4.0 billion years ago, with the oldest known rocks on Earth, the Acasta Gneiss of Canada. The geologic record we do have for the Precambrian, particularly its older part, is difficult to decipher, because many of these ancient rocks (1) have been metamorphosed and complexly deformed; (2) in many areas, they are deeply buried beneath younger rocks; and (3) they contain few fossils useful for determining relative ages.

Geologists divide the Precambrian into two eons: the Archean (4.6 to 2.5 billion years ago) and the Proterozoic (2.5 billion to 542 million years ago), which are subdivided by using prefixes such as *paleo* (old or ancient), *neo* (new or recent), and so on (Figure 18.1). So the Precambrian lasted for more than 4.0 billion years, and Earth has existed for 4.6 billion years. Consider this: suppose that one second equals one year. If this was the case and you were to count to 4.6 billion, the task would take you and your descendants more than 146 years. In short, the Precambrian is more than 88% of all geologic time, and yet we cover this lengthy interval in only a few pages.

THE ORIGIN AND EVOLUTION OF CONTINENTS

Rocks that are 3.8 to 4.0 billion years old and are thought to represent continental crust are known from several areas, including Minnesota, Greenland, Canada, and South Africa.

According to one model for the origin of continents, even older crust was thin, unstable, and composed of ultramafic igneous rock. At first, this early crust was disrupted by rising basaltic magma at ridges, and was consumed at subduction zones. A second stage in crustal evolution began when partial melting of earlier-formed basaltic crust resulted in the formation of andesitic island arcs, and partial melting of lower crustal andesites yielded granitic magmas that were emplaced in the crust. As plutons were emplaced in these island arcs, they became more like continental crust. By approximately 4.0 billion years ago, plate movements accompanied by subduction and collisions of island arcs had formed several granitic continental nuclei (Figure 18.2).

SHIELDS, PLATFORMS, AND CRATONS

Each continent has a vast area of exposed Precambrian rocks known as a **shield**. Extending outward from shields are **platforms**, consisting of Precambrian rocks beneath more recent rocks. A shield and platform collectively form a **craton**, which we can think

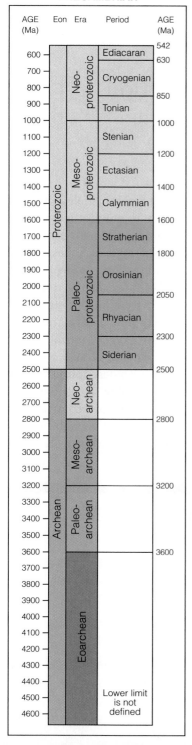

PRECAMBRIAN

Figure 18.1 The Precambrian Geologic Time Scale
This most recent version of the geologic time scale was published by the International Commission on Statigraphy (ICS) in 2004. Notice the use of the prefixes *eo* (early or dawn), *paleo* (old or ancient), *meso* (middle), and *neo* (new or recent). The age columns on the left and right sides of the time scale are in hundreds and thousands of millions of years (1,800 million years = 18 billion years, for example). See Figure 17.1 for the complete geologic time scale.

Figure 18.2 Three Stages in the Origin of Granitic Continental Crust
Andesitic island arcs formed by the partial melting of basaltic oceanic crust are intruded by granitic magmas. As a result of plate movements, island arcs collide and form larger units or cratons.

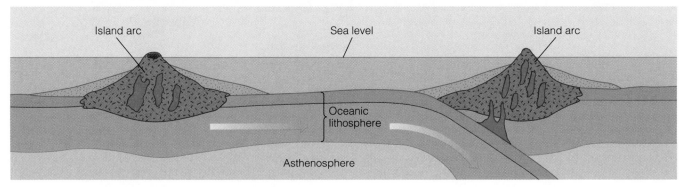

a. Two island arcs on separate plates move toward each other.

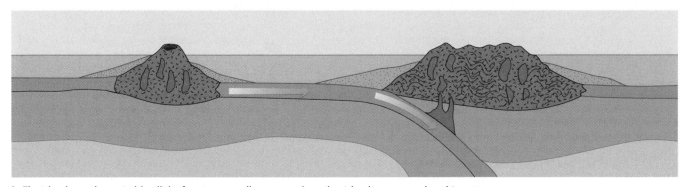

b. The island arcs shown in (a) collide, forming a small craton, and another island arc approaches this craton.

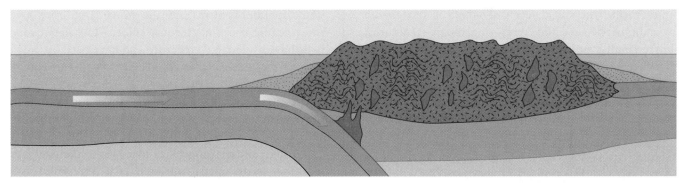

c. The island arc shown in (b) collides with the craton.

of as the stable core or nucleus of a continent (Figure 18.3). Many of the rocks within a craton have been strongly deformed, intruded by plutons, and altered by metamorphism, but they have experienced little or no deformation since the end of the Precambrian. Their stability since that time contrasts sharply with their Precambrian history of orogenic activity.

The **Canadian shield**, the exposed part of the North American craton, is a vast area of subdued topography, lakes, and exposed Archean and Proterozoic rocks, thinly covered in places by Pleistocene glacial deposits. Beyond the Canadian shield, exposures of Precambrian rocks are limited to areas of deep erosion, such as the Grand Canyon, and areas of orogeny, such as the Appalachian and Rocky Mountains.

Geologists have delineated several smaller

> **Canadian shield** The exposed part of the North American craton; mostly in Canada but also in Minnesota, Wisconsin, Michigan, and New York.

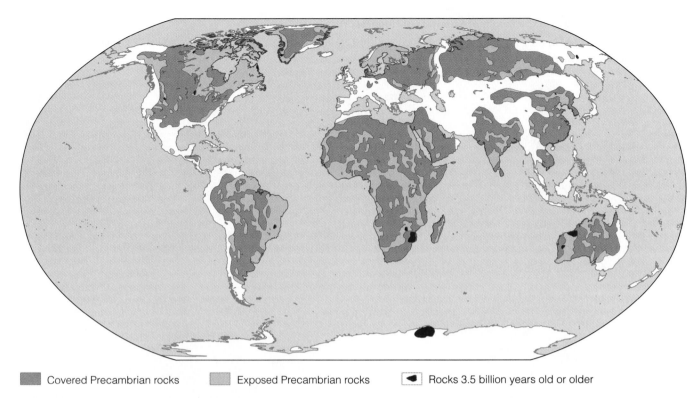

| Covered Precambrian rocks | Exposed Precambrian rocks | Rocks 3.5 billion years old or older |

Figure 18.3 Precambrian Cratons

The areas of exposed Precambrian rocks are the shields, and buried Precambrian rocks are the platforms. Shields and platforms collectively make up the cratons.

SOURCE: FROM A. M. GOODWIN, "THE MOST ANCIENT CONTINENTAL MARGINS," IN *THE GEOLOGY OF CONTINENTAL MARGINS*, BURK AND DRAKE (EDS), 1974, P. 768 (FIG 1), © SPRINGER-VERLAG.

units within the shield, each of which is recognized by radiometric ages and trends of geologic structures. These smaller units, as well as others making up the platform, are the subunits that constitute the North American craton. Each smaller unit was likely an independent minicontinent that later assembled into a large craton. The amalgamation of these units took place along deformation belts during the Paleoproterozoic.

ARCHEAN EARTH HISTORY

By far the most common rocks of Archean age are complexes consisting of granite and gneiss, as well as subordinate but reasonably common rock successions known as **greenstone belts**. An ideal greenstone belt has three major rock units: The lower and middle units are mostly volcanic rocks, and the upper unit is sedimentary. Low-grade metamorphism and the mineral chlorite give the volcanic rocks a greenish color. Most greenstone belts have a synclinal structure and have been intruded by granitic plutons and complexly folded and cut by thrust faults (Figure 18.4a).

A currently widely accepted model for the origin of greenstone belts holds that they develop in *back-arc marginal basins* that first open and then close. An early stage of extension takes place when the back-arc marginal basin opens, during which time volcanism and sedimentation take place, and finally an episode of compression occurs as it closes (Figure 18.4b,c). During closure, the rocks are intruded by plutons and metamorphosed, and the greenstone belt takes on a synclinal form as it is folded and faulted.

Undoubtedly, the present style of plate tectonics, involving the opening and closing of ocean basins, has been the primary agent of Earth evolution since at least the Paleoproterozoic. But many geologists are now convinced that some form of plate tectonics was operating during the Archean as well. However, Earth's radiogenic heat production has diminished through time, so during the Archean, when more heat

greenstone belt A linear or podlike association of igneous and sedimentary rocks; typically synclinal and consists of lower and middle volcanic units and an upper sedimentary rock unit.

Figure 18.4 Greenstone Belts and Their Origin

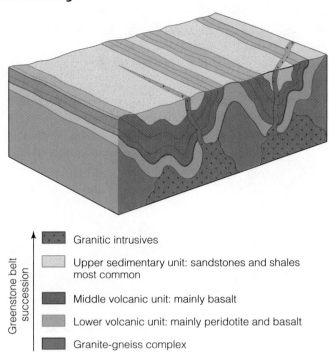

Granitic intrusives

Upper sedimentary unit: sandstones and shales most common

Middle volcanic unit: mainly basalt

Lower volcanic unit: mainly peridotite and basalt

Granite-gneiss complex

Greenstone belt succession

a. Two adjacent greenstone belts show their synclinal structure. The two lower units in greenstone belts are mostly volcanic rocks, whereas the upper unit is sedimentary.

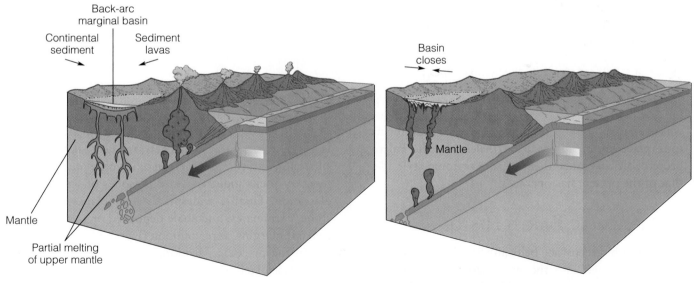

b. Basalt lavas and sediment derived from the continent and island arc fill the back-arc marginal basin.

c. Closure of the back-arc marginal basin causes compression and deformation. The evolving greenstone belt is deformed into a syncline-like structure into which granitic magmas intrude.

b. Outcrop of the Mesnard Quartzite. The crests of the ripple marks point toward the observer.

a. The Kona Dolomite. The bulbous structures are stromatolites that resulted from the activities of cyanobacteria (blue-green algae).

Figure 18.5 Paleoproterozoic Sedimentary Rocks

was available, seafloor spreading and plate movements took place more rapidly. Nevertheless, by the Paleoproterozoic and since then, a plate tectonic style like that of the present has operated.

PROTEROZOIC EARTH HISTORY

The origin of greenstone belts and granite-gneiss complexes continued into the Proterozoic but at a considerably reduced rate. Whereas most Archean rocks have been metamorphosed, many Proterozoic rocks have been little altered. Furthermore, the Proterozoic was a time of deposition of *banded iron formations*, consisting of alternating layers of iron minerals and chert; *red beds*, sandstones, siltstones, and shales with iron oxide cement; and sedimentary rocks indicating two episodes of widespread glaciation. Finally, widespread assemblages of sandstone, carbonates, and shale deposited on passive continental margins were common (Figure 18.5).

In the preceding section, we noted that Archean crust assembled through a series of island-arc and minicontinent collisions, providing the nuclei around which Proterozoic continental crust accreted. One large landmass so formed, called **Laurentia**, consisted mostly of North America and Greenland, parts of northwestern Scotland, and perhaps parts of the Baltic Shield of Scandinavia.

The first major episode of Laurentia's crustal evolution took place during the Paleoproterozoic, between

Laurentia A Proterozoic continent composed mostly of North America, Greenland, and parts of Scotland and Scandinavia.

2.0 and 1.8 billion years ago (BYA). Several major **orogens**—zones of deformed rocks—developed, many of which were metamorphosed and intruded by plutons. So this was a time of continental accretion during which collisions between Archean-age crust formed a large craton (Figure 18.6). Following the

Paleoproterozoic amalgamation of cratons, considerable accretion took place between 1.8 and 1.6 BYA in what is now the southwestern and central United States, as successively younger belts were sutured to the craton (Figure 18.6).

Between 1.6 and 1.3 BYA, extensive igneous activity—particularly the emplacement of granitic plutons and eruptions of rhyolite and ash flows—occurred that was unrelated to orogenic activity. According to one hypothesis, these rocks resulted from large-scale upwelling of magma beneath a supercontinent.

Another important event in the evolution of Laurentia, the *Grenville orogeny* in the eastern United States and Canada took place between 1.3 and 1.0 BYA (Neoproterozoic) (Figure 18.6). Grenville deformation represents the final episode of Proterozoic continental accretion of Laurentia. Contemporaneous with Grenville deformation was an episode of rifting in Laurentia, resulting in the origin of the *Midcontinent Rift*. The central part of the rift is filled with hundreds of overlapping basaltic lava flows and sedimentary rocks forming a pile several kilometers thick.

> **orogen** A linear part of Earth's crust that was or is being deformed during an orogeny.

Figure 18.6 Proterozoic Evolution of Laurentia

These three illustrations show the overall trends in the Proterozoic evolution of Laurentia, but they do not show many of the details of this long, complex episode in Earth history.

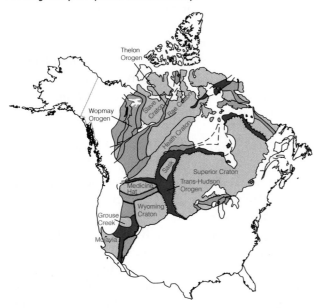

a. During the Paleoproterozoic, Archean cratons were sutured along deformation belts called orogens.

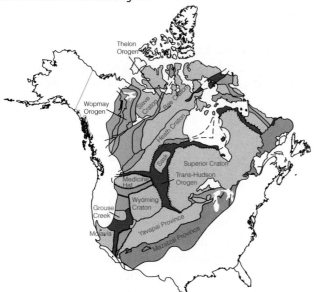

b. Laurentia grew along its southeastern margin by accretion of the Yavapai and Mazatzal provinces.

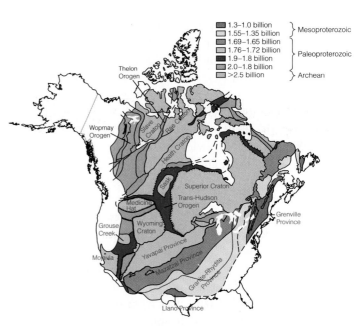

c. The last episodes in the Proterozoic accretion of Laurentia involved the origin of the Granite-Rhyolite province and the Grenville-Llano province.

LO2 The Paleozoic Geography of Earth

By the beginning of the Paleozoic Era, six major continents were present. Besides these large landmasses, geologists have also identified numerous microcontinents, such as *Avalonia*, and island arcs associated with microplates. The six major Paleozoic continents are *Baltica* (Russia west of the Ural Mountains and the major part of northern Europe), *China* (a complex area consisting of at least three Paleozoic continents that were not widely separated and are here considered to include China, Indochina, and the Malay Peninsula), *Gondwana* (Africa, Antarctica, Australia, Florida, India, Madagascar, and parts of the Middle East and southern Europe), *Kazakhstania* (a triangular continent centered on Kazakhstan but considered by some to be an extension of the Paleozoic Siberian continent), *Laurentia* (most of present North America, Greenland, northwestern Ireland, Scotland, and part of eastern Russia), and *Siberia* (Russia east of the Ural Mountains and Asia north of Kazakhstan and south of Mongolia).

In contrast to today's global geography, the Cambrian world consisted of these six continents dispersed around the globe at low tropical latitudes (Figure 18.7a). Water circulated freely among ocean basins, and the polar regions were apparently ice free. By the Late Cambrian, shallow seas had covered large areas of Laurentia, Baltica, Siberia, Kazakhstania, and China, whereas highlands were present in northeastern Gondwana, eastern Siberia, and central Kazakhstania.

During the Ordovician and Silurian periods, plate movement played a major role in the changing global geography. Gondwana moved southward during the Ordovician and began to cross the South Pole, as indicated by Upper Ordovician glacial deposits found today in the Sahara (Figure 18.7b). In contrast to the passive continental margin Laurentia exhibited during the Cambrian, an active convergent plate boundary formed along its eastern margin during the Ordovician, as indicated by the Late Ordovician *Taconic orogeny* that occurred in New England.

During the Silurian, Baltica, along with the newly attached Avalonia, moved northwestward relative to Laurentia and collided with it to form the larger continent of *Laurasia*. This collision, which closed the northern *Iapetus Ocean*, is marked by the *Caledonian orogeny*. After this orogeny, the southern part of the Iapetus Ocean still remained open between Laurentia and Avalonia-Baltica. Siberia and Kazakhstania moved from a southern equatorial position during the Cambrian to north temperate latitudes by the end of the Silurian Period.

During the Devonian, as the southern Iapetus Ocean narrowed between Laurasia and Gondwana, mountain building continued along the eastern margin of Laurasia with the *Acadian orogeny* (Figure 18.8a). Erosion of the ensuing highlands spread vast amounts of reddish fluvial sediments over large areas of northern Europe and eastern North America.

Other Devonian tectonic events, probably related to the collision of Laurentia and Baltica, include the Cordilleran *Antler orogeny* and the change from a passive continental margin to an active convergent plate boundary in the Uralian mobile belt of eastern Baltica. The distribution of reefs, evaporites, and red beds, as well as the existence of similar floras throughout the world, suggest a rather uniform global climate during the Devonian Period.

During the Carboniferous Period, southern Gondwana moved over the South Pole, resulting in extensive continental glaciation. The advance and retreat of these glaciers produced global changes in sea level that affected sedimentation patterns on the cratons. As Gondwana continued moving northward, it collided with Laurasia during the Early Carboniferous and continued suturing with it during the rest of the Carboniferous. The final phase of collision between Gondwana and Laurasia is indicated by the Ouachita Mountains of Oklahoma, formed by thrusting during the Late Carboniferous and Early Permian. Elsewhere, Siberia collided with Kazakhstania and moved toward the Uralian margin of Laurasia (Baltica), colliding with it during the Early Permian.

The assemblage of Pangaea concluded during the Permian with the completion of many of the continental collisions that began during the Carboniferous (Figure 18.8b). An enormous single ocean, *Panthalassa*, surrounded Pangaea and spanned Earth from pole to pole. Waters of this ocean probably circulated more freely than at present, resulting in more equable water temperatures.

The formation of a single landmass had climatic consequences for the terrestrial environment as well. Terrestrial Permian sediments indicate that arid and semiarid conditions were widespread over Pangaea. The mountain ranges produced by the *Hercynian*, *Alleghenian*, and *Ouachita orogenies* were high enough to create rainshadows that blocked the moist, subtropical, easterly winds—much as the southern

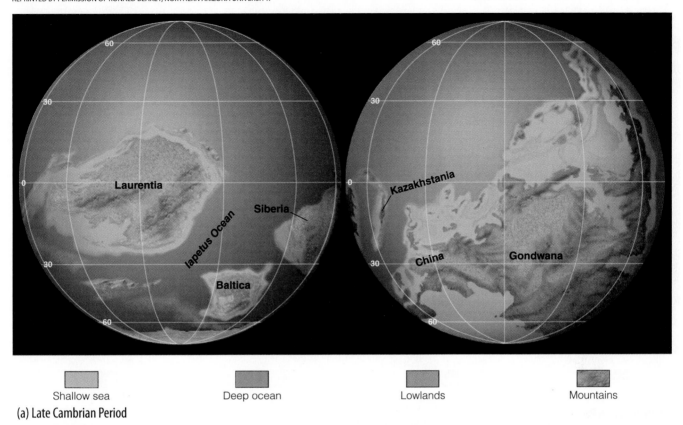

Shallow sea   Deep ocean   Lowlands   Mountains

(a) Late Cambrian Period

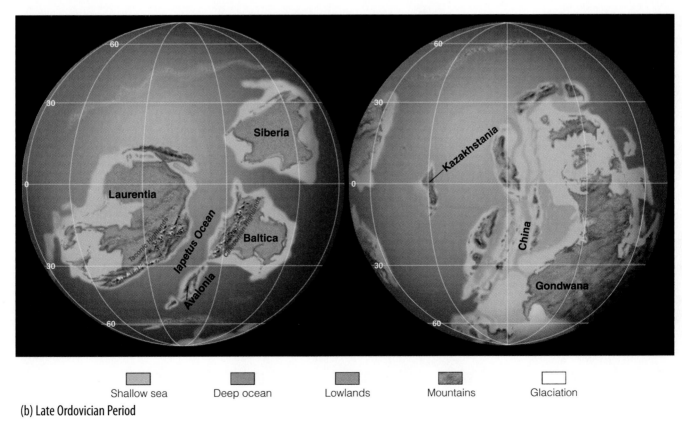

Shallow sea   Deep ocean   Lowlands   Mountains   Glaciation

(b) Late Ordovician Period

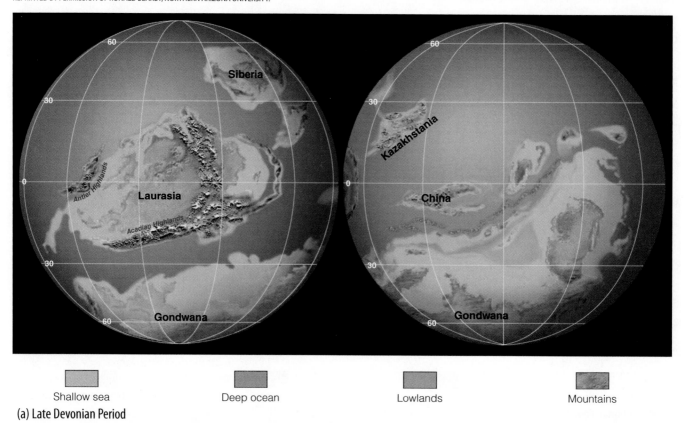

(a) Late Devonian Period

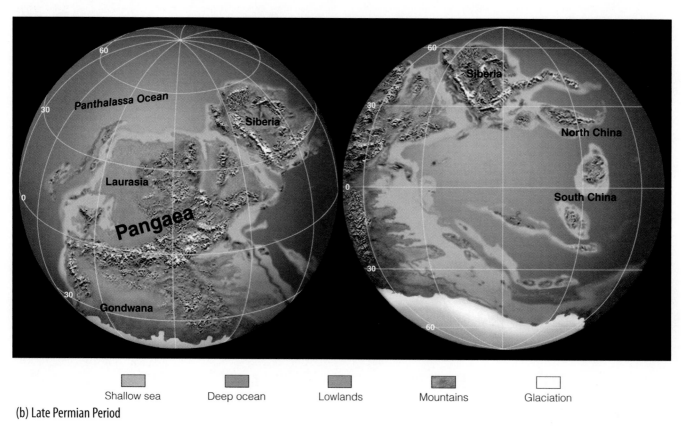

(b) Late Permian Period

Andes Mountains do in western South America today. This produced very dry conditions in North America and Europe, as evident from the extensive Permian red beds and evaporites found in western North America, central Europe, and parts of Russia. Permian coals, indicative of abundant rainfall, were mostly limited to the northern temperate belts (latitude 40 to 60 degrees north), whereas the last remnants of the Carboniferous ice sheets continued their recession.

LO3 The Paleozoic Evolution of North America

We divide the Paleozoic history of the North American craton into two parts, the first dealing with the relatively stable continental interior over which shallow seas advanced (transgressed) and retreated (regressed), and the second with the mobile belts where mountain building occurred.

Geologists commonly divide the sedimentary record of North America into six cratonic sequences. A *cratonic sequence* is a major transgressive-regressive cycle bounded by craton-wide unconformities. The transgressive phase is usually well-preserved, whereas the regressive phase of each sequence is marked by an unconformity.

THE SAUK SEQUENCE

Rocks of the **Sauk sequence** (Late Neoproterozoic–Early Ordovician) record the first major transgression onto the North American craton. During the Late Neoproterozoic and Early Cambrian, deposition of marine sediments was limited to the passive shelf areas of the Appalachian and Cordilleran borders of the craton. The craton itself was above sea level and experiencing weathering and erosion. Because North America was located in a tropical climate at this time (Figure 18.7a), and because there is no evidence of any terrestrial vegetation, weathering and erosion of the exposed Precambrian basement rocks must have proceeded rapidly.

During the Middle Cambrian, the transgressive phase of the Sauk began with shallow seas encroaching over the craton. By the Late Cambrian, the Sauk Sea had covered most of North America, leaving only a portion of the Canadian shield and a few large islands above sea level (Figure 18.9). These islands, collectively named the *Transcontinental Arch*, extended from New Mexico to Minnesota and the Lake Superior region.

THE TIPPECANOE SEQUENCE

As the Sauk Sea regressed from the craton during the Early Ordovician, a landscape of low relief emerged. The exposed rocks were predominantly limestones and dolostones that were deeply eroded because North America was still located in a tropical environment (Figure 18.7b). The resulting craton-wide unconformity marks the boundary between the Sauk and Tippecanoe sequences.

Like the Sauk sequence, deposition of the **Tippecanoe sequence** (Middle Ordovician–Early Devonian) began with a major transgression onto the craton. This transgressing sea deposited clean quartz sands over most of the craton. The Tippecanoe basal sandstones were followed by widespread carbonate deposition. The limestones were formed by calcium carbonate–secreting marine organisms such as corals and brachiopods.

As the Tippecanoe Sea gradually regressed from the craton during the Late Silurian, precipitation of evaporite minerals took place in the Appalachian, Ohio, and Michigan Basins (Figure 18.10). In the Michigan Basin alone, approximately 1,500 m of sediments was deposited, nearly half of which are halite and anhydrite.

By the Early Devonian, the regressing Tippecanoe Sea had retreated to the craton margin, exposing an extensive lowland topography. During this regression, marine deposition was initially restricted to a few interconnected cratonic basins and finally, by the end of the Tippecanoe, to only the margins surrounding the craton. As the Tippecanoe Sea regressed during the Early Devonian, the craton experienced mild deformation, resulting in the formation of many domes, arches, and

Sauk sequence A widespread association of sedimentary rocks bounded above and below by unconformities that was deposited during a latest Proterozoic to Early Ordovician transgressive-regressive cycle of the Sauk Sea.

Tippecanoe sequence A widespread body of sedimentary rocks bounded above and below by unconformities; deposited during a Middle Ordovician to Early Devonian transgressive-regressive cycle of the Tippecanoe Sea.

Figure 18.9 Paleogeography of North America During the Cambrian Period
Note the position of the Cambrian paleoequator. During this time, North America straddled the equator, as indicated in Figure 18.7a.

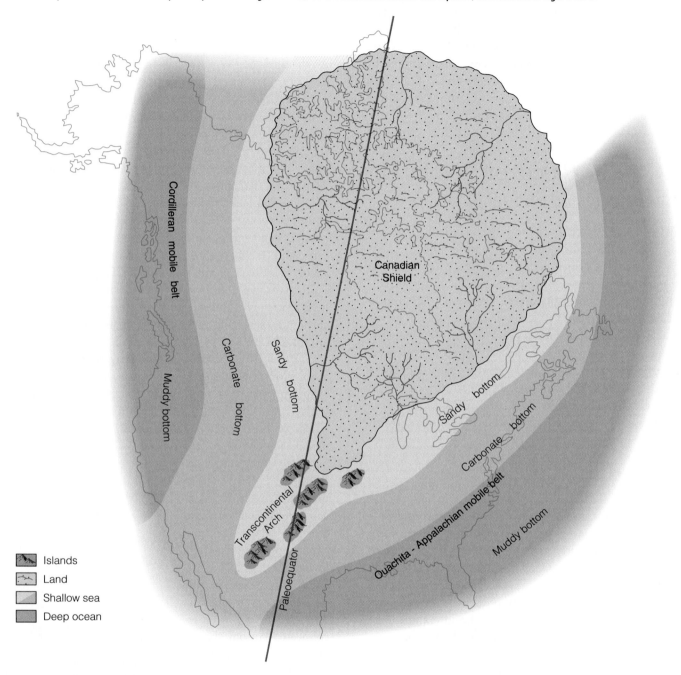

Islands
Land
Shallow sea
Deep ocean

Kaskaskia sequence
A widespread sequence of Middle Devonian to Upper Mississippian sedimentary rocks bounded above and below by unconformities; deposited during a transgressive-regressive cycle of the Kaskaskia Sea.

basins. These structures were mostly eroded during the time the craton was exposed, so that they were eventually covered by deposits from the encroaching Kaskaskia Sea.

THE KASKASKIA SEQUENCE

The boundary between the Tippecanoe sequence and the overlying **Kaskaskia sequence** (Middle Devonian–Late Mississippian) is marked by a major unconformity. As the Kaskaskia Sea transgressed over the low-relief landscape of the craton, most of the basal beds consisted of clean, well-sorted, quartz sandstones.

Figure 18.10 Paleogeography of North America During the Silurian Period
Note the development of reefs in the Michigan, Ohio, and Indiana-Illinois-Kentucky areas.

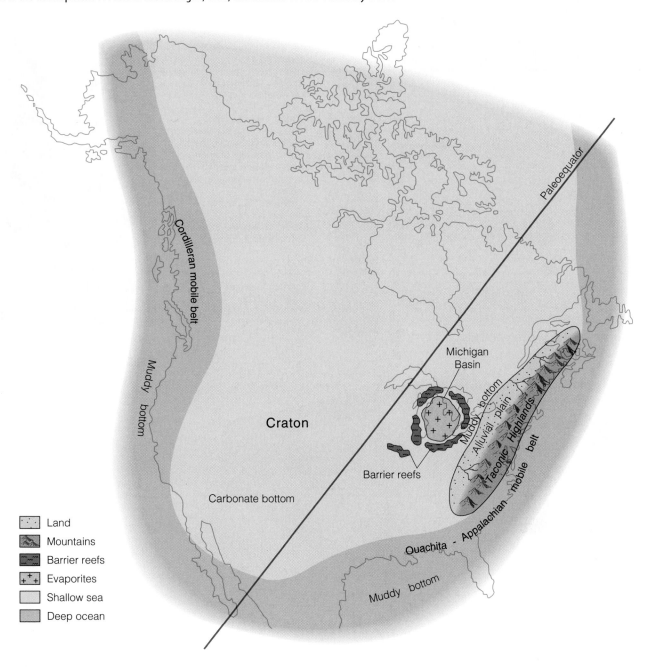

Legend:
- Land
- Mountains
- Barrier reefs
- Evaporites
- Shallow sea
- Deep ocean

Except for widespread Upper Devonian and Lower Mississippian black shales, the majority of Kaskaskian rocks are carbonates, including reefs, and associated evaporite deposits. In many other parts of the world, such as southern England, Belgium, central Europe, Australia, and Russia, the Middle and early Late Devonian epochs were times of major reef building.

During the Late Mississippian regression of the Kaskaskia Sea from the craton, carbonate deposition was replaced by vast quantities of detrital sediments. Before the end of the Mississippian, the Kaskaskia Sea had retreated to the craton margin, once again exposing the craton to widespread weathering and erosion resulting in a craton-wide unconformity at the end of the Kaskaskia sequence.

Figure 18.11 Columnar Section of a Complete Cyclothem

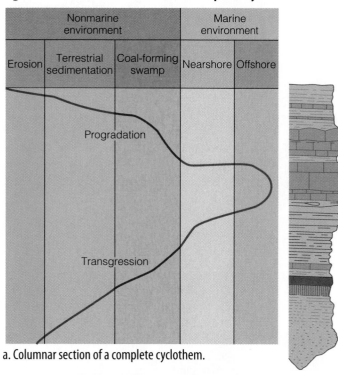

a. Columnar section of a complete cyclothem.

Disconformity
Brackish and
  nonmarine shales
Marine shales
Algal limestones with
  nearshore and brackish
  water invertebrate fossils
Limestones with offshore
  invertebrate fossils
Limestones and shale with
  offshore
  invertebrate fossils

Marine shales with nearshore
  invertebrate fossils

Coal
Underclay
Nonmarine shales and
  sandstones
Nonmarine sandstones
  Disconformity

b. Pennsylvanian coal bed, West Virginia.

COURTESY OF WAYNE E. MOORE

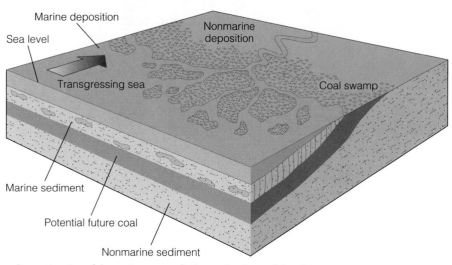

c. Reconstruction of the environment of a Pennsylvanian coal-forming swamp.

THE ABSAROKA SEQUENCE

The extensive unconformity separating the Kaskaskia and Absaroka sequences essentially divides the strata into the North American Mississippian and Pennsylvanian systems. The rocks of the **Absaroka sequence** (Late Mississippian–Early Jurassic) not only differ from those of the Kaskaskia sequence but also result from different tectonic regimes.

One characteristic feature of Pennsylvanian rocks is their repetitive pattern of alternating marine and nonmarine strata. Such rhythmically repeating sedimentary sequences are called **cyclothems** (Figure 18.11). They result from repeated alternations of marine and nonmarine environments, usually in areas of low relief. Although seemingly simple, cyclothems reflect a delicate interplay between nonmarine deltaic and shallow marine interdeltaic and shelf environments.

Cyclothems represent transgressive and regressive sequences, with an erosional surface separating one cyclothem from another. Thus, an idealized cyclothem passes upward from fluvial-deltaic deposits, through coals, to detrital shallow-water marine sediments, and finally to limestones typical of an open marine environment (Figure 18.11a).

Such repetitious sedimentation over a widespread area requires an explanation. The hypothesis currently favored by many geologists is a rise and fall of sea level related to advances and retreats of Gondwanan continental glaciers. When the Gondwanan ice sheets advanced, sea level dropped; when they melted, sea level rose. Late Paleozoic cyclothem activity on all of the cratons closely corresponds to Gondwanan glacial–interglacial cycles.

During the Pennsylvanian, the area of greatest deformation occurred in the southwestern part of the North American craton, where a series of fault-bounded uplifted blocks formed the *Ancestral Rockies* (Figure 18.12). These mountain ranges had diverse geologic histories and were not all

> **cyclothem** A vertical sequence of cyclically repeating sedimentary rocks resulting from alternating periods of marine and nonmarine deposition; commonly contain a coal bed.

Figure 18.12 Paleogeography of North America During the Pennsylvanian Period

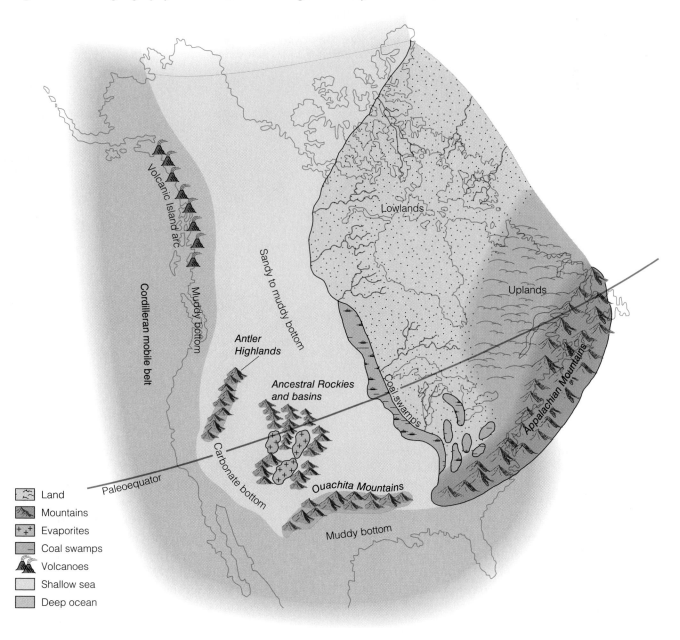

Land
Mountains
Evaporites
Coal swamps
Volcanoes
Shallow sea
Deep ocean

Figure 18.13 The Ancestral Rockies

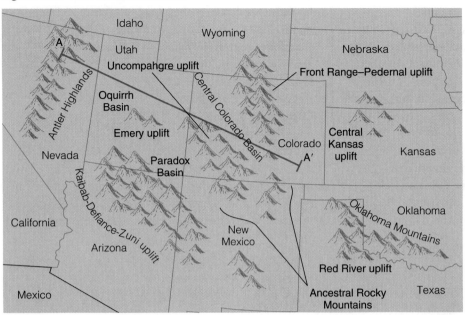

a. Location of the principal Pennsylvanian highland areas and basins of the southwestern part of the craton.

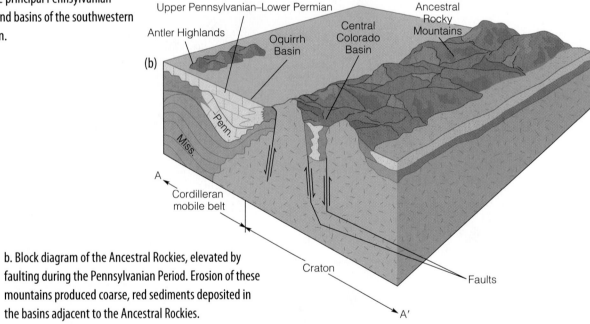

b. Block diagram of the Ancestral Rockies, elevated by faulting during the Pennsylvanian Period. Erosion of these mountains produced coarse, red sediments deposited in the basins adjacent to the Ancestral Rockies.

elevated at the same time. Uplift of these mountains, some of which were elevated more than 2 km along near-vertical faults, resulted in the erosion of the overlying Paleozoic sediments and exposure of the Precambrian igneous and metamorphic basement rocks (Figure 18.13). As the mountains eroded, tremendous quantities of coarse, red sediments were deposited in the surrounding basins, where they are preserved in many areas, such as the spectacular Garden of the Gods, Colorado.

While the various intracratonic basins were filling with sediment during the Late Pennsylvanian, the Absaroka Sea slowly began retreating from the craton. During the Middle and Late Permian, the Absaroka Sea was restricted to west Texas and southern New Mexico, forming a complex of lagoonal, reef, and open-shelf environments (Figure 18.14). By the end of the Permian Period, the Absaroka Sea had retreated from the craton, exposing continental red beds over most of the southwestern and eastern region.

Figure 18.14 Paleogeography of North America During the Permian Period

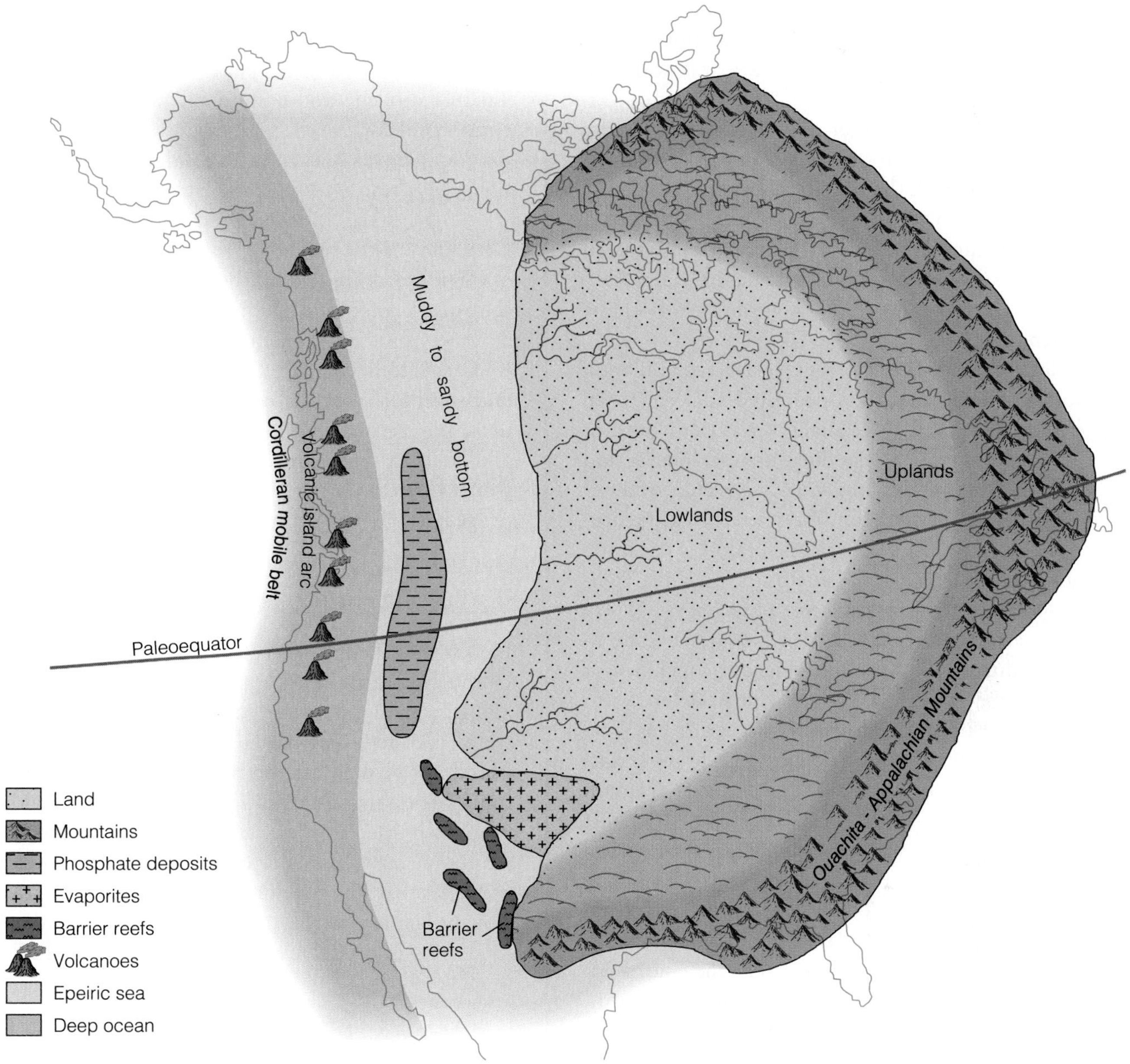

LO4 The History of the Paleozoic Mobile Belts

We now turn to the Paleozoic orogenic activity in the **mobile belts** (elongated areas of mountain-building activity along the margins of continents). The mountain building occurring during the Paleozoic Era had a profound influence on the climate and sedimentary history of the craton. In addition, it was part of the global tectonic regime that sutured the continents together, forming Pangaea by the end of the Paleozoic.

THE APPALACHIAN MOBILE BELT

Throughout Sauk time (Neoproterozoic–Early Ordovician), the Appalachian region was a broad, passive

> **mobile belt** An elongated area of deformation as indicated by folds and faults; generally adjacent to a craton.

continental margin. Sedimentation was closely balanced by subsidence as extensive carbonate deposits succeeded thick, shallow marine sands. During this time, movement along a divergent plate boundary was widening the Iapetus Ocean (Figure 18.15a).

Beginning with the subduction of the Iapetus plate beneath Laurentia (an oceanic–continental convergent plate boundary), the Appalachian mobile belt was born (Figure 18.15b). The resulting **Taconic orogeny**, named after the present-day Taconic Mountains of eastern New York, central Massachusetts, and Vermont, was the first of several orogenies to affect the Appalachian region.

A large *clastic wedge* (an extensive accumulation of mostly detrital sediments deposited adjacent to an uplifted area) formed in the shallow seas to the west of the Taconic orogeny. These deposits are thickest and coarsest nearest the highland area and become thinner and finer-grained away from the source area, eventually grading into carbonates on the craton. The clastic wedge resulting from the erosion of the Taconic Highlands is referred to as the *Queenston Delta*.

The second Paleozoic orogeny to affect Laurentia began during the Late Silurian and concluded at the end of the Devonian Period. The **Acadian orogeny** affected the Appalachian mobile belt from Newfoundland to Pennsylvania, as sedimentary rocks were folded and thrust against the craton. As with the preceding Taconic orogeny, the Acadian orogeny occurred along an oceanic–continental convergent plate boundary. As the northern Iapetus Ocean continued to close during the Devonian, the plate carrying Baltica finally collided with Laurentia, forming a continental-continental convergent plate boundary along the zone of collision (Figure 18.15c). Weathering and erosion of the Acadian Highlands produced the *Catskill Delta*, a thick clastic wedge named for the Catskill Mountains in northern New York, where it is well exposed.

Taconic orogeny An Ordovician episode of mountain building that resulted in the deformation of the Appalachian mobile belt.

Acadian orogeny An episode of Devonian deformation in the northern Appalachian mobile belt resulting from the collision of Baltica with Laurentia.

Hercynian-Alleghenian orogeny Pennsylvanian to Permian orogenic event during which the Appalachian mobile belt in eastern North America and the Hercynian mobile belt of southern Europe were deformed.

The Taconic and Acadian orogenies were part of the same major orogenic event related to the closing of the Iapetus Ocean. This event began with an oceanic–continental convergent plate boundary during the Taconic orogeny and culminated with a continental-continental convergent plate boundary during the Acadian orogeny as Laurentia and Baltica became sutured (Figure 18.15). After this, the **Hercynian-Alleghenian orogeny** began, followed by orogenic activity in the Ouachita mobile belt.

The Hercynian mobile belt of southern Europe and the Appalachian and Ouachita mobile belts of North America mark the zone along which Europe (part of Laurasia) collided with Gondwana. While Gondwana and southern Laurasia collided during the Pennsylvanian and Permian Periods, eastern Laurasia (Europe and southeastern North America) joined with Gondwana (Africa) as part of the Hercynian-Alleghenian orogeny.

THE CORDILLERAN MOBILE BELT

During the Neoproterozoic and Early Paleozoic, the Cordilleran area was a passive continental margin along which extensive continental shelf sediments were deposited. Beginning in the Middle Paleozoic, an island arc formed off the western margin of the craton. This eastward-moving island arc collided with the western border of the craton during the Late Devonian and Early Mississippian, resulting in a highland area, termed the Antler Highlands (Figure 18.12). This orogenic event, the *Antler orogeny*, was the first in a series of orogenic events to affect the Cordilleran mobile belt.

THE OUACHITA MOBILE BELT

The Ouachita mobile belt extends for approximately 2,100 km from the subsurface of Mississippi to the

Approximately 80% of the former Ouachita Mobile Belt is buried beneath a Mesozoic and Cenozoic sedimentary cover. The two major exposed areas in this region are the Ouachita Mountains of Oklahoma and Arkansas and the Marathon Mountains of Texas.

Figure 18.15 Evolution of the Appalachian Mobile Belt

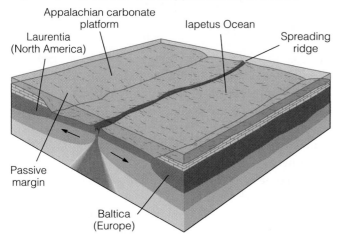

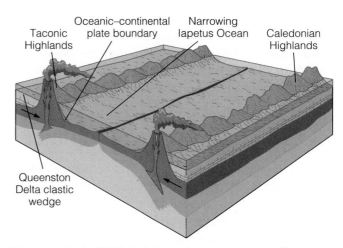

a. During the Neoproterozoic to the Early Ordovician, the Iapetus Ocean was opening along a divergent plate boundary. Both the east coast of Laurentia and the west coast of Baltica were passive continental margins where large carbonate platforms existed.

b. Beginning in the Middle Ordovician, the passive margins of Laurentia and Baltica became oceanic–continental convergent plate boundaries, resulting in orogenic activity.

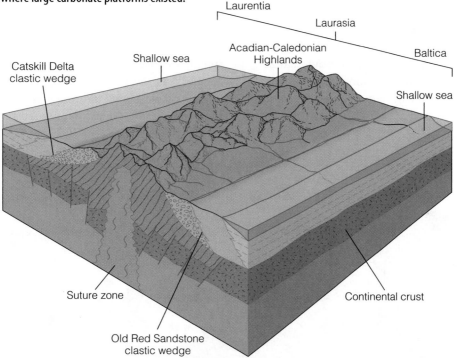

c. By the Late Paleozoic, Laurentia and Baltica collided along a continental–continental convergent plate boundary, forming a larger continental landmass, Laurasia.

Marathon region of Texas. During the Late Neoproterozoic to Early Mississippian, shallow-water detrital and carbonate sediments were deposited on a broad continental shelf, while in the deeper-water portion of the adjoining mobile belt, bedded cherts and shales were accumulating. Beginning in the Mississippian Period, the rate of sedimentation increased dramatically as the region changed from a passive continental margin to an active convergent plate boundary, marking the beginning of the **Ouachita orogeny**.

Thrusting of sediments continued throughout the Pennsylvanian and Early Permian, driven by the compressive forces

Ouachita orogeny A period of mountain building that took place in the Ouachita mobile belt during the Pennsylvanian Period.

generated along the zone of subduction as Gondwana collided with Laurasia. The collision of Gondwana and Laurasia is marked by the formation of a large mountain range, most of which eroded during the Mesozoic Era. Only the rejuvenated Ouachita and Marathon Mountains remain of this once-lofty mountain range.

LO5 The Role of Microplates

It is becoming increasingly clear that accretion along the continental margins is more complicated than the somewhat simple, large-scale plate interactions that we have described. Geologists now recognize that numerous microplates, such as Avalonia (Figure 18.7b), existed during the Paleozoic and were involved in the orogenic events that occurred during that time.

A careful examination of Paleozoic global paleogeographic maps shows numerous microplates, and their location and role during the formation of Pangaea must be taken into account. Thus, although the basic history of the formation of Pangaea during the Paleozoic remains the same, geologists now realize that microplates also played an important role and help explain some previously anomalous geologic situations.

LO6 The Breakup of Pangaea

Just as the formation of Pangaea influenced geologic and biologic events during the Paleozoic, the breakup of this supercontinent profoundly affected geologic and biologic events during the Mesozoic. The movement of continents affected global climatic and oceanic regimes, as well as the climates of individual continents.

Geologic, paleontologic, and paleomagnetic data indicate that the breakup of Pangaea took place in four general stages. The first stage involved rifting between Laurasia and Gondwana during the Late Triassic. By the end of the Triassic, the newly formed and expanding Atlantic Ocean separated North America from Africa (Figure 18.16a). This was followed by the rifting of North America from South America sometime during the Late Triassic and Early Jurassic.

The second stage in Pangaea's breakup involved rifting and movement of the various Gondwana continents during the Late Triassic and Jurassic periods. As early as the Late Triassic, Antarctica and Australia, which remained sutured together, began separating from South America and Africa, whereas India

Figure 18.16 Paleogeography of the World During the Mesozoic Era

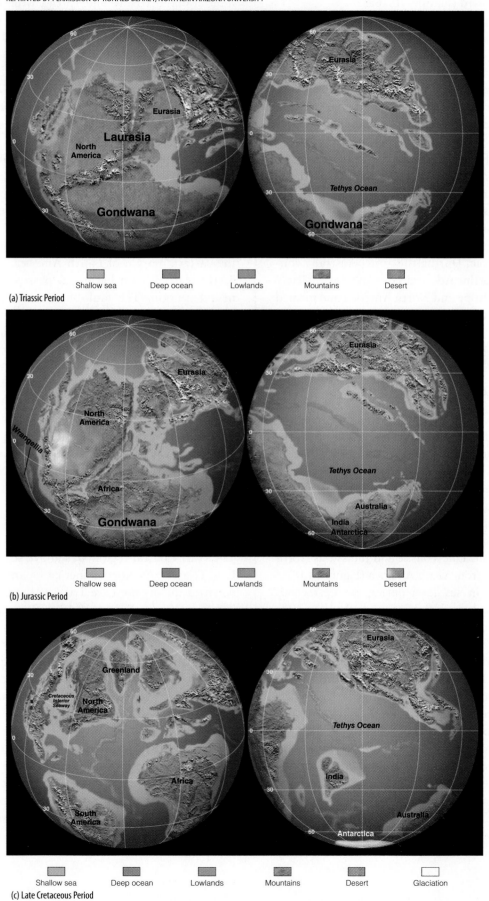

(a) Triassic Period

Shallow sea   Deep ocean   Lowlands   Mountains   Desert

(b) Jurassic Period

Shallow sea   Deep ocean   Lowlands   Mountains   Desert

(c) Late Cretaceous Period

Shallow sea   Deep ocean   Lowlands   Mountains   Desert   Glaciation

began drifting away from the Gondwana continent and moved northward (Figures 18.16a, b).

The third stage of breakup began during the Late Jurassic, when South America and Africa began to separate (Figure 18.16b). During this stage, the eastern end of the Tethys Sea began closing as a result of the clockwise rotation of Laurasia and the northward movement of Africa. This narrow Late Jurassic and Cretaceous seaway between Africa and Europe was the forerunner of the present Mediterranean Sea.

By the end of the Cretaceous, Australia and Antarctica had separated, India had nearly reached the equator, South America and Africa were widely separated, and the eastern side of what is now Greenland had begun to separate from Europe (Figure 18.16c).

The final stage in Pangaea's breakup occurred during the Cenozoic. During this time, Australia continued moving northward, and Greenland completely separated from Europe and North America and formed a separate landmass.

LO7 The Mesozoic History of North America

The beginning of the Mesozoic Era was essentially the same in terms of mountain building and sedimentation as the preceding Permian Period in North America (Figure 18.17). Terrestrial sedimentation continued over much of the craton, while block faulting and igneous activity began in the Appalachian region as North America and Africa began to separate (Figure 18.16). The newly forming Gulf of Mexico was the site of extensive evaporite deposition during the Late Triassic and Jurassic as North America separated from South America (Figure 18.18).

A global rise in sea level during the Cretaceous resulted in worldwide transgressions onto the continents such that marine deposition was continuous over much of western North America (Figure 18.19). A volcanic island arc system that formed off the western edge of the craton during the Permian was sutured to North America sometime during the Permian or Triassic. During the Jurassic, the entire Cordilleran area was involved in a series of major mountain-building episodes, resulting in the formation of the Sierra Nevada, the Rocky Mountains, and other lesser mountain ranges.

EASTERN COASTAL REGION

During the Early and Middle Triassic, coarse detrital sediments derived from the erosion of the recently uplifted Appalachians (*Alleghenian orogeny*) filled intermontane basins and spread over the surrounding areas. As erosion continued during the Mesozoic, this once-lofty mountain system was reduced to a low-lying plain.

During the Late Triassic, the first stage in the breakup of Pangaea began with North America separating from Africa and the Atlantic Ocean starting to form (Figure 18.16a). Fault-block basins developed in response to upwelling magma beneath Pangaea in a zone stretching from present-day Nova Scotia to North Carolina (Figure 18.20). Erosion of the adjacent fault-block mountains filled these basins with great quantities (up to 6,000 m) of poorly sorted, red, nonmarine detrital sediments.

As the Atlantic Ocean grew, rifting ceased along the eastern margin of North America, and this once-active plate margin became a passive, trailing continental margin. The fault-block mountains produced by this rifting continued to erode during the Jurassic and Early Cretaceous until all that was left was a large low-relief area.

The sediments produced by this erosion contributed to the growing eastern continental shelf. During the Cretaceous Period, the Appalachian region was re-elevated and once again shed sediments onto the continental shelf, forming a gently dipping, seaward-thickening wedge of rocks up to 3,000 m thick. These rocks are currently exposed in a belt extending from Long Island, New York, to Georgia.

GULF COASTAL REGION

The Gulf Coastal region was above sea level until the Late Triassic. As North America separated from South America during the Late Triassic, the Gulf of Mexico began to form (Figure 18.18). With oceanic waters flowing into this newly formed, shallow, restricted basin, conditions were ideal for evaporite formation. More than 1,000 m of evaporites were precipitated at this time, and most geologists think that these Jurassic evaporites are the source of the Cenozoic salt domes found today in the Gulf of Mexico and southern Louisiana.

By the Late Jurassic, circulation in the Gulf of Mexico was less restricted, and evaporite deposition ended. Normal marine conditions returned to the area with alternating transgressing and regressing seas. The resulting sediments were covered and buried by thousands of meters of Cretaceous and Cenozoic sediments.

Figure 18.17 Paleogeography of North America During the Triassic Period

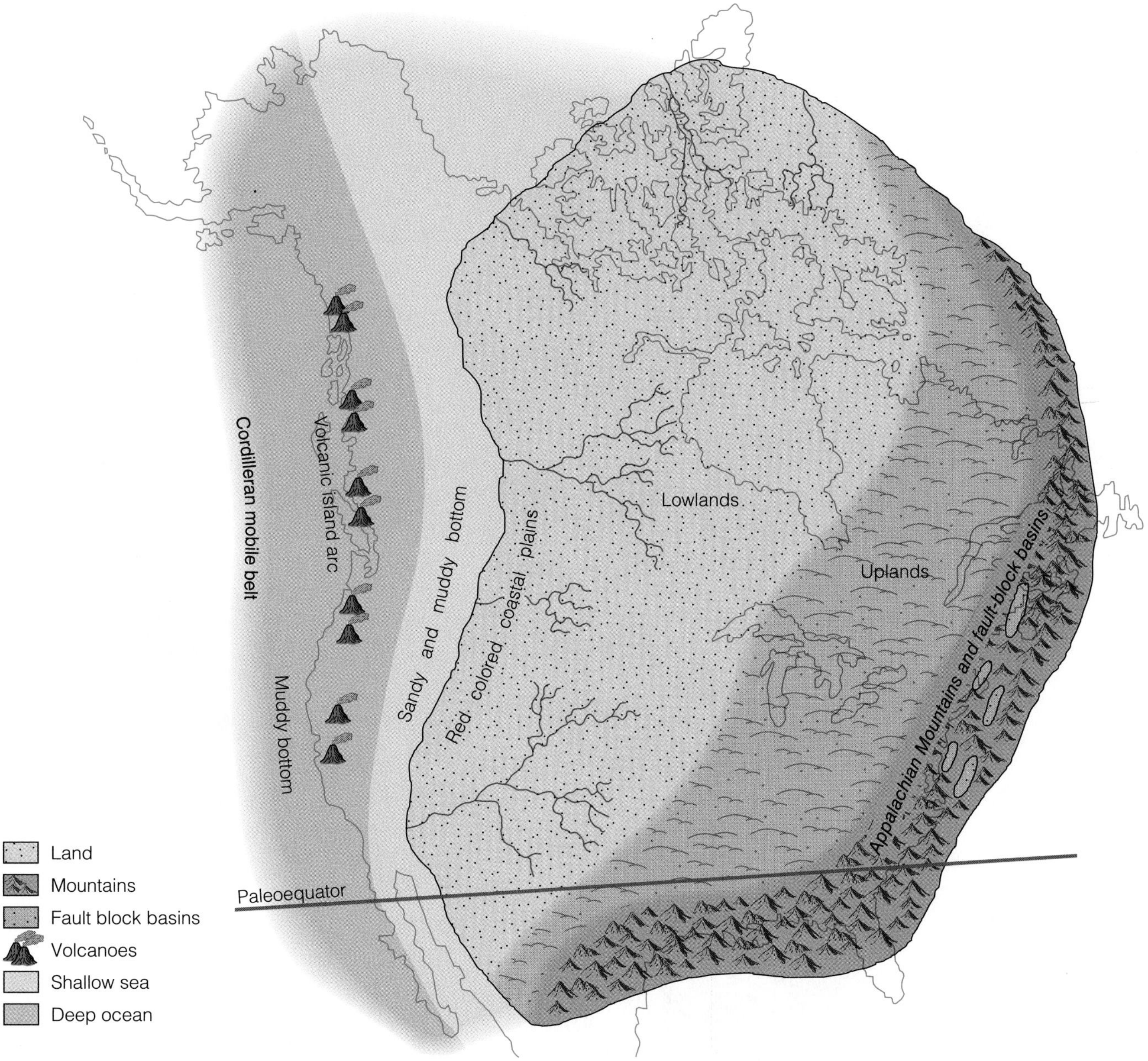

Legend:
- Land
- Mountains
- Fault block basins
- Volcanoes
- Shallow sea
- Deep ocean

Map labels: Cordilleran mobile belt, Volcanic island arc, Muddy bottom, Sandy and muddy bottom, Red colored coastal plains, Lowlands, Uplands, Appalachian Mountains and fault-block basins, Paleoequator

During the Cretaceous, the Gulf Coastal region, like the rest of the continental margin, was flooded by northward-transgressing seas forming a wide seaway that extended from the Arctic Ocean to the Gulf of Mexico (Figure 18.19).

WESTERN REGION

During the Permian, an island arc and ocean basin formed off the western North American craton (Figure 18.14), followed by subduction of an oceanic plate beneath the island arc and the thrusting of oceanic and island arc rocks eastward against the craton mar-

gin. This event, similar to the preceding Antler orogeny and known as the *Sonoma orogeny*, occurred at or near the Permian–Triassic boundary.

Following the Late Paleozoic–Early Mesozoic destruction of the volcanic island arc during the Sonoma orogeny, the western margin of North America became an oceanic–continental convergent plate boundary. During the Late Triassic, a steeply dipping subduction zone developed along the western margin of North America in response to the westward movement of North America over the Pacific plate. This newly created oceanic–continental plate boundary controlled Cordilleran tectonics for the rest of the Mesozoic

Figure 18.18 Paleogeography of North America During the Jurassic Period

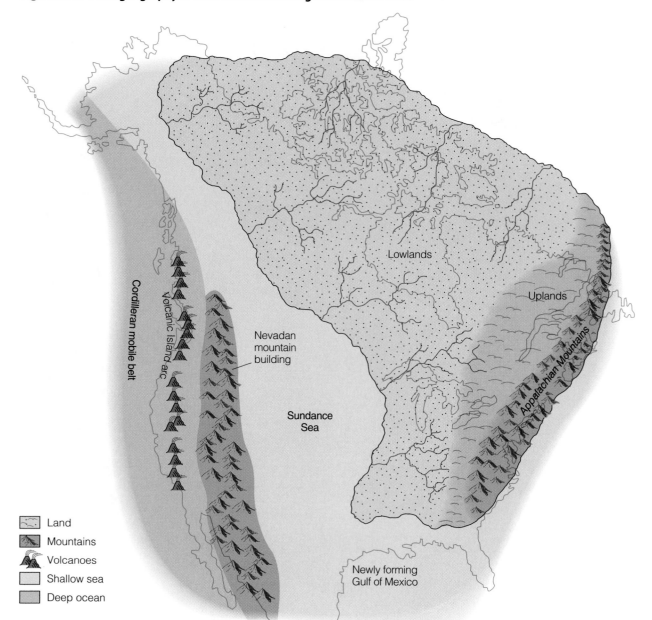

Cordilleran mobile belt

Volcanic Island arc

Nevadan mountain building

Lowlands

Uplands

Appalachian Mountains

Sundance Sea

Newly forming Gulf of Mexico

Land
Mountains
Volcanoes
Shallow sea
Deep ocean

Nevadan orogeny Late Jurassic to Cretaceous deformation that strongly affected the western part of North America.

Era and for most of the Cenozoic Era; this subduction zone marks the beginning of the modern circum-Pacific orogenic system.

The general term *Cordilleran orogeny* is applied to the mountain-building activity that began during the Jurassic and continued into the Cenozoic. The Cordilleran orogeny consisted of a series of individually named, but interrelated, mountain-building events that occurred in different regions at different times. Most of this Cordilleran orogenic activity is related to the continued westward movement of the North American plate as it overrode the Farallon plate, and its history is highly complex.

The first phase of the Cordilleran orogeny, the **Nevadan orogeny**, began during the Middle to Late Jurassic and continued into the Cretaceous as large volumes of granitic magma were generated at depth beneath the western edge of North America. These

Figure 18.19 Paleogeography of North America During the Cretaceous Period

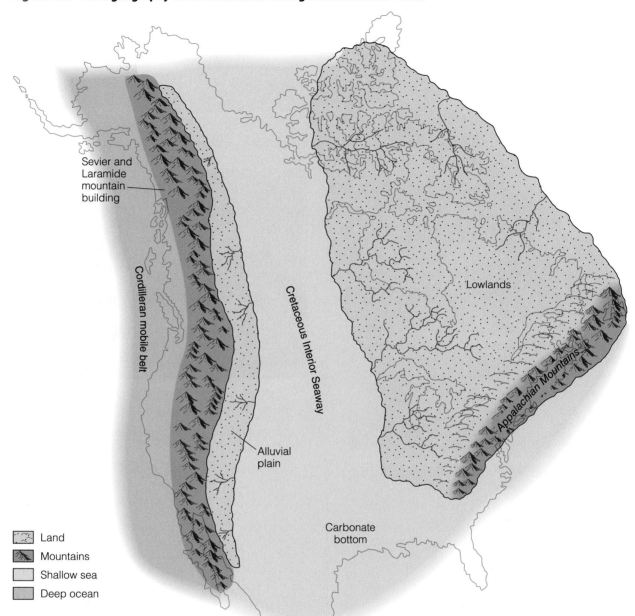

Sevier and Laramide mountain building

Cordilleran mobile belt

Cretaceous Interior Seaway

Alluvial plain

Lowlands

Appalachian Mountains

Carbonate bottom

Land

Mountains

Shallow sea

Deep ocean

granitic masses ascended as huge batholiths that are now recognized as the Sierra Nevada, southern California, Idaho, and Coast Range batholiths.

The second phase of the Cordilleran orogeny, the **Sevier orogeny**, was mostly a Cretaceous event. Subduction of the Farallon plate beneath the North American plate continued during this time, resulting in numerous overlapping, low-angle thrust faults in which blocks of older rocks were thrust eastward on top of younger rocks. This deformation produced generally north-south–trending mountain ranges stretching from Montana to western Canada.

During the Late Cretaceous to Early Cenozoic, the final pulse of the Cordilleran orogeny occurred. The **Laramide orogeny** developed east of the Sevier orogenic belt in the present-day Rocky Mountain areas of New Mexico, Colorado, and Wyoming. Most of the features of the present-day

Sevier orogeny
Cretaceous deformation that affected the continental shelf and slope areas of the Cordilleran mobile belt.

Laramide orogeny
A Late Cretaceous to Early Cenozoic episode of deformation in the area of the present-day Rocky Mountains.

Figure 18.20 North America Triassic Fault-Block Basins

a. Areas where Triassic fault-block basin deposits crop out in eastern North America.

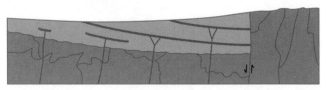

b. After the Appalachians were eroded into a low-lying plain by the Middle Triassic, fault-block basins formed as a result of Late Triassic rifting between North America and Africa.

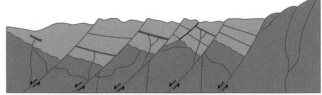

c. These valleys accumulated tremendous thickness of sediments and were themselves broken by a complex of normal faults during rifting.

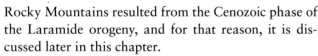

d. Palisades of the Hudson River. This sill was one of many intruded into the Newark sediments during the Late Triassic rifting that marked the separation of North America from Africa.

Rocky Mountains resulted from the Cenozoic phase of the Laramide orogeny, and for that reason, it is discussed later in this chapter.

Concurrent with the tectonism in the Cordilleran mobile belt, Early Triassic sedimentation on the western continental shelf consisted of shallow-water marine sandstones, shales, and limestones. During the Middle and Late Triassic, the western shallow seas regressed farther west, exposing large areas of former seafloor to erosion. Marginal marine and nonmarine Triassic rocks, particularly red beds, contribute to the spectacular and colorful scenery of the region.

These rocks represent a variety of depositional environments, including fluvial, deltaic, floodplain, and desert dunes. The Upper Triassic *Chinle Formation*, for example, is widely exposed throughout the Colorado Plateau and is probably most famous for its petrified wood, spectacularly exposed in Petrified

Forest National Park, Arizona. Although best known for its petrified wood, the Chinle Formation has also yielded fossils of amphibians and various reptiles, including small dinosaurs.

The Early Jurassic deposits in a large part of the western region consist mostly of clean, cross-bedded sandstones indicative of wind-blown deposits. The thickest and most prominent of these is the *Navajo Sandstone*, a widespread cross-bedded sandstone that accumulated in a coastal dune environment along the southwestern margin of the craton. The sandstone's most distinctive feature is its large-scale cross-beds, some of them more than 25 m high.

Marine conditions returned to the region during the Middle Jurassic, when a seaway called the *Sundance Sea* twice flooded the interior of western North America (Figure 18.18). The resulting deposits were largely derived from erosion of the tectonic highlands

Figure 18.21 Dinosaur Bones in Bas Relief
The north wall of the visitors' center at Dinosaur National Monument, showing dinosaur bones in bas relief, just as they were deposited 140 million years ago in the Morrison Formation.

to the west that paralleled the shoreline. These highlands resulted from intrusive igneous activity and associated volcanism that began during the Triassic.

During the Late Jurassic, a mountain chain formed in Nevada, Utah, and Idaho as a result of the deformation produced by the Nevadan orogeny. As the mountain chain grew and shed sediments eastward, the Sundance Sea began retreating northward. A large part of the area formerly occupied by the Sundance Sea was then covered by multicolored detrital sediments that comprise the *Morrison Formation*, which contains the world's richest assemblage of Jurassic dinosaur remains (Figure 18.21).

Shortly before the end of the Early Cretaceous, Arctic waters spread southward over the craton, forming a large inland sea in the Cordilleran area. By the beginning of the Late Cretaceous, this incursion joined the northward-transgressing waters from the Gulf area to create an enormous *Cretaceous Interior Seaway*, which occupied the area east of the Sevier orogenic belt (Figure 18.19). Extending from the Gulf of Mexico to the Arctic Ocean and more than 1,500 km wide at its maximum extent, this seaway effectively divided North America into two large landmasses until just before the end of the Late Cretaceous.

As the Mesozoic Era ended, the Cretaceous Interior Seaway withdrew from the craton. During this regression, marine waters retreated to the north and south, and marginal marine and continental deposition formed widespread coal-bearing deposits on the coastal plain.

LO8 Cenozoic Earth History

At 66 million years long, the Cenozoic Era is comparatively brief, constituting only 1.4% of all geologic time. Nevertheless, 66 million years is an extremely long time, certainly long enough for significant evolution of Earth and its biota. Furthermore, Cenozoic rocks are at or near the surface and have been little altered, thereby making access to and interpretation of them easier than for rocks of previous eras.

Geologists divide the Cenozoic Era into two periods. The Paleogene Period (66 to 23 million years ago) includes the Paleocene, Eocene, and Oligocene epochs, and the Neogene Period (23 million years ago to the present) includes the Miocene, Pliocene, Pleistocene, and Holocene or Recent epochs (see Figure 17.1). Although you may see the terms Tertiary Period (66–1.8 million years ago) and Quaternary Period (for the last 1.8 million years), they are no longer recommended.

Many of Earth's features have long histories, but the present distribution of land and sea and the landforms of the continents developed recently in the context of geologic time. For instance, the Appalachian Mountain region began its evolution during the Precambrian, but its present expression is largely the product of Cenozoic uplift and erosion. Likewise,

Earth's systems continue to interact, accounting for a continuously evolving planet.

distinctive landforms such as glacial valleys, badlands topography, and volcanoes of our national parks developed during the past few thousand years to several millions of years.

CENOZOIC PLATE TECTONICS AND OROGENY

The Late Triassic fragmentation of the supercontinent Pangaea (Figure 18.16a) began an episode of plate motions that continues even now. As a result, Cenozoic orogenic activity was largely concentrated in two major zones or belts, the *Alpine-Himalayan belt* and the *circum-Pacific belt.*

Within the Alpine-Himalayan orogenic belt, the *Alpine orogeny* began during the Mesozoic, but major deformation also occurred from the Eocene to Late Miocene as the African and Arabian plates moved northward against Eurasia. Deformation resulting from plate convergence formed mountains between Spain and France, the Alps of mainland Europe, and the mountains of Italy and North Africa. This orogenic belt remains geologically active.

Farther east in the Alpine-Himalayan orogenic belt, the *Himalayan orogeny* resulted from the collision of India with Asia (see Figure 10.18). Sometime during the Eocene, India's northward drift rate decreased abruptly, indicating the probable time of collision. In any event, an orogeny resulted during which two continental plates became sutured, which is why the present-day Himalayas are far inland rather than at a continental margin.

Plate subduction in the circum-Pacific orogenic belt took place throughout the Cenozoic, giving rise to orogenies in the Aleutians, the Philippines, and Japan, and along the west coasts of North, Central, and South America. The Andes Mountains in western South America, for example, formed as a result of convergence of the Nazca and South American plates (see Figure 10.17). Spreading at the East Pacific Rise and subduction of the Cocos and Nazca plates beneath Central and South America, respectively, account for continuing orogenic activity in these regions.

LO9 The North American Cordillera

The *North American Cordillera* is a complex mountainous region in western North America extending from Alaska into central Mexico (Figure 18.22). It has a long, complex geologic history involving accretion of island arcs along the continental margin, orogeny at an oceanic–continental boundary, vast outpourings of basaltic lavas, and block faulting. The most recent episode of large-scale deformation was the *Laramide orogeny*, which began during the Late Cretaceous, 85 to 90 million years ago. Like many other orogenies, it took place along an oceanic–continental boundary but was much farther inland than is typical (Figure 18.23).

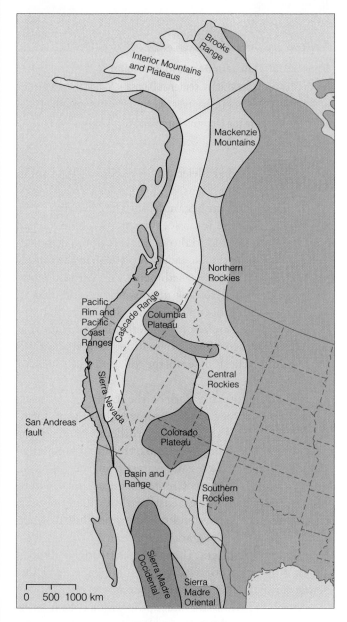

Figure 18.22 The North American Cordillera
The North American Cordillera is a complex mountainous region extending from Alaska into central Mexico. It consists of the elements shown here.

Figure 18.23 The Laramide Orogeny

The Laramide orogeny resulted when the Farallon plate was subducted beneath North America during the Late Cretaceous to Eocene periods.

SOURCE: REPRODUCED WITH PERMISSION FROM J. B. MURPHY, G. L. OPPLIGER, G. H. BRIMHALL, JR., AND A. HYNES, "MANTLE PLUMES AND MOUNTAINS," *AMERICAN SCIENTIST*, VOL. 87, NO. 2, P. 152 (MARCH–APRIL 1999).

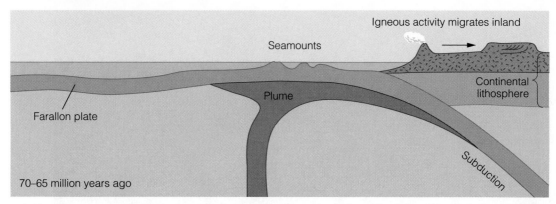

a. As North America moved westward over the Farallon plate, beneath which was the deflected head of a mantle plume, the angle of subduction decreased and magmatism shifted inland.

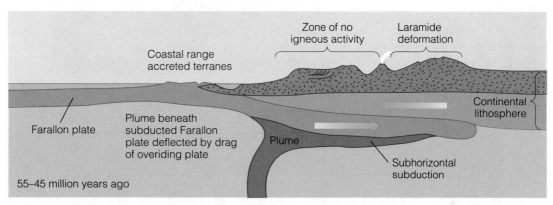

b. With nearly horizontal subduction, magmatism ceased and the continental crust was deformed mostly by vertical forces.

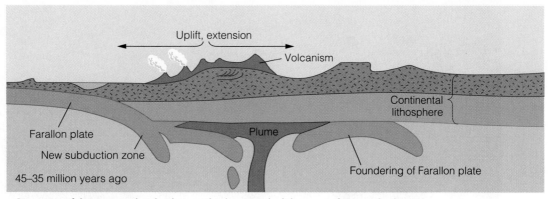

c. Disruption of the oceanic plate by the mantle plume marked the onset of renewed volcanism.

Figure 18.24 The Colorado Plateau and the Basin and Range Province

a. Mexican Hat in Utah is an erosional feature measuring about 18 m across. It is made up of Permian rocks, but its present form resulted from Cenozoic erosion.

b. The Sierra Nevada at the western margin of the Basin and Range has risen along normal faults, so that it is more than 3,000 m above the valley to the east.

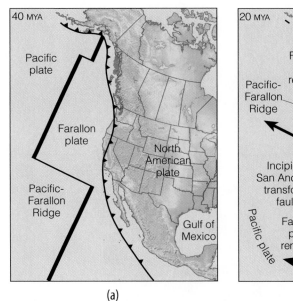
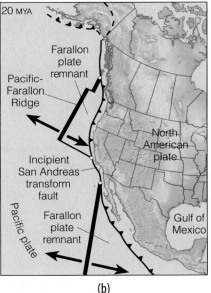
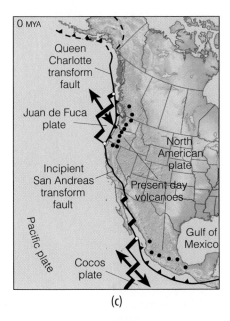

(a)   (b)   (c)

Figure 18.25 Three Stages in Westward Drift of North America

Three stages (a), (b), and (c) in the westward drift of North America and its collision with the Pacific-Farallon Ridge. As the North American plate overrode the ridge, its margin became bounded by transform faults rather than a subduction zone.

SOURCE: REPRINTED WITH PERMISSION FROM W. R. DICKINSON, "CENOZOIC PLATE TECTONIC SETTING OF THE CORDILLERAN REGION IN THE WESTERN UNITED STATES," *CENOZOIC PALEOGEOGRAPHY OF THE WESTERN UNITED STATES*, PACIFIC COAST SYMPOSIUM 3, 1979, P. 2 (FIG. 1).

The Laramide orogeny ceased about 40 million years ago, but since that time the mountain ranges formed during the orogeny were eroded, and the valleys between the ranges filled with sediments. Many of the ranges were nearly buried in their own erosional debris, and their present-day elevations are the result of renewed uplift.

In other parts of the Cordillera, the Colorado Plateau was uplifted far above sea level, but the rocks were little deformed (Figure 18.24a). In the Basin and Range Province, block faulting began during the Middle Cenozoic and continues to the present. At its western edge, the province is bounded by a large escarpment that forms the east face of the Sierra Nevada (Figure 18.24b).

In the Pacific Northwest, an area of about 200,000 km², mostly in Washington, is covered by the Cenozoic Columbia River basalts (see Figure 5.13). Issuing from long fissures, these flows overlapped to produce an aggregate thickness of about 1,000 m. Widespread volcanism also took place in Oregon, Idaho, California, Arizona, and New Mexico.

The present-day elements of the Pacific Coast section of the Cordillera developed as a result of the westward drift of North America, the partial consumption of the oceanic Farallon plate, and the collision of North America with the Pacific-Farallon Ridge (Figure 18.25). During the Early Cenozoic, the entire Pacific Coast was bounded by a subduction zone that stretched from Mexico to Alaska. Most of the Farallon plate was consumed at this subduction zone, and now only two small remnants exist: the Juan de Fuca and Cocos plates (Figure 18.25). The continuing subduction of these small plates accounts for seismicity and volcanism in the Cascade Range of the Pacific Northwest and Central America, respectively. Westward drift of the North American plate also resulted in its collision with the Pacific-Farallon Ridge and the origin of the Queen Charlotte and San Andreas transform faults (Figure 18.25).

THE CONTINENTAL INTERIOR AND THE GULF COASTAL PLAIN

The vast, shallow seas that had invaded the continents during the previous eras were largely absent during the Cenozoic. The notable exception was the Zuni Sea, which occupied a large area in the continental interior during part of the Cenozoic. Sediments derived from the Laramide highlands to the west and southwest were transported eastward and deposited in a variety of continental, transitional, and marine environments.

Figure 18.26 **Maximum Extent of Pleistocene Glaciers in North America**

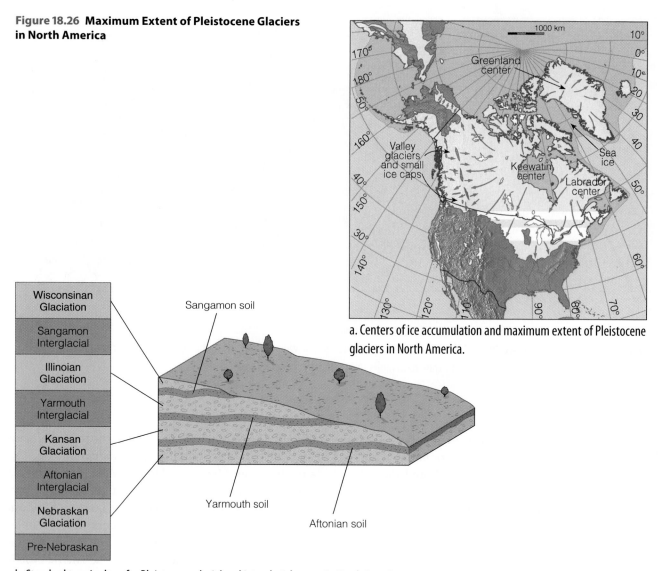

a. Centers of ice accumulation and maximum extent of Pleistocene glaciers in North America.

| Wisconsinan Glaciation |
| Sangamon Interglacial |
| Illinoian Glaciation |
| Yarmouth Interglacial |
| Kansan Glaciation |
| Aftonian Interglacial |
| Nebraskan Glaciation |
| Pre-Nebraskan |

Sangamon soil

Yarmouth soil

Aftonian soil

b. Standard terminology for Pleistocene glacial and interglacial stages in North America.

The Gulf Coast sedimentation pattern was established during the Jurassic and persisted through the Cenozoic. Much of the sediment deposited on the Gulf Coastal Plain was detrital, but in the Florida section of the coastal plain and the Gulf Coast of Mexico, significant carbonate deposition occurred. A carbonate platform was established in Florida during the Cretaceous, and carbonate deposition continues at the present.

EASTERN NORTH AMERICA

The eastern seaboard has been a passive continental margin since Late Triassic rifting separated North America from North Africa and Europe. The present distinctive topography of the Appalachian Mountains is the product of Cenozoic uplift and erosion. By the end of the Mesozoic, the Appalachian Mountains had been eroded to a plain. Cenozoic uplift rejuvenated the streams, which responded by renewed downcutting. As the streams eroded downward, they were superposed on resistant strata and cut large canyons across these strata. The distinctive topography of the Valley and Ridge Province is the product of Cenozoic erosion and preexisting geologic structures. It consists of northeast-southwest–trending ridges of resistant upturned strata and intervening valleys eroded into less-resistant strata.

PLEISTOCENE GLACIATION

We know today that the Pleistocene (Ice Age) began 1.8 million years ago and ended about 10,000 years ago. During this time, several intervals of widespread continental glaciation took place, especially in the Northern Hemisphere, each separated by warmer interglacial periods. In addition, valley glaciers were more common at lower elevations and latitudes, and many extended much farther than they do today.

As you would expect, the climatic effects responsible for Pleistocene glaciers were worldwide. Nevertheless, Earth was not as frigid as it is portrayed in movies and cartoons, nor was the onset of the climatic conditions leading to glaciation rapid. Indeed, evidence from several types of investigations indicates that the climate cooled gradually from the Eocene through the Pleistocene. Furthermore, evidence from seafloor sediments shows that 20 major warm–cold cycles have occurred during the last 2 million years.

Geologists know that, at their greatest extent, Pleistocene glaciers covered about three times as much of Earth's surface as they do now (Figure 18.26a). Like the vast ice sheets now in Greenland and Antarctica, they were probably 3 km thick. Geologists have identified four major Pleistocene glacial episodes that took place in North America—the *Nebraskan*, *Kansan*, *Illinoian*, and *Wisconsinan* (Figure 18.26b)—each named for the state in which the most southerly glacial deposits are well exposed. The three interglacial stages are named for localities of well-exposed soils and other deposits (Figure 18.26b).

The Pleistocene is not a period in the distant past; we see the effects of Pleistocene glaciation all around us in much of North America. Valley glaciation shaped many of the most majestic features of several national parks in the United States and Canada. Small valley glaciers are still active at high elevations in the Rocky Mountains, the Cascade Range, and mountains in Alaska.

CHAPTER 19
LIFE HISTORY

The fossil record of life on Earth stretches back almost 4 billion years. For most of that time span, life was single-celled bacteria and simple algae living in the seas. Complex, land-dwelling animals like this elk in Yellowstone National Park are, geologically speaking, recent phenomena indeed. More recent still are the self-aware and culturally sophisticated animals like the person who took this photo.

NATIONAL PARK SERVICE

Introduction

"We now know that extinction is the rule, not the exception."

Scientists have been seriously investigating the history of life on Earth for two centuries. Without science, our only knowledge of how the animals and plants around us changed would be from written human records. Such records are interesting, but they simply do not go far enough back in time to help us understand anything more than the current diversity of organisms and a few very recent extinctions.

By looking at geologic materials, scientists have clearly established the outlines of Earth history going back some 4 billion years. Although there is much we still do not know about the history of life, we have learned a great deal. For example, we now know that extinction is the rule, not the exception. We are sure that more than 99% of all species that have ever existed in our seas and on land are now extinct. But we have also learned that the variety and complexity of organisms have increased after episodes of mass extinction, time and time again.

This chapter has the ambitious task of summarizing what we know about the many and varied creatures that have inhabited our planet almost since Earth began. The evidence we rely on is the **fossils** that have endured through millions, even billions, of years. The quality of the *fossil record* varies considerably, depending on the types of organisms existing at a particular time and the environment in which they lived.

The preservation of any one organism as a fossil is rare, but fossils are nevertheless quite common. This apparent contradiction is easily explained when you consider that so many billions of organisms have existed during so many millions of years that if only a tiny fraction were preserved, the total number of fossils is phenomenal. In fact, fossils of many **vertebrate** animals (those with a segmented vertebral column), such as dinosaurs, are much more common than most people realize. In general, the fossil record does give a good overview of life history.

fossil The remains or traces of once-living organisms.

vertebrate Any animal that has a segmented vertebral column; includes fish, amphibians, reptiles, birds, and mammals.

stromatolite A biogenic sedimentary structure, especially in limestone, produced by the entrapment of sediment grains on sticky mats of photosynthesizing bacteria.

LEARNING OUTCOMES

After reading this unit, you should be able to do the following:

LO1 Discuss the evolution of life during the Precambrian

LO2 Review life during the Paleozoic Era

LO3 Discuss life of the Mesozoic Era

LO4 Review life during the Cenozoic Era

LO1 Precambrian Life History

Scientists had long assumed that the fossils so common in Paleozoic rocks must have had a long earlier history, but little was known about these organisms. A few enigmatic Precambrian fossils had been reported, but they were mostly dismissed as inorganic structures rather than traces of organisms. In fact, the Precambrian was once called the *Azoic*, meaning "devoid of life."

Then, during the early 1900s, Charles Walcott proposed that layered, moundlike structures from Proterozoic rocks in Ontario, Canada, were ancient reefs constructed by algae. However, paleontologists did not demonstrate until 1954 that they were actually the products of organic activity. These structures, now called **stromatolites**, still form in a few environments

Figure 19.1 Stromatolites

a. Stromatolites from the Neoproterozoic in Glacier National Park, Montana. Note the gently curved structures or layers in the rock.

b. Present-day stromatolites displaying their pillow-shaped growth in Shark Bay, Australia, one of few places where they still live.

such as Shark Bay, Australia, when sediment is trapped on sticky mats of photosynthesizing bacteria, commonly called *blue-green algae* (Figure 19.1). We now know that stromatolites are common in some Proterozoic-age rocks, but the oldest ones are in 3.3- to 3.5-billion-year-old (Paleoarchean) rocks in Australia.

The importance of stromatolites in Earth history should not be overlooked. The atmosphere of the planet during the Paleoarchean had no free oxygen (no oxygen gas), but stromatolites, like plants, produce oxygen as a by-product of photosynthesis. Slowly but surely, as the Archean and especially the Proterozoic unfolded, increasing amounts of oxygen accumulated in Earth's atmosphere.

Paleoarchean stromatolites are among the earliest fossils on Earth, but indirect evidence of life from before stromatolites comes from 3.8-billion-year-old rocks in what is now Greenland. These rocks contain small carbon spheres that may be of organic origin, but the evidence is not conclusive. All known fossils from Archean and Early Paleoproterozoic rocks are of single-celled bacteria lacking a cell nucleus that reproduced asexually, much as bacteria do today. Such cells are termed **prokaryotic cells** as opposed to **eukaryotic cells**, the latter having a cell nucleus and other internal structures not present in prokaryotic cells, and most reproduce sexually.

The origin of eukaryotic cells is one of the most important events in life history. No one doubts they were present by the Mesoproterozoic, and some evidence indicates they first evolved as long as 1.4 billion years ago. A currently popular theory (supported by evidence mostly from living organisms) holds that two or more prokaryotic cells entered into a beneficial symbiotic relationship, and the symbionts became increasingly interdependent until the unit could exist only as a whole.

The first eukaryotes (organisms composed of eukaryotic cells) were still single-celled, but by Paleoproterozoic or Mesoproterozoic time, the first multicelled organisms had made their appearance. Carbonaceous impressions of what appear to be multicelled algae are known from several areas, but the first multicelled animal fossils are found in Neoproterozoic-age rocks. Some of the oldest so far reported are from the 545- to 600-million-year-old Ediacaran fauna of Australia, consisting of multicelled animals now known from all continents except Antarctica. Some investigators think these fossils represent jellyfish, sea pens, segmented worms, and arthropods (Figure 19.2). One wormlike fossil has even been cited as a possible ancestor of the trilobites that were so common during the Early Paleozoic (Figure 19.2b). Other researchers disagree; they think these animals represent an early evolutionary development distinct from the ancestry of any present-day animals.

prokaryotic cell A cell that lacks a nucleus and organelles such as mitochondria and plastids; the cells of bacteria and cyanobacteria.

eukaryotic cell A cell with an internal membrane-bounded nucleus that contains chromosomes and other internal structures; possessed by all organisms except bacteria.

Figure 19.2 Fossils of the Ediacaran Fauna of Australia

a. *Tribrachidium heraldicum,* a possible primitive echinoderm.

b. *Spriggina floundersi,* a possible ancestor of trilobites.

The nature of the Ediacaran fossils is debated, but all scientists agree that complex, multicelled animals were not only present by the Neoproterozoic but also widely distributed. One good example is the fossil *Kimberella,* from Russia, which was perhaps a sluglike creature. In addition, tracks and trails provide compelling evidence of complex animals such as worms.

Although more fossils are being found, the Proterozoic fossil record is not very good, because the animals then present lacked durable skeletons. However, minute scraps of shell-like material and spicules, presumably from sponges, in Neoproterozoic rocks indicate that hard skeletal elements were present by then. Nevertheless, animals with durable skeletons of chitin (a complex organic substance), silica (SiO_2), and calcium carbonate ($CaCO_3$) were not abundant until the beginning of the Paleozoic Era.

LO2 Paleozoic Life History

At the beginning of the Paleozoic Era, animals with skeletons appeared rather abruptly in the fossil record. In fact, their appearance is described as an explosive development of new types of animals and is referred to as the "Cambrian explosion" by most scientists. This sudden appearance of new animals in the fossil record is rapid, however, only in the context of geologic time, having taken place over millions of years during the Early Cambrian Period.

The earliest of these animals with skeletons were *invertebrates*—that is, animals that lack a segmented vertebral column. Rather than focusing on the history of each invertebrate group (Table 19.1), we will survey the evolution of the Paleozoic marine invertebrate communities through time, concentrating on the major features and changes that took place.

MARINE INVERTEBRATES

Although almost all of the major invertebrate phyla evolved during the Cambrian Period (Table 19.1), many were represented by only a few species. Whereas trace fossils are common and echinoderms diverse, trilobites, brachiopods, and archaeocyathids (bottom-dwelling organisms that constructed reeflike structures and lived only during the Cambrian) comprised the majority of Cambrian skeletonized life. It is important to remember, however, that the fossil record is biased toward organisms with durable skeletons and that we generally know little about the soft-bodied organisms of that time. At the end of the Cambrian Period, trilobites suffered mass extinctions, and even though they persisted until the end of the Paleozoic Era, their numbers were considerably diminished.

A major transgression that began during the Middle Ordovician (Tippecanoe sequence) resulted in a widespread inundation of the craton. This vast, shallow sea, and a uniformly warm climate during this time, opened numerous new marine habitats that were soon filled by a variety of organisms, such that the Ordovician is characterized by a dramatic increase in the diversity of the total shelly fauna. The end of the Ordovician, however, was a time of mass extinctions in the marine realm. More than 100 families of marine invertebrates became extinct, and many geologists think these extinctions were the result of extensive glaciation that occurred in Gondwana at the end of the Ordovician Period (see Chapter 18).

TABLE 19.1

THE MAJOR INVERTEBRATE GROUPS AND THEIR GEOLOGIC RANGES

Phylum Protozoa	Cambrian–Recent	**Phylum Mollusca**	Cambrian–Recent
Class Sarcodina	Cambrian–Recent	Class Monoplacophora	Cambrian–Recent
Order Foraminifera	Cambrian–Recent	Class Gastropoda	Cambrian–Recent
Order Radiolaria	Cambrian–Recent	Class Bivalvia	Cambrian–Recent
Phylum Porifera	Cambrian–Recent	Class Cephalopoda	Cambrian–Recent
Class Demospongea	Cambrian–Recent	**Phylum Annelida**	Precambrian–Recent
Order Stromatoporoida	Cambrian–Oligocene	**Phylum Arthropoda**	Cambrian–Recent
Phylum Archaeocyatha	Cambrian	Class Trilobita	Cambrian–Permian
Phylum Cnidaria	Cambrian–Recent	Class Crustacea	Cambrian–Recent
Class Anthozoa	Ordovician–Recent	Class Insecta	Silurian–Recent
Order Tabulata	Ordovician–Permian	**Phylum Echinodermata**	Cambrian–Recent
Order Rugosa	Ordovician–Permian	Class Blastoidea	Ordovician–Permian
Order Scleractinia	Triassic–Recent	Class Crinoidea	Cambrian–Recent
Phylum Bryozoa	Ordovician–Recent	Class Echinoidea	Ordovician–Recent
Phylum Brachiopoda	Cambrian–Recent	Class Asteroidea	Ordovician–Recent
Class Inarticulata	Cambrian–Recent	**Phylum Hemichordata**	Cambrian–Recent
Class Articulata	Cambrian–Recent	Class Graptolithina	Cambrian–Mississippian

The mass extinction at the end of the Ordovician Period was followed by rediversification and recovery of many of the decimated groups. In fact, the Silurian and Devonian were times of major reef building in which organic reef builders diversified in new ways, building massive reefs larger than any produced during the Cambrian or Ordovician. Another mass extinction occurred near the end of the Devonian and resulted in a worldwide near-total collapse of the massive reef communities.

The Carboniferous invertebrate marine community responded to the Late Devonian extinctions in much the same way the Silurian invertebrate marine community responded to the Late Ordovician extinctions—that is, with renewed diversification. However, large organic reefs like those existing earlier in the Paleozoic virtually disappeared and were replaced by small patch reefs that flourished during the Late Paleozoic.

The Permian invertebrate marine faunas resembled those of the Carboniferous. However, they were not as widely distributed because of the restricted size of the shallow seas on the cratons and the reduced shelf space along the continental margins.

THE PERMIAN MASS EXTINCTION

The greatest recorded mass extinction to affect Earth's biota occurred at the end of the Permian Period (Figure 19.3). By the time the Permian ended, roughly 50% of all marine invertebrate families and about 90% of all marine invertebrate species became extinct. In addi-

tion, more than 65% of all amphibians and reptiles as well as nearly 33% of insects on land also became extinct.

What caused such a crisis for both marine and land-dwelling organisms? Currently, many scientists think an episode of deep-sea anoxia and increased oceanic CO_2 levels resulted in a highly stratified ocean during the Late Permian. In other words, there was very little, if any, circulation of oxygen-rich surface waters into the deep ocean. During this time, stagnant waters also covered the shelf regions, thus affecting the shallow marine fauna.

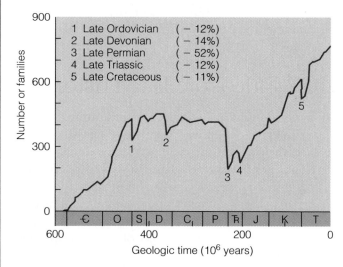

Figure 19.3 Phanerozoic Diversity of Marine Invertebrate and Vertebrate Families

Currently, many scientists think that a large-scale marine regression, coupled with climatic changes in the form of global warming (caused by an increase in carbon dioxide levels), may have been responsible for the mass extinctions recorded in the fossil record.

result in stagnant, stratified oceans, rather than a well-mixed oxygenated oceanic system.

During the Late Permian, widespread volcanism and continental fissure eruptions were also taking place. These eruptions released additional carbon dioxide into the atmosphere, thus contributing to increased climatic instability and ecologic collapse. By the end of the Permian, a near-collapse of both the marine and terrestrial ecosystem had occurred. Although the ultimate cause of such devastation is still being debated and investigated, it is safe to say that it was probably a combination of interconnected and related geologic and biologic events.

In addition, there is also evidence of increased global warming during the Late Permian. This would also contribute to a stratified global ocean, because warming of the high latitudes would significantly reduce or eliminate the down-welling of cold, dense, oxygenated waters from the polar areas into the deep oceans at lower latitudes, as occurs today. This would

VERTEBRATES

Besides the numerous invertebrate groups, vertebrates (animals with backbones) also evolved and diversified during the Paleozoic. Remains of the most primitive vertebrates, the fish, are found in Upper Cambrian marine rocks in Wyoming. These fish, referred to as *ostracoderms*, were jawless, had poorly developed fins, had an external covering of bony armor, and lived from the Early Cambrian to Late Devonian (Figure 19.4a).

Figure 19.4 Recreation of a Devonian Seafloor
a. An ostracoderm (*Hemicyclaspis*).   b. A placoderm (*Bothriolepis*).
c. An acanthodian (*Parexus*).   d. A ray-finned fish (*Cheirolepis*).

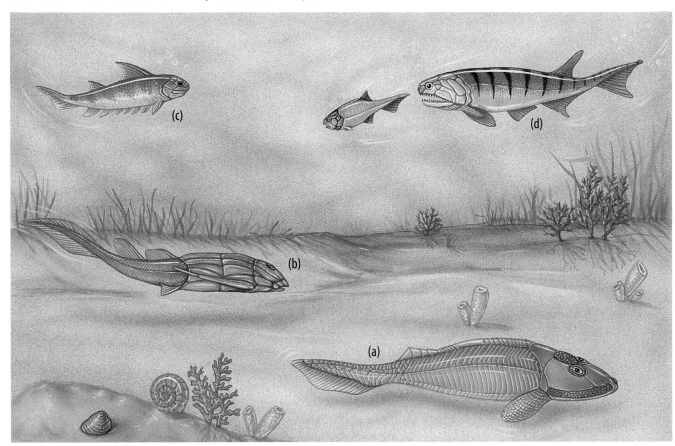

Figure 19.5 Artist's Conception of a Late Devonian Landscape in the Eastern Part of Greenland
Shown is *Ichthyostega*, an amphibian that grew to a length of about 1 m. The flora of the time was diverse, consisting of a variety of small and large seedless vascular plants.

The origin of jaws was a major evolutionary advance among primitive vertebrates. Whereas their jawless ancestors could only feed on detritus, jawed fish could chew food and become active predators, thus opening many new ecological niches.

The fossil remains of the first jawed fish are found in Lower Silurian rocks and belong to the *acanthodians*, a group of enigmatic fish characterized by large spines, scales covering much of the body, jaws, teeth, and reduced body armor (Figure 19.4c). Many scientists think the acanthodians included the probable ancestors of the present-day bony and cartilaginous fish groups.

The other jawed fish, the *placoderms* (heavily armored jawed fish), evolved during the Late Silurian. The placoderms exhibited considerable variety, including small bottom-dwellers (Figure 19.4b), as well as some of the largest marine predators that ever existed.

Among the bony fish, one group known as lobe-finned fish was particularly important, because they included the ancestor of the amphibians, the first land-dwelling vertebrate animals (Figure 19.5). In fact, the similarity between the group of lobe-finned fish known as *crosspterygians* and the earliest amphibians is striking and one of the most widely cited examples of a transition from one major group to another. However,

recent discoveries of older lobe-finned fish and newly published findings of tetrapod-like fish are filling the gaps in the evolution from fish to amphibians.

Although amphibians were the first vertebrates to live on land, having evolved by the Devonian, they were not the first land-dwelling organisms. Land plants, which probably evolved from green algae, first appeared during the Ordovician. Furthermore, insects, millipedes, spiders, and even snails invaded the land before amphibians. All of these organisms encountered several problems during the transition from water to land. The most critical obstacles for animals were drying out, reproduction, the effects of gravity, and the extraction of oxygen from the atmosphere by lungs rather than from water by gills. These problems were partly solved by some of the lobe-finned fish; they already had a backbone and limbs that could be used for support and walking on land, and lungs to extract oxygen from the atmosphere.

Amphibians were limited in colonizing the land, however, because they had to return to water to lay their gelatinous eggs and because they never completely solved the problem of drying out. In contrast, the reptiles evolved skin or scales to preserve their internal moisture and an egg in which the developing embryo is surrounded by a liquid-filled sac and provided with both food and a waste sac. The evolution of such an egg allowed vertebrates to colonize all parts of the land, because they no longer had to return to the water as part of their reproductive cycle. The oldest known reptiles evolved during the Mississippian Period and were small, agile animals.

One of the descendant groups of these early reptiles was the *pelycosaurs*, or finback reptiles, which became the dominant reptile group by the Permian (Figure 19.6). The pelycosaurs were the first land-dwelling vertebrates to become diverse and widespread. Moreover, they were the ancestors of the *therapsids*, the advanced mammal-like reptiles that gave rise to mammals during the Triassic Period.

As the Paleozoic Era came to an end, the therapsids constituted about 90% of the known reptile genera and occupied a wide range of ecological niches. The mass extinctions that decimated the marine fauna at the close of the Paleozoic had an equally great effect on the terrestrial population. By the end of the Permian, about 90% of all marine invertebrate species were extinct, compared with more than two-thirds of all amphibians and reptiles. Plants, in contrast, apparently did not experience as great a turnover as invertebrates and animals.

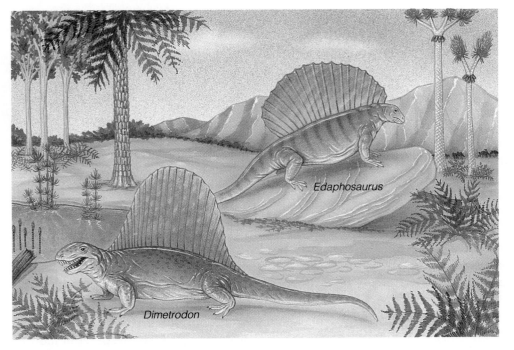

Figure 19.6 Pelycosaurs
Most pelycosaurs or finback reptiles had a characteristic sail on their back. One hypothesis explains the sail as a type of thermoregulatory device. Other hypotheses are that it was a sexual display or a device to make the reptile look more intimidating. Shown here are the carnivore *Dimetrodon* and the herbivore *Edaphosaurus*.

PLANTS

Plants encountered many of the same problems animals faced when they made the transition to land: drying out, the effects of gravity, and reproduction. Plants adapted by evolving a variety of structural features that allowed them to invade the land during the Ordovician.

The most common and widespread land plants are vascular plants, which have a tissue system of specialized cells for the movement of water and nutrients. Nonvascular plants, such as mosses and fungi, lack these specialized cells and are typically small and usually live in low, moist areas. Nonvascular plants were probably the first to make the transition to land, but their fossil record is poor.

The earliest known vascular land plants are small, leafless, Y-shaped stems from the Middle Silurian of Wales and Ireland. They are known as **seedless vascular plants**, because they did not produce seeds, nor did they have a true root system. Although these plants lived on land, they never completely solved the problem of drying out and were thus restricted to moist areas. Even their living descendants, such as ferns, are usually found in moist areas. During the Pennsylvanian Period, the seedless vascular plants became abundant and diverse because of the widespread coal-forming swamps that were ideally suited to their lifestyle.

Along with the evolution of diverse seedless vascular plants, another significant floral event occurred during the Devonian. The evolution of the seed at this time liberated vascular plants from their dependence on moist conditions and allowed them to spread over all parts of the land. The first to do so were the flowerless seed plants, or **gymnosperms**, which include the living cycads, conifers, and ginkgoes. Whereas the seedless vascular plants dominated the flora of the Pennsylvanian coal-forming swamps, the gymnosperms made up an important element of the Late Paleozoic flora, particularly in the nonswampy areas.

LO3 Mesozoic Life History

The Mesozoic Era is designated as the "Age of Reptiles," alluding to the fact that reptiles were the most common land-dwelling vertebrate animals. Although dinosaurs and their relatives evoke considerable interest, many other groups of organisms were not only present, but also quite common. Many invertebrates were common in the seas as well as on land, and vertebrates other than reptiles proliferated. For instance, mammals evolved from mammal-like reptiles during the Triassic, and birds probably evolved

seedless vascular plant A type of land plant that has specialized tissues for transporting fluids and nutrients throughout the body and that reproduces by spores rather than seeds (e.g., ferns and horsetail rushes).

gymnosperm A flowerless, seed-bearing land plant.

from small carnivorous dinosaurs during the Jurassic. Another wave of extinctions took place at the end of the Mesozoic.

© ISTOCKPHOTO.COM/ROBERT BLANDCHARD

MARINE INVERTEBRATES

Following the Permian mass extinctions, the Mesozoic was a time when marine invertebrates repopulated the seas. Among the mollusks, the clams, oysters, and snails became increasingly diverse and abundant, and the cephalopods were among the most important Mesozoic invertebrate groups. In contrast, the brachiopods never completely recovered from their near extinction and have remained a minor invertebrate group ever since. In areas of warm, clear, shallow marine waters, corals again proliferated, but these corals were of a new and more familiar type.

Single-celled animals known as *foraminifera* (Table 19.1) were also important, and they diversified tremendously during the Jurassic and Cretaceous periods. Floating or planktonic forms in particular became extremely common, but many of them became extinct at the end of the Mesozoic, and only a few types survived into the Cenozoic.

THE DIVERSIFICATION OF REPTILES

Reptile diversification began during the Mississippian Period, with the evolution of the first animals capable of laying eggs on land, which along with their skin or scales to preserve internal moisture truly freed them from an aquatic environment. From this basic stock of so-called *stem reptiles*, all other reptiles as well as birds and mammals evolved.

archosaur One of a group of animals including dinosaurs, flying reptiles (pterosaurs), crocodiles, and birds.

dinosaur Any of the Mesozoic reptiles that belong to the groups designated as ornithischians and saurischians.

Saurischia An order of dinosaurs characterized by a lizardlike pelvis; includes theropods, prosauropods, and sauropods.

Ornithischia One of the two orders of dinosaurs, characterized by a birdlike pelvis; includes ornithopods, stegosaurs, ankylosaurs, pachycephalosaurs, and ceratopsians.

bipedal Walking on two legs as a means of locomotion, as in humans, birds, and some dinosaurs.

quadrupedal Walking on four legs as a means of locomotion.

Archosaurs and the Origin of Dinosaurs

Reptiles known as **archosaurs** (*archo*, "ruling," and *sauros*, "lizard") include crocodiles, pterosaurs (flying reptiles), dinosaurs, and birds. Including such diverse animals in a single group implies that they share a common ancestor, and indeed several characteristics unite them. For instance, all have teeth set in individual sockets, except today's birds, but even the earliest birds had this feature.

All **dinosaurs** possess several shared characteristics yet differ enough for us to recognize two distinct orders: **Saurischia** and **Ornithischia**. Each order has a distinctive pelvic structure: Saurischian dinosaurs have a lizardlike pelvis and are thus called *lizard-hipped dinosaurs*, whereas Ornithischians have a birdlike pelvis and are called *bird-hipped dinosaurs* (Figure 19.7). It is now clear that they had a common ancestor much like archosaurs, known from Middle Triassic rocks in Argentina. These dinosaur ancestors were small (less than 1 m long), long-legged carnivores that walked and ran on their hind limbs, so they were **bipedal**, as opposed to **quadrupedal** animals, which move on all four limbs.

Dinosaurs

The term *dinosaur* was proposed by Sir Richard Owen in 1842 to mean "fearfully great lizard," although now "fearfully" has come to mean "terrible"—thus the characterization of dinosaurs as "terrible lizards." But of course they were not terrible, or at least no more so than animals living today, and they were not lizards. Nevertheless, dinosaurs more than any other kind of animal have inspired awe and thoroughly captured the public imagination (Figure 19.8). Their popularization in cartoons, books, and movies has unfortunately often been inaccurate and contributed to misunderstandings. For instance, many people think that all

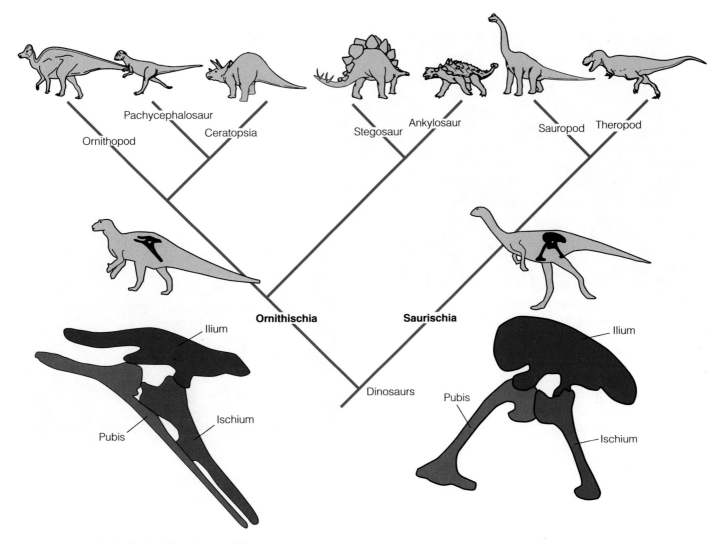

Figure 19.7 The Relationships Among Dinosaurs

Pelvises of ornithischian and saurischian dinosaurs are shown for comparison. All of the dinosaurs shown here were herbivores, except for the theropods. Note that bipedal and quadrupedal dinosaurs are found in both ornithischians and saurischians.

dinosaurs were large. It is true that *many* were large, but not all were large. In fact, dinosaurs varied from giants weighing several tens of metric tons to those no larger than a chicken.

Although various media now portray dinosaurs as more active animals, the misconception that they were lethargic beasts persists. Evidence now available indicates that some were quite active and perhaps even warm-blooded. It also appears that some species cared for their young long after hatching, a behavioral characteristic most often found in birds and mammals. Many questions, however, remain unanswered about dinosaurs, but their fossils and the rocks that contain

them are revealing more and more about their evolutionary relationships and behavior.

Among the saurischian dinosaurs, two suborders are defined: theropods and sauropods. All *theropods* were bipedal carnivores that ranged in size from tiny *Compsognathus* to comparative giants such as *Tyrannosaurus* and similar but even larger species. Some of the smaller theropods such as *Velociraptor* and its relative *Deinonychus*, with a large, sickle-shaped claw on each hind foot, likely used these claws in a slashing type of attack (Figure 19.8a). Some remarkable discoveries beginning in 1996 by Chinese paleontologists have yielded several species of small theropods with feathers.

Figure 19.8 Dinosaurs

PHOTO OF A RESTORATION AT THE CALIFORNIA ACADEMY OF SCIENCES, SAN FRANCISCO, COURTESY SUE MONROE

a. The theropod dinosaur *Deinonychus* in its probable attack posture. Restoration in the California Academy of Sciences, San Francisco.

SUE MONROE

b. Skeleton of the ceratopsian *Triceratops* in the Natural History Museum, London, England. This rhinoceros-sized herbivore was common during the Late Cretaceous of western North America.

Included among the sauropods are the giant, quadrupedal herbivores such as *Apatosaurus*, *Diplodocus*, and *Brachiosaurus*, the largest land animals of any kind. According to one estimate, *Brachiosaurus* weighed more than 75 metric tons, and partial remains indicate that even larger sauropods existed.

The diversity of ornithischians is evident from the five distinct suborders recognized: ornithopods, pachycephalosaurs, ankylosaurs, stegosaurs, and ceratopsians (Figure 19.8b). All ornithischians were herbivores, but some were bipeds, whereas others were quadrupeds. In fact, although ornithopods such as the well-known duck-billed dinosaurs were primarily bipeds, their well-developed forelimbs allowed them to walk on all fours as well. Also included among the ornithischians were the heavily armored ankylosaurs, the stegosaurs with plates on their backs as well as spikes on their tails for defense, the horned ceratopsians (Figure 19.8b) and the peculiar, domeheaded pachycephalosaurs.

Warm-Blooded Dinosaurs?

Were dinosaurs *endotherms* (warm-blooded) like today's mammals and birds, or were they *ectotherms* (cold-blooded) like all of today's reptiles? Almost everyone now agrees that some compelling evidence exists for dinosaur endothermy.

Bones of endotherms typically have numerous passageways that, when the animals are alive, contain blood vessels, but considerably fewer passageways are present in bones of ectotherms. Proponents of dinosaur endothermy note that dinosaur bones are more

Dinosaurs roamed Earth for more than 140 million years.

Figure 19.9 Flying Reptiles

a. This long-tailed pterosaur from the Jurassic of Europe had a wingspan of about 0.6 m.

b. *Pteranodon* was a short-tailed Cretaceous pterosaur with a wingspan greater than 6 m.

similar to those of living endotherms. Crocodiles and turtles have this so-called endothermic bone, yet they are ectotherms, and some small mammals have bones more typical of ectotherms. It may be that bone structure is related more to body size and growth patterns than to endothermy.

Endotherms must eat more than comparable-size ectotherms, because their metabolic rates are so much higher. Consequently, endothermic predators require large prey populations and thus constitute a much smaller proportion of the total animal population than their prey, usually only a few percent. Where data are sufficient to allow an estimate, dinosaur predators made up 3% to 5% of the total population. Nevertheless, uncertainties in the data make this argument less than convincing for many paleontologists.

Endothermy seems to be a prerequisite for having a large brain, because a complex nervous system needs a rather constant body temperature. Some dinosaurs did have large brains compared to body size, especially the small- and medium-sized carnivores, but many did not. So brain size for some dinosaurs may be a convincing argument, but even more compelling evidence for theropod endothermy comes from their probable relationship to birds, and the recent discoveries in China of theropods with feathers or a featherlike outer covering. Today, only endotherms have hair, fur, or feathers for insulation.

Good arguments for endothermy exist for several types of dinosaurs, although the large sauropods were probably not endothermic, but nevertheless were capable of maintaining a rather constant body temperature. Large animals heat up and cool down more slowly than smaller ones, because they have a small surface area compared to their volume. With their comparatively smaller surface area for heat loss, sauropods probably retained heat more effectively than their smaller relatives.

In general, a fairly good case can be made for endothermy in many theropods and in some ornithopods. Nevertheless, disagreement exists, and for some dinosaurs the question is still open.

Flying Reptiles

Paleozoic insects were the first animals to achieve flight, but the first among vertebrates were the **pterosaurs**, or flying reptiles. They were common in the skies from the Late Triassic until their extinction at the end of

pterosaur Any of the Mesozoic flying reptiles.

Figure 19.10 Mesozoic Marine Reptiles

a. Ichthyosaurs looked and probably lived much like today's porpoises.

b. Long-necked plesiosaurs were aquatic reptiles with flipper-like forelimbs. Both long-necked and short-necked plesiosaurs probably came ashore to lay their eggs.

the Cretaceous (Figure 19.9). Adaptations for flight include a wing membrane supported by an elongated fourth finger; light, hollow bones; and development of those parts of the brain associated with muscular coordination and sight. The fact that at least one pterosaur species had a coat of hair or hairlike feathers suggests that it, and perhaps all pterosaurs, were endotherms.

Most pterosaurs were no bigger than today's sparrows, robins, and crows. However, a few species had wingspans of several meters, and one Cretaceous pterosaur found in Texas had a wingspan of at least 12 m! The comparatively large pterosaurs probably took advantage of

ichthyosaur Any of the porpoiselike, Mesozoic marine reptiles.

thermal updrafts to stay airborne, mostly by soaring but occasionally by flapping their wings for maneuvering. Smaller pterosaurs probably stayed aloft by vigorously flapping their wings, just as present-day small birds do.

Marine Reptiles

The most familiar Mesozoic marine reptiles are the rather porpoiselike **ichthyosaurs** (Figure 19.10a). Most of these fully aquatic animals were about 3 m long, but one species reached about 12 m. All ichthyosaurs had a streamlined body, a powerful tail for propulsion, and flipperlike forelimbs for maneuvering.

Their numerous sharp teeth indicate they were fish eaters, although they undoubtedly preyed on other marine organisms as well. Ichthyosaurs were so completely aquatic that it is doubtful they could come onto land, so females probably retained eggs within their bodies and gave birth to live young. A few fossils with small ichthyosaurs within the appropriate part of the body cavity support this interpretation.

Another well-known group, the **plesiosaurs**, belonged to one of two subgroups: short-necked and long-necked (Figure 19.10b). Most were modest-sized animals 3.6 to 6 m long, but one species found in Antarctica measures 15 m. Short-necked plesiosaurs may have been bottom feeders, but their long-necked cousins probably used their necks in a snakelike fashion to capture fish with their numerous sharp teeth. These animals probably came ashore to lay their eggs.

BIRDS

Several fossils with feather impressions have been discovered in the Jurassic Solnhofen Limestone of Germany, but in almost every other known physical feature these fossils are most similar to small theropods. The birdlike creature known as *Archaeopteryx* retained dinosaurlike teeth, tail, brain size, and hindlimb structure but had feathers and a wishbone, characteristics typical of birds.

Most paleontologists now think that some kind of small theropod was the ancestor of birds. Even the wishbone consisting of fused clavicles so typical of birds is found in a number of theropods, and recent discoveries of theropods in China with some kind of feathery covering are further evidence of this relationship.

Two more Mesozoic fossils shed more light on bird evolution. One specimen, from China, is slightly younger than *Archaeopteryx* and has both primitive and advanced features. For instance, it retains abdominal ribs similar to those of *Archaeopteryx* and theropods, but it has a reduced tail more typical of present-day birds. Another Mesozoic bird from Spain is also a mix of primitive and advanced characteristics.

Archaeopteryx's fossil record is not good enough to resolve whether it is the ancestor of today's birds or an animal that died out without leaving descendants. Of course, that in no way diminishes the fact that it shows both reptile and bird characteristics. However, some scientists claim that fossils of two crow-sized individuals known as *Protoavis* are an even earlier bird than *Archaeopteryx*. These Late Triassic fossils have hollow bones and the breastbone structure of birds, but because no feather impressions were found, many paleontologists think these are specimens of small theropod dinosaurs.

MAMMALS

In a previous section, we briefly mentioned therapsids, or advanced mammal-like reptiles. One particular group of therapsids known as **cynodonts** was the most mammal-like of all and gave rise to mammals during the Late Triassic. This transition is especially well documented by fossils and is so gradual that classification of some fossils as either reptile or mammal is difficult.

In fact, the first mammals retained several reptilian characteristics but had mammalian features as well. A good example is their lower jaw. In typical mammals, the lower jaw is a single bone, whereas in reptiles it consists of several bones. The first mammals retained more than one bone in the lower jaw but had teeth and a jaw–skull joint characteristic of mammals. In fact, the quadrate and articular bones that formed the jaw–skull joint in reptiles were modified and became the incus and amalleus, two small bones in the mammalian middle ear. Another typical mammalian feature is occlusion, which means that the chewing teeth meet surface to surface to allow grinding, a feature also found in some advanced cynodonts.

In short, some mammalian features evolved more rapidly than others, thereby accounting for animals that have characteristics of both reptiles and mammals. Even though mammals appeared at the same time as dinosaurs, their diversity remained low, and all of them were small animals during the rest of the Mesozoic Era (Figure 19.11).

PLANTS

Triassic and Jurassic land plant communities were composed of seedless vascular plants and *gymnosperms*. Among the gymnosperms the conifers continued to diversify, and cycads that superficially resemble palms made their appearance. Both seedless vascular plants and gymnosperms are well represented in the present-day flora, but neither group is as abundant as it once was.

plesiosaur A type of Mesozoic marine reptile.

cynodont A type of advanced mammal-like reptile; the ancestor of mammals was among the cynodonts.

Figure 19.11 Mesozoic Mammals
One of the earliest mammals, the triconodont *Triconodon*.

SOURCE: ADAPTED WITH PERMISSION FROM J. BENES, *PREHISTORIC ANIMALS AND PLANTS*, 1979, PP. 174–175, ILLUSTRATION BY ZDENEK BURIAN, PUBLISHED BY HIPPOCRENE BOOKS, INC.

The long dominance of seedless plants and gymnosperms ended during the Early Cretaceous, when many were replaced by **angiosperms**, or flowering plants. Recent studies have identified both fossil and living gymnosperms that show close relationships to angiosperms. Since they first evolved, angiosperms have adapted to nearly every terrestrial habitat from mountains to deserts. Some have even adapted to shallow coastal waters. Their reproduction involving flowers to attract animal pollinators and the evolution of enclosed seeds largely accounts for their success. They now account for about 96% of all vascular plant species.

CRETACEOUS MASS EXTINCTIONS

The mass extinctions at the close of the Mesozoic were second in magnitude only to those at the end of the Paleozoic. Casualties of the Mesozoic extinctions include dinosaurs, flying reptiles, marine reptiles, and several kinds of marine creatures such as ammonites.

angiosperm Any of the vascular plants that have flowers and seeds; the flowering plants.

A hypothesis for extinctions that has become popular since 1980 is based on a discovery at the Cretaceous–Paleogene boundary in Italy—a clay layer 2.5 cm thick with an abnormally high concentration of the platinum-group element iridium. Since this discovery, high iridium concentrations have been identified at many other Cretaceous–Paleogene boundary sites. The significance of this discovery lies in the fact that iridium is rare in crustal rocks but occurs in much higher concentrations in some meteorites. Several investigators proposed a meteorite impact to explain this iridium anomaly and further postulated that the impact of a large meteorite, perhaps 10 km in diameter, set in motion a chain of events that led to extinctions.

The meteorite-impact scenario goes something like this: On impact, about 60 times the mass of the meteorite was blasted from Earth's crust high into the atmosphere, and the heat generated at impact started raging fires that added more particulate matter to the atmosphere. Sunlight was blocked for several months, causing a temporary cessation of photosynthesis; food chains collapsed and extinctions followed. In addition, with sunlight greatly diminished, Earth's surface temperatures were drastically reduced and could have added to the biologic stress.

Some now claim that a probable impact site has been found centered on the town of Chicxulub on the Yucatán Peninsula of Mexico. The structure is about 180 km in diameter and lies beneath layers of sedimentary rock.

Most geologists now concede that a large meteorite impact occurred, but we also know that vast outpourings of lava were taking place in what is now India. Perhaps these eruptions brought about detrimental atmospheric changes. Furthermore, the vast, shallow seas that covered large parts of the continents had mostly withdrawn by the end of the Cretaceous, and the mild, equable Mesozoic climates became harsher and more seasonal by the end of that era.

LO4 Cenozoic Life History

Even though we emphasize the evolution of mammals in this section, you should be aware of other important life events. Flowering plants continued to dominate land plant communities, the present-day groups of birds evolved during the Early Paleogene, and some marine invertebrates continued to diversify, eventually giving rise to today's marine fauna.

Mammals coexisted with dinosaurs for more than 100 million years, yet their Mesozoic fossil record indicates that they were not abundant, diverse, or very large. Extinctions at the end of the Mesozoic eliminated dinosaurs and some of their relatives, thereby creating the adaptive opportunities that mammals quickly exploited. The Age of Mammals, as the Cenozoic Era is commonly called, had begun.

Marine Invertebrates and Phytoplankton

Survivors of the Mesozoic extinctions populated the Cenozoic marine ecosystem. Especially abundant among the invertebrates were the foraminifera, radiolarians, corals, bryozoans, mollusks, and echinoids. Only a few species of phytoplankton survived into the Cenozoic, but those that did flourished and diversified. Marine and freshwater plants called *diatoms*, with silica skeletons, were particularly abundant.

Foraminifera were also abundant, and their shells accumulated to form thick limestones, some of which the ancient Egyptians used to construct the Sphinx and the Pyramids. Corals once again became the dominant reef builders, and just as during the Mesozoic, bivalves and gastropods were major components of the marine invertebrate communities.

Diversification of Mammals

Among living mammals, **monotremes**, such as the platypus, lay eggs, whereas marsupials and placentals give birth to live young. **Marsupial mammals** are born in an immature, almost embryonic condition, and then develop further in their mother's pouch. **Placental mammals** have no shelled egg, but they have developed a placenta within the uterus across which nutrients and oxygen are carried from the mother to the developing embryo. As a result, placental mammals are much more fully developed than marsupials before birth.

A measure of the success of placental mammals is that more than 90% of all mammals, fossil and living, are placental. Judging from the fossil record, monotremes have never been very common; the only living ones are platypuses and spiny anteaters of the Australian region. Marsupials have been more successful in terms of number of species and geographic distribution, but even they have been largely restricted to South America and the Australian region.

Although mammals first appeared during the Triassic, a major adaptive radiation began during the Paleocene and continued throughout the Cenozoic. Several groups of Paleocene mammals are *archaic*, meaning that they were holdovers from the Mesozoic Era or that they did not give rise to any of today's mammals (Figure 19.12). But also among these mammals were the first rodents, rabbits, **primates**, carnivores, and hoofed mammals. However, even these had not yet become clearly differentiated from their ancestors, and the differences between herbivores and carnivores were slight. Most were small; large mammals were not present until the Late Paleocene, and the first giant terrestrial mammals did not appear until the Eocene.

Diversification continued during the Eocene, when several more types of mammals appeared, but if we could go back and visit this time, we probably would not recognize many of these animals. Some would be vaguely familiar, but the ancestors of horses, camels, rhinoceroses, and elephants would bear little resemblance to their living descendants. By Oligocene time, all of the orders of existing mammals were present, but diversification continued as more familiar families and genera appeared. Miocene and Pliocene mammals were mostly mammals that we could readily identify, although a few unusual types still existed (Figure 19.13).

monotreme Any of the egg-laying mammals; includes only the platypus and spiny anteater of the Australian region.

marsupial mammal Any of the pouched mammals such as kangaroos and wombats that give birth to their young in a very immature state; most common in Australia.

placental mammal Any of the mammals that have a placenta to nourish the developing embryo; most living and fossil mammals.

primate Any of the mammals that belong to the order Primates; characteristics include large brain, stereoscopic vision, and grasping hands.

Figure 19.12 Paleocene Mammals

The archaic mammalian fauna of the Paleocene Epoch included such animals as (1) *Protictus*, an early carnivore; (2) insectivores; (3) *Ptilodus*; and (4) *Pantolambda*, which stood about 1 m tall.

CENOZOIC MAMMALS

Mammals arose from mammal-like reptiles known as *cynodonts* during the Late Triassic. Following the Mesozoic extinctions, they began an adaptive radiation and soon became the most abundant land-dwelling vertebrates. Now more than 4,000 species exist, ranging from tiny shrews to elephants and whales.

Numerous groups of mammals evolved during the Cenozoic, and some such as camels and horses and their relatives have excellent fossil records. Camels evolved from small, four-toed ancestors and were particularly abundant in North America, where most of their evolutionary history is recorded. They became extinct in North America during the Pleistocene, but not before some species migrated to South America and Asia. Horses and their living relatives, rhinoceroses and tapirs, also evolved from small Early Cenozoic ancestors (Figure 19.14). Horses and rhinoceroses were common in North America, and like camels, they died out here but survived in the Old World.

Whales provide an excellent example of how our knowledge of life history is constantly improving. Fossil whales have always been common, but until several years ago, fossils linking fully aquatic animals with land-dwelling ancestors were largely absent. It turns out that this transition took place in a part of the world where the fossil record was poorly known. Now, however, a number of fossils are available, showing that whales evolved from land-dwelling ancestors during the Eocene.

PLEISTOCENE FAUNAS

One remarkable aspect of the Cenozoic history of mammals is that so many very large species existed during the Pleistocene Epoch. In North America, there were mastodons and mammoths, giant bison, huge

cooler temperatures of the Pleistocene. Large animals have proportionately less surface area compared to their volume, and therefore retain heat more effectively than do small animals.

At the end of the Pleistocene, almost all of the large terrestrial mammals of North America, South America, and Australia became extinct. Extinctions also occurred on the other continents, but they had considerably less impact. These extinctions were modest by comparison to earlier ones, but they were unusual in that they affected mostly large terrestrial mammals. The debate over the cause of this extinction continues between those who think the large mammals could not adapt to the rapid climatic changes at the end of the Ice Age and those who think these mammals were killed off by human hunters, a hypothesis known as *prehistoric overkill.*

PRIMATE EVOLUTION

Several evolutionary trends in the order Primates help to define the order, some of which are related to its *arboreal*, or tree-dwelling, ancestry. These include changes in the skeleton and mode of locomotion, an increase in brain size, a shift toward smaller, fewer, and less specialized teeth, and the evolution of stereoscopic vision and grasping hands with opposable thumbs. Not all of these trends took place in every primate group, nor did they evolve at the same rate in each group.

Figure 19.13 Mural Showing Pliocene Mammals of the Western North America Grasslands

The animals shown are (1) *Amybeledon*, a shovel-tusked mastodon; (2) *Teleoceras*, a short-legged rhinoceros; (3) *Cranioceras*, a horned, hoofed mammal; (4) a rodent; (5) a rabbit; (6) *Merycodus*, an extinct pronghorn; (7) *Synthetoceras*, a hoofed mammal with a horn on its snout; and (8) *Pliohippus*, a one-toed grazing horse.

COURTESY OF THE SMITHSONIAN INSTITUTION, RECONSTRUCTION PAINTING OF TELEOCERAS MAMMALS BY JAY H. MATTERNES © 1982

ground sloths, giant camels, and beavers nearly 2 m long (Figure 19.15). Australia and Europe also had giant mammals. In addition to mammals, giant birds up to 3.5 m tall and weighing 585 kg existed in New Zealand, Madagascar, and Australia.

Many smaller mammals were also present, but the major evolutionary trend in mammals was toward large body size. Perhaps this was an adaptation to the

The *prosimians*, or lower primates, include the lemurs, lorises, tarsiers, and tree shrews and are the oldest primate lineage, with a fossil record extending back to the Paleocene. Sometime during the Late Eocene, the *anthropoids*, or higher primates— which include monkeys, apes, and humans—evolved from a prosimian lineage, and by the Oligocene, the anthropoids were a well-established group. The

Figure 19.14 Evolution of Horses
Simplified diagram showing some of the trends from the earliest known horse to the one-toed grazing horses of the present. Trends shown include increase in size, reduction in the number of toes and lengthening of the legs, and development of high-crowned teeth with complex chewing surfaces. Notice that *Merychippus* had three toes, whereas *Pliohippus* had only one. Another evolutionary lineage of horses, not shown here, led to the now-extinct three-toed browsers.

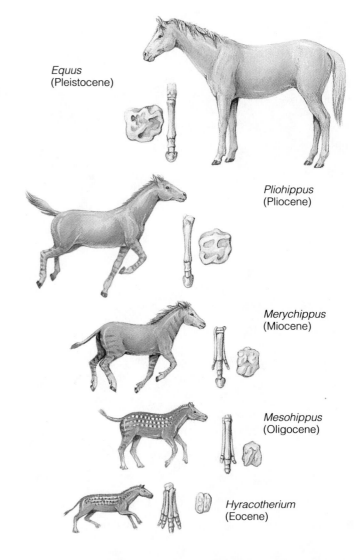

Equus
(Pleistocene)

Pliohippus
(Pliocene)

Merychippus
(Miocene)

Mesohippus
(Oligocene)

Hyracotherium
(Eocene)

Figure 19.15 Large Pleistocene Mammals
Among the diverse Pliocene and Pleistocene mammals of Florida were 6-m-long giant sloths and armored mammals known as glyptodonts that weighed more than 2 metric tons. Horses, camels, elephants, carnivores, and a variety of rodents were also common.

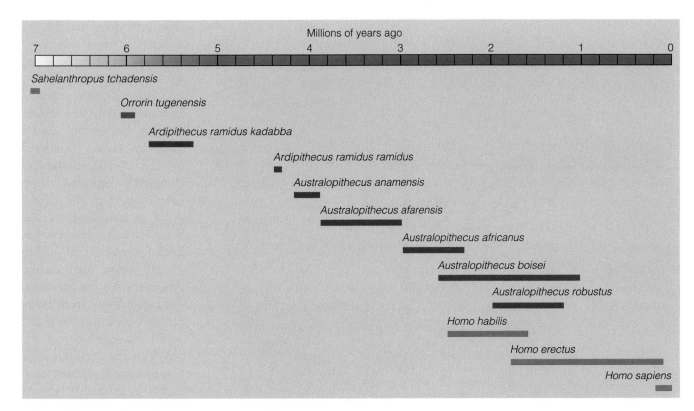

Figure 19.16 Fossil Record of Hominids
The geologic ranges for the commonly accepted species of hominids.

hominoids consist of the two families of apes and one of humans, and the extinct ancestors of both. The hominoid lineage diverged from Old World monkeys sometime before the Miocene.

During the Miocene, Africa collided with Europe, which changed Africa's climate and allowed animals to migrate between the two landmasses. Two apelike groups evolved during this time that ultimately gave rise to present-day hominoids. Members of one of these groups most likely was the ancestor of all later hominoids, the group that includes modern humans and our ancestors. The **hominids** (family Hominidae), the primate family that includes present-day humans and their extinct ancestors, have a fossil record extending back almost 7 million years (Figure 19.16). Several features distinguish them from other primates. Hominids are bipedal; that is, they have an upright posture and walk on two legs rather than four. In addition, they show a trend toward a large and complex brain. Other features include a reduced face and small canine teeth, omnivorous feeding, increased manual dexterity, and the use of sophisticated tools.

Australopithecine is a collective term for all members of the genus *Australopithecus*, which currently includes five species. Many paleontologists accept the evolutionary scheme in which *Australopithecus anamensis*, the oldest known australopithecine, is ancestral to *A. afarensis* (Figure 19.17), who in turn is ancestral to *A. africanus* and the genus *Homo*, as well as the side branch of australopithecines represented by *A. robustus* and *A. boisei*.

The earliest member of our own genus *Homo* is *Homo habilis*, which lived from 2.5 to 1.6 million years ago. Evolving from *H. habilis* was *H. erectus*, a widely distributed species that migrated from Africa during the Pleistocene Epoch. The archaeological record indicates that *H. erectus* made tools, used fire, and lived in caves, an advantage for those living in more northerly climates.

Debate still surrounds the transition from *H. erectus* to our own species, *Homo sapiens*. Paleoanthropologists are split into two camps. One view supports the "out of Africa" idea; that is, the idea that early modern humans evolved from a single woman in Africa, whose

hominoid An abbreviated term for Hominoidea, the superfamily that includes apes and humans.

hominid Abbreviated form of Hominidae, the family to which humans belong; bipedal primates include *Australophithecus* and *Homo*.

Figure 19.17 *Australopithecus afarensis*
Recreation of a Pliocene landscape showing *Australopithecus afarensis,* an early hominid species, gathering and eating fruits and seeds.

chiefly in caves and hutlike structures, which also contain a variety of specialized stone tools and weapons.

About 30,000 years ago, humans closely resembling modern Europeans moved into the region inhabited by the Neanderthals and completely replaced them. *Cro-Magnons*, the name given to successors of the Neanderthals in France, lived from about 35,000 to 10,000 years ago; during this period, the development of art and technology far exceeded anything the world had seen before.

With the appearance of Cro-Magnons, human evolution has become almost entirely cultural rather than biological. This means that although we are becoming rapidly more sophisticated in terms of the technology and knowledge we can bring to bear on our problems, we are not necessarily changing physically as a species.

offspring migrated—perhaps as recently as 100,000 years ago—to populate Europe and Asia, driving the earlier hominid population to extinction.

The alternative explanation, the "multiregional" hypothesis, maintains that early modern humans did not have an isolated origin in Africa, but rather that they established separate populations throughout Europe and Asia. Occasional contact and interbreeding between these populations enabled our species to maintain its overall cohesiveness, while still preserving the regional differences in people we see today. At this time, not enough evidence points to which, if either, theory is correct, and the emergence of *H. sapiens* from ancestral *H. erectus* is still not resolved.

Probably the most famous of all fossil humans are the *Neanderthals*, who inhabited Europe and the Near East from about 200,000 to 30,000 years ago. Based on specimens from more than 100 sites, we now know that Neanderthals were not much different from us, only more robust. Neanderthals were the first humans to move into truly cold climates. Their remains are found

The largest single cultural change for our species occurred not with the computer revolution but about 10,000 to 15,000 years ago when our forefathers learned to farm rather than to rely strictly on a hunter-gatherer lifestyle. The agricultural revolution occurred at several different places around the world at roughly the same time, as the last cold period of the Pleistocene was ending. It is what made settled life in towns possible—and eventually civilization as we know it.

We are still expanding the area of the planet that we have under the plow. This fact is just one aspect of our increasing impact on the globe. We now consume fossil fuels originally created in the Paleozoic and Mesozoic eras, and we mine mineral resources that date back all the way to the Proterozoic.

Many species in Earth history have changed the planet, from the humble stromatolites forward in time, but no species is changing the planet as rapidly as we humans are now doing. On the bright side, no species up until our own has been able to understand the changes it imposes on Earth—still our only home.

Index

What's Inside: Physical geology and historical geology; the scientific method; theory; hypothesis; geology and human experience; current geologic and environmental issues; origins of the universe and solar system; the planet Earth; plate tectonic theory; the rock cycle; igneous, sedimentary, and metamorphic rocks; the theory of organic evolution; the geologic time scale; and the principle of uniformatarianism.

Learning Objectives

Upon completion of this material, your students should understand the following.

- Geology is the study of Earth.

- Earth is a complex system of interconnected components that interact and affect one another in various ways.

- Theories are based on the scientific method and can be tested by experiments and/or observations.

- Geology plays an important role in human experience and affects us both as individuals and members of society and nation-states.

- The universe is thought to have originated approximately 14 billion years ago with a big bang. The solar system and planets evolved 4.6 billion years ago from a turbulent, rotating cloud of material surrounding the embryonic Sun.

- Earth consists of three concentric layers—core, mantle, and crust—and this orderly division formed during Earth's early history.

- Plate tectonics is the unifying theory of geology and this theory revolutionized the science.

- The rock cycle illustrates the interrelationships between Earth's internal and external processes and shows how and why the three major rock groups are related.

- The theory of organic evolution provides the conceptual framework for understanding the history of life.

- An appreciation of geologic time and the principle of uniformitarianism is central to understanding the evolution of Earth and its biota.

- Geology is an integral part of our lives.

Chapter Outline

Enrichment Topics

Topic 1. Scale of the Solar System. Use a football field analogy to convey to students the size of the solar system and relative distances of the planets to the Sun.

Put the Sun on the goal line.
Mercury is on the 1 foot line
Venus is on the 2 foot line
Earth is on the 1 yard line
Mars is on the 1 1/2 yard line
Jupiter is on the 5 yard line
Saturn is on the 10 yard line
Uranus is on the 20 yard line
Neptune is on the 30 yard line
Pluto is on the 40 yard line.

On this same scale, the nearest star would be 500 miles away.

Lecture Suggestions

1. A survey of the classroom and its contents illustrates our dependence on mineral substances (metal and plastic furniture, cement block or sheetrock walls, floors with vinyl tile or synthetic carpet, glass in light fixtures and windows). Further discussion could involve mineral fuels for heating, lighting, and air-conditioning. Even natural fibers and materials in students' clothing (wool, cotton, silk, leather) depend on proper soil and the availability of fresh water.

2. Almost every day there are stories in the newspapers and on radio and television that are relevant to geology—about volcanic eruptions, earthquakes, floods, landslides, subsidence or collapse of old mines, water quality, pollution, waste (especially toxic) disposal. Start a file of these clips and introduce them at the appropriate places in the course. It can be very effective to display a printed article as you discuss that article. Or suggest that students bring in articles they come across or summaries of information from newscasts.

3. Point out ways in which people employ the scientific method in drawing conclusions in their daily lives. Extract from these examples the elements and sequence of thought embodied by the scientific method. What are the facts? What is the explanation? Contrast this method with conclusions based on the supernatural. Where does faith come in? Is faith important in science? If so, where?

Key Terms

4. Teach students what a planet is by discussing why Pluto is no longer considered a planet. Encourage a class discussion on whether Pluto should regain its planetary status or remain as a lesser "dwarf planet."

5. As Thomas Kuhn has proposed, the geosynclinal theory represents perhaps the clearest example of how a reigning theory is questioned and is eventually discarded for another. In particular, emphasize the largely descriptive nature of geology and its hypotheses prior to plate tectonic theory as an analogy to the initial and latter stages of the scientific method. A discussion of plate tectonic theory could aid in illustrating the nature of the scientific method, the development of scientific theories, and the day-to-day business of doing science, as well as the elements and history of the theory itself. However, you may want to postpone this discussion until Chapter 2: Plate Tectonics: A Unifying Theory.

6. The rock cycle is really the rock recycle: any rock can become any other type of rock. The rock cycle is a description of the processes by which rocks and materials (such as magma or sediments) are endlessly transformed from one state to another.

7. When covering the principle of uniformitarianism, ask a number of students to each give one example of this principle drawn from their daily experience.

Consider This

1. Why is plate tectonics a theory and not a fact? What portions of it are theory, and what are fact?

2. Why is plate tectonics called the unifying theory of geology? How could the distribution of volcanoes, earthquakes, mountain ranges, and mineral deposits be explained without it?

3. Why can the rock cycle be considered a part of plate tectonics? Does the fact that the rock cycle involves the hydrologic cycle and atmospheric processes separate it from plate tectonics? How are hydrologic and atmospheric processes also a part of plate tectonics?

4. Speculate on the implications of plate tectonics with respect to world politics. What does plate tectonics have to do with global politics? How much better or worse might things be if Pangaea had never broken apart?

5. How would understanding earth history be different without the principle of uniformitarianism?

What's Inside: Continental drift hypothesis and evidence; Earth's magnetic field; paleomagnetism; magnetic reversals; seafloor spreading; plate tectonic theory; the three types of plate boundaries; and the driving mechanism of plate tectonics.

CHAPTER 2 | PREP CARD

Plate Tectonics: A Unifying Theory

Learning Objectives

Upon completion of this material, your students should understand the following.

- Plate tectonics is the unifying theory of geology and has revolutionized geology.

- The hypothesis of continental drift was based on considerable geologic, paleontologic, and climatologic evidence.

- The hypothesis of seafloor spreading accounts for continental movement, and the idea that thermal convection cells provide a mechanism for plate movement.

- The three types of plate boundaries are divergent, convergent, and transform. Along these boundaries, new plates are formed, consumed, or slide past one another.

- Interaction along plate boundaries accounts for most of Earth's earthquake and volcanic activity.

- The rate of movement and motion of plates can be calculated in several ways.

- Some type of convective heat system is involved in plate movement.

- Plate movement affects the distribution of natural resources.

- Plate movement affects the distribution of the world's biota and has influenced evolution.

Chapter Outline

Introduction 23

LO1   Early Ideas about Continental Drift 24
  Alfred Wegener and the Continental Drift
    Hypothesis 24

LO2   What Is the Evidence for Continental Drift? 25
  Continental Fit 25
  Similarity of Rock Sequences and Mountain Ranges 25
  Glacial Evidence 26
  Fossil Evidence 27

LO3   Earth's Magnetic Field 28

LO4   Paleomagnetism and Polar Wandering 30

LO5   Magnetic Reversals and Seafloor Spreading 31

Enrichment Topics

Topic 1. Pre- and Post-Pangaea Supercontinents. Pangaea, which started splitting up about 200 million years ago, was not the first supercontinent. Rodinia was together about 1 billion years ago, and there is evidence for a supercontinent as many as 3 billion years ago. How does the geologic history of Rodinia parallel that of Pangaea? What is the geologic evidence that the continent was once together and then broke up? The Burke Museum of Natural History and Culture, University of Washington, 2006. http://www.washington.edu/burkemuseum/ geo_history_wa/Dance%20of%20the%20Giant%20Continents.htm

Topic 2. Crazy Ideas in Science. Alfred Wegener's continental drift idea turned out to be true. Some of the other "Crazy Theories That Turned Out to Be True" involve ice ages and the asteroid impact theory for the dinosaur extinction. *Christian Science Monitor*, November 9, 2004. http://www.csmonitor.com/2004/1109/ p18s02-hfks.html

Topic 3. Visualizing Seafloor Spreading. Seismic reflection, used to image the mid-Atlantic ridge, reveals a variety of rock types. Bonatti et al., *Nature* 423: 499–505, 2003. http://www.nature.com/ nature/journal/v423/n6939/images/nature01594-f2.2.jpg.

Topic 4. Seeing Tectonic Plate Movement in Action. Global Positioning System (GPS) receivers are used to determine precise geodetic position measurements each day to chronicle plate movements. Horizontal movement vectors, which are the result of plate tectonic motions, are shown on the map, and motions for various locations can be graphed. http:// sideshow.jpl.nasa.gov/mbh/series.html

LECTURE SUGGESTIONS

1. Relative rates of motion between tectonic plates are a difficult concept to grasp for many beginning geology students. Conduct a demonstration in which members of the class are designated as specific plates and plate margins that move with a set of tape measures.

2. Why was the concept of seafloor spreading necessary for continental drift to be accepted? How could scientists ignore the overwhelming evidence that the continents could move over the face of Earth?

3. Demonstrate the relationship between hot spots and surface volcanic chains with a piece of paper and a lighted

Key Terms

Glossopteris flora 24
continental drift 24
Pangaea 24
magnetism 28
magnetic field 28
paleomagnetism 30
Curie point 30
magnetic reversal 31
seafloor spreading 31
thermal convection cell 32
magnetic anomaly 32
plate tectonic theory 33
divergent plate boundary 34
convergent plate boundary 36
oceanic–oceanic plate boundary 36
oceanic–continental plate boundary 38
continental–continental plate boundary 38
transform plate boundary 39
transform fault 39
hot spot 40

match. As you move the paper over the match a burn trace is left with the oldest burn at the farthest distance from the match. Be careful!

4. Use this idea to explain mid ocean ridge spreading. Many students in introductory geology generally understand that new crust is created at the mid oceanic ridge (MOR) by separation of plates, but some of the subtleties can escape them. For example, the bilateral symmetry of the paleomagnetic reversal pattern, or the migration of the ridge axis can be confusing. A demonstration using students can help clear up the situation.
   - Depending on the class size, all or part of the class may be used. Classrooms that have a central aisle are ideal. The volunteers, about 20 to 30 students, gather in the central aisle. They are referred to as the magma in the chamber underlying the central ridge axis. They are told to shuffle slowly to the front of the class in pairs, and for each pair to separate, one going left and the other to the right upon reaching the front of the room. When they reach the front of the room, they are to hold their arms up if the instructor's arms are up, or leave them down if the instructor's arms are down.
   - In this way, the plates grow at the front of the room, as the students diverge from each other at the ridge axis and are subsequently replaced by other students emerging from the magma chamber.
   - Hands up are normal polarity rocks and hands down are reverse polarity rocks. The bilateral symmetry of the paleomagnetic pattern and progression of rock age from oldest (first to come out) to youngest (last to come out) should now be obvious to most students. Having the students take a small step toward the audience for each minute that they are part of the new crust will generate the subsidence that is expressed as ridge topography. By varying the rate at which the students walk and separate, various paleomagnetic reversal pattern widths and ridge topographies (East Pacific Rise vs. Mid Atlantic Ridge) can be generated.

5. Note that the direction of plate movement at any given point along a spreading ridge or subduction zone must be perpendicular to that spreading ridge or subduction zone.

6. Stress the differences between continental and oceanic crust and between the lithosphere and the asthenosphere. Note especially the rigid behavior, lower density, and brittle, highly fractured nature of the continental crust.

7. A pot of boiling soup and its surface crust is a useful analogy for describing the process of convection cell motion.

CONSIDER THIS

1. Why is plate tectonics given the status of theory while continental drift attained only the status of hypothesis?
2. Because plate movements have been directly measured, is plate tectonics a theory or a fact? Why?
3. Compare and contrast continental drift and the plate tectonic theory. What does plate tectonics have that continental drift does not?
4. What is the origin of the term "transform" in transform fault?
5. Why are transform faults associated with spreading ridges at approximately right angles to the ridge? Why don't transform faults intersect subduction zones as well?
6. Why can spreading ridges never be directly connected to subduction zones, but instead the two must be linked by a transform fault?

What's Inside: Matter, elements, atoms, bonding, and compounds; crystalline solids; mineral composition; ferromagnesian and nonferromagnesian silicates and other mineral groupings; physical properties of minerals; rock-forming minerals; processes for forming minerals; natural resources and reserves.

Learning Objectives

Upon completion of this material, your students should understand the following.

- All matter, including minerals, is made up of atoms that bond to form elements and compounds.

- Geologists have a specific definition for the term mineral.

- You can distinguish minerals from other naturally occurring and manufactured substances.

- Minerals are incredibly varied, yet only a few are particularly common.

- Geologists use physical properties such as color, hardness, and density to identify minerals.

- Minerals originate in various ways and under varied conditions.

- Some minerals, designated rock-forming minerals, are particularly common in rocks, whereas others are found in small amounts.

- Some minerals and rocks are important natural resources that are essential to industrialized societies.

Chapter Outline

Enrichment Topics

Topic 1. Theory of the Atom. Although the theory of matter as a composite of extremely small, independent particulate units known as atoms originated with Greek philosophers, it is only recently that scientists have been able to observe and photograph, and even manipulate single atoms and their movement using the scanning electron microscope. Thus, until the invention of the electron microscope, the theory of the atom and all theory based on it were deduced from observation and experiment, the subject never having been observed. This is an example of how evidence can be used to create hypotheses and theories.

Topic 2. Chemical Composition of the "Average" Rock. It is interesting to look at the volume percentages of the components of Earth's crust. The volume percentages of the eight most common elements are:
Oxygen = 93.8 percent
Silicon = 0.9 percent
Potassium = 1.8 percent
Aluminum = 0.5 percent
Sodium = 1.3 percent
Iron = 0.4 percent
Calcium = 1.0 percent
Magnesium = 0.3 percent

Note that these percentages add up to 100.0 percent. This means that all other elements combined, including some economically important ones, are present in just miniscule traces.

Topic 3. A Gold Rush in Greek Mythology. Strabo, an ancient Greek historian and geographer, wrote that the inhabitants of Colchis, a country on the eastern shore of the Black Sea, used to separate gold carried down from the mountains in streams. They did this by running the stream sediments, in water, through troughs lined with fleeces. The gold stuck to the fleeces while the accompanying sediment was carried away. These fleeces were then hung to dry, after which the gold could be knocked off, like beating a carpet. According to Greek mythology, tales of the "Golden Fleece" inspired Jason and his crew on the Argo, and these Argonauts (the Greek, *nautikos* means sailor) sailed in search of it.

Topic 4. Mega Crystals. In "the Sistine Chapel of Crystals," gypsum crystals up to 50 feet (15 m) long are found. They are the largest natural crystals on Earth! How did these crystals form? Unlike most caves, these caves are hot! Why are

Key Terms

mineral 47
element 48
atom 48
nucleus 48
proton 48
neutron 48
electron 49
electron shell 49
atomic number 49
atomic mass number 49
bonding 49
compound 49
ion 50
ionic bond 50
covalent bond 51
crystalline solid 52
crystal 52
silicate 55
silica 55
silica tetrahedron 55
ferromagnesian silicate 57
nonferromagnesian silicate 57
carbonate mineral 57
luster 58
cleavage 59
hardness 60
specific gravity 60
density 61
rock 61
rock-forming mineral 61
resource 63
reserve 63

they hot? There is a plan to open the cave for tourists. How can this be done without destroying this beautiful natural resource? http://www.crystalinks.com/mexicocrystals.html

Lecture Suggestions

1. Cleavage planes are probably the most difficult mineral property for students to recognize. Point out that they can be recognized by naturally fractured and flat surfaces which all reflect or "wink" light at the same time when turned in the light. Also point out that although surfaces that are parallel to one another have the same cleavage plane, a sample need not show evidence of more than one cleavage surface in order to qualify it as a cleavage plane.

2. To demonstrate that cleavage within a crystal continues down to a very small scale (theoretically all the way to an atomic scale), examine a piece of mica or a crystal of halite and point out that the fracture lines that can be seen indicate the presence of additional planes of weakness within the crystal. Furthermore, one could hammer a halite crystal until the fragments are microscopic, and the three cleavage planes would still give each fragment a cubic shape.

3. Use a broken sample of crystalline quartz to illustrate the difference between crystal faces and cleavage planes. The quartz crystal faces are flat, but when broken, the crystal fractures like glass; it does not possess a cleavage. Compare this to a halite or calcite crystal, both of which do break along flat cleavage planes. Point out that crystal faces are naturally flat, reflective surfaces but differ from cleavage planes in that the former result from a crystal having grown into an open space, whereas the latter, which is a result of inter-atomic bonding, is seen after fracturing.

4. While the students are examining the quartz crystals, point out the fine, parallel striations that are on the crystal faces. These striations, which are perpendicular to the long axis of the original crystal, are always present on natural quartz crystals and develop during growth. Note that many "quartz crystals" sold as jewelry, art objects, or trinkets do not exhibit these striations. These commercially available crystals, evidently, have either been ground and polished, removing the lines and possibly reshaping the faces, or they might not be real quartz at all. If these "crystals" are actually glass, how could the student tell?

5. Note that streak is a more constant property for a mineral than the overall color, which can vary. Compare the streaks of different forms of the mineral hematite (Fe_2O_3). If you can, obtain samples which are earthy, sub-metallic, and metallic (specular hematite or specularite). They should all exhibit the same "dried blood" red streak, though their overall colors, as well as lusters, are different. The name hematite, incidentally, is from the Greek *haema,* meaning blood (the same root is found in hemoglobin, hemophiliac, hematology) in allusion to the color of the streak.

Consider This

1. Can a rock be composed of only one mineral? Why or why not? If so, give an example.
2. Can a native element be a mineral? Can a native element be a rock? Why or why not? If yes, give examples of each case.
3. Why don't some minerals, such as quartz, break along cleavage planes, whereas other minerals, such as halite, always do?
4. What properties of a mineral might give it special powers, such as the ability to heal? Because there are none, what is the reason that so many people think that minerals do possess powers?
5. Why are physical properties more important than chemical composition in determining whether or not a mineral is valued as a gemstone?
6. List at least 20 common items that have been derived or fabricated from mineral resources.
7. Give some examples of how searching for and controlling (or trying to control) mineral resources have played a major role in human history.

CHAPTER 4 | PREP CARD

Igneous Rocks and Intrusive Igneous Activity

Learning Objectives

Upon completion of this material, your students should understand the following.

- With few exceptions magma is composed of silicon and oxygen with lesser amounts of several other chemical elements.

- Temperature and composition are the most important controls on the mobility of magma and lava.

- Most magma originates within Earth's upper mantle or lower crust at or near divergent or convergent plate boundaries.

- Several processes bring about chemical changes in magma, so magma may evolve from one kind into another.

- All igneous rocks form when magma or lava cools and crystallizes, or by the consolidation of pyroclastic materials ejected during explosive eruptions.

- Geologists use texture and composition to classify igneous rocks.

- Intrusive igneous bodies called plutons form when magma cools below Earth's surface. The origin of the largest plutons is not fully understood.

Chapter Outline

Enrichment Topics

Topic 1. Igneous Rock Classification. There are lots of ways to classify igneous rocks. Grab a handful of igneous rocks, preferably with their chemical analyses, and explore some of those ways—color-texture, modal, normative. http://csmres.jmu.edu/geollab/Fichter/IgnRx/IgHome.html

Topic 2. Devil's Tower. Geologists do not agree on whether Devil's Tower is the neck of a long-dead volcano or an igneous intrusion. How could geologists tell the difference? http://www.nps.gov/archive/deto/geology.htm

Lecture Suggestions

1. Bowen's reaction series can be compared to changing two brick walls into two new walls. In one case—the discontinuous series—the bricks are individually removed, and a new wall is simultaneously built from the bricks but in a new pattern. The constituents of olivine, for example, are thus recreated into the new structure of pyroxene. In the second case—the continuous series—bricks are individually removed from the wall and replaced by different bricks having a different color but retaining the original pattern. For plagioclase, an original constituent (calcium) is replaced by a new constituent (sodium) but the structure is never torn down.

2. Bowen's reaction series is a very powerful tool for explaining more than one facet of rock formation. The series helps us understand why some minerals are typically found together in igneous rocks (e.g., olivine and pyroxene with calcium plagioclase; or quartz, biotite, and orthoclase with sodium plagioclase), but others are not (e.g., olivine and pyroxene with sodium plagioclase and orthoclase). Bowen's reaction series also explains temperature relations (why mafic magmas are hotter than felsic ones). It can also be seen why a mafic magma, by giving up heat, can partially melt material to form felsic magma, but the reverse isn't possible. Later, when weathering is discussed, Bowen's reaction series will help to understand the greater stability at Earth's surface of the lower temperature minerals (e.g., quartz) and lesser stability of the higher temperature ones (e.g., olivine).

3. Crystal settling may be explained by an analogy of a series of salts that precipitate as saltwater evaporates while the remaining water becomes increasingly saturated in

Key Terms

magma 66
lava 66
lava flow 66
pyroclastic materials 66
igneous rock 66
volcanic (extrusive igneous) rock 66
plutonic (intrusive igneous) rock 66
felsic magma 66
mafic magma 66
intermediate magma 67
viscosity 67
magma chamber 68
Bowen's reaction series 68
hot spot 71
mantle plume 71
crystal settling 71
assimilation 72
country rock 72
magma mixing 72
aphanitic texture 73
phaneritic texture 73
porphyritic texture 74
vesicle 74
pyroclastic (fragmental) texture 74
pluton 78
concordant pluton 79
discordant pluton 79
dike 79
sill 79
laccolith 79
volcanic pipe 80
volcanic neck 80
batholith 80
stock 80
stoping 81

the remaining salts. Precipitation from a solution and crystallization of a magma are not the same thing.

4. Dust off the lava lamp you relegated to the attic and bring it to class. You can use it to illustrate points about how densities vary with temperature, how gravity is the driving force in a rising magma, and what shapes are assumed by the rising bodies as they deform and shoulder aside the other material. You may want to note that unlike the lava lamp example, plutons, once they ascend, do not later descend back into the same rocks through which they rose.

Consider This

1. If small bodies such as dikes (e.g., basalt dikes) and sills cool rapidly and are therefore usually fine grained, how is it possible for pegmatites, having crystals several feet in length, to form in these small bodies?
2. Batholiths are thought to have been emplaced and solidified several kilometers underground, yet we can see them exposed in mountain ranges today. What has happened to all of the rocks that must have been above the batholiths?
3. How can a single volcano erupt distinctly felsic composition lavas in one event, and distinctly mafic lavas in another?

CHAPTER 5 | PREP CARD

Volcanoes and Volcanism

Learning Objectives

Upon completion of this material, your students should understand the following.

- In addition to lava flows, erupting volcanoes eject pyroclastic materials, especially ash and gases.

- Geologists identify the basic types of volcanoes by their eruptive style, composition, and shape.

- Although all volcanoes are unique, most are identified as shield volcanoes, cinder cones, or composite volcanoes.

- Volcanoes characterized as lava domes tend to erupt explosively and are therefore dangerous.

- Active volcanoes in the United States are found in Hawaii, Alaska, and the Cascade Range of the Pacific Northwest.

- Eruptions in Hawaii and Alaska are commonplace, but only two eruptions occurred in the continental United States during the 1900s and one in 2004. Canada has had no eruptions in recorded modern times.

- Some eruptions yield vast sheets of lava or pyroclastic materials rather than forming volcanoes.

- Geologists have devised the volcanic explosivity index as a measure of an eruption's size.

- Most volcanoes are located in belts at or near divergent and convergent plate boundaries.

- Some volcanoes are carefully monitored to help geologists anticipate eruptions.

Chapter Outline

Enrichment Topics

Topic 1. Volcanism and Pollutants. In the years it erupts, the biggest single source of air pollution in Washington State is Mount St. Helens, putting out between 50 and 250 tons of sulfur dioxide a day. SO_2 is a component of smog and also a contributor to acid rain. With their pollution control devices, all of the state's industries combined produce about 120 tons a day. Mount Etna in Italy can produce 100 times more SO_2 than Mount St. Helens. In all though, SO_2 emissions from volcanoes account for only about 15 million tons a year but emissions from human activities account for about 200 million tons. Volcanoes also produce the noxious gases hydrogen sulfide and hydrogen chloride. Volcanoes also emit CO_2, a greenhouse gas that contributes to rising global temperatures: Mount St. Helens between 500 and 1,000 tons a day and Mount Etna around 25 million tons. "Mount St. Helens Volcano is Washington State's No. 1 Air Polluter," *The Seattle Times*, 2004.

Topic 2. Undersea Eruption. An undersea eruption was caught on camera, producing 22 million cubic meters of new lava along the East Pacific Rise, one of the mid-ocean ridges. The lava buried seismometers that were placed along the ridge to monitor earthquake activity. Images show that the eruption was right along the ridge crest, just as predictions suggest new oceanic crust forms. http://dsc.discovery.com/news/2007/12/04/undersea-volcano.html

Topic 3. Plateau Basalts and Mass Extinctions. The Deccan Traps, a region of plateau basalts that erupted about 65 million years ago, may be the cause of the extinction of the dinosaurs and other contemporaneous Late Cretaceous faunas rather than the favored asteroid impact hypothesis. The argument is that the climatic effects of such large scale eruptions would be sufficient to cause extinctions. http://www.geosociety.org/news/pr/07-59.htm; http://geoweb.princeton.edu/people/keller/deccan/deccan.html

Topic 5. Predicting Eruptions. On January, 14, 1993, scientists meeting to discuss the science of predicting volcanic eruptions took a field trip, led by volcanologist Stanley Williams, into the crater of Galeras volcano in the Columbian Andes. After 40 years of dormancy, Galeras had come to life in March 1989 and erupted on July 16, 1992, when the lava dome blew apart and ash rose 3.5 miles into the air. After that, Galeras was quiet. During their trip into the crater, Williams and his colleagues made measurements they hoped would someday help them predict volcanic eruptions. As they were about to descend, tons of red- and white-hot ash and rock

Key Terms

volcanism 84
lava tube 85
pahoehoe 87
aa 87
columnar jointing 87
pillow lava 87
volcanic ash 89
volcano 91
crater 91
caldera 91
shield volcano 91
cinder cone 92
**composite volcano
(stratovolcano)** 93
lahar 93
lava dome 93
nuée ardente 95
basalt plateau 95
fissure eruption 95
pyroclastic sheet deposit 96
circum-Pacific belt 96
Mediterranean belt 96
Cascade Range 96
**volcanic explosivity index
(VEI)** 99
volcanic tremor 100

were thrown from the crater's mouth. The eruption was small and brief, but it was enough to kill six of the scientists and three hikers who were in the crater that day. Williams himself was seriously injured. The disaster allowed scientists to learn more about eruptions and since then, volcanologists have predicted several eruptions by monitoring earthquakes, the location of magma within a volcano, and the swelling in the flanks of a volcano. Between 1979 and 2000, 23 volcanologists died while working on volcanoes. Yet the importance of this work is underscored by the approximately 30,000 people who were killed by volcanoes in the past 20 years. Dr. Williams has written a book about the experience, *Surviving Galeras*, 2001.

Topic 6. Volcanism and Antarctic Ice Loss. Climate change may cause the West Antarctic ice sheet to melt, which will raise global sea level. The recent discovery of evidence of a volcanic eruption taking place beneath the ice is a big concern because a volcanic eruption beneath the ice will cause melting from the bottom, as well as melting from warmer temperatures at the top. Add to that the possibility that melting beneath the ice sheet can cause water to lubricate the location where the ice meets the underlying rock and cause the ice sheet to slip into the sea faster. http://sciencenow.sciencemag.org/cgi/content/full/2008/122/3

Lecture Suggestions

1. Differential viscosity can also be illustrated in this manner. Place three small samples of honey or molasses in separate beakers. Put one sample in a bath of ice water and another in a bath of hot water before class and let them sit until they warm or cool significantly. By pouring the three samples simultaneously, you can show the students how temperature has a significant effect on the viscosity of a given substance. They may remember this better if you remind them of the old phrase, "as slow as molasses in January."

2. Is the inside of Earth entirely molten? Try this simple demonstration. Bring two identical looking eggs to class, one of them hard boiled, the other uncooked. During the discussion of magma sources, spin the hard cooked egg on its long axis. Then, you may invite (or select) two students to come forward, give each of them one of the eggs, and ask each to spin his/her egg (not too close to the table edge). One student (with the cooked egg) should be able to do this, without much trouble. The other student will find it impossible. Some of the students by now may have guessed at the difference between the eggs, or you can elicit the answer. You can then point out the analogy with Earth, which spins very well (with a little wobble) on its axis.

3. Don't forget the comparison of dissolved gases in a magma coming out of solution as the magma rises to the surface (and pressure is released) to what happens when a bottle of club soda is opened. You can even bring a bottle to class, shake it up, and while pointing it at the class, ask what will happen if you open the top. Even when you don't go through with it, they will get the point. It's more fun when you "erupt" the bottle, though.

Consider This

1. Why are felsic magmas more viscous than mafic magmas?
2. Where are the most hotspots found? Are they found everywhere on oceanic plates? Are they found on continental plates? How does a mantle plume penetrate the thickness of the continental lithosphere to create an eruption?
3. Do volcanic eruptions vary in frequency on a human timescale? Are there certain years or decades in which eruptions are more common?
4. What different kinds of hazards can people living near volcanoes expect? Why do people live on and near known or potentially active volcanoes? What are some benefits which volcanoes may provide?

What's Inside: Mechanical and chemical weathering; regolith; soil horizons; soil erosion; designations of sedimentary particles; depositional environments; lithification; detrital and chemical sedimentary rocks; sedimentary facies; and resources found in sedimentary rocks.

CHAPTER 6 | PREP CARD
Weathering, Soil, and Sedimentary Rocks

Learning Objectives

Upon completion of this material, your students should understand the following.

- Weathering yields the raw materials for both soils and sedimentary rocks.

- Some weathering processes bring about physical changes in Earth materials with no change in composition, whereas others result in compositional changes.

- A variety of factors are important in the origin and evolution of soils.

- Soil degradation involves any loss of soil productivity that results from erosion, chemical pollution, or compaction.

- Sediments are deposited as aggregates of loose solids that may become sedimentary rocks if they are compacted and/or cemented.

- Geologists use texture and composition to classify sedimentary rocks.

- A variety of features preserved in sedimentary rocks are good indicators of how the original sediment was deposited.

- Most evidence of prehistoric life in the form of fossils is found in sedimentary rocks.

Chapter Outline

Enrichment Topics

Topic 1. Soils and a Positive Feedback Mechanism for Global Warming. Soils are rich with carbon, which microbes eat. Warmer temperatures are causing microbes to decompose soil carbon faster and release it into the atmosphere. This increases atmospheric CO_2, which increases global temperatures and increases the rate at which microbes decompose soil carbon. This is called a positive feedback mechanism for global warming. There are ways to increase the storage of carbon in the soil, but the effect would be small compared to the potential effect of the release of carbon from soil due to warmer temperatures. The only way to reduce the loss of soil carbon is to reduce greenhouse gas emissions into the atmosphere. http://www.csmonitor.com/2006/0421/p09s01-coop.html

Topic 2. Soil Erosion and Coral Reefs. Corals cannot exist in muddy water because sediment clogs their feeding apparatus and inhibits oxygen diffusion. Extensive deforestation in tropical areas is resulting in dramatic increases in the rates of soil erosion. This, in turn, is threatening the health and diversity of coral reefs, which exhibit incredible biodiversity and are nurseries for important commercial fish species. It is estimated that excess sediments due to deforestation have killed 75 percent of the coral reefs in the Philippines and in the Caribbean Sea off of Costa Rica. Australia's Great Barrier Reef receives five to ten times as much sediment as it did before the arrival of Europeans.

Topic 3. Philippine Logging Ban. The Philippine government has taken a drastic step to reducing soil erosion. In 2004, the Philippine government banned logging because soil erosion on deforested slopes brought about massive landslides that killed hundreds or thousands of people during storms. In February 2006, for example, a massive landslide killed more than 1,000 people and buried portions of the town of Guinsaugon. So many slopes have been deforested that it will be a long while before landslides are reduced in number and severity. http://news.bbc.co.uk/2/hi/asia-pacific/4068477.stm, http://news.bbc.co.uk/2/hi/asia-pacific/4723770.stm

Topic 4. Modern Agriculture and Soil Loss. Intensive farming greatly increases the erosion rate of farmland. Plowing loosens the soil and monoculture farming (planting only one crop) leaves it exposed for part of the year. Without protection, the soil is vulnerable to rain, wind, and gravity. Loose soil exposed to

Key Terms

the elements is easily and rapidly eroded. Usually the soil that erodes is the nutrient-rich topsoil. Artificial fertilizers must be added to make up for losses of soil nutrients or the land will not be as productive. About one-third of the world's farmland is currently eroding at 7 to 100 times the rate that soil is forming. China has the highest rate of soil loss, with a national annual average of about 18 tons per acre (40 metric tons per hectare). In some locations, soil is lost at 25 tons per acre (48 metric tons per hectare) per year. If every location lost soil at that rate, the entire world's topsoil would be eroded in 150 years. Since the arrival of Europeans, the United States has lost about one-third of its topsoil.

Lecture Suggestions

1. Point out the importance of soils to everyday activities. For example, if a student has a hamburger for lunch, everything which he/she eats depended on soil: the lettuce, tomato, and wheat for the bun were grown in soils, and the beef cattle had to eat plants grown in soil. If the hamburger was wrapped in aluminum foil, the aluminum came from bauxite.
2. Stress that detrital sedimentary rocks are classified primarily according to particle size, not composition. Note that sand is not a compositional term indicating quartz particles, but a size category.
3. Perform a simple demonstration of how graded beds form using a sturdy glass or clear plastic jar. Into the jar, place various sizes of sediment—some small gravel, coarse sand, fine sand, clay. Then, partially fill the jar with water. Thoroughly shake the jar then set it on the table, allowing the sediment to settle in a graded fashion. This shows how graded bedding can be used to determine the top direction, and tells something about the environment of deposition.
4. To illustrate how certain depositional environments are recognized by geologists, offer a few sets of sedimentary rock types, textures, structures, and types of fossils (e.g., mud cracks, raindrop prints, and fossil plants or vertebrates) and have the students determine which depositional environments are most likely represented by the collective evidence.

Consider This

1. The ecosystem encompasses the interactions among the lithosphere, hydrosphere, atmosphere, and biosphere. Soils are very different in different ecosystems. What role do they play in deserts, forests, tundra, and other ecosystems?
2. How did the rates of weathering and erosion change after the evolution of land plants?
3. Our attitudes toward the land and our "place" in nature largely determine how land is used. What are the three general attitudes toward land use commonly found within this and other societies?
4. What are some of the ways in which the rate of soil erosion can be reduced?
5. Is soil a renewable or nonrenewable resource?
6. What effect might increased soil erosion have on global warming?
7. If a sedimentary facies is deposited during a transgression or regression, is that facies of the same age everywhere? If so, why? If not, how might age equivalence within the facies be demonstrated?
8. Has the abundance of some sedimentary rock types changed over the duration of Earth's existence?
9. What types of sedimentary rocks are evidence of arid conditions and of tropical climates?
10. If fossils did not exist, would geologists have discovered that Earth has a history longer than recorded human history?

What's Inside: Agents of metamorphism including heat, pressure, and fluid activity; contact, dynamic, and regional metamorphism; metamorphic grade; foliated texture; nonfoliated texture; metamorphic zones; metamorphic facies; plate tectonics and metamorphism; and natural resources found in metamorphic rocks.

CHAPTER 7 | PREP CARD
Metamorphism and Metamorphic Rocks

Learning Objectives

Upon completion of this material, your students should understand the following.

- Metamorphic rocks result from the transformation of other rocks by various processes occurring beneath Earth's surface.

- Heat, pressure, and fluid activity are the three agents of metamorphism.

- Contact, dynamic, and regional metamorphism are the three types of metamorphism.

- Metamorphic rocks are typically divided into two groups, foliated and nonfoliated, primarily on the basis of texture.

- Metamorphic rocks with a foliated texture include slate, phyllite, schist, gneiss, amphibolite, and migmatite.

- Metamorphic rocks with a nonfoliated texture include marble, quartzite, greenstone, hornfels, and anthracite.

- Metamorphic rocks can be grouped into metamorphic zones based on the presence of index materials that form under specific temperature and pressure conditions.

- The successive appearance of particular metamorphic minerals indicates increasing or decreasing metamorphic intensity.

Chapter Outline

Enrichment Topics

Topic 1. Useful Garnets. Garnets are useful as gemstones and for industrial purposes, but they are also useful for geologists. Garnets are important minerals for geothermaobarometry, the measure of the previous temperature and pressure history of a metamorphic rock, because temperature-time histories are preserved in the compositional zonations that are typical of their growth pattern. A garnet with no zonation was likely homogenized by diffusion, which tells geologists something of the time-temperature history of the host rock. Garnets are also useful for determining the metamorphic facies of a rock. Rocks that had basalt composition often contain eclogite garnet. Pyrope is formed at high pressures, which means that a pyrope-containing peridotite formed at great depth, possibly as deep as 100 km. Silica-rich garnets are formed at even greater depths; if they are found as inclusions within diamonds they likely originated at 300 to 400 km depth in the crust.

Lecture Suggestions

1. When describing metamorphic rocks, it is important to stress that: a) foliation results from the preferred orientation of minerals; b) schist appears to be layered because its principle minerals (biotite and muscovite) are oriented with their flat cleavage surfaces parallel to one another; and c) the banded foliation of gneiss can be distinguished from other foliation types and from stratification by its coarse crystals and the differences in mineral composition from one layer to another.

2. Point out that foliation in a metamorphic rock does not have any necessary relationship to any preexisting layering, such as stratification, which may have been present in the parent rock. You could use a thick stack of cards (perhaps 5"x7" size) on the sides of which you rule a series of parallel lines to represent stratification. Be sure the "stratification" is at a distinct angle to the layered edges of the cards, which will represent "foliation."

3. Note that metamorphic rocks can form from any parent rock type, even another metamorphic rock. Make frequent reference to the parent rock type(s) from which a given metamorphic rock formed. Make clear that although parent rock type is easily identified in some metamorphic rocks such as quartzite and marble, the parent rock types of others often cannot be identified without a whole-rock

Key Terms

chemical analysis of the rock's composition (e.g., gneiss can form from impure sandstones or from granitic igneous rocks).

4. Review the rock cycle, focusing on metamorphism and metamorphic rocks, and emphasize the link of metamorphism to plate tectonics.

5. Note that nonfoliated rock types also exhibit recrystallization, which usually results in increased crystal size, but often without changes in mineral or chemical composition.

6. Stress that although heat is involved in the metamorphism of rocks, the heat never, by definition, is great enough to melt all or even a significant portion of the parent rock.

7. Note that new minerals can form in metamorphic rocks without introduction of new or additional elements, because the rock is transformed as a new equilibrium adjustment to increased pressure and temperature is reached. When hot fluids are involved, however, these may introduce new or additional elements and remove others. Thus, in some metamorphic rocks, new minerals are formed by additions and losses of various elements.

8. When discussing the causes and tectonic contexts of metamorphism, stress that low-temperature, high-pressure metamorphism is usually associated with disruptions of the crust such as folds and faults which result from compressional forces.

Consider This

1. Why can a glacier be considered a metamorphic rock?

2. How reasonable is it to assume that any metamorphic rock must be older than the igneous or sedimentary rock found in the same region?

3. Are diamonds the product of the metamorphism of anthracite coal?

4. What type(s) of metamorphism are likely to occur along continental-continental convergent plate boundaries?

What's Inside: Elastic rebound theory; seismology; the focus and epicenter; the circum-Pacific belt; the Mediterranean-Asiatic belt; P-waves; S-waves; R-waves; L-waves; locating an earthquake's epicenter; the Modified Mercalli Intensity Scale; the Richter Magnitude Scale; the destructive effects of earthquakes; predicting earthquakes; Earth's interior; the core; the mantle; Earth's internal heat; and the crust.

CHAPTER 8 | PREP CARD

Earthquakes and Earth's Interior

Learning Objectives

Upon completion of this material, your students should understand the following.

- Energy is stored in rocks and is released when they fracture, thus producing various types of waves that travel outward in all directions from their source.

- Most earthquakes take place in well-defined zones at transform, divergent, and convergent plate boundaries.

- An earthquake's epicenter is found by analyzing earthquake waves at no fewer than three seismic stations.

- Intensity is a qualitative assessment of the damage done by an earthquake.

- The Richter Magnitude Scale and Moment Magnitude Scale are used to express the amount of energy released during an earthquake.

- Great hazards are associated with earthquakes, such as ground-shaking, fire, tsunami, and ground failure.

- Efforts by scientists to make accurate, short-term earthquake predictions have thus far met with only limited success.

- Geologists use seismic waves to determine Earth's internal structure.

- Earth has a central core overlain by a thick mantle and a thin outer layer of crust.

- Earth possesses considerable internal heat that continuously escapes at the surface.

Chapter Outline

Enrichment Topics

Topic 1. What a Unit Means. This chart, from the USGS, is useful for students who want to check on what a unit or partial unit change in magnitude means for the displacement and energy of an earthquake. http://neic.usgs.gov/neis/eqlists/eqstats.html

Magnitude Change	Ground Motion Change (Displacement)	Energy Change
1.0	10.0 times	about 32 times
0.5	3.2 times	about 5.5 times
0.3	2.0 times	about 3 times
0.1	1.3 times	about 1.4 times

Topic 2. Hills Magnify Ground Shaking. Seismologists have known for a long time that hills amplify ground motions from earthquakes. New models show that hills that are topped by soil experience ground motions more than 80 percent higher than on the nearby plain. The high-frequency waves, those that cause the most damage, are magnified as much as 10 times. The waves get trapped in the thin layer of soil and are reinforced as they bounce between its upper and lower boundaries. These findings have important implications for building construction on hills. *Science News*, May 4, 2002.

Topic 3. More about the Boxing Day Tsunami. On December 26, 2004, the most massive earthquake in 40 years, measuring 9.0 on the Richter Magnitude Scale, struck 60 miles off the northwestern coast of Sumatra. The quake released the stresses that build up as the India Plate subducts into the Sunda Trench beneath the overriding Burma plate. The earthquake was the result of thrust faulting and it had a shallow focus. About 1,200 km of the plate boundary slipped, and the average displacement on the fault was about 15 meters. The seafloor above the fault was likely uplifted several meters and it was this vertical motion that displaced the trillions of tons of water above it and generated the tsunami. Several aftershocks measuring about 5.0 on the Richter Magnitude Scale followed. The tsunami sped across the sea surface, barely noticeable in open water, but grew as it went up the shoreline. Within one hour, the wave slammed into Sumatra and within two, it reached India. The tsunami reached Africa about eight hours later. The tsunami struck in separate waves, the third being the most massive. In all, eleven countries around the Indian Ocean suffered damage. Wave heights were greatest near the earthquake epicenter and decreased with distance, ranging from over 33 feet (10 m) tall at Sumatra down to

Key Terms

13 feet (4 m) at Sri Lanka, Thailand, and Somalia The death toll from the tsunami is around 230,000 making it the most deadly earthquake in recorded history. About 1.2 million people were left homeless.

Topic 4. Imaging Earth's Interior with Tomography. There are many beautiful images available on the internet that show the results of seismic tomography studies. To find some, Google "seismic tomography" and choose images. Here's one in which scientists have drawn in features of a subduction zone: http://www.whoi.edu/cms/images/lstokey/2005/1/v42n2-detrick2en_5301.jpg

And here's another showing blobs in the mantle: http://www.geo.cornell.edu/geology/classes/Geo101/graphics/s12fsl.jpg

Lecture Suggestions

1. The difference between P- and S-wave motions can be effectively illustrated using a Slinky.
2. Check out the earthquake database of the National Earthquake Information Center, maintained by the U.S. Geological Survey in Golden, Colorado (http://wwwneic.cr.usgs.gov/). Collect data on numbers and magnitudes of earthquakes that have occurred in the past week or two. The class will be interested to see how many quakes occur every day.
3. Demonstrate density differences among iron (meteorite, if available), peridotite, basalt, and granite to suggest how concentric layers in Earth's interior may have developed.
4. Bring to class samples of the interior of Earth: granite and basalt for the two types of crust, peridotite for the mantle, and if possible, an iron-nickel meteorite for the core. Point out that these are not entirely accurate representation.

Consider This

1. What tests could you design to determine whether mantle convection takes place only in the upper portion or if it involves the entire mantle?
2. Suggest a reason why, as seismic tomography has indicated, there are large pockets of hot material beneath the interiors of continents rather than along the spreading ridges, and why there are zones of cold rock that encircle and extend inward beneath the continents on their Pacific margins.

What's Inside: Historic oceanic investigations; methods of investigating the ocean floor; composition of the oceanic crust; continental margins; volcanism at continental margins; oceanic trenches; oceanic ridges; submarine hydrothermal vents; seafloor sedimentation; and natural resources found in the ocean.

CHAPTER 9 | PREP CARD

The Seafloor

Learning Objectives

Upon completion of this material, your students should understand the following.

- Scientists use echo sounding, seismic profiling, sampling, and observations from submersibles to study the largely hidden seafloor.

- Oceanic crust is thinner and compositionally less complex than continental crust.

- The margins of continents consist of a continental shelf and slope and in some cases, a continental rise with adjacent abyssal plains. The elements that make up a continental margin depend on the geologic activity that takes place in these marginal areas.

- Although the seafloor is flat and featureless in some places, it also has ridges, trenches, seamounts, and other features.

- Geologic activities at or near divergent and convergent plate boundaries account for distinctive seafloor features such as submarine volcanoes and deep-sea trenches.

- Most seafloor sediment comes from the weathering and erosion of continents and oceanic islands and from the shells of tiny marine organisms.

- Organisms in warm, shallow seas build wave-resistant structures known as reefs.

Chapter Outline

Enrichment Topics

Topic 1. Our Dying Coral Reefs. Coral reefs are known as the "rainforests of the sea" because they harbor such an incredible abundance and diversity of life. These spectacular and beautiful ecosystems are home to more than one-fourth of all marine plant and animal species. Reefs are built of tiny coral polyps that construct calcium carbonate ($CaCO_3$) shells around their bodies. The coral polyps enjoy a mutually beneficial relationship with minute algae called zooxanthellae; the photosynthetic algae supply oxygen and food to the corals, and the corals provide a home and nutrients (their wastes) for the zooxanthellae. The algae give the coral their bright colors of pink, yellow, blue, purple, and green. Coral are very temperature sensitive; if the water gets too hot, the coral eject their zooxanthellae, giving up their means of producing food. Without the algae, the coral turn white, a phenomenon called coral bleaching. Sometimes zooxanthellae move back in when conditions improve, and sometimes a heat tolerant species of algae will take its place. But if the reef goes without algae for too long, the coral starves and the reef dies. Coral reefs may recover from one bleaching event, but multiple events can kill them. Dr. Clive Wilkinson, coordinator of the Global Coral Reef Monitoring Network, blames the current upsurge in coral bleaching on rising seawater temperatures due to global warming. There are many other problems damaging or killing coral reefs including excess sedimentation, pollution, and insensitive tourists. According to Wilkinson's report, *Status of Coral Reefs of the World: 2004*, 20 percent of coral reefs are severely damaged and unlikely to recover, and another 24 percent are at imminent risk of collapse.

Topic 2. Life at the Very Bottom of the Sea. The Challenger Deep is the deepest spot in the ocean, a hole at the bottom of the Marianas Trench–10,896 meters below the sea surface. Amazingly, Japanese researchers have found life in the sediments at these great depths. The organisms are foraminifera, single-celled creatures that are common at the sea surface, but rare in the deep sea. The Challenger Deep species are mostly new; rather than being multi-chambered and having a shell like surface forms, they are soft-walled. DNA analyses of these critters suggest that they are a primitive form dating from the Precambrian and are the ancestors of the more complex, more modern forms. The organisms probably evolved from populations that were able to adapt to steady increases in pressure as the Challenger Deep developed

Key Terms

seismic profiling 177
ophiolite 178
continental margin 179
continental shelf 179
continental slope 180
continental rise 180
turbidity current 180
submarine fan 180
submarine canyon 180
active continental margin 181
passive continental margin 182
abyssal plain 183
oceanic trench 183
oceanic ridge 184
submarine hydrothermal vent 184
black smoker 184
seamount 186
guyot 186
aseismic ridge 187
pelagic clay 187
ooze 188
reef 189
Exclusive Economic Zone (EEZ) 190

during the past 6 to 9 million years. *National Geographic News*, February 3, 2005. http://news.nationalgeographic.com/news/2005/02/0203_050203_deepest.html.

Topic 3. Hydrothermal Vent Life. Hydrothermal vents are incredibly harsh environments for life. The fluids are very acidic (with a pH as low as 2.8); they emit poisonous gases and toxic metals. Fluid temperatures can be up to 750°F (400°C) but grade into frigid seawater. Hydrothermal vents are the only mid- or deep-sea communities that produce their own food. Too deep and dark for photosynthesis, they are the only communities on Earth in which chemosynthesis provides the bulk of the food energy. Hydrothermal vent communities support a typical food web, in which some animals, like snails, eat the bacteria and some, like fish, consume the animals that eat the bacteria. Other animals live symbiotically with the bacteria. In one symbiotic vent relationship, chemosynthetic bacteria live within the tissues of giant tube worms (*Riftia pachyptila*); they provide the worms with a constant source of food and are given shelter in return. Life forms differ between oceans and vents. In all, the vents are home to more than 300 species, nearly all of them unique to vent sites, such as giant clams (*Calyptogena magnifica*) and eyeless shrimp (*Rimicaris exoculata*). Exploring descriptions and photos of vent life is an interesting way to draw students in to this strange world.

Lecture Suggestions

1. If a globe made before the mid-1960s is available, the fact that the seafloors have no information plotted, like the "terra incognita" of earlier times, may serve to illustrate the amount of knowledge that has been acquired about the seafloor and how important its exploration was to geology and the formulation of plate tectonic theory.
2. Oceanographic studies have commonly been advanced by strategic and geopolitical concerns. The echo sounder, used to determine ocean depths and ultimately to map the oceans, was developed during World War II to provide better underwater information on submarines. During the Cold War era, the submarine became a principle component of offensive and defensive planning, and improved maps for all the ocean floors became imperative. These improved maps formed the basis for other studies and led to the recognition of the relationships among such things as oceanic ridges, trenches, island arcs, and the major fractures now known as transform faults.
3. The relationships between sea-floor spreading and sea level may be illustrated with an aquarium and two blocks of rock, one larger and representing the greater volume of hot rock extruded at the mid-ocean ridges during intervals of rapid spreading, and the other, smaller one representing the cooler mid-ocean ridges extruded during times of lower spreading rates. By placing one block and then the other in the aquarium tank, you can illustrate the different displacement of water.
4. There has been a tremendous amount of work done lately on the environmental problems facing the oceans. Bringing in some of these topics can add relevancy and immediacy to a lecture.

Consider This

1. Spreading ridges worldwide have elevations of 2,500 m, ± 200 m. Sea level is controlled in part by the spreading rates as well as the constant rate of cooling of spreading ridge basalts. Suggest a way to determine the depth of the Atlantic Ocean basin at any time since the rifting of Pangaea.

CHAPTER 10 | PREP CARD

Deformation, Mountain Building, and the Continents

Learning Objectives

Upon completion of this material, your students should understand the following.

- Rock deformation involves changes in the shape or volume or both of rocks in response to applied forces.

- Geologists use several criteria to differentiate among geologic structures such as folds, joints, and faults.

- Correctly interpreting geologic structures is important in human endeavors such as constructing highways and dams, choosing sites for power plants, and finding and extracting some resources.

- Deformation and the origin of geologic structures are important in the origin and evolution of mountains.

- Most of Earth's large mountain systems formed, and in some cases, continue to form, at or near the three types of convergent plate boundaries.

- Terranes have special significance in mountain building.

- Earth's continental crust, and especially mountains, stands higher than adjacent crust because of its composition and thickness.

Chapter Outline

Enrichment Topics

Topic 1. Devastating Indian Earthquakes. Plate tectonics is responsible for earthquakes in India in which tens of thousands of people die. In January 2001, more than 10,000 people were killed in an intraplate earthquake on the Indian subcontinent. The Indian subcontinent is driving ever northward into Asia, so even where the plates don't meet, weak spots are prone to earthquakes. India has suffered five such earthquakes since 1965; the January 2001 quake was on an ancient rift that originated when India separated from Gondwana 150 million years ago. At an even greater risk is the northeastern edge of the plate where it is colliding with Asia, and the strain is building quickly. *Science*NOW, January 29, 2001.

Topic 2. Birth of a Mountain Range, But When? There are two camps in the debate on when the Sierra Nevada Mountains rose— one says that it was between 40 and 60 million years ago when a subducted ocean plate slid beneath the continent and raised it up; the other thinks that it occurred between 3 and 5 million years ago when a large chunk of crust broke off from beneath the continent, melted, became buoyant, and caused the range to rise. One way of trying to determine when the range rose is to look at hydrogen isotopes of ancient raindrops to distinguish the height of the cloud from which the drop fell. This technique favors the first explanation, that the Sierra Nevada rose about 50 million years ago, but the results are still controversial. *Science*NOW, July 6, 2006.

Lecture Suggestions

1. A large sample of "silly putty" can be used to illustrate how a given material can respond differently to different stresses or stresses applied at different rates. If a ball of the material is dropped from a short distance onto a table top, it will bounce; and although a small amount of the stress, resulting from the impact of the ball with the surface, is accommodated in a plastic manner (leaving a flat spot on the ball), most of the stress has been accommodated in an elastic fashion. The material can then be deformed plastically by squeezing it (compression) or stretching it out (tension), or simply by letting gravity pull on the ball as it sits on the table. Fracture can result, of course, if the material is pulled (by tension or shearing) too fast. Point out that even though "silly putty" is not a rock, rocks can respond in similar fashion under the right circumstances.

Key Terms

2. Strike and dip are difficult to visualize and need to be demonstrated in class. To illustrate strike, as well as apparent and true dip, stack some books and prop them at an angle. Let the binding's trace be the strike and the cover be the bed's dip surface. Take a pencil and orient it on the dip surface (cover) so that it is parallel to strike (the binding). Now slowly rotate the pencil so its eraser end remains fixed on the dip surface and its point moves away from the dip surface in a horizontal plane, until it lies perpendicular to the strike (binding edge). Notice that the distance between the pencil point and the dip surface (book cover) increases from zero (when the pencil is parallel to strike) to a maximum value, when it is perpendicular to the strike. The angle between the eraser and the pencil's projection (or its shadow from an overhead light) on the cover will increase from zero to some maximum value. These are dip or inclination angles, and the maximum value is that of the true dip.

3. Students are likely to remember hanging wall and foot wall if they are told that the hanging wall is characterized as that wall from which a geologist could only hang (as opposed to walk), and the footwall as that wall which would be beneath a geologist's feet.

4. Once they can recognize the hanging wall and footwall, it may help the students remember the types of dip-slip faults from the acronym FUN. This stands for "Footwall (moved) Up = Normal" fault. Obviously, the other type is HUR ("Hanging wall Up = Reverse").

5. Ask students to compare and contrast the types of geologic features, structures, and activity that occur on continental-continental, oceanic-continental, and oceanic-oceanic convergent plate boundaries. Also compare and contrast the geologic features, structures, and activity along divergent, convergent, and transform plate boundaries.

6. The terms and structures of anticlines and synclines can be more readily understood if it is noted that "cline" means slope and "anti" means opposite or away from, whereas "syn" means together or toward.

7. A three-layered peanut butter and jelly sandwich, cut so as to illustrate a fault, can be used to demonstrate the types of forces and resulting structures which form along convergent, divergent, and transform boundaries.

Consider This

1. What will happen to the western coast of California in 30 or 40 million years as the result of movements along the San Andreas fault?

2. If orogenesis is typically associated with the active margins of continents, and the Rocky Mountains are relatively young and not the product of suturing of two continents, how can the mid-continent location of the Rocky Mountains be explained by plate tectonic theory?

3. What type of forces and geologic structures dominate the northern Rocky Mountains? Which dominate the middle Rocky Mountains?

What's Inside: Description of mass wasting; shear strength; major factors causing mass wasting; mass movements (rate of movement, type of movement, type of material); rockfalls; slumps and rock slides; five types of flows; complex movements; and reducing and eliminating the damages of mass wasting.

Learning Objectives

Upon completion of this material, your students should understand the following.

- It is important to understand the different types of mass wasting because mass wasting affects us all and causes significant destruction.

- Factors such as slope angle, weathering and climate, water content, vegetation, and overloading are interrelated, and all contribute to mass wasting.

- Mass movements can be triggered by such factors as overloading, soil saturation, and ground shaking.

- Mass wasting is categorized as either rapid mass movements or slow mass movements,

- The different types of rapid mass movements are rockfalls, slumps, rock slides, mudflows, debris flows, and quick clays; each type has recognizable characteristics.

- The different types of slow mass movements are earthflows, solifluction, and creep; each type has recognizable characteristics.

- People can minimize the effects of mass wasting by conducting geologic investigations of an area and stabilizing slopes to prevent and ameliorate movement.

Chapter Outline

Enrichment Topics

Topic 1. Beauty versus Death. In January 2005, a mudslide roared down a slope in La Conchita, California, fatally burying 10 people. Two slides had struck in March 1995, fortunately killing no one. Geologists warn that further slides are inevitable and may even be larger. The reason is that the ground underneath that part of California rises at a rate of 15 feet every 1,000 years and the slopes are already unstable. When there is excess rain, a slide can result. The slides are so fast that no warning system could be designed to avert tragedy. Whereas some residents of the 160 homes in the town have packed up, citing the excess danger, others say the town's natural beauty is worth the risk. Indeed, many other locations in California are prone to landslides, floods, and wildfires. Why should this one be abandoned? But even if a way can be found to protect the town, who should pay? And what if the cost of saving the town is worth more than the town itself? As it is, the new landslide made short work of a $400,000 retaining wall built after the 1995 slide.

Topic 2. Better than Water. Whereas water is always cited as the best erosional agent, humans have water beat, hands down. It has been estimated that the amount of material moved by human activities globally, including excavation for housing, road building, and mining, totals about 30 Gt/yr (1 Gt = 109 tons or 1012 kg). This compares to about 24 Gt/yr long-distance sediment transfer by rivers (including the material contributed by agriculture), 4.3 Gt/yr moved by glaciers, and .6 Gt/yr moved downslope by mass wasting (creep and slope wash). Humans can also exacerbate water erosion, by loosening the soil. http://soils.usda.gov/use/worldsoils/mapindex/eh2orisk.html and *GSA Today*, September, 1994.

Topic 3. Fires and Mass Wasting. The devastating fires that strike southern California every few years increase the risk of soil erosion, earthflows, debris flows, and mud flows. Mass wasting becomes more probable in such circumstances because these brushfires leave a waxy residue from the grasses as a covering on the ground surface. Thus, when heavy rainfall occurs, water cannot readily percolate into the soil. Sheet and rill runoff soaks into these soils along natural fissures in the soil and openings created by creep. When groundwater is concentrated in this manner, the slope's stability is significantly reduced, allowing for the potential for mass wasting. http://landslides.usgs.gov/research/wildfire/

Key Terms

Lecture Suggestions

1. When mass wasting is discussed, many students can relate to their own experiences building sand castles at the beach. Ask them whether they could build a steeper pile (or attain a greater angle of repose) with wet or dry sand. What happens to the sand castle structure as it loses moisture and the surface tension between the particles diminishes? Finally, note how the addition of too much water turns the sand into a slurry which cannot be piled up but will flow instead.

2. Compare the stability of a variety of materials by pouring samples of soil, fine sand, coarse gravel, etc. into piles. Discuss the factors that affect the angle of repose. Encourage students to look at piles of gravel, sand, etc. that they might see near cement factories or at construction sites. What can they tell about the material from the shape of the piles?

Consider This

1. If you purchased property and planned to build your home on a slope:
   A. describe the ideal set of geologic conditions that you would look for.
   B. describe the precautions you would take if your study revealed the possibility of one of the following:
      i. slumps
      ii. rock slides
      iii. mudflows
      iv. debris flows
      v. earthflows
      vi. creep

CHAPTER 12 | PREP CARD
Running Water

Learning Objectives

Upon completion of this material, your students should understand the following.

- Running water, one part of the hydrologic cycle, does considerable geologic work.

- Water is continuously cycled from the oceans to land and back to the oceans.

- Running water transports large quantities of sediment and deposits sediment in or adjacent to braided and meandering rivers and streams.

- Alluvial fans (on land) and deltas (in a standing body of water) are deposited when a stream's capacity to transport sediment decreases.

- Flooding is a natural part of stream activity that takes place when a channel receives more water than it can handle.

- The several types of structures to control floods are only partly effective.

- Rivers and streams continuously adjust to changes.

- The concept of a graded stream is an ideal, although many rivers and streams approach the graded condition.

- Most valleys form and change in response to erosion by running water coupled with other geologic processes such as mass wasting.

Chapter Outline

Enrichment Topics

Topic 1. Big River in the Dry Western U.S. The Colorado River originates high in the Rocky Mountains of Colorado, fed year-round by snowmelt, rain, and groundwater. As the river travels across the parched lands of Utah, Arizona, and into Mexico, evaporation far exceeds precipitation. The Colorado River provides water to rapidly growing desert cities in California (Los Angeles), Nevada (Las Vegas), and Arizona (Phoenix and Tucson), primarily from the dams at Lake Mead and Lake Powell. Scientists have recognized that, as the desert southwest suffers through a continuing drought, dry times may be more typical of the weather pattern of the region than were the wet years when the river water was initially allocated. In response to the drier climate and to the likelihood that climate change will likely cause further rainfall reductions in the region, states have come together to agree to new water allocations. In April 2007, seven western states and the federal government agreed that if water allocations were cut, the reductions would be shared equally among the states, something that could happen as soon as 2010. The conservation group "Living Rivers" recommends limiting growth in the region due to lack of water, but water managers recognize that these states will continue growing at a brisk pace.

Topic 2. Rift Zones, River Valleys, and Drainage Basins. Divergent boundaries that form as the result of doming and rifting of the crust typically proceed to rift along some, but not all, of the rifts which are formed. The remaining failed rifts are typically oriented at right angles to the margins of the continents which form by rifting. Once the margins of the continents cool and subside, these failed rifts become the axes of great river systems. Examples include the Mississippi, Niger, and Amazon Rivers.

Topic 3. Power at a Price. The Three Gorges Dam, the world's largest dam, on China's Yangtze River is scheduled to become fully operational in about 2011. At more than 600 feet high (183 m) and 1.5 miles (2.4 km) wide, the dam will create a reservoir hundreds of feet deep and nearly 400 miles long. The turbines will create as much electricity as 18 nuclear power plants or 15 coal burning plants. Part of the reason for the project is to end the Yangtze's massive floods, which have killed more than one million people in the past century. The dam's cost, though, is great. The rising waters will flood 395 square miles (632 square km) of land, displacing 1.2 million people from nearly 500 cities, towns, and villages along the river. Magnificent scenery and irreplaceable architectural and

Key Terms

archeological sites will be lost including ancestral burial sites, temples dating back hundreds of years, and even fossil locales. More than 300,000 farmers will need to be relocated, but because the Yangtze's floodplains are among the most fertile land in China, the replacement land is unlikely to be as productive. Alterations of river conditions will further endanger baiji dolphins (*Lipotes vexillifer*) and finless porpoises (*Neophocaena phocaenoides*), among other creatures. The price tag for the dam may be as high as $50 billion.

Topic 4. History of the Clean Water Act. During the 1950s and 1960s, water quality declined, because not all cities had sewage treatment plants, not all pollutants were treated in them, and many new chemicals were being put into the water. Intensive agriculture increased the runoff of nutrients, herbicides, and insecticides. Industrialization and urbanization intensified. The precipitous decline in water quality came to a head in 1969 when the Cuyahoga River, a tributary of Lake Erie, caught fire as it flowed through downtown Cleveland, Ohio. This spurred Congress to pass the Clean Water Act of 1972, which established air quality standards, set emissions limits, empowered state and federal government with enforcement, and increased funding for air pollution research. The law, amended in 1990, now regulates 189 toxic air pollutants, oversees alternative fuels, and also restricts the pollutants that contribute to acid rain and stratospheric ozone depletion. The amended law introduced a market-based system for controlling the emissions that cause acid rain, known as cap-and-trade. The goal is to restore and maintain the cleanliness of the nation's waters for recreation, fishing, and wildlife.

Lecture Suggestions

1. When discussing graded streams, diagram the factors that act to increase rates of erosion and deposition in terms of a sine-cosine curve balanced about an axis of equilibrium, with deposition represented by positive values and erosion represented by negative values. Point out that a stable equilibrium is never reached, but instead, all streams are in a constant flux about such an equilibrium state.
2. Drainage basin types are readily recognized if geologic maps are used to illustrate their configurations. If the essentials of geologic maps can be briefly explained, the relationships among geologic structures, drainage basin patterns, and landforms can be easily understood. The same can be demonstrated for incised and superposed streams.

Consider This

1. If some submarine canyons extend seaward from stream mouths, is sea level really an ultimate base level?
2. All physical systems tend toward a state of homeostasis or dynamic equilibrium. Is the dynamic equilibrium of a graded stream ever actually reached?
3. If deposition is inhibited during times of glacial advance (less meltwater and sediment load) and enhanced during time of glacial retreat (more meltwater carrying greater sediment loads), how might a study of the number of and stratigraphy of stream terraces in the lower Mississippi Valley be a gauge of the number of glacial advances and retreats?
4. Explain why many streams in the Rocky Mountains cut directly across mountain ranges.
5. The Army Corps of Engineers commonly straightens and otherwise modifies the course and dimensions of streams. What are some of the effects that the straightening of a stream channel could have?
6. If you were a city planner in a community through which a stream flowed, what information would you need in order to determine where, or if, a development project should be located in the vicinity of a stream? Consider a variety of stream and drainage basin types.

What's Inside: Groundwater and the hydrologic cycle; porosity; permeability; the water table; zone of aeration; zone of saturation; the downward movement of groundwater; springs; water wells; artesian systems; karst topography; caves and caverns; the effects of modifying the groundwater system; and hydrothermal activity.

CHAPTER 13 | PREP CARD

Groundwater

Learning Objectives

Upon completion of this material, your students should understand the following.

- Groundwater is one reservoir of the hydrologic cycle and accounts for approximately 22 percent of the world's supply of freshwater.

- Porosity and permeability are largely responsible for the amount, availability, and movement of groundwater.

- The water table separates the zone of aeration from the underlying zone of saturation and is a subdued replica of the overlying land surface.

- Groundwater moves downward because of the force of gravity.

- In an artesian system, groundwater is confined and builds up high hydrostatic pressure.

- Groundwater is an important agent of both erosion and deposition and is responsible for karst topography and a variety of cave features.

- Modifications of the groundwater system may result in a lowering of the water table, saltwater incursion, subsidence, and contamination.

Chapter Outline

Enrichment Topics

Topic 1. Life in a Groundwater Reservoir. It may seem impossible, but organisms have been found in some porous groundwater aquifers and karst features despite that there is no light, little oxygen, and scant nutrients. Called stygobionts, the creatures occupy the dark spaces between sand grains. Stygobionts are colorless, eyeless, and have worm-shaped bodies for easier movement. They move and metabolize more slowly and have longer life spans than typical aquatic organisms. Without light for photosynthesis, food energy comes into the ecosystem as organic matter from the surface. The food web is simple— microbes, such as bacteria and fungi, consume the organic matter and serve as food for somewhat larger invertebrate predators. Because food is scarce, most stygobionts will eat whatever they can find, a good adaptation to have in an extreme environment.

Topic 2. Bangladesh and Arsenic. Well-meaning aid organizations wanted to save rural Bangladeshis from the sewage-contaminated surface water they once drank that annually killed about 250,000 children. To gain access to the underlying groundwater, they drilled more than 14 million tube wells, although the local people called the groundwater "devil's water." It turned out that the aquifers lying beneath Bangladesh contain high arsenic concentrations, and now about 29 percent of all wells in Bangladesh contain water with arsenic levels higher than the guidelines established by the World Health Organization. The result is that as many as 20,000 Bangladeshis die of arsenic poisoning each year. Arsenic in low levels poisons slowly, and the seriousness of the situation was not immediately realized. Over time, the damage done by arsenic poisoning can be undone with clean water and nutritious foods, but most of the victims are too poor to have access to these commodities. Arsenic can be filtered out of the water, but many rural communities cannot afford the treatment technology. Research is being done on less expensive systems that exploit the affinity of arsenic for the magnetic mineral magnetite. When added to a bucket of well water, this mineral attracts the arsenic, which is then collected with a magnet and discarded. In the meantime, alternative water supplies are being constructed in the most affected areas.

Topic 3. Groundwater Mining in Other Regions of the World. Groundwater is being used unsustainably in many poor and developing nations. In Africa, 20 percent of annual groundwater use is unsustainable; in Jordan and Yemen,

Key Terms

the rate is 30 percent; and in Israel, 15 percent. In northern China, water tables are dropping nearly 5 feet (1.5 m) a year. In Tamil Nadu, India, the water table has dropped by 100 feet (30 m) in 30 years and many of the aquifers there have run dry. Some of the world's largest cities, including Beijing, Buenos Aires, Dhaka, and Lima depend heavily on groundwater. So much groundwater is being pumped from beneath Mexico City that parts of the city have sunk approximately 30 feet (9 m) since the 1900s.

Topic 4. Bacteria and Pollution. In the summer of 1999, high levels of coliform bacteria caused officials to shut down Huntington State Beach, south of Los Angeles. Although the source of the bacteria is unknown, some scientists suggest that the bacteria came from groundwater flow. Because coliform bacteria indicates the presence of feces, and other sources of the bacteria were ruled out, this would suggest that the groundwater is contaminated with fecal matter. *Science*Now, May 25, 2004.

Topic 5. The Ogallala Aquifer. The Midwestern United States is the breadbasket of America. For more than six decades, farming in the region has depended on irrigation water from the Ogallala Aquifer. The aquifer is a 15- to 30-foot-thick (5 to 10 m) layer of permeable sand and gravel that lies between 4.5 and 30 feet (1.5 and 10 m) below the surface. The aquifer stretches for 800 miles (1,300 km) through South Dakota, Wyoming, Nebraska, Colorado, Kansas, Oklahoma, Texas, and New Mexico. The Ogallala Aquifer irrigates more than 14 million acres (57,000 square km) of land and provides an estimated one-third of all irrigation water in the United States. Besides agriculture, water from the aquifer is used for the region's cities and industries. On average, water is being pumped at eight times the rate that it is being recharged, and the water table is dropping, in some areas as much as 3 to 5 feet (90 to 150 cm) per year. Some hydrologists estimate that one-fourth of the aquifer's original supply of water will be depleted by 2020, and it may be completely spent in areas where the aquifer is shallow. Estimates for the aquifer's remaining lifespan vary from 60 to 220 years in different areas. If the aquifer is eventually exhausted, water will need to be brought in from elsewhere, at enormous cost, or the productivity of this farmland will diminish greatly.

Lecture Suggestions

1. Point out how an artesian system is one where water is able to flow uphill because it has been confined and is under pressure. This is not unlike the water distribution system that may be in the classroom, where water flows up and out the faucet because it has been confined in the pipes and is under pressure, which is maintained usually by storing the water at a higher elevation in a tank.
2. A sample of sand from a swimming pool filter can lead to a discussion of porosity and the ability of an aquifer to remove particulate matter from groundwater. Ask students why sand is used and whether just any sand will do. What are the qualities of the grains that would make them suitable for use in a filter? Another natural filter that has applications for its ability to remove particles as small as 1 micron from water is diatomaceous earth. Ask the students to suggest other natural materials that might be useful in this manner.

Consider This

1. Can groundwater exist in igneous and metamorphic rocks?
2. How might the existence of hot springs in Georgia, Arkansas, and the Black Hills be explained?
3. If you were a hydrologist employed by a construction company planning a large development project in an arid region, along an ocean coast, and downstream from a large city and a landfill, what factors would you need to consider and what data would you need in order to plan a sufficient water supply for the new development?

CHAPTER 14 | PREP CARD

Glaciers and Glaciation

Learning Objectives

- Upon completion of this material, your students should understand the following.

- Moving bodies of ice on land known as glaciers cover about 10 percent of Earth's land surface.

- During the Pleistocene Epoch (Ice Age), glaciers were much more widespread than they are now.

- Water frozen in glaciers constitutes one reservoir in the hydrologic cycle.

- In any area with a yearly net accumulation of snow, the snow is first converted to granular ice known as firn and eventually into glacial ice.

- The concept of the glacial budget is important to understanding the dynamics of any glacier.

- Glaciers move by a combination of basal slip and plastic flow, but several factors determine their rates of movement and under some conditions, they may move rapidly.

- Glaciers effectively erode, transport, and deposit sediment, thus accounting for the origin of several distinctive landforms.

Chapter Outline

Enrichment Topics

Topic 1. Can a Continental Glacier Flow Uphill? When a glacier reaches a great thickness in the zone of accumulation, the elevation difference between that region and the zone of wastage produces a slope on which gravity acts to "pull" the glacial ice down the "hill" of its own creation. The underlying ground surface may locally slope uphill, as long as the upper slope of the glacier continues downward.

Topic 2. Can a Glacial Stream Flow Uphill? There is at least one place in the Sierra Nevada Mountains where a stream composed of melt water beneath a glacier flowed uphill. The pressure of the ice weighing down on the bottom of the hill was greater than the pressure of the ice at the top of the hill—so much greater that the uphill force was greater than the force of gravity and the water flowed uphill.

Topic 3. Snowball Earth. A controversial hypothesis suggests that during 10 million years of the Precambrian, Earth was covered with ice up to 1 km thick during which time nearly all life was wiped out. Computer models show that if the polar ice caps grow beyond a certain point, runaway freezing occurs. Evidence for the snowball hypothesis includes signs of past glaciation in places that should have been too hot. But if snowball Earth existed, how did it escape the ice? One possibility is that volcanoes with tops above the ice pumped greenhouse gases into the atmosphere. With the water cycle at a standstill because the water was all in ice, there was no rain and the greenhouse gases remained in the atmosphere. Eventually there was intense global warming and a big thaw about 600 million years ago. With the planet freed from ice, complex organisms could then evolve.

Topic 4. Ice Cores and the Paleoenvironment. The most powerful window into past climate is the ice contained in glaciers and ice caps. To collect an ice core, scientists drill a hollow pipe into an ice sheet or glacier. The cores taken from the Greenland and Antarctic ice caps supply data that span long periods of time. Two Greenland cores go back 100,000 years, whereas one Antarctic core goes back 420,000 years through four glacial cycles. The European Project for Ice Coring in Antarctica, (EPICA) about two miles (3,190 m) long, has cut through eight glacial cycles covering 740,000 years and is still being drilled. Gases and particles trapped in snowfall can be analyzed by examining ice layers. These substances represent atmospheric conditions at the time the snow fell. Scientists can analyze CO_2 in the gases to determine the concentration

Key Terms

of that greenhouse gas at the time and to ascertain its source, whether from volcanic eruptions or burning fossil fuels, for example. The presence of beryllium-10 in the ice core is evidence of the strength of solar radiation. Ash indicates a volcanic eruption; dust, an expansion of deserts; and pollen, the types of plants that were on the planet at the time. The amount of pollen found in the ice layer may be an indicator of the amount of precipitation that fell. Paleoclimatologists can discern air temperature at the time the snow fell by measuring the ratios of different isotopes of oxygen and hydrogen. These isotope ratios also reveal global sea level.

Topic 5. Arctic Warming. Recent rising temperatures have not been spread evenly around Earth with temperatures in the high latitudes warming the most dramatically. During the past two decades, temperatures have risen as much as 8°F (4.4°C) in some Arctic locations. Satellite mapping of the extent of Arctic sea ice in September shows a 20 percent drop off since 1979. The loss of arctic sea ice, for example, is destroying the habitat that is needed by polar bears so that polar bear populations are in significant decline and their mean body weight is decreasing. Warmer temperatures have caused the populations of ringed seals, the bears' main food, to decline. The future does not look bright for the Arctic—a global average temperature increase of 5°F (2.8°C) means a one or two degree increase at the equator but a 12°F (6.7°C) increase at the North Pole. Forests will expand into the tundra, sea ice and glaciers will continue to melt, and Arctic ecosystems will be lost.

Lecture Suggestions

1. The deposition of till in ground moraines and terminal/recessional moraines can be likened to a lawn fertilized by a spreader. When moving, the fertilizer is evenly spread across the ground. However, when stationary and with the chute open, the fertilizer will accumulate in a ridge-like pile.
2. Basal slip may be explained by analogy with the phenomenon that permits ice-skaters to glide over ice. The weight of the skater is concentrated on the skate blade, and this pressure is great enough to melt the ice beneath. It is this thin layer of water that permits basal slip of the skate and thus the skater. Similarly the weight of 40 m or more of ice will cause some basal melting and thus basal slip.
3. A model planetarium would be useful for demonstrating the variations in orbital parameters on which the Milankovitch theory is based.

Consider This

1. Both valley and continental glaciers produce depositional landforms. However, valley glaciers carry sediment within all levels as well as on their surface, whereas continental glaciers carry sediment only in the lower levels and at their base. How then do continental glaciers produce such stratified drift deposits as eskers and kames, whereas valley glaciers do not?
2. Suggest a plausible relationship between the Milankovitch cycles and the production/removal of atmospheric carbon dioxide.
3. Predictions of global warming project 0.9 to 3.6°F (0.5 to 2°C) increase in the global average annual temperature by the middle of the 21st century if nothing is done to reduce greenhouse gas emissions. Such an increase in temperature would result in some melting of glaciers and an increase in sea level of between 7 and 32 inches (18 and 81 cm) by the end of the century (IPCC report, 2007).
    a. What regions of the world would be most severely affected by such a rise in sea level?
    b. What effects would such a sea level rise have on the groundwater supplies of coastal cities?
    c. What effects would such a sea level rise have on sediment deposition at the mouths of large rivers or depositional environments such as the Mississippi Delta?
    d. What effects would such a sea level rise have on coastal erosion?

What's Inside: Desertification; sediment transport by wind; suspended load; bed load; saltation; wind erosion; four major dune types; loess; air-pressure belts; the Coriolis effect; distribution of deserts; semiarid regions; arid regions; characteristics of deserts; weathering in deserts; running water in deserts; and important desert landforms.

CHAPTER 15 | PREP CARD

The Work of Wind and Deserts

Learning Objectives

Upon completion of this material, your students should understand the following.

- Wind transports sediment and modifies the landscape through the processes of abrasion and deflation.

- Dunes and loess are the result of wind depositing material.

- Dunes form when wind flows over and around an obstruction.

- The four major dune types are barchan, longitudinal, transverse, and parabolic.

- Loess is formed from wind-blown silt and clay and is derived from three main sources—deserts, Pleistocene glacial outwash deposits, and river floodplains in semiarid regions.

- The global pattern of air-pressure belts and winds is responsible for Earth's atmospheric circulation patterns.

- Deserts are dry, receive less than 25 cm of rain per year, have high evaporation rates, typically have poorly developed soils, and are mostly or completely devoid of vegetation.

- The majority of deserts are found in the dry climates of the low and middle latitudes.

- Deserts have many distinctive landforms produced by both wind and running water.

Chapter Outline

Enrichment Topics

Topic 1. Desertification in the Sahel. Although desertification is sometimes caused by a decrease in rainfall, human activities are often to blame. Human activities, such as deforestation, intensive agriculture, and overgrazing, degrade the soil. In developing countries, the soil nutrients that were taken up by trees or animals are burned for heat and to cook food, and so the nutrients are lost and the region is unable to recover its fertility. A significant example of desertification has been in the semiarid region south of the Sahara Desert of Africa. Traditionally, the Sahel was populated by nomadic tribes, who maintained low numbers and migrated frequently, following the rains and being careful not to overgraze an area. In recent times, the nomadic tribes have been constrained to remain in one spot due to the enforcement of national boundaries and the initiation of intensive agriculture. The use of groundwater and the relative abundance of rain in the 1960s allowed the region's population to grow. This proved disastrous when the worst drought of the century hit from 1968 to 1973. During that time, nearly 250,000 people and 3.5 million cattle died of starvation, and the Sahara expanded southward by about 100 miles (160 km). Among the western United States, sub-Saharan Africa, the Middle East, western Asia, parts of Central and South America, and Australia, more than 100,000 square miles (350,000 sq. km.) of land are lost to desertification each year. A 2004 United Nations report warns that one-third of the world's land surface is at risk for desertification and that one-fifth of the world's population is threatened by the impacts of this land change.

Topic 2. Climatic Change in the Sahara. A desert is not always a desert, and the Sahara has not always been arid. In keeping with glacial cycles, the Sahara is sometimes humid and lush. The last humid cycle lasted from about 5,000 to 10,000 years ago, when the region was inhabited by hunter-gatherers. Some locations were relatively densely populated. As the climate became more arid, people congregated around the remaining lakes and rivers or moved south into wetter regions. Egyptian civilization came about because people congregated in the Nile River valley. http://www.thenakedscientists.com/HTML/content/interviews/interview/584/

Key Terms

Lecture Suggestions

1. An interesting topic is the different conditions that lead to the different types of deserts and the different types of vegetation that are found in each. Also, the adaptation that animals and plants have to desert conditions is a fascinating topic.

2. Global warming and the effects warmer temperatures will have on all aspects of the natural and human environment are timely and interesting topics. There are many excellent articles in *New York Times* and other sources.

Consider This

1. Explain the sequence of events and interrelationships of human modification, soil development, erosion, rainfall, and temperature which could transform the Brazilian rainforest into a desert.

What's Inside: Changing shorelines; waves; tides; wind-generated waves; shoreline erosion and common features of shoreline erosion; shoreline deposition and the shoreline depositional features; the nearshore sediment budget; depositional, erosional, submergent, and emergent coasts; storm waves and coastal flooding; Hurricane Katrina; problems resulting from rising sea level; and methods currently used to combat rising sea level.

Learning Objectives

Upon completion of this material, your students should understand the following.

- Wind-generated waves and their associated nearshore currents effectively modify shorelines by erosion and deposition.

- The gravitational attraction by the Moon and Sun and Earth's rotation are responsible for the rhythmic daily rise and fall of sea level known as tides.

- Seacoasts and lakeshores are both modified by waves and nearshore currents, but seacoasts also experience tides, which are insignificant in even the largest lakes.

- Distinctive erosional and depositional landforms such as wave-cut platforms, spits, and barrier islands are found along shorelines.

- The concept of a nearshore sediment budget considers equilibrium, losses, and gains in the amount of sediment in a coastal area.

- Several types of coasts are recognized based on criteria such as deposition and erosion and fluctuations in sea level.

Chapter Outline

Enrichment Topics

Topic 1. Submarine Canyons. Submarine canyons are like canyons on land, some as long as the Grand Canyon. They form from gravity driven water and sediments that occur almost as landslides, known as turbidity currents. In 2004, scientists "witnessed" a turbidity current when they left some scientific instruments in Monterey Canyon off of California. The canyon is 20 km offshore with a depth of 525 meters. One instrument was later found 110 m down canyon and another 400 meters down canyon. A remotely-operated vehicle ended up buried beneath a 70 cm layer of sediment 550 meters down the canyon. One instrument that was initially placed 10 km up the canyon was also swept downslope. The scientists calculated that if the turbidity current deposited 70 cm of sediment along the length of the canyon, the total material would have been 2.2 million cubic centimeters. Five other turbidity currents were measured in the canyon in the following three years. On the other coast, the Hudson Canyon off of New York, which is on a passive margin, receives only a small amount of sedimentation. The canyon was carved and received most of its sediment during the last ice age. *Science News*, January 1, 2005.

Topic 2. Rip Currents. Rip currents form when nearshore currents move along the shore in opposite directions; where they meet, there is excess water that rushes out to sea. They are found on any beach with breaking waves and can extend out from the shoreline through the surf zone and past the breakers. More than 100 people are killed due to rip currents in the U.S. each year, and over 80 percent of rescues performed by surf beach lifeguards are on people stuck in rip currents. If you are caught in a rip current, do not try to swim in to shore; you will just become exhausted and will get pulled out to sea. Do swim parallel to the shore until you no longer feel that you are being dragged offshore. Then you should be able to safely swim inland. This may be the most practical bit of information you will learn in this course!

Topic 3. Coastal Zones. The coastal zone is a highly productive environment that includes the continental shelves, estuaries, bays, and lagoons. Although these waters comprise less than 10 percent of Earth's surface, they account for about 25 percent of the global biological production and store 90 percent of the global sedimentary accumulation of organic carbon. Nearly 50 percent of the U.S. population inhabits coastal counties that make up 10 percent of the nation's land

Key Terms

area and account for at least 30 percent of the nation's gross national product. As human activity in this region has increased, the environmental effects have been especially accelerated, including losses of habitat and living resources. *EOS*, May 25, 1994.

Topic 4. How Do Scientists Learn About the Seafloor? Learning about the seafloor is difficult due to the great distances (horizontal and vertical) and tremendous amounts of water that get in the way. Research ships travel out to sea for days to months, carrying equipment that can sample sediment, rocks, and make maps. A bathymetric map shows the 3-dimensional geographic features of the seafloor on a 2-dimensional map. A device towed behind a ship sends out sound waves that strike the bottom then return. Because the speed of sound waves is known, the amount of time it takes for a wave to make a round trip allows scientists to calculate the distance to the object. When this information is compiled, a map of the seafloor emerges. A gravity corer is a hollow tube deployed on a cable that accelerates through the water. When it reaches the bottom the tube slices into the sediment so that a nicely layered sample is collected inside. Seafloor rocks are collected with a dredge, a giant rectangular bucket that is dragged behind the ship. Submersibles are small diving crafts that are not attached to the mother ship. The battery-operated submersible *Alvin* can descend to more than 13,000 feet (4,000 m) carrying a pilot and two passengers who collect samples, take photos and video, and make accurate descriptions of what they see. Submersibles offer scientists their best view of the deep ocean, but they are potentially dangerous and expensive. Because remotely operated vehicles (ROVs) carry no human passengers, they can visit dangerous locations. The ROV *Jason*, which photographed rooms inside the *Titanic* and *Argo*, has been dangerously close to hydrothermal vents.

Lecture Suggestions

1. The dynamic nature of beaches can be illustrated by comparing old postcards of beach resorts with more recent photos. Likewise, old maps, navigational charts, or aerial photographs can be used to demonstrate the extent to which a shoreline may change in a relatively short period of time.
2. An interesting discussion can be held by asking students to consider the predicament of sailors in the days of wooden ships. What could they tell about the sea bottom by looking at the shoreline in unfamiliar waters? What sorts of obstacles might they encounter after a major storm?
3. Suggest that students consider how the early voyagers across the Pacific Ocean, those who travelled from Asia and Indonesia to settle in Poynesia, Micronesia, and Melanesia managed to navigate across great expanses of open seas. They had no compasses or other instruments, and certainly no maps, yet they were very successful. Aside from using stars, they also used wave refraction caused by distant, unseen land masses.
4. Compare photos of beaches from before and after the December 26, 2005, tsunami. This will illustrate examples of beach erosion.

Consider This

1. A one-foot (30 cm) sea level rise in Florida would cause the loss of 100 feet (30 m) of beaches. What will this do for the flat-lying state? How will this affect the damage storms can do as storm surge is added to higher sea level?
2. If you were a real estate developer, what geologic factors should you consider before constructing homes or motels on a barrier island? Under what circumstances would you proceed with such a project? If you were a potential buyer, what questions would you ask the developer?
3. Given the difficulty of preventing beach and coastal wetland erosion and of restoring beaches, suggest several solutions to these important problems.

What's Inside: Early concepts of geologic time and the age of Earth; James Hutton; relative dating methods; correlating rock units; absolute dating methods; development of the geologic time scale; and the relationship between geologic time and climate change.

CHAPTER 17 | PREP CARD

Geologic Time: Concepts and Principles

Learning Objectives

Upon completion of this material, your students should understand the following.

- The concept of geologic time and its measurements have changed throughout human history.

- The principle of uniformitarianism is fundamental to geology.

- Relative dating—placing geologic events in a sequential order—provides a means to interpret geologic history.

- The three types of unconformities—disconformities, angular unconformities, and nonconformities—are erosional surfaces separating younger from older rocks and represent significant intervals of geologic time for which we have no record at a particular location.

- Time equivalency of rock units can be demonstrated by various correlation techniques.

- Absolute dating methods are used to date geologic events in terms of years before present.

Chapter Outline

Enrichment Topics

Topic 1. Uniformitarianism and Catastrophism. Uniformitarianism states that the natural processes we see operating today are the same as those that operated in the past and so we can learn about the past by observing the present. Uniformitarianism recognizes that Earth changes slowly over vast periods of time. Prior to the acceptance of uniformitarianism was catastrophism, the idea that Earth was formed by sudden, short-lived catastrophic events. Catastrophism was linked to religion; Noah's flood was a prime example of a catastrophic event, and it contrasted entirely with the idea of epeiric seas slowly encroaching over the land over a long period of time. To distinguish their ideas from the prevailing religious views, scientists rejected catastrophism entirely, and uniformitarianism as a scientific philosophy held for about two centuries. But late in the 20th century, scientists recognized that catastrophes have shaped our planet's history. There is no better example than the asteroid impact many think caused the extinction at the end of the Mesozoic. Today, scientists combine the uniformitarianism and catastrophism into something that makes more sense as a whole. Because we see catastrophes happen (fortunately, none as bad as a massive asteroid impact), they can be a part of a uniformitarian view. The present is the key to the past and sometimes both time periods can include catastrophes.

Topic 2. Climate Change in Earth's Past. Climate change was not invented in the modern era; climate has varied tremendously in Earth's past. Paleoclimatologists are piecing together the story of the Paleocene-Eocene Thermal Maximum (PETM), mostly from the chemistry of forams collected from ocean sediment cores. After the end of the Cretaceous Period 66 million years ago, temperatures rose until they became so high they triggered an even greater warming event around 55 million years ago, the PETM. About half of the warming, 3.6°F (2°C), took place over no more than a few hundred years and the rest over less than 5,000 years. Sea-surface temperatures increased by between 9° and 14°F (5° and 8°C), and the deep sea warmed dramatically as well. Scientists think that the most likely explanation for the warming is that a load of methane flooded the atmosphere. The most likely source of such vast amounts of methane is the methane hydrate deposits buried in seafloor sediments. According to this scenario, the PETM was triggered when ocean temperatures rose high enough that the hydrate structure melted and released the

Key Terms

methane trapped inside. The high temperatures of the PETM had many consequences. Warm surface water caused ocean currents to switch direction. Because warm water cannot hold as much gas as cold water, oceanic oxygen levels were very low. In the atmosphere, methane broke down and formed CO_2, which then formed carbonic acid that rained down and caused the carbonate shells of many organisms to dissolve. High acidity and low oxygen caused 50 percent of deep-sea forams and possibly other deep sea animals to die out. Although there was no mass extinction, the abundance of some life forms and/or their evolutionary pathways altered. Modern mammals from rodents to primates first evolved and flourished. The PETM lasted for about 200,000 years, likely ending when all the available methane had been released into the atmosphere. Over time, the CO_2 the methane broke down into was sequestered in forests and plankton and dissolved into the oceans.

Topic 3. The Oldest Known Rocks. The Acasta gneiss in northwest Canada near Great Slave Lake has been dated at 4.03 billion years of age. This age is about 100 million years older than the previous "record-holder" from western Greenland. The newly discovered rocks are of granitic composition, indicating that differentiation of continental and oceanic crust had begun before this time in Earth history because granitic rocks could not have formed directly from oceanic crust. These early continental crust samples were dated first using standard techniques applied to uranium-lead isotopic pairs contained within zircon crystals. More refined measurements were made with a sensitive high-mass-resolution ion microprobe—a machine which allows analysis of the best-preserved portions of a single crystal and is more precise than other techniques. Isolated zircon crystals dated at nearly 4.3 billion years of age have been known for some years. However, these crystals were derived from sandstones, not their granitic parent rocks. http://pubs.usgs.gov/gip/geotime/age.html

Lecture Suggestions

1. Discuss whether the occurrence of such events as asteroid impacts, which suddenly and globally disrupt ecosystems and the strata and sediments of a large area, invalidate the principle of uniformitarianism.
2. Note how uniformitarianism must be an accepted principle in other sciences, even though it may not be stipulated.
3. Evaluate the assertion that uniformitarianism requires that any geologic or evolutionary phenomenon which is not observable today cannot have happened in the past.
4. Consider whether or not the principle of fossil succession depends upon the existence of organic evolution.

Consider This

1. Does an unconformity encompass the same duration of time everywhere it occurs?
2. Much negative attention is often given to the sources of error and the limits of uncertainty that are given in accompaniment of any absolute age. Why is such attention unwarranted in regard to the sources of error, and why is such attention misleading when aimed at the limits of uncertainty for any given date?
3. Does the fact that much of geology is a historical science distinguish it from other sciences in any significant way that affects the confidence one can have in its conclusions? Why or why not? Is it actually any different from the other sciences in its methodology?
4. Why can fossils be used to demonstrate the age equivalence of geographically separated and (often) lithologically dissimilar strata?
5. Is there any single region on Earth that has a complete rock record?

What's Inside: Shields, platforms, and cratons; Laurentia; orogeny; six major Paleozoic continents; the assemblage of Pangaea; the six cratonic sequences of the North American craton; mountain building and the Paleozoic mobile belts; microplates; the four-step breakup of Pangaea; the Mesozoic history of North America; Cenozoic plate tectonics; the North American Cordillera; and Pleistocene glaciation.

Learning Objectives

Upon completion of this material, your students should understand the following.

- The societal implications for studying Earth systems and subsystems over geologic time.
- The distinction between a craton and a shield, and how they form.
- How the craton of North America evolved during the Precambrian.
- The six major continents that existed at the beginning of the Paleozoic Era.
- The movements of Laurentia and Gondwana during the Early Paleozoic.
- The formation of Laurasia and location of Gondwana during the Late Paleozoic.
- How Pangaea had been formed from Laurentia and Gondwana during the Permian.
- How cratonic sequences reflect widespread transgressions and regressions.
- The extent, character and significance of the successive records of the Sauk, Tippecanoe, and Kaskaskia sequences.
- The Absaroka sequence, cyclothems, and the formation of coals during the Pennsylvanian.
- The timing and significance of the Taconic orogeny.
- The Acadian orogeny, how its character changed, and the formation of Pangaea.
- The four stages in the breakup of Pangaea.
- The timing and initial site of the separation of North America from Africa.
- The evolution of the Gulf Coastal region during the Jurassic and Cretaceous.
- The general evolution of the western region of North America during the Mesozoic.
- The importance of the Cenozoic Alpine-Himalayan and circum-Pacific orogenic belts.
- The nature and general evolution of the North American Cordillera.

Enrichment Topics

Topic 1. Recent Basin and Range Faulting, Idaho. In 1983, the Borah fault segment of the Lost River Range in central Idaho ruptured causing a magnitude 7.2 earthquake in a remote area of Idaho. Remarkable eyewitness accounts by local elk hunters described how the fault scrap formed in a matter of seconds with a roar and a cloud of dust that seemed to dash across the landscape. Seconds later, a 25-mi (40 km) long surface rupture formed. Almost more remarkable was the groundwater activity that occurred in the area. On the valley floor at the epicenter, water spouts erupted from fissures in rock like fire hydrants, shooting water and debris several hundred feet across a road. Dry sand was blown from the alluvial floor, followed by water gushing several feet high for many hours. As many as 40 craters, some the size of Olympic-sized swimming pools, were left after the water ceased to flow. In Yellowstone Park, 1,550 miles (2,500 km) away, Old Faithful became less faithful, some geysers ceased to erupt and some new springs opened. The Borah Fault earthquake gave geologists a new understanding not only of basin and range faulting but also the associated effects on groundwater systems. *Natural History*, June, 1986.

Lecture Suggestions

1. Discuss how geologists working in the Appalachians have managed to unravel such a complicated geologic story. An interesting account of the development of geologic thought about the Appalacian Mountain chain is given in John McPhee's book, *In Suspect Terrane*.
2. Two other books by the same author, John McPhee, provide useful information and anecdotes for lectures about the tectonics of western North America, *Basin and Range*, and *Rising from the Plain*.
3. The breakup of Pangaea is the vital part of Earth's story during the Mesozoic. Emphasize that the story is still ongoing. Mountain building events that began in western North America are still happening.
4. Remind students that if they could travel through time, they would only feel reasonably comfortable during the last 65 million years of Earth's history. Most of the continents were relatively close to their present positions, and animals and plants, though strange, were somewhat familiar.

- The cause of and result of the collision of North American with the Pacific-Farallon Ridge.
- The cause of the present topography of the Appalachian Mountains.
- The Pleistocene climatic cycles of North America.
- The mineral resources dominating during different periods of geologic time.

Chapter Outline

Consider This

1. Have students try to list the types of data that geologists might collect to piece together ancient mountain ranges that are now separated by vast oceans. How can geologists be confident that the Appalachian, Caledonian, Alleghenian and Hercynian orogenies were all closely related in time and space?
2. Does a transgression or regression encompass the same duration of time everywhere it occurs?
3. Knowing what events occurred with the formation of Pangaea in the Paleozoic, speculate on what might happen should the Atlantic and/or the Pacific oceans close in the future. What types of new mountain ranges might form? What would happen to the old orogenic zones?
4. How would the climate of western North America have been influenced by microplate accretion? By Cenozoic mountain building?
5. How much extension has taken place in the Basin and Range during extensional tectonics? How much shortening took place during Sevier compressional tectonics? How could a geologist calculate these figures?

Key Terms

CHAPTER 19 | PREP CARD

Life History

Learning Objectives

Upon completion of this material, your students should understand the following.

- Which life forms appeared during the Precambrian.

- When marine invertebrates with durable skeletons first appeared.

- The groups which dominated the marine invertebrate communities of the Cambrian and Ordovician Periods.

- The faunas dominant during the Silurian through Permian Periods and the mass extinctions at the end of the Permian.

- When the fish appeared, some of the forms which developed, and their relationship to the amphibians.

- Why the reptiles were able to colonize all parts of the land; the dominant reptile groups during the Permian.

- The problems both plants and animals had to overcome in the transition from water to land.

- The probable ancestor of terrestrial vascular plants, early development of these plants, and later evolution of gymnosperms then angiosperms.

- The general development of the dinosaurs from initial evolution through reaching maximum abundance and diversity.

- The basis for recognizing two distinct orders of dinosaurs and what these orders are.

- What pterosaurs, ichthyosaurs, and plesiosaurs were.

- When birds and earliest mammals evolved.

- The effect of the mass extinctions at the end of the Cretaceous, and a possible hypothesis to explain the extinctions.

- Why mammals became dominant during the Cenozoic.

- When primates and hominids evolved and how hominids differ from other mammal groups.

- Why there may have been so many very large species of mammals during the Pleistocene.

Enrichment Topics

Topic 1. Plateau Basalts and Mass Extinctions. The Deccan Traps, a region of plateau basalts that erupted about 65 million years ago, is the leading contender for an alternative to the asteroid impact hypothesis for the extinction of the dinosaurs and other contemporaneous Late Cretaceous faunas. The argument is that the climatic effects of such large scale eruptions would be sufficient to cause extinctions. A statistical analysis of the correlation between the ages of the Deccan flood eruptions and the end of the Cretaceous Period has shown that they apparently match with a statistical difference of zero—i.e., within the limits of error the two ages are the same. This does not necessarily mean that all episodes of massive volcanism lead to extinctions, or that all extinctions are the result of volcanism. It has even been proposed that major impact by an extraterrestrial body may have initiated these flood basalts. *Geology Today*, March–April, p. 51, 1994; *Geophysical Research Letters*, Vol. 20, p. 1399, 1993 .

Topic 2. More on the Deccan Traps. Recent studies indicate that Deccan volcanism lasted longer than previously thought, from at least 67 million years ago (m.y.a.) to less than 63 m.y.a. Initiation of Deccan volcanic activity thus predates the Cretaceous-Tertiary boundary, dated at ~66 m.y.a. The mass extinction could thus be a combined effect of Deccan volcanism and bolide impact. *EOS* August 2, 1994.

Lecture Suggestions

1. Read Stephan J. Gould's book *Wonderful Life* for excellent commentaries on science, evolution and extinction. Gould discusses contingencies in the geologic record or the "what if" factor. The fauna of the Burgess Shale contained many species that may have been from phyla that became extinct. Had those phyla survived and the ones that gave rise to modern life gone extinct, what would this world be like today?

2. Lead the students in a discussion about the role that serendipity plays in science. For example, Walcott never had the opportunity to advance his research on the Burgess Shale because of other commitments. Would he have interpreted them differently with sufficient time to do thorough research? Was he limited by scientific vision at that time?

Chapter Outline

Key Terms

3. It is important that students understand the concept of co-evolution. What cause and effect relationships determined the evolution of life'? Discuss the concept of the food chain and levels of predators. What does each level on the food chain do for the organisms on the levels above and below?

4. There is much good information written at the introductory level about the Cretaceous mass extinctions. Have the students divide into three groups and research the following theories: (1) meteorite impact, (2) voluminous volcanism, (3) their time on Earth was over.

Consider This

1. Consider the important concept of extinction. Given the number of species which have become extinct through geologic time is it really important for humans to try to save every endangered species?

2. Is there any way to determine which species should be protected and which may not be important in the long run? Are humans too focused on "charismatic megafauna"—a.k.a. big, pretty animals—rather than on species which may be more important to protect?

3. Why was it necessary for plants to populate the land before animals could do so? What types of animals might have been able to inhabit the land without plants being there?

4. Why are the best dinosaur fossil localities where they are? Why aren't dinosaur fossils found in other parts of the country which have Mesozoic rocks?

5. Formulate a hypothesis as to what would happen on Earth in the event of a meteorite impact such as that suggested for the end of the Cretaceous Period. What is the likelihood that we would survive? What habitats would be destroyed? Would there be global warming or cooling? What is the likelihood of such an event (remember the 1994 collision of Shoemaker-Levy with Jupiter)?

6. Is the term "Age of Mammals" really a fair assessment of the Cenozoic Era? What might be a better name?

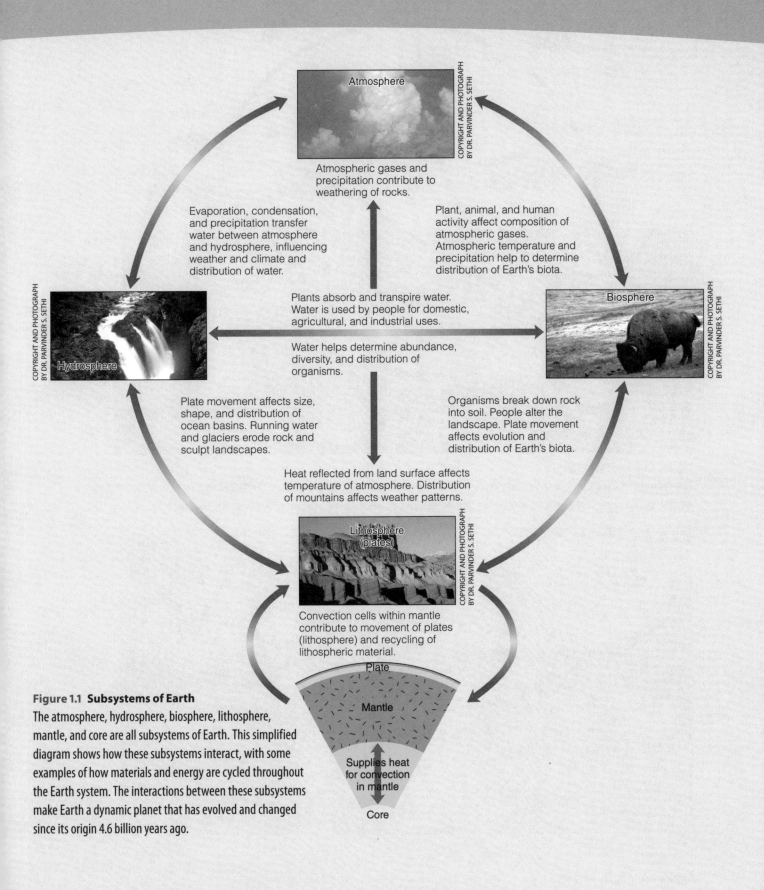

Atmosphere

COPYRIGHT AND PHOTOGRAPH BY DR. PARVINDER S. SETHI

Atmospheric gases and precipitation contribute to weathering of rocks.

Evaporation, condensation, and precipitation transfer water between atmosphere and hydrosphere, influencing weather and climate and distribution of water.

Plant, animal, and human activity affect composition of atmospheric gases. Atmospheric temperature and precipitation help to determine distribution of Earth's biota.

COPYRIGHT AND PHOTOGRAPH BY DR. PARVINDER S. SETHI

Plants absorb and transpire water. Water is used by people for domestic, agricultural, and industrial uses.

Water helps determine abundance, diversity, and distribution of organisms.

Biosphere

COPYRIGHT AND PHOTOGRAPH BY DR. PARVINDER S. SETHI

Hydrosphere

Plate movement affects size, shape, and distribution of ocean basins. Running water and glaciers erode rock and sculpt landscapes.

Organisms break down rock into soil. People alter the landscape. Plate movement affects evolution and distribution of Earth's biota.

Heat reflected from land surface affects temperature of atmosphere. Distribution of mountains affects weather patterns.

Lithosphere (plates)

COPYRIGHT AND PHOTOGRAPH BY DR. PARVINDER S. SETHI

Convection cells within mantle contribute to movement of plates (lithosphere) and recycling of lithospheric material.

Plate

Mantle

Supplies heat for convection in mantle

Core

Figure 1.1 Subsystems of Earth

The atmosphere, hydrosphere, biosphere, lithosphere, mantle, and core are all subsystems of Earth. This simplified diagram shows how these subsystems interact, with some examples of how materials and energy are cycled throughout the Earth system. The interactions between these subsystems make Earth a dynamic planet that has evolved and changed since its origin 4.6 billion years ago.

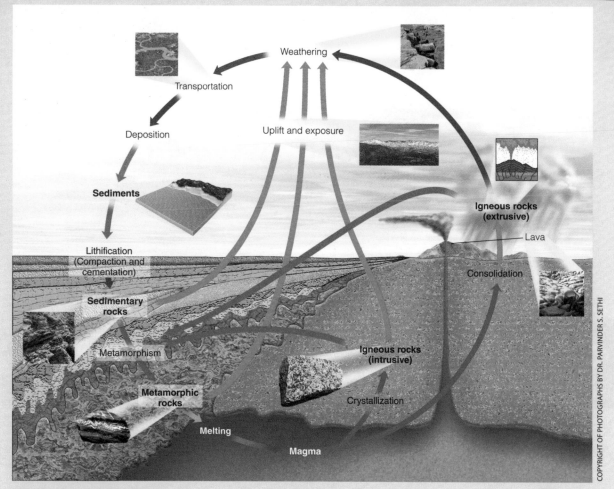

Figure 1.14 The Rock Cycle

This cycle shows the interrelationships between Earth's internal and external processes and how the three major rock groups are related. An ideal cycle includes the events on the outer margin of the cycle, but interruptions, indicated by internal arrows, are common.

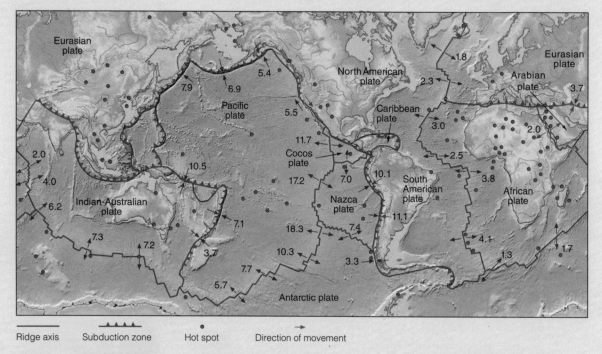

Figure 2.11 Earth's Plates

A world map showing Earth's plates, their boundaries, their relative motion and rates of movement in centimeters per year, and hot spots.